AS-Level
Mathematics

AS Maths is seriously tricky — no question about that.
To do well, you're going to need to <u>revise properly</u> and <u>practise hard</u>.

This book has <u>thorough notes</u> on everything in modules C1, C2, S1, M1 and D1.
It'll help you learn the stuff you need and take you <u>step-by-step</u> through loads of examples.

It's got practice questions... lots of them. For <u>every topic</u> there are warm-up and exam-style
questions. Plus there are <u>two full practice exams</u> at the end of each module.

And of course, we've done our best to make the whole thing vaguely entertaining for you.

Complete Revision and Practice
Exam Board: Edexcel

Contents

Contents

Editors:
Mary Falkner, Josephine Gibbons, Paul Jordin, Sharon Keeley-Holden, Simon Little, Ali Palin, Andy Park, David Ryan, Lyn Setchell, Caley Simpson, Jane Towle, Dawn Wright

Contributors:
Andy Ballard, Charley Darbishire, Dave Harding, Claire Jackson, Tim Major, Mark Moody, Garry Rowlands, Mike Smith, Claire Thompson, Julie Wakeling, Kieran Wardell, Chris Worth

Proofreaders:
Mona Allen, Alastair Duncombe, Paul Garrett, Helen Greaves, Glenn Rogers

Published by CGP

ISBN: 978 1 84762 580 9

Printed by Elanders Ltd, Newcastle upon Tyne.

Based on the classic CGP style created by Richard Parsons.

A Few Definitions and Things

Yep, this is a pretty dull way to start a book. A list of definitions. But at least it gets it out of the way right at the beginning — it would be a bit mean of me to try and sneak it in halfway through and hope you wouldn't notice.

Polynomials

POLYNOMIALS are expressions of the form $a + bx + cx^2 + dx^3 + \dots$

$5y^3 + 2y + 23$ ← Polynomial in the variable y.

$1 + x^2$

$z^4 + 3z - z^2 - 1$ ← Polynomial in the variable z.

An expression is made up of <u>terms</u>.

E.g: z^4, $3z$, $-z^2$ and -1

x, y and z are always **VARIABLES**
They're usually what you solve equations to find. They often have more than one possible value.

Letters like a, b, c are always **CONSTANTS**
Constants never change. They're fixed numbers — but can be represented by letters. π is a good example. You use the symbol π, but it's just a number = 3.1415...

Functions

FUNCTIONS take a value, do something to it, and output another value.

$f(x) = x^2 + 1$ ← function f takes a value, squares it and adds 1.

$g(x) = 2 - \sin 2x$ ← function g takes a value (in degrees), doubles it, takes the sine of it, then takes the value away from 2.

You can plug values into a function — just replace the variable with a certain number.

$f(-2) = (-2)^2 + 1 = 5$

$f(0) = (0)^2 + 1 = 1$

$f(252) = (252)^2 + 1 = 63505$

$g(-90°) = 2 - \sin(-180°) = 2 - 0 = 2$

$g(0°) = 2 - \sin 0° = 2 - 0 = 2$

$g(45°) = 2 - \sin 90° = 2 - 1 = 1$

Exam questions use functions all the time. They generally don't have that much to do with the actual question. It's just a bit of terminology to get comfortable with.

Multiplication and Division

There's three different ways of showing **MULTIPLICATION:**

1) with good old-fashioned "times" signs (×):

$f(x) = (2x \times 6y) + (2x \times \sin x) + (z \times y)$

The multiplication signs and the variable x are easily confused.

2) or sometimes just use a little dot:

$f(x) = 2x.6y + 2x.\sin x + z.y$

Dots are better for long expressions — they're less confusing and easier to read.

3) but you often don't need anything at all:

$f(x) = 12xy + 2x\sin x + zy$

And there's three different ways of showing **DIVISION:**

1) $\dfrac{x + 2}{3}$

2) $(x + 2) \div 3$

3) $(x + 2)/3$

Equations and Identities

This is an **IDENTITY:**

$x^2 - y^2 \equiv (x + y)(x - y)$

But this is an **EQUATION:**

$y = x^2 + x$

Make up any values you like for x and y, and it's always true. The left-hand side always equals the right-hand side.

The difference is that the identity's true for all values of x and y, but the equation's only true for certain values.

NB: If it's an identity, use the $\equiv$ sign instead of =.

This has at most two possible solutions for each value of y. e.g. if $y = 0$, x can only be 0 or -1.

Laws of Indices

You use the laws of indices a helluva lot in maths — when you're integrating, differentiating and ...er... well loads of other places. So take the time to get them sorted <u>now</u>.

Three mega-important *Laws of Indices*

You <u>must</u> know these three rules. I can't make it any clearer than that.

$$a^m \times a^n = a^{m+n}$$

If you <u>multiply</u> two numbers — you <u>add</u> their powers.

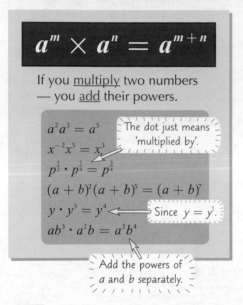

$a^2 a^3 = a^5$

The dot just means 'multiplied by'.

$x^{-2} x^5 = x^3$

$p^{\frac{1}{2}} \cdot p^{\frac{1}{4}} = p^{\frac{3}{4}}$

$(a+b)^2 (a+b)^5 = (a+b)^7$

$y \cdot y^3 = y^4$ ← Since $y = y^1$.

$ab^3 \cdot a^2 b = a^3 b^4$

Add the powers of *a* and *b* separately.

$$\frac{a^m}{a^n} = a^{m-n}$$

If you <u>divide</u> two numbers — you <u>subtract</u> their powers.

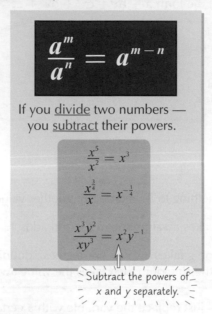

$\dfrac{x^5}{x^2} = x^3$

$\dfrac{x^{\frac{3}{4}}}{x} = x^{-\frac{1}{4}}$

$\dfrac{x^3 y^2}{xy^3} = x^2 y^{-1}$

Subtract the powers of *x* and *y* separately.

$$(a^m)^n = a^{mn}$$

If you have a <u>power</u> to the <u>power of something else</u> — <u>multiply</u> the powers together.

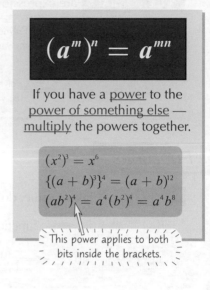

$(x^2)^3 = x^6$

$\{(a+b)^3\}^4 = (a+b)^{12}$

$(ab^2)^4 = a^4 (b^2)^4 = a^4 b^8$

This power applies to both bits inside the brackets.

Other important stuff about *Indices*

You can't get very far without knowing this sort of stuff. Learn it — you'll definitely be able to use it.

$$a^{\frac{1}{m}} = \sqrt[m]{a}$$

You can write <u>roots</u> as powers...

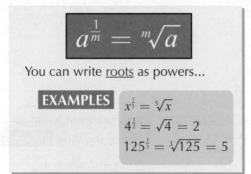

EXAMPLES

$x^{\frac{1}{5}} = \sqrt[5]{x}$

$4^{\frac{1}{2}} = \sqrt{4} = 2$

$125^{\frac{1}{3}} = \sqrt[3]{125} = 5$

$$a^{\frac{m}{n}} = \sqrt[n]{a^m} = \left(\sqrt[n]{a}\right)^m$$

A power that's a <u>fraction</u> like this is the <u>root of a power</u> — or the <u>power of a root</u>.

It's often easier to work out the root first, then raise it to the power.

EXAMPLES

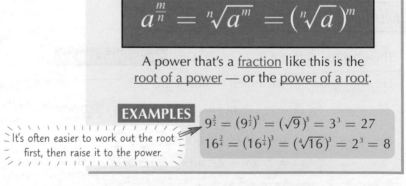

$9^{\frac{3}{2}} = (9^{\frac{1}{2}})^3 = (\sqrt{9})^3 = 3^3 = 27$

$16^{\frac{3}{4}} = (16^{\frac{1}{4}})^3 = (\sqrt[4]{16})^3 = 2^3 = 8$

$$a^{-m} = \frac{1}{a^m}$$

A <u>negative</u> power means it's on the bottom line of a fraction.

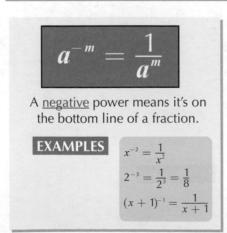

EXAMPLES

$x^{-2} = \dfrac{1}{x^2}$

$2^{-3} = \dfrac{1}{2^3} = \dfrac{1}{8}$

$(x+1)^{-1} = \dfrac{1}{x+1}$

$$a^0 = 1$$

This works for <u>any</u> number or letter.

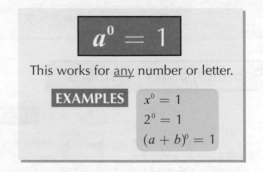

EXAMPLES

$x^0 = 1$

$2^0 = 1$

$(a+b)^0 = 1$

Indices, indices — de fish all live indices...

What can I say that I haven't said already? Blah, blah, important. Blah, blah, learn these. Blah, blah, use them all the time. Mmm, that's about all that needs to be said really. So I'll be quiet and let you get on with what you need to do.

Surds

A surd is a number like $\sqrt{2}$, $\sqrt[3]{12}$ or $5\sqrt{3}$ — one that's written with the $\sqrt{\ }$ sign. They're important because you can give <u>exact</u> answers where you'd otherwise have to round to a certain number of decimal places.

Surds are sometimes the only way to give an Exact Answer

Put $\sqrt{2}$ into a calculator and you'll get something like 1.414213562...
But square 1.414213562 and you get 1.999999999.

And no matter how many decimal places you use, you'll never get <u>exactly</u> 2.
The only way to write the exact, spot-on value is to <u>use surds</u>.

So, as you're not allowed a calculator for your C1 exam, leave your answer as a <u>surd</u>.

There are basically Three Rules for using Surds

There are three <u>rules</u> you'll need to know to be able to use surds properly. Check out the 'Rules of Surds' box below.

EXAMPLES (i) Simplify $\sqrt{12}$ and $\sqrt{\dfrac{3}{16}}$.   (ii) Show that $\dfrac{9}{\sqrt{3}} = 3\sqrt{3}$.   (iii) Find $(2\sqrt{5} + 3\sqrt{6})^2$.

(i) <u>Simplifying</u> surds means making the number in the $\sqrt{\ }$ sign <u>smaller</u>, or getting rid of a <u>fraction</u> in the $\sqrt{\ }$ sign.

$$\sqrt{12} = \sqrt{4 \times 3} = \sqrt{4} \times \sqrt{3} = 2\sqrt{3}$$
$$\sqrt{\frac{3}{16}} = \frac{\sqrt{3}}{\sqrt{16}} = \frac{\sqrt{3}}{4}$$
Using $\sqrt{\dfrac{a}{b}} = \dfrac{\sqrt{a}}{\sqrt{b}}$.

Using $\sqrt{ab} = \sqrt{a}\sqrt{b}$.

(ii) For questions like these, you have to write a number (here, it's 3) as $3 = (\sqrt{3})^2 = \sqrt{3} \times \sqrt{3}$

$$\frac{9}{\sqrt{3}} = \frac{3 \times 3}{\sqrt{3}} = \frac{3 \times \sqrt{3} \times \sqrt{3}}{\sqrt{3}} = 3\sqrt{3}$$
Cancelling $\sqrt{3}$ from the top and bottom lines.

(iii) Multiply surds very <u>carefully</u> — it's easy to make a silly mistake.

$$(2\sqrt{5} + 3\sqrt{6})^2 = (2\sqrt{5} + 3\sqrt{6})(2\sqrt{5} + 3\sqrt{6})$$
$$= (2\sqrt{5})^2 + 2 \times (2\sqrt{5}) \times (3\sqrt{6}) + (3\sqrt{6})^2$$
$$= (2^2 \times \sqrt{5}^2) + (2 \times 2 \times 3 \times \sqrt{5} \times \sqrt{6}) + (3^2 \times \sqrt{6}^2)$$
$$= 20 + 12\sqrt{30} + 54$$
$$= 74 + 12\sqrt{30}$$

$= 4 \times 5 = 20$
$= 9 \times 6 = 54$
$= 12\sqrt{5}\sqrt{6} = 12\sqrt{30}$

Rules of Surds
There's not really very much to remember.
$$\sqrt{ab} = \sqrt{a}\sqrt{b}$$
$$\sqrt{\frac{a}{b}} = \frac{\sqrt{a}}{\sqrt{b}}$$
$$a = (\sqrt{a})^2 = \sqrt{a}\sqrt{a}$$

Remove surds from fractions by Rationalising the Denominator

Surds are pretty darn complicated.

So they're the last thing you want at the bottom of a fraction.

But have no fear — <u>Rationalise the Denominator</u>...

Yup, you heard... (it means getting rid of the surds from the bottom of a fraction).

EXAMPLE Rationalise the denominator of $\dfrac{1}{1 + \sqrt{2}}$

Multiply the top and bottom by the denominator (but change the sign in front of the surd).
$$\frac{1}{1 + \sqrt{2}} \times \frac{1 - \sqrt{2}}{1 - \sqrt{2}}$$
$$\frac{1 - \sqrt{2}}{(1 + \sqrt{2})(1 - \sqrt{2})} = \frac{1 - \sqrt{2}}{1^2 + \sqrt{2} - \sqrt{2} - \sqrt{2}^2}$$
$$\frac{1 - \sqrt{2}}{1 - 2} = \frac{1 - \sqrt{2}}{-1} = -1 + \sqrt{2}$$

This works because:
$(a + b)(a - b) = a^2 - b^2$

Surely the pun is mightier than the surd...

You'll need to work with surds in your <u>non-calculator C1 exam</u>, as roots are nigh on impossible (well, very tricky) to work out without a calculator. Learn the rules in the box so you can write them down without thinking — then get <u>loads</u> of practice.

Multiplying Out Brackets

In this horrific nightmare that is AS-level maths, you need to manipulate and simplify expressions <u>all the time</u>.

Remove brackets by **Multiplying** them out

Here are the basic types you have to deal with. You'll have seen them before. But there's no harm in reminding you, eh?

<u>Multiply Your Brackets Here — we do all shapes and sizes</u>

Single Brackets

$$a(b + c + d) = ab + ac + ad$$

Double Brackets

$$(a + b)(c + d) = ac + ad + bc + bd$$

Squared Brackets

$$(a + b)^2 = (a + b)(a + b) = a^2 + 2ab + b^2$$

Use the middle stage until you're comfortable with it. Just <u>never</u> make this <u>mistake</u>: $(a + b)^2 = a^2 + b^2$

Long Brackets

Write it out again with <u>each term</u> from one bracket separately multiplied by the <u>other bracket</u>.

$$(x + y + z)(a + b + c + d)$$
$$= x(a + b + c + d) + y(a + b + c + d) + z(a + b + c + d)$$

Then <u>multiply out each</u> of these <u>brackets</u>, one at a time.

Single Brackets

Multiply all the terms inside the brackets by the bit outside — separately.

$$3xy(x^2 + 2x - 8)$$

All the stuff in the brackets now needs sorting out. Work on each bracket separately.

$$(3xy \times x^2) + (3xy \times 2x) + (3xy \times (-8))$$

I've put brackets round each bit to make it easier to read.

$$(3x^3y) + (6x^2y) + (-24xy)$$

Multiply the numbers first, then put the x's and other letters together.

$$3x^3y + 6x^2y - 24xy$$

Squared Brackets

Either write it as two brackets and multiply it out...

$$(2y^2 + 3x)^2$$

$$(2y^2 + 3x)(2y^2 + 3x)$$

The dot just means 'multiplied by' — the same as the × sign.

$$2y^2.2y^2 + 2y^2.3x + 3x.2y^2 + 3x.3x$$

From here on it's simplification — nothing more, nothing less.

$$4y^4 + 6xy^2 + 6xy^2 + 9x^2$$

$$4y^4 + 12xy^2 + 9x^2$$

...or do it in one go.

$$\underset{a^2}{(2y^2)^2} + \underset{2ab}{2(2y^2)(3x)} + \underset{b^2}{(3x)^2}$$

$$4y^4 + 12xy^2 + 9x^2$$

Long Brackets

$$(2x^2 + 3x + 6)(4x^3 + 6x^2 + 3)$$

Each term in the first bracket has been multiplied by the second bracket.

$$2x^2(4x^3 + 6x^2 + 3) + 3x(4x^3 + 6x^2 + 3) + 6(4x^3 + 6x^2 + 3)$$

Now multiply out each of these brackets.

$$(8x^5 + 12x^4 + 6x^2) + (12x^4 + 18x^3 + 9x) + (24x^3 + 36x^2 + 18)$$

Then simplify it all...

$$8x^5 + 24x^4 + 42x^3 + 42x^2 + 9x + 18$$

Go forth and multiply out brackets...

OK, so this is obvious, but I'll say it anyway — if you've got 3 or more brackets together, multiply them out 2 at a time. Then you'll be turning a really hard problem into two easy ones. You can do that loads in maths. In fact, writing the same thing in different ways is what maths is about. That and sitting in classrooms with tacky 'maths can be fun' posters...

Taking Out Common Factors

Common factors need to be hunted down, and taken outside the brackets. They are a danger to your exam mark.

Spot those **Common Factors**

A bit which is in each term of an expression is a <u>common factor</u>.

Spot Those Common Factors

$$2x^3z + 4x^2yz + 14x^2y^2z$$

Look for any bits that are in each term.

<u>Numbers:</u> there's a common factor of 2 here because 2 divides into 2, 4 and 14.

<u>Variables:</u> there's at least an x^2 in each term and there's a z in each term.

So there's a common factor of $2x^2z$ in this expression.

And Take Them Outside a Bracket

If you spot a common factor you can "<u>take it out</u>":

$$2x^2z(x + 2y + 7y^2)$$

Write the common factor outside a bracket.

and put what's left of each term inside the bracket.

Afterwards, always <u>multiply back out</u> to check you did it right:

Check by Multiplying Out Again

$$2x^2z(x + 2y + 7y^2) = 2x^3z + 4x^2yz + 14x^2y^2z$$

But it's not just numbers and variables you need to look for...

Brackets:

$$(y + a)^2(x - a)^3 + (x - a)^2$$

$(x-a)^2$ is a common factor
— it comes out to give:

$$(x - a)^2((y + a)^2(x - a) + 1)$$

Look for **Common Factors** when **Simplifying Expressions**

EXAMPLE Simplify... $(x + 1)(x - 2) + (x + 1)^2 - x(x + 1)$

There's an $(x + 1)$ factor in each term, so we can take this out as a common factor (hurrah).

$$(x + 1)\{(x - 2) + (x + 1) - x\}$$

The terms inside the big bracket are the old terms with an $(x + 1)$ removed.

At this point you should check that this multiplies out to give the original expression.
(You can just do this in your head, if you trust it.)

Then simplify the big bracket's innards:

$$(x + 1)(x - 2 + x + 1 - x)$$
$$= (x + 1)(x - 1)$$
$$= x^2 - 1$$

Get this answer by multiplying out the two brackets (or by using the "difference of two squares").

Bored of spotting trains or birds? Try common factors...

You'll be doing this business of taking out common factors a lot — so get your head round this. It's just a case of looking for things that are in all the different terms of an expression, i.e. <u>bits they have in common</u>. And if something's in all the different terms, save yourself some time and ink, and write it once — instead of two, three or more times.

Algebraic Fractions

No one likes fractions. But just like Mondays, you can't put them off forever. Face those fears. Here goes...

The first thing you've got to know about fractions:

You can just add the stuff on the top lines because the bottom lines are all the same.

$$\frac{a}{x} + \frac{b}{x} + \frac{c}{x} \equiv \frac{a+b+c}{x}$$

x is called a common denominator — a fancy way of saying 'the bottom line of all the fractions is x.'

Add fractions by putting them over a **Common Denominator**...

Finding a common denominator just means 'rewriting some fractions so all their bottom lines are the same'.

EXAMPLE Simplify $\frac{1}{2x} - \frac{1}{3x} + \frac{1}{5x}$

You need to rewrite these so that all the bottom lines are equal. What you want is something that all these bottom lines divide into.

Put It over a Common Denominator

30 is the lowest number that 2, 3 and 5 go into. So the common denominator is $30x$.

$$\frac{15}{30x} - \frac{10}{30x} + \frac{6}{30x}$$

Always check that these divide out to give what you started with.

$$\frac{15 - 10 + 6}{30x} = \frac{11}{30x}$$

...even **horrible** looking ones

Yep, finding a common denominator even works for those fraction nasties — like these:

EXAMPLE Simplify $\frac{3}{x+2} + \frac{5}{x-3}$

Find the Common Denominator

Take all the individual 'bits' from the bottom lines and multiply them together. Only use each bit once unless something on the bottom line is squared.

The individual 'bits' here are $(x+2)$ and $(x-3)$.

$$(x+2)(x-3)$$

Put Each Fraction over the Common Denominator

Make the denominator of each fraction into the common denominator.

$$\frac{3(x-3)}{(x+2)(x-3)} + \frac{5(x+2)}{(x+2)(x-3)}$$

Multiply the top and bottom lines of each fraction by whatever makes the bottom line the same as the common denominator.

Combine into One Fraction

Once everything's over the common denominator you can just add the top lines together.

$$= \frac{3(x-3) + 5(x+2)}{(x+2)(x-3)}$$

All the bottom lines are the same — so you can just add the top lines.

All you need to do now is tidy up the top.

$$= \frac{3x - 9 + 5x + 10}{(x+2)(x-3)} = \frac{8x+1}{(x+2)(x-3)}$$

So much prettier now all the terms are <u>together</u>. Simple.

Well put me over a common denominator and pickle my walrus...

Adding fractions — turning lots of fractions into one fraction. Sounds pretty good to me, since it means you don't have to write as much. Better do it carefully, though — otherwise you can watch those marks shoot straight down the toilet.

Simplifying Expressions

I know this is basic stuff but if you don't get really comfortable with it you <u>will</u> make silly mistakes. You will.

Cancelling stuff on the top and bottom lines

Cancelling stuff is good — because it means you've got rid of something, and you don't have to write as much.

EXAMPLE Simplify $\dfrac{ax + ay}{az}$

You can do this in two ways. Use whichever you prefer — but make sure you understand the ideas behind both.

Factorise — then Cancel

$$\frac{ax + ay}{az} = \frac{a(x + y)}{az}$$

Factorise the top line.

Cancel the 'a'. $\dfrac{\cancel{a}(x + y)}{\cancel{a}z} = \dfrac{x + y}{z}$

Split into Two Fractions — then Cancel

$$\frac{ax + ay}{az} = \frac{ax}{az} + \frac{ay}{az}$$

This is an okay thing to do — just think what you'd get if you added these.

$$= \frac{\cancel{a}x}{\cancel{a}z} + \frac{\cancel{a}y}{\cancel{a}z} = \frac{x}{z} + \frac{y}{z}$$

This answer's the same as the one from the first box — honest. Check it yourself by adding the fractions.

Simplifying complicated-looking **Brackets**

EXAMPLE Simplify the expression $(x - y)(x^2 + xy + y^2)$

There's only one thing to do here.... Multiply out those brackets!

$$(x - y)(x^2 + xy + y^2) = x(x^2 + xy + y^2) - y(x^2 + xy + y^2)$$

Multiplying each term in the first bracket by the second bracket.

$$= (x^3 + x^2y + xy^2) - (x^2y + xy^2 + y^3)$$

Multiplying out each of these two brackets.

$$= x^3 + x^2y + xy^2 - x^2y - xy^2 - y^3$$

Don't forget these become minus signs because of the minus sign in front of the bracket.

And then the x^2y and the xy^2 terms disappear...

$$= x^3 - y^3$$

Sometimes you just have to do **Anything** you can think of and **Hope**...

Sometimes it's not easy to see what you're supposed to do to simplify something.
When this happens — just do anything you can think of and see what 'comes out in the wash'.

EXAMPLE Simplify $4x + \dfrac{4x}{x + 1} - 4(x + 1)$

There's nothing obvious to do — so do what you can. Try adding them as fractions...

$$4x + \frac{4x}{x + 1} - 4(x + 1) = \frac{(x + 1) \times 4x}{x + 1} + \frac{4x}{x + 1} - \frac{(x + 1) \times 4x(x + 1)}{x + 1}$$

The common denominator is $(x+1)$.

$$= \frac{4x^2 + 4x + 4x - 4(x + 1)^2}{x + 1}$$

Still looks horrible. So work out the brackets — but don't forget the minus signs.

$$= \frac{4x^2 + 4x + 4x - 4x^2 - 8x - 4}{x + 1}$$

$$= -\frac{4}{x + 1}$$

Aha — everything disappears to leave you with this. And this is definitely simpler than it looked at the start.

Don't look at me like that...

Choose a word, any word at all. Like "Simple". Now stare at it. Keep staring at it. Does it look weird? No? Stare a bit longer. Now does it look weird? Yes? Why is that? I don't understand.

C1 Section 1 — Practice Questions

So that was the first section. Now, before you get stuck into Section Two, test yourself with these questions. Go on.
If you thought this section was a doddle, you should be able to fly through them...

Warm-up Questions

1) Pick out the constants and the variables from the following equations:
   a) $(ax + 6)^2 = 2b + 3$
   b) $f(x) = 12a + 3b^3 - 2$
   c) $y = \dfrac{-b \pm \sqrt{b^2 - 4ac}}{2a}$
   d) $\dfrac{dy}{dx} = x^2 + ax + 2$

2) What symbol should be used instead of the equals sign in identities?

3) Which of these are identities (i.e. true for all variable values)?
   A $(x + b)(y - b) = xy + b(y - x) - b^2$ B $(2y + x^2) = 10$
   C $a^2 - b^2 = (a - b)(a + b)$ D $a^3 + b^3 = (a + b)(a^2 - ab + b^2)$

4) Simplify these:
   a) $x^3.x^5$ b) $a^7.a^8$ c) $\dfrac{x^8}{x^2}$ d) $(a^2)^4$ e) $(xy^2).(x^3yz)$ f) $\dfrac{a^2b^4c^6}{a^3b^2c}$

5) Work out the following:
   a) $16^{\frac{1}{2}}$ b) $8^{\frac{1}{3}}$ c) $16^{\frac{3}{4}}$ d) x^0 e) $49^{-\frac{1}{2}}$

6) Find exact answers to these equations:
   a) $x^2 - 5 = 0$ b) $(x + 2)^2 - 3 = 0$

7) Simplify: a) $\sqrt{28}$ b) $\sqrt{\dfrac{5}{36}}$ c) $\sqrt{18}$ d) $\sqrt{\dfrac{9}{16}}$

8) Show that a) $\dfrac{8}{\sqrt{2}} = 4\sqrt{2}$, and b) $\dfrac{\sqrt{2}}{2} = \dfrac{1}{\sqrt{2}}$

9) Find $(6\sqrt{3} + 2\sqrt{7})^2$

10) Rationalise the denominator of: $\dfrac{2}{3 + \sqrt{7}}$

11) Remove the brackets and simplify the following expressions:
    a) $(a + b)(a - b)$
    b) $(a + b)(a + b)$
    c) $35xy + 25y(5y + 7x) - 100y^2$
    d) $(x + 3y + 2)(3x + y + 7)$

12) Take out the common factors from the following expressions:
    a) $2x^2y + axy + 2xy^2$
    b) $a^2x + a^2b^2x^2$
    c) $16y + 8yx + 56x$
    d) $x(x - 2) + 3(2 - x)$

13) Put the following expressions over a common denominator:
    a) $\dfrac{2x}{3} + \dfrac{y}{12} + \dfrac{x}{5}$ b) $\dfrac{5}{xy^2} - \dfrac{2}{x^2y}$ c) $\dfrac{1}{x} + \dfrac{x}{x + y} + \dfrac{y}{x - y}$

14) Simplify these expressions:
    a) $\dfrac{2a}{b} - \dfrac{a}{2b}$ b) $\dfrac{2p}{p + q} + \dfrac{2q}{p - q}$ c) "A bird in the hand is worth two in the bush"

C1 Section 1 — Practice Questions

The warm-up questions should have been a <u>walk in the park</u>, so now for the main event — <u>exam-style questions</u>. The questions on this page are like the ones you'll get <u>in your exam</u>, so make sure you can do them before moving on.

Exam Questions

1   a)   Write down the value of $27^{\frac{1}{3}}$.

(1 mark)

      b)   Find the value of $27^{\frac{4}{3}}$.

(2 marks)

2   Simplify

      a) $\left(5\sqrt{3}\right)^2$

(1 mark)

      b) $\left(5 + \sqrt{6}\right)\left(2 - \sqrt{6}\right)$

(2 marks)

3   Given that $10000\sqrt{10} = 10^{k}$, find the value of k.

(3 marks)

4   Express $\dfrac{5 + \sqrt{5}}{3 - \sqrt{5}}$ in the form $a + b\sqrt{5}$, where a and b are integers.

(4 marks)

5   Factorise completely

$$2x^4 - 32x^2.$$

(3 marks)

6   Write

$$\frac{x + 5x^3}{\sqrt{x}}$$

   in the form $x^m + 5x^n$, where m and n are constants.

(2 marks)

7   Show that

$$\frac{\left(5 + 4\sqrt{x}\right)^2}{2x}$$

   can be written as $\dfrac{25}{2}x^{-1} + Px^{-\frac{1}{2}} + Q$, and find the value of the integers P and Q.

(3 marks)

Sketching Quadratic Graphs

If a question doesn't seem to make sense, or you can't see how to go about solving a problem, try drawing a <u>graph</u>. It sometimes helps if you can actually <u>see</u> what the problem is, rather than just reading about it.

Sketch the graphs of the following quadratic functions:

① $y = 2x^2 - 4x + 3$ ② $y = 8 - 2x - x^2$

Quadratic graphs are **Always** u-shaped or n-shaped

 A The first thing you need to know is whether the graph's going to be u-shaped or n-shaped (upside down). To decide, look at the <u>coefficient of x^2</u>.

$y = 2x^2 - 4x + 3$

- The coefficient of x^2 here is <u>positive</u>... ...so the graph's u-shaped. → **+ve**

$y = 8 - 2x - x^2$

- The coefficient of x^2 here is <u>negative</u>... ...so the graph's upside down (n-shaped) → **−ve**

B Now find the places where the graph crosses the <u>axes</u> (both the y-axis and the x-axis).

(i) Put $x = 0$ to find where it meets the <u>y-axis</u>.

$y = 2x^2 - 4x + 3$

$y = (2 \times 0^2) - (4 \times 0) + 3$ so $y = 3$ ← That's where it crosses the y-axis.

(ii) Solve $y = 0$ to find where it meets the <u>x-axis</u>.

$2x^2 - 4x + 3 = 0$

$b^2 - 4ac = -8 < 0$ ← You could use the formula. But first check $b^2 - 4ac$ to see if $y = 0$ has any roots.

So it has no solutions, and doesn't cross the x-axis.

For more info, see page 16.

(i) Put $x = 0$.

$y = 8 - 2x - x^2$

$y = 8 - (2 \times 0) - 0^2$ so $y = 8$

(ii) Solve $y = 0$.

$8 - 2x - x^2 = 0$ ← This equation factorises easily...

$\Rightarrow (2 - x)(x + 4) = 0$

$\Rightarrow x = 2$ or $x = -4$

- The minimum or maximum of the graph is always at $x = \frac{-b}{2a}$
- The maximum value is <u>halfway</u> between the roots — the graph's symmetrical.

 C Finally, find the <u>minimum</u> or <u>maximum</u> (i.e. the <u>vertex</u>).

Since $y = 2(x - 1)^2 + 1$ ← By <u>completing the square</u> (see page 13).

the minimum value is $y = 1$, which occurs at $x = 1$.

The maximum value is at $x = -1$

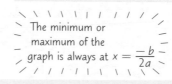 So the maximum is $y = 8 - (2 \times -1) - (-1)^2$

i.e. the graph has a maximum at the point (−1, 9).

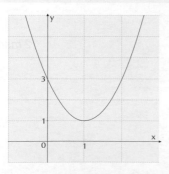

Sketching Quadratic Graphs

A) <u>up or down</u> — decide which direction the curve points in.

B) <u>axes</u> — find where the curve crosses them.

C) <u>max / min</u> — find the vertex.

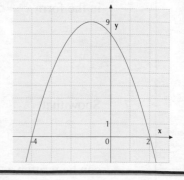

Van Gogh, Monet — all the greats started out sketching graphs...

So there are <u>three steps</u> here to learn. Simple enough. You can do the third step (finding the max/min point) by either a) completing the square, which is covered a bit later, or b) using the fact that the graph's symmetrical — so once you've found the points where it crosses the x-axis, the point halfway between them will be the max/min. It's all laughs here...

Factorising a Quadratic

Factorising a quadratic means putting it into <u>two brackets</u> — and is useful if you're trying to draw a graph of a quadratic or solve a quadratic equation. It's pretty easy if $a = 1$ (in $ax^2 + bx + c$ form), but can be a real pain otherwise.

$$x^2 - x - 12 = (x - 4)(x + 3)$$

Factorising's not so bad when a = 1

EXAMPLE Solve $x^2 - 8 = 2x$ by factorising.

A Put into $ax^2 + bx + c = 0$ Form

$x^2 - 2x - 8 = 0$ ⟵ So $a = 1$, $b = -2$, $c = -8$.
Write down the two brackets with x's in: $x^2 - 2x - 8 = (x\quad)(x\quad)$

B Find the Two Numbers

Find two numbers that <u>multiply</u> together to make c but which also <u>add</u> or <u>subtract</u> to give b (you can ignore any minus signs for now).

1 and 8 multiply to give 8 — and add / subtract to give 9 and 7.
2 and 4 multiply to give 8 — and add / subtract to give 6 and 2.

This is the value for b you're after — so this is the right combination: 2 and 4.

C Find the Signs

Now all you have to do is put in the plus or minus signs.

It must be +2 and −4 because 2 × (−4) = −8 and 2 + (−4) = 2 − 4 = −2

If c is negative, then the signs must be different.

$x^2 - 2x - 8 = (x\quad 4)(x\quad 2)$

$x^2 - 2x - 8 = (x + 2)(x - 4)$

D Solve the Equation

All you've done so far is to factorise the equation — you've still got to solve it.

$(x + 2)(x - 4) = 0$

$\Rightarrow x + 2 = 0$ or $x - 4 = 0$

Don't forget this last step. The factors aren't the answer.

$\Rightarrow x = -2$ or $x = 4$

Factorising Quadratics

A) **Rearrange the equation into the standard $ax^2 + bx + c$ form.**

B) **Write down the two brackets:**
$(x\quad)(x\quad)$

C) **Find two numbers that multiply to give 'c' and add / subtract to give 'b' (ignoring signs).**

D) **Put the numbers in the brackets and choose their signs.**

Another Example...

EXAMPLE Solve $x^2 + 4x - 21 = 0$ by factorising.

This equation is already in the standard format — you can write down the brackets straight away.

$x^2 + 4x - 21 = (x\quad)(x\quad)$

This is the value of 'b' you're after — 3 and 7 are the right numbers.

1 and 21 multiply to give 21 — and add / subtract to give 22 and 20.
3 and 7 multiply to give 21 — and add / subtract to give 10 and 4.

$x^2 + 4x - 21 = (x + 7)(x - 3)$

And solving the equation to find x gives... $\Rightarrow x = -7$ or $x = 3$

Scitardauq Gnisirotcaf — you should know it backwards...

Factorising quadratics — this is <u>very</u> basic stuff. You've really got to be comfortable with it. If you're even slightly rusty, you need to practise it until it's second nature. Remember why you're doing it — you don't factorise simply for the pleasure it gives you — it's so you can <u>solve</u> quadratic equations. Well, that's the theory anyway...

Factorising a Quadratic

It's not over yet...

Factorising a quadratic when a ≠ 1

These can be a real pain. The basic method's the same as on the previous page — but it can be a bit more awkward.

EXAMPLE   Factorise $3x^2 + 4x - 15$

A

Write Down Two Brackets

As before, write down two brackets — but instead of just having x in each, you need two things that will multiply to give $3x^2$.

It's got to be $3x$ and x here.

$$3x^2 + 4x - 15 = (3x \quad)(x \quad)$$

B

The Fiddly Bit

You need to find two numbers that multiply together to make 15 — but which will give you $4x$ when you multiply them by x and $3x$, and then add / subtract them.

$(3x \quad 1)(x \quad 15) \Rightarrow x$ and $45x$ which then add or subtract to give $46x$ and $44x$.

$(3x \quad 15)(x \quad 1) \Rightarrow 15x$ and $3x$ which then add or subtract to give $18x$ and $12x$.

$(3x \quad 3)(x \quad 5) \Rightarrow 3x$ and $15x$ which then add or subtract to give $18x$ and $12x$.

This is the value you're after — so this is the right combination.

$(3x \quad 5)(x \quad 3) \Rightarrow 5x$ and $9x$ which then add or subtract to give $14x$ and $4x$.

C

Add the Signs

You know the brackets must be like these... $\Longrightarrow (3x \quad 5)(x \quad 3) = 3x^2 + 4x - 15$

So all you have to do is put in the plus or minus signs.

'c' is negative — that means the signs in the brackets are different.

You've only got two choices — if you're unsure, just multiply them out to see which one's right.

$$(3x + 5)(x - 3) = 3x^2 - 4x - 15$$

or...

$$(3x - 5)(x + 3) = 3x^2 + 4x - 15 \quad \Longleftarrow \text{So it's this one.}$$

Sometimes it's best just to **Cheat** and use the **Formula**

Here are two final points to bear in mind:

1)   It <u>won't</u> always factorise.

2)   Sometimes factorising is so <u>messy</u> that it's easier to just use the quadratic formula...

So if the question doesn't tell you to factorise, don't assume it will factorise.
And if it's something like this thing below, don't bother trying to factorise it...

EXAMPLE   Solve $6x^2 + 87x - 144 = 0$

This <u>will</u> actually factorise, but there are 2 possible bracket forms to try.
$(6x \quad)(x \quad)$ or $(3x \quad)(2x \quad)$ And for each of these, there are 8 possible ways of making 144 to try.

And you can quote me on that...

"He who can properly do quadratic equations is considered a god."   "Quadratic equations are the music of reason."
Plato   James J Sylvester

Completing the Square

Completing the Square is a handy little trick that you should <u>definitely</u> know how to use.
It can be a bit fiddly — but it gives you <u>loads</u> of information about a quadratic really quickly.

Take any old quadratic and put it in a **Special Form**

Completing the square can be really confusing. For starters, what does "Completing the Square" <u>mean</u>?
<u>What</u> is the square? <u>Why</u> does it need completing? Well, there is <u>some</u> logic to it:

1)   The <u>square</u> is something like this: $(x + \text{something})^2$ It's basically the factorised equation (with the factors
both the same), but there's something missing...

2) ...So you need to '<u>complete</u>' it by adding a
number to the square to make it equal to the $(x + \text{something})^2 + d$
original equation.

You'll start with something like this...   ...sort the x-coefficients...   ...and you'll end up with something like this.

$$2x^2 + 8x - 5 \implies 2(x + 2)^2 + ? \implies 2(x + 2)^2 - 13$$

Lovely!

Make completing the square a bit **Easier**

There are only a few stages to completing the square — if you can't be bothered trying to understand it,
just <u>learn how to do it</u>. But I reckon it's worth spending a bit more time to get your head round it <u>properly</u>.

 A

| Take Out a Factor of 'a' |
— take a factor of a out
of the x^2 and x terms.

$f(x) = 2x^2 + 3x - 5$ ← This is in the form $ax^2 + bx + c$

This '2' is an 'a'.

$f(x) = 2\left(x^2 + \frac{3}{2}x\right) - 5$ ← Check that the bracket multiplies out to what you had before.

This is $\frac{b}{a}$

B | Rewrite the Bracket | — rewrite the bracket as one bracket squared.

The number in the brackets is <u>always</u> $\frac{b}{2a}$
half the old number in front of the x.

$f(x) = 2\left(x + \frac{3}{4}\right)^2 + d$

d is a number you have to find to make the new form equal to the old one.

Don't forget the 'squared' sign.

C | Complete the Square | — find d.

To do this, <u>make the old
and new equations equal
each other</u>...

$$2\left(x + \frac{3}{4}\right)^2 + d = 2x^2 + 3x - 5$$

...and you can find d.

$$2x^2 + 3x + \frac{9}{8} + d = 2x^2 + 3x - 5$$

The x^2 and x bits are the same on
both sides, so they can disappear.

$$\frac{9}{8} + d = -5$$

$$\Rightarrow d = -\frac{49}{8}$$

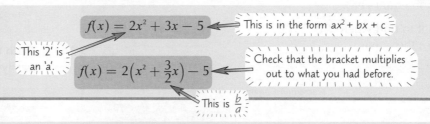

Completing the Square

A)   **THE BIT IN THE BRACKETS
IS ALWAYS —** $a\left(x + \frac{b}{2a}\right)^2$

B)   **CALL THE NUMBER
AT THE END d —** $a\left(x + \frac{b}{2a}\right)^2 + d$

C)   **MAKE THE TWO FORMS
EQUAL —** $ax^2 + bx + c = a\left(x + \frac{b}{2a}\right)^2 + d$

D | So the Answer is: |

$$f(x) = 2x^2 + 3x - 5 = 2\left(x + \frac{3}{4}\right)^2 - \frac{49}{8}$$

Complete your square — it'd be root not to...

Remember — you're basically trying to write the expression as one bracket squared, but it doesn't quite work. So you have
to add a number (d) to make it work. It's a bit confusing at first, but once you've learnt it you won't forget it in a hurry.

Completing the Square

Once you've completed the square, you can very quickly say <u>loads</u> about a quadratic function. And it all relies on the fact that a squared number can <u>never</u> be less than zero... <u>ever</u>.

Completing the square can sometimes be Useful

This is a quadratic written as a completed square. As it's a quadratic function and the coefficient of x^2 is positive, it's a u-shaped graph.

This is a square — it can never be negative. The smallest it can be is O.

$$f(x) = 3x^2 - 6x - 7 = 3(x-1)^2 - 10$$

A

Find the Minimum — make the bit in the brackets equal to zero.

When the squared bit is zero, $f(x)$ reaches its minimum value. This means the graph reaches its lowest point.

$$f(x) = 3(x-1)^2 - 10$$

This number here is the minimum.

$$f(1) = 3(1-1)^2 - 10$$

$f(1)$ means using $x = 1$ in the function

$$f(1) = 3(0)^2 - 10 = -10$$

So the minimum is −10, when $x = 1$

B

Where Does f(x) Cross the x-axis? — i.e. find x.

Make the completed square function equal zero.

$$3(x-1)^2 - 10 = 0$$

Solve it to find where $f(x)$ crosses the x-axis.

$$\Rightarrow (x-1)^2 = \frac{10}{3}$$

da-de-dah ... rearranging again.

$$\Rightarrow x - 1 = \pm\sqrt{\frac{10}{3}}$$

$$\Rightarrow x = 1 \pm \sqrt{\frac{10}{3}}$$

So $f(x)$ crosses the x-axis when...

$$x = 1 + \sqrt{\frac{10}{3}} \quad \text{or} \quad x = 1 - \sqrt{\frac{10}{3}}$$

These notes are all about graphs with positive coefficients in front of the x^2. But if the coefficient is negative, then the graph is flipped upside down (n-shaped, not u-shaped).

With this information, you can easily sketch the graph...

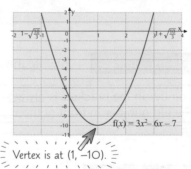

$$f(x) = 3x^2 - 6x - 7$$

Vertex is at (1, −10).

Some functions don't have Real Roots

By completing the square, you can also quickly tell if the graph of a quadratic function ever crosses the x-axis. It'll only cross the x-axis if the function changes sign (i.e. goes from positive to negative or vice versa). Take this function...

Find the Roots

$$f(x) = x^2 + 4x + 7$$

$$f(x) = (x+2)^2 + 3$$

This number's positive.

The smallest this bit can be is zero (at $x = -2$).

$(x + 2)^2$ is never less than zero so $f(x)$ is never less than three.

This means that:

a) $f(x)$ can <u>never</u> be negative.
b) The graph of $f(x)$ <u>never</u> crosses the x-axis.

If the coefficient of x^2 is negative, you can do the same sort of thing to check whether $f(x)$ ever becomes positive.

Don't forget — two wrongs don't make a root...

You'll be pleased to know that that's the end of me trying to tell you how to do something you probably really don't want to do. Now you can push it to one side and run off to roll around in a bed of nettles... much more fun.

The Quadratic Formula

Unlike factorising, the quadratic formula <u>always</u> works... no ifs, no buts, no butts, no nothing...

The **Quadratic Formula** — a reason to be cheerful, but careful...

If you want to solve a quadratic equation $ax^2 + bx + c = 0$,
then the answers are given by this formula:

$$x = \frac{-b \pm \sqrt{b^2 - 4ac}}{2a}$$

The formula's a godsend — but use the power wisely...

> If any of the coefficients (i.e. if *a*, *b* or *c*) in your quadratic equation are negative — be <u>especially</u> careful.

> Always take things nice and <u>slowly</u> — don't try to rush it.

> It's a good idea to write down what *a*, *b* and *c* are <u>before</u> you start plugging them into the formula.

> There are a couple of minus signs in the formula — which can catch you out if you're not paying <u>attention</u>.

I shall teach you the ways of the **Formula**

EXAMPLE: Solve the quadratic equation $3x^2 - 4x = 8$, leaving your answer in surd form.

> The mention of surds is a <u>big</u> clue that you should use the formula.

 Rearrange the Equation

Get the equation in the standard $ax^2 + bx + c = 0$ form.

$3x^2 - 4x = 8$

$3x^2 - 4x - 8 = 0$

 Find a, b and c

Write down the coefficients a, b and c — making sure you don't forget minus signs.

$3x^2 - 4x - 8 = 0$

$a = 3 \quad b = -4 \quad c = -8$

 Stick Them in the Formula

Very carefully, plug these numbers into the formula. It's best to write down each stage as you do it.

$$x = \frac{-b \pm \sqrt{b^2 - 4ac}}{2a}$$

$$x = \frac{-(-4) \pm \sqrt{(-4)^2 - 4 \times 3 \times (-8)}}{2 \times 3}$$

$$x = \frac{4 \pm \sqrt{16 + 96}}{6}$$

$$x = \frac{4 \pm \sqrt{112}}{6}$$

> Simplify your answer as much as possible, using the rules of surds (see page 3).

$$x = \frac{2 \pm 2\sqrt{7}}{3}$$

> The $\pm$ sign means that we have two different expressions for x — which you get by replacing the $\pm$ with + and −.

$$x = \frac{2 + 2\sqrt{7}}{3} \quad \text{or} \quad x = \frac{2 - 2\sqrt{7}}{3}$$

Using this magic formula, I shall take over the world... ha ha ha...

Okay, maybe it's not <u>quite</u> that good... but it's really important. So learn it properly — which means spending enough time until you can just say it out loud the whole way through, with no hesitations. Or perhaps you could try singing it as loud as you can to the tune of your favourite cheesy song. Sha-la-la-la-la-la-la-ha... La-di-da... Sha-la-la-la-la-la-la-ha...

The Quadratic Formula

By using part of the quadratic formula, you can quickly tell if a quadratic equation has two solutions, one solution, or no solutions at all. Tell me more, I hear you cry...

How Many Roots? *Check the b² – 4ac bit...*

$$x = \frac{-b \pm \sqrt{b^2 - 4ac}}{2a}$$

When you try to find the roots of a quadratic function, this bit in the square-root sign ($b^2 - 4ac$) can be positive, zero, or negative. It's <u>this</u> that tells you if a quadratic function has two roots, one root, or no roots.

The $b^2 - 4ac$ bit is called the <u>discriminant</u>.

<u>Because</u> — if the discriminant is positive, the formula will give you two different values — when you add or subtract the $\sqrt{b^2 - 4ac}$ bit.

<u>But</u> if it's zero, you'll only get one value, since adding or subtracting zero doesn't make any difference.

<u>And</u> if it's negative, you don't get any (real) values because you can't take the square root of a negative number.

Well, not in C1. In some areas of maths, you can actually take the square root of negative numbers and get 'imaginary' numbers. That's why we say no 'real' roots — because there are 'imaginary' roots!

It's good to be able to picture what this means:

A root is just the value of x when $y = 0$, so it's where the graph touches or crosses the x-axis.

$b^2 - 4ac > 0$	$b^2 - 4ac = 0$	$b^2 - 4ac < 0$
Two roots	One root	No roots

So the graph crosses the x-axis twice and these are the roots:

The graph just touches the x-axis from above (or from below if the x^2 coefficient is negative).

The graph doesn't touch the x-axis at all.

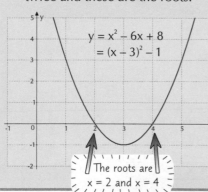

$y = x^2 - 6x + 8$
$= (x - 3)^2 - 1$

The roots are $x = 2$ and $x = 4$

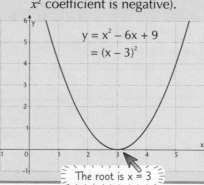

$y = x^2 - 6x + 9$
$= (x - 3)^2$

The root is $x = 3$

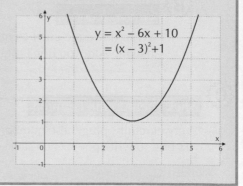

$y = x^2 - 6x + 10$
$= (x - 3)^2 + 1$

Identify *a*, *b* and *c* to find the **Discriminant**

The <u>first</u> thing you have to do when you're given a quadratic is to <u>work out</u> what a, b and c are. Make sure you get them the <u>right way round</u> — it's easy to get mixed up if the quadratic's in a <u>different order</u>.

EXAMPLE   Find the discriminant of $15 - x - 2x^2$. How many real roots does $15 - x - 2x^2$ have?

First, identify a, b and c: $a = -2$, $b = -1$ and $c = 15$ (NOT $a = 15$, $b = -1$ and $c = -2$).

Then put these values into the formula for the discriminant:

$b^2 - 4ac = (-1)^2 - (4 \times -2 \times 15) = 1 + 120 = 121.$

The discriminant is > 0, so $15 - x - 2x^2$ has two distinct real roots.

ha ha ha ha haaaaaa...   ha ha ha...   ha ha ha...   ha ha ha........

The Quadratic Formula

The discriminant often comes up in exam questions — but sometimes they'll be sneaky and not actually tell you that's what you have to find. Any question that mentions roots of a quadratic will probably mean that you need to find the discriminant.

a, *b* and *c* might be Unknown

In exam questions, you're often given a <u>quadratic</u> where one or more of *a*, *b* and *c* are given in terms of an <u>unknown</u> (usually *k*, but sometimes *p* or *q*). This means that you'll end up with an <u>equation</u> or <u>inequality</u> for the discriminant <u>in terms of the unknown</u> — you might have to <u>solve</u> it to find the <u>value</u> or <u>range of values</u> of the unknown.

EXAMPLE Find the range of values of *k* for which: a) f(*x*) has 2 distinct roots, b) f(*x*) has 1 root, c) f(*x*) has no real roots, where f(*x*) = $3x^2 + 2x + k$.

First, decide what *a*, *b* and *c* are: $a = 3, b = 2, c = k$

Then work out what the discriminant is: $b^2 - 4ac = 2^2 - 4 \times 3 \times k$ $= 4 - 12k$

These calculations are exactly the same. You don't need to do them if you've done a) because the only difference is the (in)equality symbol.

a) <u>Two distinct roots</u> means:

$b^2 - 4ac > 0 \Rightarrow 4 - 12k > 0$
$\Rightarrow 4 > 12k$
$\Rightarrow k < \frac{1}{3}$

b) <u>One root</u> means:

$b^2 - 4ac = 0 \Rightarrow 4 - 12k = 0$
$\Rightarrow 4 = 12k$
$\Rightarrow k = \frac{1}{3}$

c) <u>No roots</u> means:

$b^2 - 4ac < 0 \Rightarrow 4 - 12k < 0$
$\Rightarrow 4 < 12k$
$\Rightarrow k > \frac{1}{3}$

You might have to Solve a Quadratic Inequality to find *k*

When you put your values of *a*, *b* and *c* into the formula for the <u>discriminant</u>, you might end up with a <u>quadratic inequality</u> in terms of *k*. You'll have to solve this to find the range of values of *k* — there's more on this on p. 23.

EXAMPLE The equation $kx^2 + (k + 3)x + 4 = 0$ has two distinct real solutions. Show that $k^2 - 10k + 9 > 0$, and find the set of values of *k* which satisfy this inequality.

Identify *a*, *b* and *c*: $a = k, b = (k + 3)$ and $c = 4$

Then put these values into the formula for the discriminant:

$b^2 - 4ac = (k + 3)^2 - (4 \times k \times 4) = k^2 + 6k + 9 - 16k = k^2 - 10k + 9$.

The original equation has two distinct real solutions, so the discriminant must be > 0.

So $k^2 - 10k + 9 > 0$.

Now, to find the set of values for *k*, you have to factorise the quadratic:

$k^2 - 10k + 9 = (k - 1)(k - 9)$.

The solutions of this equation are *k* = 1 and *k* = 9. From the graph, you can see that this is a u-shaped quadratic which is > 0 when

$k < 1$ or when $k > 9$.

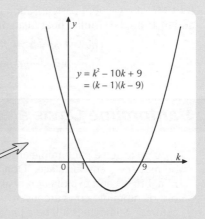

$y = k^2 - 10k + 9$
$= (k - 1)(k - 9)$

I'll try not to discriminate...

Don't panic if you're not sure how to solve quadratic inequalities — they're covered in more detail on p. 23. Chances are you'll get a discriminant question in the exam, so you need to know what to do. Although it might be tempting to hide under your exam desk and hope it doesn't find you, there's no escaping these questions — so get practising.

Cows

The stuff on this page isn't strictly on the syllabus. But I've included it anyway because I reckon it's really important stuff that you ought to know.

There are loads of Different Types of Cows

Dairy Cattle

Every day a dairy cow can produce up to 128 pints of milk — which can be used to make 14 lbs of cheese, 5 gallons of ice cream, or 6 lbs of butter.

The Jersey
The Jersey is a small breed best suited to pastures in high rainfall areas. It is kept for its creamy milk.

Advantages
1) Can produce creamy milk until old age.
2) Milk is the highest in fat of any dairy breed (5.2%).
3) Fairly docile, although bulls can't be trusted.

Disadvantages
1) Produces less milk than most other breeds.

The Holstein-Friesian
This breed can be found in many areas. It is kept mainly for milk.

Advantages
1) Produce more milk than any breed.
2) The breed is large, so bulls can be sold for beef.

Disadvantages
1) Milk is low in fat (3.5%).

Beef Cattle

Cows are sedentary animals who spend up to 8 hours a day chewing the cud while standing still or lying down to rest after grazing. Getting fat for people to eat.

The Angus
The Angus is best suited to areas where there is moderately high rainfall.

Advantages
1) Early maturing.
2) High ratio of meat to body weight.
3) Forages well.
4) Adaptable.

The Hereford
The Hereford matures fairly early, but later than most shorthorn breeds. All Herefords have white faces, and if a Hereford is crossbred with any other breed of cow, all the offspring will have white or partially white faces.

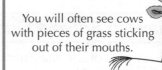

Advantages
1) Hardy.
2) Adaptable to different feeds.

Disadvantages
1) Susceptible to eye diseases.

Milk comes from Cows

> This is <u>really</u> important — try not to forget it.

Milk is an emulsion of butterfat suspended in a solution of water (roughly 80%), lactose, proteins and salts. Cow's milk has a specific gravity of around 1.03.

It's pasteurised by heating it to 63° C for 30 minutes. It's then rapidly cooled and stored below 10° C.

Louis Pasteur began his experiments into 'pasteurisation' in 1856. By 1946, the vacuum pasteurisation method had been perfected, and in 1948, UHT (ultra heat-treated) pasteurisation was introduced.

$$cow + grass = fat\ cow$$
$$fat\ cow + milking\ machine \Rightarrow milk$$

You will often see cows with pieces of grass sticking out of their mouths.

SOME IMPORTANT FACTS TO REMEMBER:
• A newborn calf can walk on its own an hour after birth.
• A cow's teeth are only on the bottom of her mouth.
• While some cows can live up to 40 years, they generally don't live beyond 20.

Pantomime Cows aren't Real

If you go to see a pantomime around Christmas time, you may see a cow on stage. Don't get concerned about animal rights and exploitation of animals — it's not a real cow. Pantomime cows are just two people wearing a cow costume. Sometimes it'll be a pantomime horse instead.

The Cow
The cow is of the bovine ilk;
One end is moo,
the other, milk.
— Ogden Nash

Famous Cows and Cow Songs

Famous Cows
1) Ermintrude from the Magic Roundabout.
2) The Laughing Cow.
3) Other TV commercial cows — Anchor, Dairylea
4) The cow that jumped over the moon.
5) Greek mythology was full of gods turning themselves and their girlfriends into cattle.

Cows in Pop Music
1) Boom Boom Cow — Black Eyed Peas
2) Saturday Night at the Moo-vies — The Drifters
3) I Kissed a Cow — Katy Perry
4) Take a Cow — Rihanna
5) Cows Don't Lie — Shakira

Where's me Jersey — I'm Friesian...

Cow-milking — an underrated skill, in my opinion. As Shakespeare once wrote, 'Those who can milk cows are likely to get pretty good grades in maths exams, no word of a lie'. Well, he probably would've written something like that if he was into cows. And he would've written it because cows are helpful when you're trying to work out what a question's all about — and once you know that, you can decide the best way forward. And if you don't believe me, remember the saying of the ancient Roman Emperor Julius Caesar — 'If in doubt, draw a cow'.

Factorising Cubics

Factorising a quadratic function is okay — but you might also be asked to <u>factorise a cubic</u> (something with x^3 in it). And that takes a bit more time — there are more steps, so there are more chances to make mistakes.

Factorising a cubic given **One Factor**

$$f(x) = 2x^3 + x^2 - 8x - 4$$

Factorising a cubic means exactly what it meant with a quadratic — putting brackets in. When they ask you to factorise a cubic equation, they'll usually tell you one of the factors.

EXAMPLE Given that $(x+2)$ is a factor of $f(x) = 2x^3 + x^2 - 8x - 4$, express $f(x)$ as the product of three linear factors.

① The first step is to find a quadratic factor. So write down the factor you know, along with another set of brackets.

$$(x + 2)(\qquad) = 2x^3 + x^2 - 8x - 4$$

Put the x^2 bit in this new set of brackets. These have to <u>multiply together</u> to give you this.

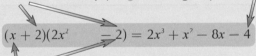

$$(x + 2)(2x^2 \qquad - 2) = 2x^3 + x^2 - 8x - 4$$

Factorising Cubics

1) **Write down the factor $(x-k)$.**

2) **Put in the x^2 term.**

3) **Put in the constant.**

4) **Put in the x term by comparing the number of x's on both sides.**

5) **Check there are the same number of x^2's on both sides.**

6) **Factorise the quadratic you've found — if that's possible.**

② Find the number for the second set of brackets. These have to <u>multiply together</u> to give you this.

$$(x + 2)(2x^2 \qquad - 2) = 2x^3 + x^2 - 8x - 4$$

③ These multiplied give you $-2x$, but there's $-8x$ in $f(x)$ — so you need an 'extra' $-6x$. And that's what this $-3x$ is for.

$$(x + 2)(2x^2 - 3x - 2) = 2x^3 + x^2 - 8x - 4$$

You only need $-3x$ because it's going to be multiplied by 2 — which makes $-6x$.

If every term in the cubic contains an 'x' (i.e. $ax^3 + bx^2 + cx$) then just take out x as your first factor before factorising the remaining quadratic as usual.

④ Before you go any further, check that there are the same number of x^2's on <u>both</u> sides.

$4x^2$ from here...

$$(x + 2)(2x^2 - 3x - 2) = 2x^3 + x^2 - 8x - 4$$

...and $-3x^2$ from here...   ...add together to give this x^2.

If this is okay, factorise the quadratic into two linear factors.

$$(2x^2 - 3x - 2) = (2x + 1)(x - 2)$$

<u>And so...</u> $2x^3 + x^2 - 8x - 4 = (x + 2)(2x + 1)(x - 2)$

If you wanted to solve a cubic, you'd do it exactly the same way — put it in the form $ax^3 + bx^2 + cx + d = 0$ and factorise.

I love the smell of fresh factorised cubics in the morning...

Factorising cubics is exactly the same as learning to unicycle... It's impossible at first. But when you finally manage it, it's really easy from then onwards and you'll never forget it. Probably. To tell the truth, I can't unicycle at all. So don't believe a word I say.

C1 Section 2 — Practice Questions

Mmmm, well, quadratic equations — not exactly designed to make you fall out of your chair through laughing so hard, are they? But (and that's a huge 'but') they'll get you <u>plenty of marks</u> come that fine morning when you march confidently into the exam hall — if you know <u>what you're doing</u>. Time for some <u>practice questions</u> methinks...

Warm-up Questions

1) Factorise the following expressions. While you're doing this, <u>sing a jolly song</u> to show how much you enjoy it.
   a) $x^2 + 2x + 1$,   b) $x^2 - 13x + 30$,   c) $x^2 - 4$,   d) $3 + 2x - x^2$,
   e) $2x^2 - 7x - 4$,   f) $5x^2 + 7x - 6$.

2) Solve the following equations. And <u>sing verse two</u> of your jolly song.
   a) $x^2 - 3x + 2 = 0$,   b) $x^2 + x - 12 = 0$,   c) $2 + x - x^2 = 0$,   d) $x^2 + x - 16 = x$,
   e) $3x^2 - 15x - 14 = 4x$,   f) $4x^2 - 1 = 0$,   g) $6x^2 - 11x + 9 = 2x^2 - x + 3$.

3) Rewrite these quadratics by <u>completing the square</u>. Then state their <u>maximum</u> or <u>minimum</u> value and the <u>value of x</u> where this occurs. Also, say if and where they <u>cross the x-axis</u> — just for a laugh, like.
   a) $x^2 - 4x - 3$,   b) $3 - 3x - x^2$,   c) $2x^2 - 4x + 11$,   d) $4x^2 - 28x + 48$.

4) How many <u>roots</u> do these quadratics have? <u>Sketch</u> their graphs.
   a) $x^2 - 2x - 3 = 0$,   b) $x^2 - 6x + 9 = 0$,   c) $2x^2 + 4x + 3 = 0$.

5) Solve these quadratic equations, leaving your answers in <u>surd form</u> where necessary.
   a) $3x^2 - 7x + 3 = 0$,   b) $2x^2 - 6x - 2 = 0$,   c) $x^2 + 4x + 6 = 12$.

   Have a peek at p.23 for help on solving a quadratic inequality.

6) If the quadratic equation $x^2 + kx + 4 = 0$ has <u>two roots</u>, what are the possible values of k?

The warm-up questions will only get you as far as the <u>third floor</u>. And you'll have to use the stairs.
To get all the way to the <u>top floor</u>, you need to take the <u>express lift</u> that is this lovely set of exam questions...

Exam Questions

1   The equation $x^2 + 2kx + 4k = 0$, where k is a non-zero integer, has equal roots.

   Find the value of k.

   (4 marks)

2   The equation $px^2 + (p + 3)x + 4 = 0$ has 2 distinct real solutions for x (p is a constant).

   a)   Show that $p^2 - 10p + 9 > 0$.

   (3 marks)

   b)   Hence find the range of possible values for p.

   (4 marks)

C1 Section 2 — Practice Questions

Two exam questions are <u>never enough</u> — so here are a few more...

3   Given that

$$5x^2 + nx + 14 \equiv m(x + 2)^2 + p,$$

find the values of the integers m, n and p.

(3 marks)

4   a)   Rewrite $x^2 - 12x + 15$ in the form $(x - a)^2 + b$, for integers a and b.

(2 marks)

      b)   (i)   Find the minimum value of $x^2 - 12x + 15$.

(1 mark)

            (ii) State the value of x at which this minimum occurs.

(1 mark)

5   a)   Use the quadratic formula to solve the equation $x^2 - 14x + 25 = 0$.
          Leave your answer in simplified surd form.

(3 marks)

      b)   Sketch the curve of $y = x^2 - 14x + 25$, giving the coordinates of the
          point where the curve crosses the x- and y-axis.

(3 marks)

      c)   Hence solve the inequality $x^2 - 14x + 25 \leq 0$.

(1 mark)

6   a)   (i)   Express $10x - x^2 - 27$ in the form $-(m - x)^2 + n$, where m and n are integers.

(2 marks)

            (ii) Hence show that $10x - x^2 - 27$ is always negative.

(1 mark)

      b)   (i)   State the coordinates of the maximum point of the curve $y = 10x - x^2 - 27$.

(2 marks)

            (ii) Sketch the curve, showing where the curve crosses the y-axis.

(2 marks)

Linear Inequalities

Solving <u>inequalities</u> is very similar to solving equations. You've just got to be really careful that you keep the inequality sign pointing the <u>right</u> way.

Find the ranges of x that satisfy these inequalities:

(i) $x - 3 < -1 + 2x$ (ii) $8x + 2 \geq 2x + 17$ (iii) $4 - 3x \leq 16$ (iv) $36x < 6x^2$

Sometimes the inequality sign *Changes Direction*

Like I said, these are pretty similar to solving equations — because whatever you do to one side, you have to do to the other. But multiplying or dividing by <u>negative</u> numbers <u>changes</u> the direction of the inequality sign.

> <u>Adding</u> or <u>subtracting</u> doesn't change the direction of the inequality sign

> <u>Multiplying</u> or <u>dividing</u> by something <u>positive</u> doesn't affect the inequality sign

EXAMPLE   If you <u>add</u> or <u>subtract</u> something from both sides of an inequality, the inequality sign <u>doesn't</u> change direction.

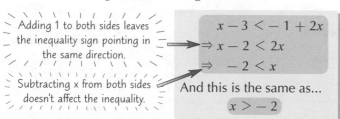

Adding 1 to both sides leaves the inequality sign pointing in the same direction.

Subtracting x from both sides doesn't affect the inequality.

$$x - 3 < -1 + 2x$$
$$\Rightarrow x - 2 < 2x$$
$$\Rightarrow -2 < x$$

And this is the same as...
$$x > -2$$

EXAMPLE   Multiplying or dividing both sides of an inequality by a <u>positive</u> number <u>doesn't</u> affect the direction of the inequality sign.

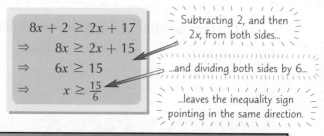

$$8x + 2 \geq 2x + 17$$
$$\Rightarrow \quad 8x \geq 2x + 15$$
$$\Rightarrow \quad 6x \geq 15$$
$$\Rightarrow \quad x \geq \frac{15}{6}$$

Subtracting 2, and then 2x, from both sides...

...and dividing both sides by 6...

...leaves the inequality sign pointing in the same direction.

But *Change* the inequality if you *Multiply* or *Divide* by something *Negative*

But multiplying or dividing both sides of an inequality by a <u>negative</u> number <u>changes</u> the direction of the inequality.

EXAMPLE

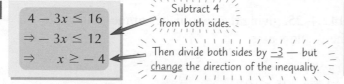

$$4 - 3x \leq 16$$
$$\Rightarrow -3x \leq 12$$
$$\Rightarrow \quad x \geq -4$$

Subtract 4 from both sides.

Then divide both sides by −3 — but <u>change</u> the direction of the inequality.

> The <u>reason</u> for the sign changing direction is because it's just the same as swapping everything from one side to the other:
> $$-3x \leq 12 \Rightarrow -12 \leq 3x \Rightarrow x \geq -4$$

Don't divide both sides by *Variables* — like x and y

You've got to be really careful when you divide by things that <u>might</u> be negative — well basically, don't do it.

EXAMPLE $36x < 6x^2$

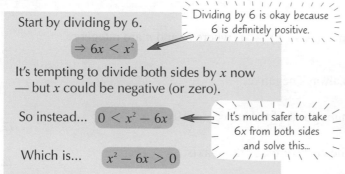

Start by dividing by 6.
$$\Rightarrow 6x < x^2$$

It's tempting to divide both sides by x now — but x could be negative (or zero).

So instead... $0 < x^2 - 6x$

Which is... $x^2 - 6x > 0$

Dividing by 6 is okay because 6 is definitely positive.

It's much safer to take 6x from both sides and solve this...

> ### *Two Types of Inequality Sign*
> **There are two kinds of inequality sign:**
> **Type 1:** $<$ — less than
> $>$ — greater than
> **Type 2:** $\leq$ — less than or equal to
> $\geq$ — greater than or equal to
> **Whatever type the question uses — use the same kind all the way through your answer.**

See the next page for more on solving quadratic inequalities.

So no one knows we've arrived safely — splendid...

So just remember — inequalities are just like normal equations except that you have to reverse the sign when multiplying or dividing by a negative number. And <u>don't</u> divide both sides by variables. (You should know not to do this with normal equations anyway because the variable could be <u>zero</u>.) OK — lecture's over.

Quadratic Inequalities

With quadratic inequalities, you're best off drawing the graph and taking it from there.

Draw a **Graph** to solve a **Quadratic** inequality

EXAMPLE   Find the ranges of x which satisfy these inequalities:

① $-x^2 + 2x + 4 \geq 1$

② $2x^2 - x - 3 > 0$

First rewrite the inequality with <u>zero</u> on one side.

$$-x^2 + 2x + 3 \geq 0$$

Then <u>draw</u> the graph of: $y = -x^2 + 2x + 3$

So find where it crosses the x-axis (i.e. where $y = 0$):

$$-x^2 + 2x + 3 \Rightarrow x^2 - 2x - 3 = 0$$
$$\Rightarrow (x + 1)(x - 3) = 0$$
$$\Rightarrow x = -1 \text{ or } x = 3$$

And the coefficient of x^2 is negative, so the graph is n-shaped. So it looks like this:

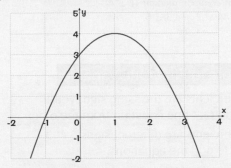

You're interested in when this is <u>positive or zero</u>, i.e. when it's above the x-axis.

From the graph, this is when x is <u>between –1 and 3</u> (including those points). So your answer is...

$$-x^2 + 2x + 4 \geq 1 \text{ when } -1 \leq x \leq 3.$$

This one already has zero on one side, so <u>draw</u> the graph of $y = 2x^2 - x - 3$.

Find where it crosses the x-axis:

$$2x^2 - x - 3 = 0$$
$$\Rightarrow (2x - 3)(x + 1) = 0$$
$$\Rightarrow x = \tfrac{3}{2} \text{ or } x = -1$$

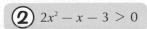
Factorise it to find the roots.

And the coefficient of x^2 is positive, so the graph is u-shaped. And looks like this:

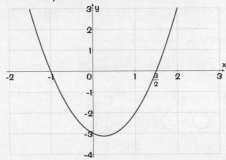

You need to say when this is <u>positive</u>. Looking at the graph, there are two parts of the x-axis where this is true — when x is <u>less than –1</u> and when x is <u>greater than 3/2</u>. So your answer is:

$$2x^2 - x - 3 > 0 \text{ when } x < -1 \text{ or } x > \tfrac{3}{2}.$$

EXAMPLE (REVISITED)   On the last page you had to solve $36x < 6x^2$.

$$36x < 6x^2$$
equation 1 $\Longrightarrow$ $6x < x^2$
$$\Rightarrow 0 < x^2 - 6x$$

So draw the graph of

$$y = x^2 - 6x = x(x - 6)$$

And this is <u>positive</u> when $x < 0$ or $x > 6$.

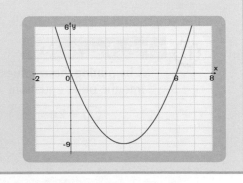

If you divide by x in equation 1, you'd only get half the solution — you'd miss the $x < 0$ part.

That's nonsense — I can see perfectly...

Call me sad, but I reckon these questions are pretty cool. They look a lot more difficult than they actually are and you get to draw a picture. Wow! When you do the graph, the important thing is to find where it crosses the x-axis (you don't need to know where it crosses the y-axis) and make sure you draw it the right way up. Then you just need to decide which bit of the graph you want. It'll either be the range(s) of x where the graph is below the x-axis or the range(s) where it's above. And this depends on the inequality sign.

Simultaneous Equations

Solving simultaneous equations means finding the answers to two equations <u>at the same time</u> — i.e. finding values for x and y for which both equations are true. And it's one of those things that you'll have to do <u>again and again</u> — so it's definitely worth practising them until you feel <u>really confident</u>.

① $3x + 5y = -4$
② $-2x + 3y = 9$

This is how simultaneous equations are usually shown. It's a good idea to label them as equation ① and equation ② — so you know which one you're working with.

But they'll look different sometimes, maybe like this. Make sure you rearrange them as '$ax + by = c$'.

$4 + 5y = -3x$
$-2x = 9 - 3y$

rearrange as
$ax + by = c$

$3x + 5y = -4$
$-2x + 3y = 9$

Solving them by *Elimination*

Elimination is a lovely method. It's really quick when you get the hang of it — you'll be doing virtually all of it in your head.

EXAMPLE:

① $3x + 5y = -4$
② $-2x + 3y = 9$

To get the x's to match, you need to multiply the first equation by 2 and the second by 3:

①×2 $\quad 6x + 10y = -8$
②×3 $\quad -6x + 9y = 27$

Add the equations together to eliminate the x's.

①+② $\quad 19y = 19$
$\quad\quad\quad y = 1$

So y is 1. Now stick that value for y into one of the equations to find x:

$y = 1$ in ① $\Rightarrow 3x + 5 = -4$
$\quad\quad\quad\quad\quad 3x = -9$
$\quad\quad\quad\quad\quad\quad x = -3$

So the solution is $x = -3$, $y = 1$.

A Match the Coefficients

Multiply the equations by numbers that will make either the x's or the y's match in the two equations. (Ignoring minus signs.)

Go for the lowest common multiple (LCM). e.g. LCM of 2 and 3 is 6.

B Eliminate to Find One Variable

If the coefficients are the <u>same</u> sign, you'll need to <u>subtract</u> one equation from the other.

If the coefficients are <u>different</u> signs, you need to <u>add</u> the equations.

C Find the Variable You Eliminated

When you've found one variable, put its value into one of the original equations so you can find the other variable.

But you should always...

D Check Your Answer

...by putting these values into the other equation.

② $-2x + 3y = 9$
$\quad\quad x = -3$
$\quad\quad y = 1$

$-2 \times (-3) + 3 \times 1 = 6 + 3 = 9$

If these two numbers are the same, then the values you've got for the variables are right.

Elimination Method

1) **Match the coefficients**

2) **Eliminate and then solve for one variable**

3) **Find the other variable (that you eliminated)**

4) **Check your answer**

Eliminate your social life — do AS-level maths

This is a fairly basic method that won't be new to you. So make sure you know it. The only possibly tricky bit is <u>matching the coefficients</u> — work out the lowest common multiple of the coefficients of x, say, then multiply the equations to get this number in front of each x.

Simultaneous Equations with Quadratics

Elimination is great for simple equations. But it won't always work. Sometimes one of the equations has not just x's and y's in it — but bits with x^2 and y^2 as well. When this happens, you can <u>only</u> use the <u>substitution</u> method.

Use Substitution if one equation is **Quadratic**

EXAMPLE: $\quad -x + 2y = 5 \quad$ ⓛ $\longleftarrow$ The <u>linear</u> equation — with only x's and y's in.
$\qquad\qquad x^2 + y^2 = 25 \quad$ ⓠ $\longleftarrow$ The <u>quadratic</u> equation — with some x^2 and y^2 bits in.

Rearrange the <u>linear equation</u> so that either x or y is on its own on one side of the equals sign.

$$\text{ⓛ} \ -x + 2y = 5$$
$$\Rightarrow x = 2y - 5$$

Substitute this expression into the <u>quadratic equation</u>...

$$\text{Sub into ⓠ:} \quad x^2 + y^2 = 25$$
$$\Rightarrow (2y - 5)^2 + y^2 = 25$$

...and then rearrange this into the form $ax^2 + bx + c = 0$, so you can solve it — either by <u>factorising</u> or using the <u>quadratic formula</u>.

$$\Rightarrow (4y^2 - 20y + 25) + y^2 = 25$$
$$\Rightarrow 5y^2 - 20y = 0$$
$$\Rightarrow 5y(y - 4) = 0$$
$$\Rightarrow y = 0 \ \text{or} \ y = 4$$

One Quadratic and One Linear Eqn

1) **Isolate variable in linear equation**
   Rearrange the linear equation to get either x or y on its own.

2) **Substitute into quadratic equation** — to get a quadratic equation in just one variable.

3) **Solve to get values for one variable** — either by factorising or using the quadratic formula.

4) **Stick these values in the linear equation** — to find corresponding values for the other variable.

Finally put both these values back into the <u>linear equation</u> to find corresponding values for x:

When $y = 0$: $-x + 2y = 5$ ⓛ
$$\Rightarrow x = -5$$

When $y = 4$: $-x + 2y = 5$ ⓛ
$$\Rightarrow -x + 8 = 5$$
$$\Rightarrow x = 3$$

So the solutions to the simultaneous equations are: $x = -5$, $y = 0$ and $x = 3$, $y = 4$.

As usual, <u>check your answers</u> by putting these values back into the original equations.

Check Your Answer

$x = -5$, $y = 0$: $-(-5) + 2 \times 0 = 5$ ✓
$\qquad\qquad\qquad (-5)^2 + 0^2 = 25$ ✓

$x = 3$, $y = 4$: $-(3) + 2 \times 4 = 5$ ✓
$\qquad\qquad\qquad 3^2 + 4^2 = 25$ ✓

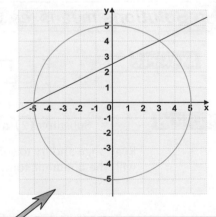

$y = x^2$ — a match-winning substitution...

The quadratic equation above is actually a <u>circle</u> about the origin with radius 5. (Don't worry, you don't need to know about circles till C2.) The linear equation is just a standard straight line. So what you're actually finding here are the two points where the line passes through the circle. And these turn out to be (–5, 0) and (3, 4). See the graph. (I thought you might appreciate seeing a graph that wasn't a line or a parabola for a change.)

Geometric Interpretation

When you have to interpret something <u>geometrically</u>, you have to draw a picture and 'say what you see'.

Two Solutions — Two points of Intersection

EXAMPLE

$y = x^2 - 4x + 5$ —— ①
$y = 2x - 3$ —— ②

SOLUTION

Substitute expression for y from ② into ①: $\quad 2x - 3 = x^2 - 4x + 5$

Rearrange and solve:
$$x^2 - 6x + 8 = 0$$
$$(x - 2)(x - 4) = 0$$
$$x = 2 \text{ or } x = 4$$

In ② gives:
$$x = 2 \Rightarrow y = 2 \times 2 - 3 = 1$$
$$x = 4 \Rightarrow y = 2 \times 4 - 3 = 5$$

There are 2 pairs of solutions: $x = 2, y = 1$ and $x = 4, y = 5$

Geometric Interpretation

So from solving the simultaneous equations, you know that the graphs meet in <u>two places</u> — the points (2, 1) and (4, 5).

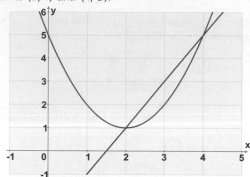

One Solution — One point of Intersection

EXAMPLE

$y = x^2 - 4x + 5$ —— ①
$y = 2x - 4$ —— ②

SOLUTION

Substitute ② in ①: $\quad 2x - 4 = x^2 - 4x + 5$

Rearrange and solve:
$$x^2 - 6x + 9 = 0$$
$$(x - 3)^2 = 0$$
$$x = 3$$

Double root — i.e. you only get 1 solution from the quadratic.

In Equation ② gives:
$$y = 2 \times 3 - 4$$
$$y = 2$$

There's 1 solution: $x = 3, y = 2$

Geometric Interpretation

Since the equations have only one solution, the two graphs only meet at one point — (3, 2). The straight line is a <u>tangent</u> to the curve.

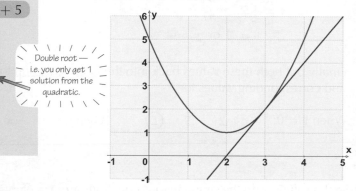

No Solutions means the Graphs Never Meet

EXAMPLE

$y = x^2 - 4x + 5$ —— ①
$y = 2x - 5$ —— ②

SOLUTION

Substitute ② in ①: $\quad 2x - 5 = x^2 - 4x + 5$

Rearrange and try to solve with the quadratic formula:
$$x^2 - 6x + 10 = 0$$
$$b^2 - 4ac = (-6)^2 - 4 \cdot 10$$
$$= 36 - 40 = -4$$

$b^2 - 4ac < 0$, so the quadratic has no roots.
So the simultaneous equations have no solutions.

Geometric Interpretation

The equations have no solutions — the graphs never meet.

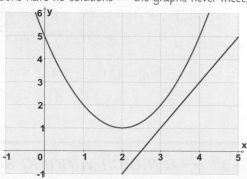

Geometric Interpretation? Frankly my dear, I don't give a damn...

There are some lovely simultaneous equation practice questions on the next page.

C1 Section 3 — Practice Questions

What's that I hear you cry? You want practice questions — and <u>lots of them</u>. Well, it just so happens I've got a <u>few here</u>. For quadratic inequalities, my advice is, 'if you're not sure, draw a picture — even if it's not accurate'. And as for simultaneous equations — well, just <u>don't rush them</u>.

Warm-up Questions

1) Solve a) $7x - 4 > 2x - 42$,   b) $12y - 3 \leq 4y + 4$,   c) $9y - 4 \geq 17y + 2$.

2) Find the <u>ranges of x</u> that satisfy these inequalities:
   a) $x + 6 < 5x - 4$,   b) $4x - 2 > x - 14$,   c) $7 - x \leq 4 - 2x$

3) Find the <u>ranges of x</u> that satisfy the following inequalities. (And watch that you use the <u>right kind</u> of inequality sign in your answers.)
   a) $3x^2 - 5x - 2 \leq 0$,   b) $x^2 + 2x + 7 > 4x + 9$,   c) $3x^2 + 7x + 4 \geq 2(x^2 + x - 1)$.

4) Find the ranges of x that satisfy these <u>jokers</u>:
   a) $x^2 + 3x - 1 \geq x + 2$,   b) $2x^2 > x + 1$,   c) $3x^2 - 12 < x^2 - 2x$

5) Solve these sets of simultaneous equations:
   a) $3x - 4y = 7$ and $-2x + 7y = -22$,   b) $2x - 3y = \frac{11}{12}$ and $x + y = -\frac{7}{12}$

6) Find where possible (and that's a bit of a <u>clue</u>) the solutions to these sets of simultaneous equations. <u>Interpret</u> your answers <u>geometrically</u>.
   a) $y = x^2 - 7x + 4$ b) $y = 30 - 6x + 2x^2$ c) $x^2 + 2y^2 - 3 = 0$
       $2x - y - 10 = 0$ $y = 2(x + 11)$ $y = 2x + 4$

7) <u>A bit trickier</u> — find where the following lines <u>meet</u>:
   a) $y = 3x - 4$ and $y = 7x - 5$
   b) $y = 13 - 2x$ and $7x - y - 23 = 0$
   c) $2x - 3y + 4 = 0$ and $x - 2y + 1 = 0$

I know, I know. Those questions <u>weren't enough</u> for you. Not to worry, there are plenty more — and the next set are <u>exam-style</u> questions. Try to contain your excitement.

Exam Questions

1   For the inequalities below, find the set of values for x:

    a) $3x + 2 \leq x + 6$,

(2 marks)

    b) $20 - x - x^2 > 0$,

(4 marks)

    c) $3x + 2 \leq x + 6$ and $20 - x - x^2 > 0$.

(1 mark)

C1 Section 3 — Practice Questions

The world is full of inequality and injustice. We need to <u>put a stop to it now</u>. The first thing we need to do is...
Oh, sorry, that's not what they meant by "solve the inequality". And I'd just come up with a solution for world peace.

2   Solve the inequalities:

   a) $3 \leq 2p + 5 \leq 15$,

   (3 marks)

   b) $q^2 - 9 > 0$.

   (4 marks)

3   a)   Factorise $3x^2 - 13x - 10$.

   (1 mark)

   b)   Hence, or otherwise, solve $3x^2 - 13x - 10 \leq 0$.

   (3 marks)

4   Find the coordinates of intersection for the following curve and line:
$$x^2 + 2y^2 = 36, \quad x + y = 6$$

   (6 marks)

5   The curve C has equation $y = -x^2 + 3$ and the line l has equation $y = -2x + 4$.

   a)   Find the coordinates of the point (or points) of intersection of C and l.

   (4 marks)

   b)   Sketch the graphs of C and l on the same axes, clearly showing
      where the graphs intersect the x- and y- axes.

   (5 marks)

6   The line l has equation $y = 2x - 3$ and the curve C has equation $y = (x + 2)(x - 4)$.

   a)   Sketch the line l and the curve C on the same axes, showing the coordinates
      of the x- and y- intercepts.

   (5 marks)

   b)   Show that the x-coordinates of the points of intersection of l and C satisfy the equation
      $x^2 - 4x - 5 = 0$.

   (2 marks)

   c)   Hence, or otherwise, find the points of intersection of l and C.

   (4 marks)

Coordinate Geometry

Welcome to geometry club... nice — today I shall be mostly talking about straight lines...

Finding the equation of a line **Through Two Points**

If you get through your exam without having to find the equation of a line through two points, I'm a Dutchman.

EXAMPLE Find the equation of the line that passes through the points (–3, 10) and (1, 4), and write it in the forms:

$$y - y_1 = m(x - x_1)$$

$$y = mx + c$$

$$ax + by + c = 0$$

— where a, b and c are integers.

You might be asked to write the equation of a line in <u>any</u> of these forms — but they're all similar.
Basically, if you find an equation in one form — you can easily <u>convert</u> it into either of the others.

The **Easiest** to find is $y - y_1 = m(x - x_1)$...

Point 1 is (–3, 10) and Point 2 is (1, 4).

Label the Points Label Point 1 as (x_1, y_1) and Point 2 as (x_2, y_2).

Point 1 — $(x_1, y_1) = (-3, 10)$

Point 2 — $(x_2, y_2) = (1, 4)$

It doesn't matter which way round you label them.

Find the Gradient Find the <u>gradient</u> of the line m — this is $m = \frac{y_2 - y_1}{x_2 - x_1}$.

$$m = \frac{4 - 10}{1 - (-3)} = \frac{-6}{4} = -\frac{3}{2}$$

Be careful here, y goes on the top, x on the bottom.

Write Down the Equation <u>Write down</u> the equation of the line, using the coordinates x_1 and y_1 — this is just $y - y_1 = m(x - x_1)$.

$x_1 = -3$ and $y_1 = 10 \Longrightarrow$

$$y - 10 = -\frac{3}{2}(x - (-3))$$

$$y - 10 = -\frac{3}{2}(x + 3)$$

...and **Rearrange** this to get the other two forms:

For the form $y = mx + c$, take everything except the y over to the right.

$$y - 10 = -\frac{3}{2}(x + 3)$$

$$\Rightarrow y = -\frac{3}{2}x - \frac{9}{2} + 10$$

$$\Rightarrow y = -\frac{3}{2}x + \frac{11}{2}$$

To find the form $ax + by + c = 0$, take everything over to one side — and then get rid of any fractions.

Multiply the whole equation by 2 to get rid of the 2's on the bottom line.

$$y = -\frac{3}{2}x + \frac{11}{2}$$

$$\Rightarrow \frac{3}{2}x + y - \frac{11}{2} = 0$$

$$\Rightarrow 3x + 2y - 11 = 0$$

Equations of Lines

1) **LABEL** the points (x_1, y_1) and (x_2, y_2).

2) **GRADIENT** — find it and call it m.

3) **WRITE DOWN THE EQUATION** using $y - y_1 = m(x - x_1)$

4) **CONVERT** to one of the other forms, if necessary.

If you end up with an equation like $\frac{3}{2}x - \frac{4}{3}y + 6 = 0$, where you've got a 2 and a 3 on the bottom of the fractions — multiply everything by the <u>lowest common multiple</u> of 2 and 3, i.e. 6.

There ain't nuffink to this geometry lark, Mister...

This is the sort of stuff that looks hard but is actually pretty easy. Finding the equation of a line in that first form really is a piece of cake — the only thing you have to be careful of is when a point has a <u>negative coordinate</u> (or two). In that case, you've just got to make sure you do the subtractions properly when you work out the gradient. See, this stuff ain't so bad...

Coordinate Geometry

This page is based around two really important facts that you've got to know — one about <u>parallel lines</u>, one about <u>perpendicular lines</u>. It's really a page of unparalleled excitement...

Two more lines...

Line l_1
$3x - 4y - 7 = 0$
$y = \frac{3}{4}x - \frac{7}{4}$

Line l_2
$x - 3y - 3 = 0$
$y = \frac{1}{3}x - 1$

...and two points...

Point A $(3, -1)$
Point B $(-2, 4)$

Parallel lines have equal *Gradient*

That's what makes them parallel — the fact that the gradients are the same.

EXAMPLE Find the line parallel to l_1 that passes through the point A $(3, -1)$.

Parallel lines have the <u>same gradient</u>.

The original equation is this: $\quad y = \frac{3}{4}x - \frac{7}{4}$

So the new equation will be this: $\quad y = \frac{3}{4}x + c$

We just need to find c.

We know that the line passes through A, so at this point x will be 3, and y will be -1.

Stick these values into the equation to find c.

$$-1 = \frac{3}{4} \times 3 + c$$

$$\Rightarrow c = -1 - \frac{9}{4} = -\frac{13}{4}$$

So the equation of the line is... $\quad y = \frac{3}{4}x - \frac{13}{4}$

And if you're only given the $ax + by + c = 0$ form it's even easier:

The <u>original</u> line is: $\quad 3x - 4y - 7 = 0$

So the <u>new</u> line is: $\quad 3x - 4y - k = 0$

Then just use the values of x and y at the point A to find k...

$$3 \times 3 - 4 \times (-1) - k = 0$$

$$\Rightarrow 13 - k = 0$$

$$\Rightarrow k = 13$$

So the equation is: $\quad 3x - 4y - 13 = 0$

The gradient of a *Perpendicular* line is: *−1 ÷ the Other Gradient*

Finding <u>perpendicular</u> lines (or '<u>normals</u>') is just as easy as finding parallel lines — as long as you remember the gradient of the perpendicular line is <u>−1 ÷ the gradient of the other one</u>.

EXAMPLE Find the line perpendicular to l_2 that passes through the point B $(-2, 4)$.

l_2 has equation: $\quad y = \frac{1}{3}x - 1$

So if the equation of the new line is $y = mx + c$, then

$$m = -1 \div \frac{1}{3}$$

$$\Rightarrow m = -3$$

Since the gradient of a perpendicular line is: −1 ÷ the other one.

Also... $\quad 4 = (-3) \times (-2) + c$

$$\Rightarrow c = 4 - 6 = -2$$

Putting the coordinates of B(−2, 4) into y = mx + c.

So the equation of the line is...

$$y = -3x - 2$$

Or if you start with: $\quad l_2 \quad x - 3y - 3 = 0$

To find a perpendicular line, swap these two numbers around, and change the sign of <u>one of them</u>. (So here, 1 and −3 become 3 and 1.)

So the new line has equation...

$$3x + y + d = 0$$

Or you could have used −3x − y + d = O.

But... $\quad 3 \times (-2) + 4 + d = 0$

$$\Rightarrow d = 2$$

Using the coordinates of point B.

And so the equation of the <u>perpendicular</u> line is...

$$3x + y + 2 = 0$$

Wowzers — parallel lines on the same graph dimension...

This looks more complicated than it actually is. All you're doing is finding the equation of a straight line through a <u>certain point</u> — the only added complication is that you have to find the gradient first. And there's another way to remember how to find the gradient of a normal — just remember that the gradients of perpendicular lines multiply together to make −1.

Curve Sketching

A picture speaks a thousand words... and <u>graphs</u> are what pass for pictures in maths. They're dead useful in getting your head round tricky questions, and time spent learning how to sketch graphs is time well spent.

The graph of **y = kxⁿ** is a different shape for different **k** and **n**

Usually, you only need a <u>rough</u> sketch of a graph — so just knowing the basic shapes of these graphs will do.

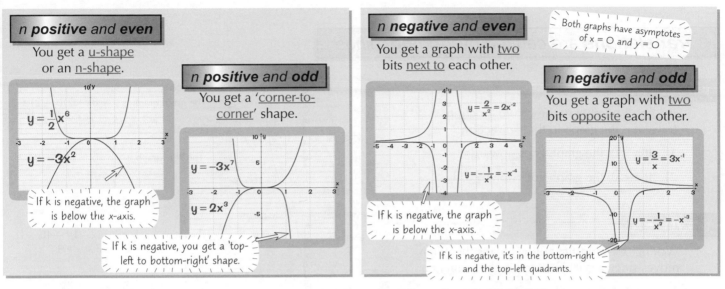

An <u>asymptote</u> of a curve is a <u>line</u> which the curve gets <u>infinitely close</u> to, but <u>never touches</u>.

If you know the **Factors** of a cubic — the graph's easy to **Sketch**

A cubic function has an x^3 term in it, and all cubics have '<u>bottom-left to top-right</u>' shape — or a '<u>top-left to bottom-right</u>' shape if the coefficient of x^3 is <u>negative</u>.

If you know the <u>factors</u> of a cubic, the graph is easy to sketch — just find where the function is <u>zero</u>.

EXAMPLE Sketch the graphs of the following <u>cubic</u> functions.

(i) $f(x) = x(x-1)(2x+1)$ (ii) $g(x) = (1-x)(x^2-2x+2)$ (iii) $h(x) = (x-3)^2(x+1)$ (iv) $m(x) = (2-x)^3$

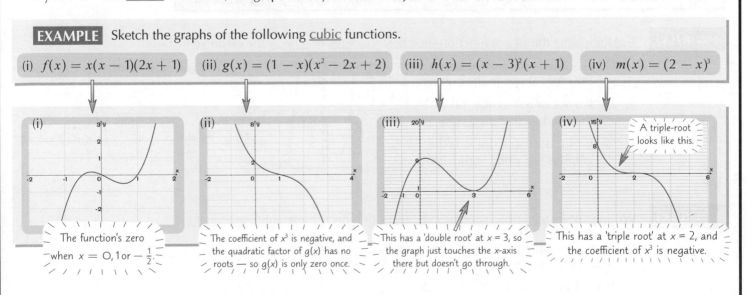

Graphs, graphs, graphs — you can never have too many graphs...

It may seem like a lot to remember, but graphs can really help you get your head round a question — a quick sketch can throw a helluva lot of light on a problem that's got you completely stumped. So being able to draw these graphs won't just help with an actual graph-sketching question — it could help with loads of others too. Got to be worth learning.

Graph Transformations

Suppose you start with any old function f(x). Then you can transform (change) it in three ways — by translating it, stretching or reflecting it.

$y = f(x)$

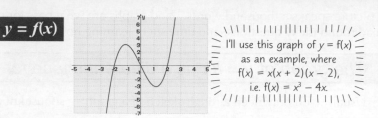

I'll use this graph of $y = f(x)$ as an example, where $f(x) = x(x + 2)(x - 2)$, i.e. $f(x) = x^3 - 4x$.

Translations are caused by Adding things

$y = f(x) + a$

Adding a number to the whole function translates the graph in the y-direction.

1) If a > 0, the graph goes upwards.

2) If a < 0, the graph goes downwards.

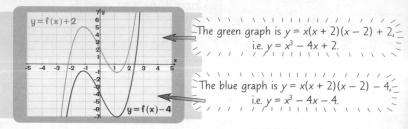

The green graph is $y = x(x + 2)(x - 2) + 2$, i.e. $y = x^3 - 4x + 2$.

The blue graph is $y = x(x + 2)(x - 2) - 4$, i.e. $y = x^3 - 4x - 4$.

$y = f(x + a)$

Writing 'x + a' instead of 'x' means the graph moves sideways ("translated in the x-direction").

1) If a > 0, the graph goes to the left.

2) If a < 0, the graph goes to the right.

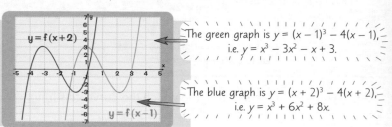

The green graph is $y = (x - 1)^3 - 4(x - 1)$, i.e. $y = x^3 - 3x^2 - x + 3$.

The blue graph is $y = (x + 2)^3 - 4(x + 2)$, i.e. $y = x^3 + 6x^2 + 8x$.

Stretches and Reflections are caused by Multiplying things

$y = af(x)$

Multiplying the whole function stretches, squeezes or reflects the graph vertically.

1) Negative values of 'a' reflect the basic shape in the x-axis.

2) If a > 1 or a < -1 (i.e. |a| > 1) the graph is stretched vertically.

3) If -1 < a < 1 (i.e. |a| < 1) the graph is squashed vertically.

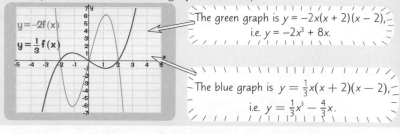

The green graph is $y = -2x(x + 2)(x - 2)$, i.e. $y = -2x^3 + 8x$.

The blue graph is $y = \frac{1}{3}x(x + 2)(x - 2)$, i.e. $y = \frac{1}{3}x^3 - \frac{4}{3}x$.

$y = f(ax)$

Writing 'ax' instead of 'x' stretches, squeezes or reflects the graph horizontally.

1) Negative values of 'a' reflect the basic shape in the y-axis.

2) If a > 1 or a < -1 (i.e. if |a| > 1) the graph is squashed horizontally.

3) If -1 < a < 1 (i.e. if |a| < 1) the graph is stretched horizontally.

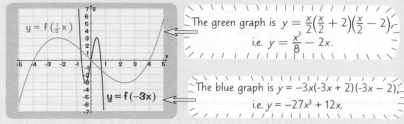

The green graph is $y = \frac{x}{2}(\frac{x}{2} + 2)(\frac{x}{2} - 2)$, i.e. $y = \frac{x^3}{8} - 2x$.

The blue graph is $y = -3x(-3x + 2)(-3x - 2)$, i.e. $y = -27x^3 + 12x$.

C1 Section 4 — Practice Questions

There you go then... a section on <u>various geometrical things</u>. And in a way it was quite exciting, I'm sure you'll agree. Though as you're probably aware, we mathematicians take our excitement from wherever we can get it. Anyway, I'll leave you alone now to <u>savour</u> these practice questions. <u>Don't skip the warm-up</u> — you don't want to hurt yourself...

Warm-up Questions

1) Find the <u>equations</u> of the <u>straight lines</u> that pass through the points

   a) $(2, -1)$ and $(-4, -19)$,   b) $\left(0, -\frac{1}{3}\right)$ and $\left(5, \frac{2}{3}\right)$.

   Write each of them in the forms
   i) $y - y_1 = m(x - x_1)$,
   ii) $y = mx + c$,
   iii) $ax + by + c = 0$, where a, b and c are integers.

2) a) The line l has equation $y = \frac{3}{2}x - \frac{2}{3}$. Find the equation of the lovely, cuddly line <u>parallel to l</u>, passing through the point with coordinates $(4, 2)$. Name this line <u>Lilly</u>.

   b) The line m (whose name is actually Mike) passes through the point $(6, 1)$ and is <u>perpendicular</u> to $2x - y - 7 = 0$. What is the equation of m?

3) The coordinates of points R and S are $(1, 9)$ and $(10, 3)$ respectively. Find the equation of the line <u>perpendicular</u> to RS, passing through the point $(1, 9)$.

4) It's lovely, lovely <u>curve-sketching time</u> — so draw rough sketches of the following curves:
   a) $y = -2x^4$,   b) $y = \frac{7}{x^2}$,   c) $y = -5x^3$,   d) $y = -\frac{2}{x^5}$.

5) Admit it — you <u>love</u> curve-sketching. We all do — and like me, you probably can't get enough of it. So more power to your elbow, and sketch these <u>cubic graphs</u>:

   a) $y = (x - 4)^3$,   b) $y = (3 - x)(x + 2)^2$,

   c) $y = (1 - x)(x^2 - 6x + 8)$,   d) $y = (x - 1)(x - 2)(x - 3)$.

6) Right — now it's time to get serious. Put your <u>thinking head on</u>, and use the graph of $f(x)$ to sketch what these graphs would look like after they've been '<u>transformed</u>'.

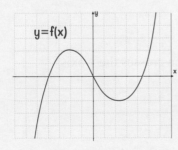

   a) $y = f(ax)$,   where   (i) $a > 1$,

                                 (ii) $0 < a < 1$,

   b) $y = af(x)$,   where   (i) $a > 1$,

                                 (ii) $0 < a < 1$,

   c)   (i) $y = f(x + a)$,   (ii) $y = f(x - a)$, where $a > 0$,

   d)   (i) $y = f(x) + a$,   (ii) $y = f(x) - a$, where $a > 0$.

C1 Section 4 — Practice Questions

The excitement continues on this page, with a thrilling selection of the finest exam questions money can buy. If you're asked to sketch a graph in your exam, it's worth taking a bit of time. They won't expect your graph to be 100% accurate, but just make sure you've got the general shape right and that it crosses the axes in the right places.

Exam Questions

1 The line PQ has equation $4x + 3y = 15$.

   a) Find the gradient of PQ.
   (2 marks)

   b) The point R lies on PQ and has coordinates $(3, 1)$. Find the equation of the line which passes through the point R and is perpendicular to PQ, giving your answer in the form $y = mx + c$. *(3 marks)*

2 The curve C has the equation
$$y = (2x + 1)(x - 2)^2.$$

   Sketch C, clearly showing the points at which the curve meets the x- and y- axes.
   (4 marks)

3

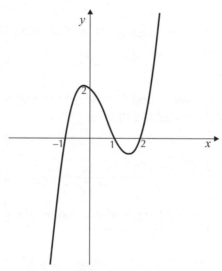

Figure 1

   Figure 1 shows a sketch of the function $y = f(x)$. The function crosses the x-axis at $(-1, 0)$, $(1, 0)$ and $(2, 0)$, and crosses the y-axis at $(0, 2)$.

   On separate diagrams, sketch the following:

   a) $y = f\left(\frac{1}{2}x\right)$.
   (3 marks)

   b) $y = f(x - 4)$.
   (2 marks)

   On each diagram, label any known points of intersection with the x- or y- axes.

C1 Section 4 — Practice Questions

Worry ye not, I'm not going to leave you wanting more — here are a few more exam questions for you to sink your teeth into.

4 a) Sketch the curve $y = f(x)$, where $f(x) = x^2 - 4$, showing clearly the points of intersection with the x- and y- axes.

(2 marks)

b) Describe fully the transformation that transforms the curve $y = f(x)$ to the curve $y = -2f(x)$.

(2 marks)

c) The curve $y = f(x)$ is translated vertically two units upwards. State the equation of the curve after it has been transformed, in term of $f(x)$.

(1 mark)

5 The line l passes through the point S $(7, -3)$ and has gradient -2.

a) Find an equation of l, giving your answer in the form $y = mx + c$.

(3 marks)

b) The point T has coordinates $(5, 1)$. Show that T lies on l.

(1 mark)

6

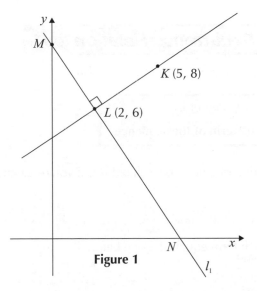

Figure 1

The points L and K have coordinates $(2, 6)$ and $(5, 8)$ respectively. The line l_1 passes through the point L and is perpendicular to the line LK, as shown in **Figure 1**.

a) Find an equation for l_1 in the form $ax + by + c = 0$, where a, b, and c are integers.

(4 marks)

The line l_1 intersects the y-axis at the point M and the x-axis at the point N.

b) Find the coordinates of M.

(2 marks)

c) Find the coordinates of N.

(2 marks)

Sequences

A sequence is a list of numbers that follow a <u>certain pattern</u>. Sequences can be <u>finite</u> or <u>infinite</u> (infinity — oooh), and they're usually generated in one of two ways. And guess what? You have to know everything about them.

A **Sequence** can be defined by its nth Term

You almost definitely covered this stuff at GCSE, so <u>no excuses</u> for mucking it up.

The point of all this is to show how you can work out any <u>value</u> (<u>the nth term</u>) from its <u>position</u> in the sequence (n).

> **EXAMPLE**   Find the nth term of the sequence 5, 8, 11, 14, 17, ...
>
> | 1st | 2nd | 3rd | 4th | 5th |
> |------|------|------|------|------|
> | 5 | 8 | 11 | 14 | 17 |
>
> +3   +3   +3   +3
>
> Each term is <u>3 more</u> than the one before it. That means that you need to start by <u>multiplying n by 3</u>.
>
> Take the first term (where $n = 1$). If you multiply n by 3, you still have to <u>add 2</u> to get 5.
>
> The same goes for $n = 2$. To get 8 you need to multiply n by 3, then add 2.
> Every term in the sequence is worked out exactly the same way.
>
> So n^{th} term is $3n + 2$.

You can define a sequence by a **Recurrence Relation** too

Don't be put off by the fancy name — recurrence relations are pretty <u>easy</u> really.

> The main thing to remember is:
>
> a_k **just means the kth term of the sequence**

The <u>next term</u> in the sequence is a_{k+1}. You need to describe how to <u>work out</u> a_{k+1} if you're given a_k.

> **EXAMPLE**   Find the recurrence relation of the sequence 5, 8, 11, 14, 17, ...
>
> From the example above, you know that each term equals the one before it, plus 3.
>
> This is written like this: $a_{k+1} = a_k + 3$
>
> So, if k = 5, $a_k = a_5$ which stands for the 5th term, and $a_{k+1} = a_6$ which stands for the 6th term.
>
> In everyday language, $a_{k+1} = a_k + 3$ means that the sixth term equals the fifth term plus 3.
>
> <u>BUT</u> $a_{k+1} = a_k + 3$ on its own <u>isn't enough</u> to describe 5, 8, 11, 14, 17, ...
>
> For example, the sequence 87, 90, 93, 96, 99, ... <u>also</u> has each term being 3 more than the one before.
>
> The description needs to be more <u>specific</u>, so you've got to <u>give one term</u> in the sequence, as well as the recurrence relation. You almost always give the <u>first value</u>, a_1.
>
> Putting all of this together gives 5, 8, 11, 14, 17, ... as $a_{k+1} = a_k + 3$, $a_1 = 5$.

Arithmetic Progressions

Right, you've got basic sequences tucked under your belt now — time to step it up a notch (sounds painful).
When the terms of a sequence progress by adding a fixed amount each time, this is called an arithmetic progression.

It's all about Finding the nth Term

The first term of a sequence is given the symbol a. The amount you add each time is called the common difference, or d. The position of any term in the sequence is called n.

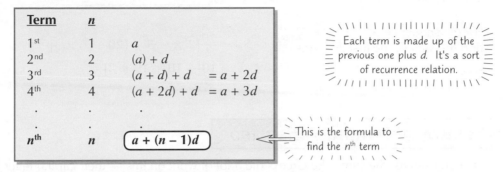

Term	$\underline{n}$		
1st	1	a	
2nd	2	$(a) + d$	
3rd	3	$(a + d) + d$	$= a + 2d$
4th	4	$(a + 2d) + d$	$= a + 3d$
.	.	.	
.	.	.	
.	.	.	
n^{th}	n	$a + (n - 1)d$	

Each term is made up of the previous one plus d. It's a sort of recurrence relation.

This is the formula to find the nth term

EXAMPLE   Find the 20th term of the arithmetic progression 2, 5, 8, 11, ... and find the formula for the nth term.

Here $a = 2$ and $d = 3$.

To get d, just find the difference between two terms next to each other — e.g. $11 - 8 = 3$

So 20th term $= a + (20 - 1)d$
$= 2 + 19 \times 3$
$= 59$

The general term is the nth term, i.e. $a + (n - 1)d$
$= 2 + (n - 1)3$
$= 3n - 1$

A Series is when you Add the Terms to Find the Total

S_n is the total of the first n terms of the arithmetic progression:

$$S_n = a + (a + d) + (a + 2d) + (a + 3d) + \ldots + (a + (n - 1)d)$$

There's a really neat version of the same formula too:

$$S_n = n \times \frac{(a + l)}{2}$$

The l stands for the last value in the progression. You work it out as $l = a + (n - 1)d$

Nobody likes formulas, so think of it as the average of the first and last terms multiplied by the number of terms.

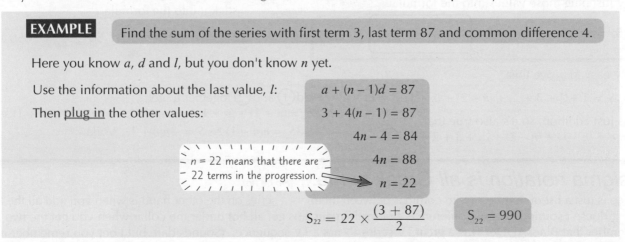

EXAMPLE   Find the sum of the series with first term 3, last term 87 and common difference 4.

Here you know a, d and l, but you don't know n yet.

Use the information about the last value, l: $a + (n - 1)d = 87$

Then plug in the other values: $3 + 4(n - 1) = 87$

$$4n - 4 = 84$$

$n = 22$ means that there are 22 terms in the progression.

$$4n = 88$$
$$n = 22$$

$$S_{22} = 22 \times \frac{(3 + 87)}{2} \qquad S_{22} = 990$$

Arithmetic Series and Sigma Notation

They **Won't** always give you the **Last Term**...

...but don't panic — there's a formula to use when the <u>last term is unknown</u>. But you knew I'd say that, didn't you?

You know $l = a + (n - 1)d$ and $S_n = n \times \dfrac{(a + l)}{2}$.

$$S_n = \frac{n}{2}[2a + (n - 1)d]$$

Plug l into S_n and rearrange to get the formula in the box:

EXAMPLE For the sequence –5, –2, 1, 4, 7, ... find the sum of the first 20 terms.

So $a = -5$ and $d = 3$.
The question says $n = 20$ too.

$S_{20} = \dfrac{20}{2}[2 \times -5 + (20 - 1) \times 3]$
$= 10[-10 + 19 \times 3]$
$S_{20} = 470$

There's **Another** way of **Writing Series**, too

So far, the letter S has been used for the sum. The Greeks did a lot of work on this — their capital letter for S is Σ or <u>sigma</u>. This is used today, together with the general term, to mean the <u>sum</u> of the series.

EXAMPLE

Starting with $n = 1$... Find $\displaystyle\sum_{n=1}^{15}(2n + 3)$ *...and ending with $n = 15$* This means you have to find the sum of the <u>first 15 terms</u> of the series with n^{th} term $2n + 3$.

The first term ($n = 1$) is 5, the second term ($n = 2$) is 7, the third is 9, ... and the last term ($n = 15$) is 33.

In other words, you need to find $5 + 7 + 9 + ... + 33$. This gives $a = 5$, $d = 2$, $n = 15$ and $l = 33$.

You know all of a, d, n and l, so you can use either formula:

$S_n = n\dfrac{(a + l)}{2}$
$S_{15} = 15\dfrac{(5 + 33)}{2}$
$S_{15} = 15 \times 19$
$S_{15} = 285$

It makes no difference which method you use.

$S_n = \dfrac{n}{2}[2a + (n - 1)d]$
$S_{15} = \dfrac{15}{2}[2 \times 5 + 14 \times 2]$
$S_{15} = \dfrac{15}{2}[10 + 28]$
$S_{15} = 285$

Use **Arithmetic Progressions** to add up the **First n Whole Numbers**

The <u>sum of the first n natural numbers</u> looks like this:

$$S_n = 1 + 2 + 3 + ... + (n - 2) + (n - 1) + n$$

So $a = 1$, $l = n$ and also $n = n$.
Now just plug those values into the formula:

Natural numbers are just positive whole numbers.

$S_n = n \times \dfrac{(a + l)}{2}$ $\Longrightarrow$ $\boxed{S_n = \dfrac{1}{2}n(n + 1)}$

EXAMPLE

Add up all the whole numbers from 1 to 100.

Sounds pretty hard, but all you have to do is stick it into the formula:

$S_{100} = \frac{1}{2} \times 100 \times 101$. So $S_{100} = 5050$

It's pretty easy to <u>prove</u> this:

1) Say, $S_n = 1 + 2 + 3 + ... + (n - 2) + (n - 1) + n$ ①

2) ① is just addition, so it's also true that:
$S_n = n + (n - 1) + (n - 2) + ... + 3 + 2 + 1$ ②

3) Add ① and ② together to get:
$2S_n = (n + 1) + (n + 1) + (n + 1) + ... + (n + 1) + (n + 1) + (n + 1)$
$\Rightarrow 2S_n = n(n + 1) \Rightarrow S_n = \frac{1}{2}n(n + 1)$. Voilà.

This sigma notation is all Greek to me... *(Ho ho ho)*

A <u>sequence</u> is just a list of numbers (with commas between them) — a <u>series</u> on the other hand is when you add all the terms together. It doesn't sound like a big difference, but mathematicians get all hot under the collar when you get the two mixed up. Remember that Black**ADD**er was a great TV <u>series</u> — not a TV sequence. (Sounds daft, but I bet you remember it now.)

C1 Section 5 — Practice Questions

Well there's not a whole heap to learn in this section, but it's still important you work through these revision questions just to make sure you're as happy as a <u>ferret in a trouser shop</u> with Sequences and Series. <u>Dead easy</u> marks to pick up in the exam — so you'd feel pretty daft if you didn't learn this stuff. Come on then, <u>get practising</u>...

Warm-up Questions

1) Find the <u>nth term</u> for the following sequences:

   a) 2, 6, 10, 14, ...

   b) 0.2, 0.7, 1.2, 1.7, ...

   c) 21, 18, 15, 12, ...

   d) 76, 70, 64, 58, ...

2) Describe the sequence 32, 37, 42, 47, ... using a recurrence relation.

3) Find the <u>last term</u> of the sequence that <u>starts with</u> 3, has a <u>common difference</u> of 0.5 and has <u>25 terms</u>.

4) Find the <u>common difference</u> in a sequence that <u>starts with</u> –2, <u>ends with</u> 19 and has <u>29 terms</u>.

5) Find the <u>sum</u> of the series that <u>starts with</u> 7, <u>ends with</u> 35 and has <u>8 terms</u>.

6) Find the <u>sum</u> of the series that <u>begins with</u> 5, 8, ... and <u>ends with</u> 65.

7) A series has <u>first term</u> 7 and <u>5th term</u> 23.

   Find: a) the common difference,   b) the 15th term, and   c) the sum of the first 10 terms.

8) A series has seventh term 36 and tenth term 30. Find the <u>sum of the first five terms</u> and the <u>nth term</u>.

9) Find $\displaystyle\sum_{n=1}^{20}(3n-1)$

10) Find $\displaystyle\sum_{n=1}^{10}(48-5n)$

Exam Questions

1. A sequence is defined by the recurrence relation: $h_{n+1} = 2h_n + 2$ when $n \geq 1$.

   a) Given that $h_1 = 5$, find the values of h_2, h_3, and h_4.

   (3 marks)

   b) Calculate the value of $\displaystyle\sum_{r=3}^{6} h_r$.

   (3 marks)

C1 Section 5 — Practice Questions

More <u>lovingly crafted</u> questions, wrought from the <u>finest numbers</u> and <u>quality equations</u> for our valued customers...

2   A sequence $a_1, a_2, a_3, \ldots$ is defined by $a_1 = k$, $a_{n+1} = 3a_n + 11$, $n \geq 1$, where k is a constant.

    a)   Show that $a_4 = 27k + 143$.

(3 marks)

    b)   Find the value of k, given that $\sum_{r=1}^{4} a_r = 278$.

(3 marks)

3   Ned has 15 cuboid pots that need filling with soil. Each pot is taller than the one before it. The different capacities of his 15 pots form an arithmetic sequence with first term (representing the smallest pot) a ml and the common difference d ml. The 7th pot is 580 ml and he will need a total of exactly 9525 ml of soil to fill all of them.

    Find the value of a and the value of d.

(7 marks)

4   The first term of an arithmetic sequence is 22 and the common difference is –1.1.

    a)   Find the value of the 31st term.

(2 marks)

    b)   If the kth term of the sequence is 0, find k.

(2 marks)

    c)   The sum to n terms of the sequence is S_n.
        Find the value of n at which S_n first becomes negative.

(4 marks)

5   David's personal trainer has given him a timetable to improve his upper-body strength, which gradually increases the amount of push-ups David does each day.

    The timetable for the first four days is shown below:

Day:	Mon	Tue	Wed	Thur
Number of push-ups:	6	14	22	30

    a)   Find an expression, in terms of n, for the number of push-ups he will have to do on day n.

(3 marks)

    b)   David follows his exercise routine for 10 days. Calculate how many push-ups he has done in total over the 10 days.

(3 marks)

    His personal trainer recommends that David takes a break from his exercise routine when he has done a cumulative total of 2450 push-ups. Given that David completes his exercises on day k, but reaches the recommended limit part-way through day $(k + 1)$,

    c)   Show that k satisfies $(2k - 49)(k + 25) < 0$.

(3 marks)

    d)   Find the value of k.

(2 marks)

Differentiation

Brrrrrr... differentiation is a bad one — it really is. Not because it's that hard, but because it comes up all over the place in exams. So if you don't know it perfectly, you're asking for trouble. <u>Differentiation</u> is a great way to work out <u>gradients</u> of graphs. You take a function, differentiate it, and you can quickly tell <u>how steep</u> a graph is. It's magic.

Use this formula to differentiate *Powers of x*

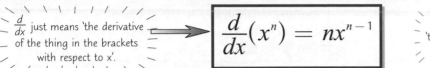

$\frac{d}{dx}$ just means 'the derivative of the thing in the brackets with respect to x'.

$$\frac{d}{dx}(x^n) = nx^{n-1}$$

<u>Derivative</u> just means 'the thing you get when you differentiate something'.

Equations are much easier to differentiate when they're written as <u>powers of x</u> — like writing $\sqrt{x}$ as $x^{\frac{1}{2}}$.

When you've done this, you can use the formula (the thing in the red box above) to differentiate the equation.

Use the differentiation formula...

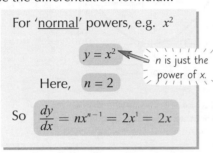

For '<u>normal</u>' powers, e.g. x^2

$$y = x^2$$

n is just the power of x.

See page 2 for more on negative powers.

Here, $n = 2$

So $\frac{dy}{dx} = nx^{n-1} = 2x^1 = 2x$

For <u>negative</u> powers, e.g. $\frac{1}{x^2} = x^{-2}$

$$y = \frac{1}{x^2} = x^{-2}$$

Remember to rewrite the equation as a <u>power of x</u>...

Here, $n = -2$

...then use the formula to find the derivative.

So $\frac{dy}{dx} = nx^{n-1} = -2x^{-3} = -\frac{2}{x^3}$

For <u>fractional</u> powers, e.g. $\sqrt{x} = x^{\frac{1}{2}}$

$$y = \sqrt{x} = x^{\frac{1}{2}}$$

Write the square root as a <u>power</u> of x...

$\frac{dy}{dx}$ can sometimes be written as $f'(x)$.

$$n = \frac{1}{2}$$

...and use that very <u>same</u> formula.

$$\frac{dy}{dx} = \frac{1}{2}x^{-\frac{1}{2}} = \frac{1}{2\sqrt{x}}$$

Power Laws:
Differentiation's much easier if you know the Power Laws really well. Like knowing that
$$x^1 = x \text{ and } \sqrt{x} = x^{\frac{1}{2}}$$
See page 2 for more info.

A constant always differentiates to O — see below.

Differentiate each term in an equation *Separately*

This formula is better than cake — even better than that really nice sticky black chocolate one from that place in town.

Even if there are loads of terms in the equation, it doesn't matter. Differentiate each bit separately and you'll be fine.

Here are a couple of examples...

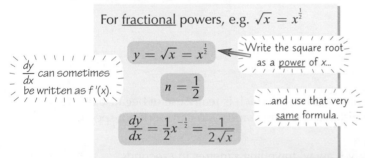

If there's a number in front of the function...

$$y = 3\sqrt{x} = 3x^{\frac{1}{2}}$$

...multiply the derivative by the same number.

$$\frac{dy}{dx} = 3\left(\frac{1}{2}x^{-\frac{1}{2}}\right)$$

i.e. $\frac{dy}{dx} = \frac{3}{2} \times x^{-\frac{1}{2}} = \frac{3}{2\sqrt{x}}$

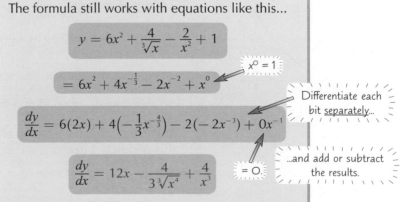

The formula still works with equations like this...

$$y = 6x^2 + \frac{4}{3\sqrt{x}} - \frac{2}{x^2} + 1$$

$$= 6x^2 + 4x^{-\frac{1}{3}} - 2x^{-2} + x^0$$

$x^0 = 1$

Differentiate each bit <u>separately</u>...

$$\frac{dy}{dx} = 6(2x) + 4\left(-\frac{1}{3}x^{-\frac{4}{3}}\right) - 2(-2x^{-3}) + 0x^{-1}$$

$= O$

...and add or subtract the results.

$$\frac{dy}{dx} = 12x - \frac{4}{3\sqrt[3]{x^4}} + \frac{4}{x^3}$$

Dario Gradient — differentiating Crewe from the rest...

If you're going to bother doing maths, you've got to be able to differentiate things. Simple as that. But luckily, once you can do the simple stuff, you should be all right. Big long equations are just made up of loads of simple little terms, so they're not really that much harder. Learn the formula, and make sure you can use it by practising all day and all night forever.

Differentiation

Differentiation is what you do if you need to find a gradient. Excited yet?

Differentiate to find Gradients...

EXAMPLE Find the gradient of the graph $y = x^2$ at $x = 1$ and $x = -2$...

You need the gradient of the graph of...

$$y = x^2$$

So differentiate this function to get...

$$\frac{dy}{dx} = 2x$$

Now when $x = 1$, $\frac{dy}{dx} = 2$,

And so the gradient of the graph at $x = 1$ is 2.

And when $x = -2$, $\frac{dy}{dx} = -4$,

So the gradient of the graph at $x = -2$ is –4.

Use differentiation to find the gradient of a <u>curve</u> — which is the same as the gradient of the <u>tangent</u> at any given point.

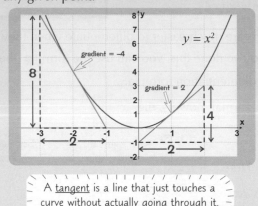

A <u>tangent</u> is a line that just touches a curve without actually going through it.

...which tell you Rates of Change...

So, you've differentiated an equation and found the gradient at a point — which is really useful because this tells you the rate of change of the curve at that point (e.g. from distance vs. time graphs you can work out speed).

EXAMPLE A sports car pulls off from a junction and drives away, travelling s metres in t seconds. For the first 10 seconds, its path can be described by the equation $s = 2t^2$.

Find: a) the speed of the car after 8 seconds and b) the car's acceleration during this period.

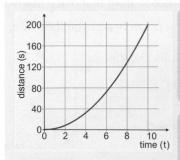

a) You can work out the speed by calculating the gradient of the curve $s = 2t^2$.

Differentiate to give: $\frac{ds}{dt} = 4t$ When $t = 8$, $\frac{ds}{dt} = 32$

So, the car is travelling at 32 ms^{-1} after 8 seconds.

b) Acceleration is the rate that speed (v) changes (i.e. it is the gradient of $v = 4t$).

So differentiate again to find the acceleration: $\frac{d^2s}{dt^2} = 4$

This means that the car's acceleration during this period is 4 ms^{-2}

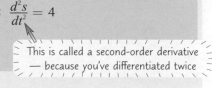

This is called a second-order derivative — because you've differentiated twice

Help me Differentiation — You're my only hope...

There's not much hard maths on this page — but there are a couple of very important ideas that you need to get your head round pretty darn soon. Understanding that <u>differentiating gives the gradient</u> of the graph is more important than washing regularly — AND THAT'S IMPORTANT. The other thing on the page you need to know is that the gradient tells you the rate of change of a function — which is also vital when working out what a question is after.

Finding Tangents and Normals

What's a tangent? Beats me. Oh no, I remember, it's one of those thingies on a curve. Ah, yes... I remember now...

Tangents **Just** touch a curve

To find the equation of a tangent or a normal to a curve, you first need to know its <u>gradient</u> —
so differentiate. Then complete the line's equation using the <u>coordinates</u> of one point on the line.

> **EXAMPLE** Find the tangent to the curve $y = (4 - x)(x + 2)$ at the point (2, 8).

Tangents and Normals...

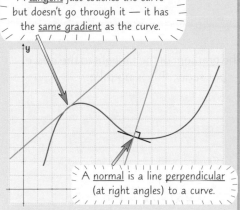

A <u>tangent</u> just touches the curve but doesn't go through it — it has the <u>same gradient</u> as the curve.

A <u>normal</u> is a line <u>perpendicular</u> (at right angles) to a curve.

To find the curve's (and the tangent's) <u>gradient</u>, first write the equation in a <u>form</u> you can differentiate...

$$y = 8 + 2x - x^2$$

...and then <u>differentiate</u> it.

$$\frac{dy}{dx} = 2 - 2x$$

The <u>gradient</u> of the tangent will be the gradient of the curve at $x = 2$.

At $x = 2$, $\dfrac{dy}{dx} = -2$,

So the tangent has <u>equation</u>,

$$y - y_1 = -2(x - x_1)$$

*in $y - y_1 = m(x - x_1)$ form.
See page 29.*

And since it passes through the <u>point</u> (2, 8), this becomes

$$y - 8 = -2(x - 2), \text{ or } y = -2x + 12.$$

You can also write it in $y = mx + c$ form.

Normals are at **Right Angles** to a curve

> **EXAMPLE** Find the normal to the curve $y = \dfrac{(x + 2)(x + 4)}{6\sqrt{x}}$ at the point (4, 4).

There's more info on parallel and perpendicular lines on p. 30.

Write the equation of the curve in a <u>form</u> you can differentiate.

$$y = \frac{x^2 + 6x + 8}{6x^{\frac{1}{2}}} = \frac{1}{6}x^{\frac{3}{2}} + x^{\frac{1}{2}} + \frac{4}{3}x^{-\frac{1}{2}}$$

Dividing everything on the top line by everything on the bottom line.

<u>Differentiate</u> it...

$$\frac{dy}{dx} = \frac{1}{6}\left(\frac{3}{2}x^{\frac{1}{2}}\right) + \frac{1}{2}x^{-\frac{1}{2}} + \frac{4}{3}\left(-\frac{1}{2}x^{-\frac{3}{2}}\right)$$

$$= \frac{1}{4}\sqrt{x} + \frac{1}{2\sqrt{x}} - \frac{2}{3\sqrt{x^3}}$$

Find the <u>gradient</u> at the point you're interested in. At $x = 4$,

$$\frac{dy}{dx} = \frac{1}{4} \times 2 + \frac{1}{2 \times 2} - \frac{2}{3 \times 8} = \frac{2}{3}$$

Because the gradient of the <u>normal</u> multiplied by the gradient of the <u>curve</u> must be −1.

So the <u>gradient</u> of the <u>normal</u> is $-\dfrac{3}{2}$.

And the <u>equation</u> of the normal is $y - y_1 = -\dfrac{3}{2}(x - x_1)$.

Finally, since the normal goes through the <u>point</u> (4, 4), the equation of the
normal must be $y - 4 = -\dfrac{3}{2}(x - 4)$, or after rearranging, $y = -\dfrac{3}{2}x + 10$.

Finding Tangents and Normals

1) **Differentiate the function.**

2) **Find the gradient, m, of the tangent or normal. This is,**
   for a <u>tangent</u>: the gradient of the curve

   for a <u>normal</u>: $\dfrac{-1}{\text{gradient of the curve}}$

3) **Write the equation of the tangent or normal in the form**
   $y - y_1 = m(x - x_1)$, or $y = mx + c$.

4) **Complete the equation of the line using the coordinates of a point on the line.**

Repeat after me... "I adore tangents and normals..."

Examiners can't stop themselves saying the words 'Find the tangent...' and 'Find the normal...'. They love the words.
These phrases are music to their ears. They can't get enough of them. I just thought it was my duty to tell you that.
And so now you know, you'll definitely be wanting to learn how to do the stuff on this page. Of course you will.

C1 Section 6 — Practice Questions

That's what <u>differentiation</u> is all about. Yes, there are <u>fiddly things</u> to remember — but overall, it's not as bad as all that. And just think of all the <u>lovely marks</u> you'll get if you can answer questions like these in the exam...

Warm-up Questions

1) An easy one to start with. Write down the <u>formula</u> for differentiating <u>any power</u> of x.

2) <u>Differentiate</u> these functions with respect to x:
   a) $y = x^2 + 2$,
   b) $y = x^4 + \sqrt{x}$,
   c) $y = \dfrac{7}{x^2} - \dfrac{3}{\sqrt{x}} + 12x^3$

3) What's the <u>connection</u> between the <u>gradient of a curve</u> at a point and the <u>gradient of the tangent</u> to the curve at the same point? (That sounds like a joke in need of a punchline — but sadly, this is no joke.)

4) Find the <u>gradients</u> of these <u>graphs</u> at $x = 2$:

a)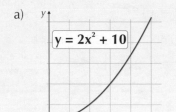

b)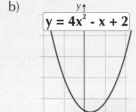

c)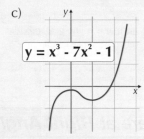

5) 1 litre of water is poured into a bowl.
   The <u>volume (v)</u> of water in the bowl (in ml) is defined by the <u>function</u>: $v = 17t^2 - 10t$
   Find the <u>rate</u> at which water is poured into the bowl when <u>$t = 4$ seconds</u>.

6) Yawn, yawn. Find the equations of the <u>tangent</u> and the <u>normal</u> to the curve $y = \sqrt{x^3} - 3x - 10$ at $x = 16$.

7) Show that the lines $y = \dfrac{x^3}{3} - 2x^2 - 4x + \dfrac{86}{3}$ and $y = \sqrt{x}$ <u>both go through</u> the point (4, 2), and are <u>perpendicular</u> at that point. Good question, that — <u>nice and exciting</u>, just the way you like 'em.

Exam Questions

1 Given that $y = x^7 + \dfrac{2}{x^3}$, find:

   a) $\dfrac{dy}{dx}$

   (2 marks)

   b) $\dfrac{d^2y}{dx^2}$

   (2 marks)

C1 Section 6 — Practice Questions

I know, I know, one page of differentiation questions just isn't enough to quench your maths thirst.
Have a few more glugs of the sweet, sweet, algebraic nectar. Mmm, numbers...

Exam Questions

2   The curve C is given by the equation $y = 2x^3 - 4x^2 - 4x + 12$.

    a)   Find $\dfrac{dy}{dx}$.

(2 marks)

    b)   Write down the gradient of the tangent to the curve at the point where $x = 2$.

(1 mark)

    c)   Hence or otherwise find an equation for the normal to the curve at this point.

(3 marks)

3   Find the gradient of the curve $y = \dfrac{1}{\sqrt{x}} + \dfrac{1}{x}$ at the point $\left(4, \dfrac{3}{4}\right)$.

(5 marks)

4   a)   Show that the equation $\dfrac{x^2 + 3x^{\frac{3}{2}}}{\sqrt{x}}$ can be written in the form $x^p + 3x^q$,
        and state the values of p and q.

(3 marks)

    b)   Now let $y = 3x^3 + 5 + \dfrac{x^2 + 3x^{\frac{3}{2}}}{\sqrt{x}}$. Find $\dfrac{dy}{dx}$, giving each coefficient in its simplest form.

(4 marks)

5   The curve C is given by the equation $y = mx^3 - x^2 + 8x + 2$, for a constant m.

    a)   Find $\dfrac{dy}{dx}$.

(2 marks)

    The point P lies on C, and has the x-value 5. The normal to C at P is parallel to the line given by the equation $y + 4x - 3 = 0$.

    b)   Find the gradient of curve C at P.

(3 marks)

    Hence or otherwise, find:

    c)   (i)   the value of m.

(3 marks)

        (ii)   the y-value at P.

(2 marks)

6   For $x \geq 0$ and $y \geq 0$, x and y satisfy the equation $2x - y = 6$.

    a)   If $W = x^2y^2$, show that $W = 4x^4 - 24x^3 + 36x^2$.

(2 marks)

    b)   (i)   Show that $\dfrac{dW}{dx} = k(2x^3 - 9x^2 + 9x)$, and find the value of the integer k.

(4 marks)

        (ii)   Find the value of $\dfrac{dW}{dx}$ when $x = 1$.

(1 mark)

    c)   Find $\dfrac{d^2W}{dx^2}$ and give its value when $x = 1$.

(2 marks)

Integration

Integration is the 'opposite' of differentiation — and so if you can differentiate, you can be pretty confident you'll be able to integrate too. There's just one extra thing you have to remember — the constant of integration...

You need the constant because there's **More Than One** right answer

When you integrate something, you're trying to find a function that returns to what you started with when you differentiate it. And when you add the constant of integration, you're just allowing for the fact that there's more than one possible function that does this...

This means the integral of 2x with respect to x.

$$\int 2x\, dx =$$

$$x^2 - 207.253$$
$$x^2 - 1$$
$$x^2$$
$$x^2 + \pi$$

If you differentiate any of these functions, you get the thing on the left — they're all possible answers.

So the answer to this integral is actually...

$$\int 2x\, dx = x^2 + C$$

The 'C' just means 'any number'. This is the constant of integration.

You only need to add a constant of integration to indefinite integrals like these ones. Definite integrals are just integrals with limits (or little numbers) next to the integral sign, but you don't need to know about these in this module.

Up the power by **One** — then **Divide** by it

The formula below tells you how to integrate any power of x (except x^{-1}).

This is an indefinite integral — it doesn't have any limits (numbers) next to the integral sign.

$$\int x^n\, dx = \frac{x^{n+1}}{n+1} + C$$

You can't do this to $\frac{1}{x} = x^{-1}$. When you increase the power by 1 (to get zero) and then divide by zero — you get big problems.

In a nutshell, this says:

> To integrate a power of x: (i) Increase the power by one — then divide by it.
>
> and   (ii) Stick a constant on the end.

EXAMPLES   Use the integration formula...

(1) For 'normal' powers,

$$\int x^3\, dx = \frac{x^4}{4} + C$$

Increase the power to 4... ...and then divide by 4.

(2) For negative powers,

$$\int \frac{1}{x^3}\, dx = \int x^{-3}\, dx$$
$$= \frac{x^{-2}}{-2} + C$$
$$= -\frac{1}{2x^2} + C$$

Increase the power by 1 to –2... ...and then divide by –2.

(3) For fractional powers,

$$\int \sqrt[3]{x^4}\, dx = \int x^{\frac{4}{3}}\, dx$$
$$= \frac{x^{\frac{7}{3}}}{(7/3)} + C$$
$$= \frac{3\sqrt[3]{x^7}}{7} + C$$

Add 1 to the power... ...then divide by this new power.

(4) And for complicated looking stuff...

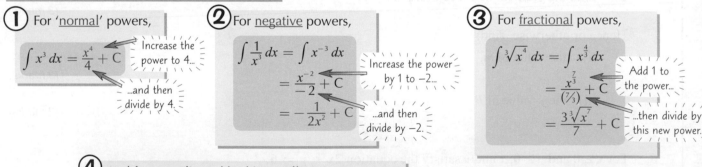

$$\int \left(3x^2 - \frac{2}{\sqrt{x}} + \frac{7}{x^2}\right) dx = \int \left(3x^2 - 2x^{-\frac{1}{2}} + 7x^{-2}\right) dx$$
$$= \frac{3x^3}{3} - \frac{2x^{\frac{1}{2}}}{(1/2)} + \frac{7x^{-1}}{-1} + C$$
$$= x^3 - 4\sqrt{x} - \frac{7}{x} + C$$

Do each of these bits separately.

CHECK YOUR ANSWERS:
You can check you've integrated properly by differentiating the answer — you should end up with the thing you started with.

Indefinite integrals — joy without limits...

This integration lark isn't so bad then — there's only a couple of things to remember and then you can do it no problem. But that constant of integration catches loads of people out — it's so easy to forget — and you'll definitely lose marks if you do forget it. You have been warned. Other than that, there's not much to it. Hurray.

Integration

By now, you're probably aware that maths isn't something you do unless you're a bit of a <u>thrill-seeker</u>.
You know, sometimes they even ask you to find a curve with a certain derivative that goes through a certain point.

You sometimes need to find the **Value** *of the* **Constant of Integration**

When they tell you something else about the curve in addition to its derivative, you can work out the value of that <u>constant of integration</u>. Usually the something is the coordinates of one of the points the curve goes through.

Really Important Bit...

When you differentiate y, you get $\frac{dy}{dx}$.
And when you integrate $\frac{dy}{dx}$, you get y
(if you ignore the constant of integration).

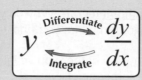

EXAMPLE The curve $f(x)$ goes through the point $(2, 8)$ and $f'(x) = 6x(x - 1)$.
Find $f(x)$.

$f'(x)$ is just another way of saying dy/dx. When you integrate $f'(x)$ you get $f(x)$ and when you differentiate $f(x)$ you get $f'(x)$.

You know the derivative $f'(x)$ and need to find the function $f(x)$ — so <u>integrate</u>.

<u>Remember:</u>
Even if you <u>don't</u> have any extra information about the curve — you still have to add a <u>constant</u> when you work out an integral <u>without limits</u>.

$f'(x) = 6x(x - 1) = 6x^2 - 6x$

So integrating both sides gives...

$f(x) = \int (6x^2 - 6x)dx$
$\Rightarrow f(x) = \frac{6x^3}{3} - \frac{6x^2}{2} + C$
$\Rightarrow f(x) = 2x^3 - 3x^2 + C$

Don't forget the constant of integration.

Check this is correct by differentiating it and making sure you get what you started with.

$f(x) = 2x^3 - 3x^2 + C = 2x^3 - 3x^2 + Cx^0$
$f'(x) = 2(3x^2) - 3(2x^1) + C(0x^{-1})$
$f'(x) = 6x^2 - 6x$

A constant always differentiates to zero.

So this function's got the correct derivative — but you haven't finished yet.

You now need to <u>find C</u> — and you do this by using the fact that it goes through the point $(2, 8)$.

$f(x) = 2x^3 - 3x^2 + C$

Putting $x = 2$ and $y = 8$ in the above equation gives...

$8 = (2 \times 2^3) - (3 \times 2^2) + C$
$\Rightarrow 8 = 16 - 12 + C$
$\Rightarrow C = 4$

So the answer you need is this one:

$f(x) = 2x^3 - 3x^2 + 4$

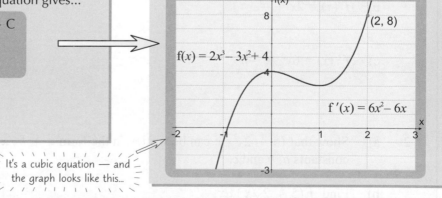

It's a cubic equation — and the graph looks like this...

Maths and alcohol don't mix — so never drink and derive...

That's another page under your belt and — go on, admit it — there was nothing too horrendous on it. If you can do the stuff from the previous page and then substitute some numbers into an equation, you can do everything from this page too. So if you think this is boring, you'd be right. But if you think it's much harder than the stuff before, you'd be wrong.

C1 Section 7 — Practice Questions

Integration is pretty much the <u>opposite of differentiation</u>. So if you can differentiate, then chances are you can integrate as well. Which brings us (kind of) neatly on to <u>these questions</u>. Come and have a go, if you think you're <u>hard enough</u>...

Warm-up Questions

1) Write down the <u>steps</u> involved in <u>integrating a power</u> of x.

2) What's an <u>indefinite integral</u>? Why do you have to add a <u>constant</u> of integration when you find an indefinite integral?

3) How can you <u>check</u> whether you've <u>integrated</u> something properly? (<u>Without</u> asking someone else.)

4) <u>Integrate</u> these: a) $\int 10x^4\,dx$,   b) $\int (3x + 5x^2)\,dx$,   c) $\int (x^2(3x + 2))\,dx$

5) Work out the <u>equation</u> of the curve that has <u>derivative</u> $\frac{dy}{dx} = 6x - 7$ and goes through the <u>point</u> (1, 0).

6) a) Find the <u>equation</u> of the curve that has <u>derivative</u> $\frac{dy}{dx} = 3x^3 + 2$ and goes through the <u>point</u> (1, 0).

   b) How would you <u>change</u> the equation if the curve had to go through the point (1, 2) instead?

   (Don't start the whole question again.)

My dad used to call exams 'umbrellas' because he thought it sounded <u>less scary</u>. So here are some <u>umbrella-style</u> questions. Also useful if it's raining...

Exam Questions

1   Find $f(x)$ in each case below. Give each term in its simplest form.

   a) $f'(x) = x^{-\frac{1}{2}} + 4 - 5x^3$

   (3 marks)

   b) $f'(x) = 2x + \dfrac{3}{x^2}$

   (2 marks)

   c) $f'(x) = 6x^2 - \dfrac{1}{3\sqrt{x}}$

   (2 marks)

2   a)   Show that $(5 + 2\sqrt{x})^2$ can be written in the form $a + b\sqrt{x} + cx$, stating the values of the constants a, b and c.

   (3 marks)

   b)   Find $\int (5 + 2\sqrt{x})^2\,dx$.

   (3 marks)

C1 Section 7 — Practice Questions

You need to be aiming to get <u>all of these right</u>. But if you do make a mistake, then it's not the end of the world — just <u>re-read the relevant part</u> of the section and then <u>have another go</u>. And keep doing this until you don't make any mistakes. Then you can feel ready to take on any integration exam questions that Core 1 might throw at you.

Exam Questions

3   The curve C has the equation $y = f(x)$, $x > 0$. $f'(x)$ is given as $2x + 5\sqrt{x} + \dfrac{6}{x^2}$.

A point P on curve C has the coordinates $(3, 7)$.

Find $f(x)$, giving your answer in its simplest form.

(6 marks)

4   $f'(x) = \dfrac{1}{\sqrt{36x}} - 2\left(\sqrt{\dfrac{1}{x^3}}\right)$ where $x > 0$

a)   Show that $f'(x) = Ax^{-\frac{1}{2}} - Bx^{-\frac{3}{2}}$ and give the values of A and B.

(3 marks)

b)   The curve C is given by $y = f(x)$ and goes through point $P\,(1, 7)$.
Find the equation of the curve.

(4 marks)

5   Curve C has equation $y = f(x)$, $x \neq 0$, where the derivative is given by $f'(x) = x^3 - \dfrac{2}{x^2}$.

The point $P\,(1, 2)$ lies on C.

a)   Find an equation for the tangent to C at the point P, giving your answer
in the form $y = mx + c$, where m and c are integers.

(4 marks)

b)   Find $f(x)$.

(5 marks)

6   $f'(x) = (x - 1)(3x - 1)$ where $x > 0$

a)   The curve C is given by $f(x)$ and goes through point $P\,(3, 10)$. Find $f(x)$.

(6 marks)

b)   The equation for the normal to C at the point P can be written in the form $y = \dfrac{a - x}{b}$
where a and b are integers. Find the values of a and b.

(4 marks)

General Certificate of Education
Advanced Subsidiary (AS) and Advanced Level

Core Mathematics C1 — Practice Exam One

Time Allowed: 1 hour 30 min

Calculators may **not** be used for this exam

There are 75 marks available for this paper.

1 a) Write down the exact value of $36^{-\frac{1}{2}}$.

(2 marks)

   b) Simplify $\dfrac{a^6 \times a^3}{\sqrt{a^4}} \div a^{\frac{1}{2}}$.

(3 marks)

2 a) Express $(5\sqrt{5} + 2\sqrt{3})^2$ in the form $a + b\sqrt{c}$, where a, b and c are integers to be found.

(4 marks)

   b) Rationalise the denominator of $\dfrac{10}{\sqrt{5} + 1}$.

(3 marks)

3 Find $\dfrac{dy}{dx}$ for each of the following:

   a) $y = x^2$

(1 mark)

   b) $y = 3x^4 - 2x$

(2 marks)

   c) $y = (x^2 + 4)(x - 2)$

(3 marks)

4 Solve the following simultaneous equations:

$$y + x = 7 \quad \text{and} \quad y = x^2 + 3x - 5$$

(5 marks)

5 The line AB is part of the line with equation $y + 2x - 5 = 0$.
   A is the point with coordinates $(1, 3)$ and B is the point with coordinates $(4, k)$.

   a) Find the value of k.

(1 mark)

   b) What is the equation of the line perpendicular to AB, that passes through A?

(4 marks)

6 a) Solve the inequality $4x + 7 > 7x + 4$.

(2 marks)

   b) Find the values of k, such that $(x - 5)(x - 3) > k$, for all possible values of x.

(3 marks)

7 Calculate $\int (4x^3 + 6x + 3)\, dx$.

(4 marks)

8 a) Find the coordinates of the point A, when A lies at the intersection of the lines l_1 and l_2, and when the equations of l_1 and l_2 respectively are $x - y + 1 = 0$ and $2x + y - 8 = 0$.

(3 marks)

   b) The points B and C have coordinates $(6, -4)$ and $(-\frac{4}{3}, -\frac{1}{3})$ respectively, and D is the midpoint of AC. Find the equation of the line BD in the form $ax + by + c = 0$, where a, b and c are integers.

(6 marks)

   c) Show that the triangle ABD is a right-angled triangle.

(3 marks)

9 a) Express $x^2 - 6x + 5$ in the form $(x + a)^2 + b$.

(2 marks)

   b) Factorise the expression $x^2 - 6x + 5$.

(2 marks)

   c) Hence sketch the graph of $y = x^2 - 6x + 5$, clearly indicating where it cuts the axes.

(3 marks)

10 A curve that passes through the point $(0, 0)$ has derivative $\dfrac{dy}{dx} = 3x^2 + 6x - 4$.

   a) Find the equation of the curve.

(4 marks)

   b) Express y as a product of 3 linear factors.

(3 marks)

11 The equation $x^2 - 4x + (k - 1) = 0$, where k is a constant, has no real roots.
Find the set of possible values of k.

(4 marks)

12 An arithmetic series has first term a and common difference d. The value of the 12th term is 79, and the value of the 16th term is 103.

   a) From the information given above, find two equations in terms of a and d, then solve them to find the values of a and d.

(4 marks)

S_n is the sum of the first n terms of the series.

   b) Find a simplified expression for S_n in terms of n, using your values of a and d from part a) above.

(3 marks)

   c) Calculate the value of S_{15}.

(1 mark)

General Certificate of Education
Advanced Subsidiary (AS) and Advanced Level

Core Mathematics C1 — Practice Exam Two

Time Allowed: 1 hour 30 min

Calculators may **not** be used for this exam

There are 75 marks available for this paper.

1 a) Simplify $(\sqrt{3} + 1)(\sqrt{3} - 1)$. *(2 marks)*

   b) Rationalise the denominator of the expression $\dfrac{\sqrt{3}}{\sqrt{3} + 1}$. *(3 marks)*

2 a) Express $\dfrac{x^2 + 2x}{\sqrt{x}}$ in the form $x^m + 2x^n$, where m and n are constants to be found. *(2 marks)*

   b) Find $\dfrac{dy}{dx}$ for $y = \dfrac{x^2 + 2x}{\sqrt{x}} + 3x^3 - x$ (where $x > 0$). *(4 marks)*

3 A curve has equation $y = f(x)$, where $\dfrac{dy}{dx} = 4(1 - x)$.
   The curve passes through the point A, with coordinates $(2, 6)$.

   a) Find the equation of the curve. *(4 marks)*

   b) Sketch the graph of $y = f(x)$, showing where it crosses the axes. *(4 marks)*

4 a) Express $x^2 - 7x + 17$ in the form $(x - m)^2 + n$, where m and n are constants.
       Hence state the maximum value of $f(x) = \dfrac{1}{x^2 - 7x + 17}$. *(4 marks)*

   b) Find the possible values of k if the equation $g(x) = 0$ is to have only one root,
       where $g(x)$ is given by $g(x) = 3x^2 + kx + 12$. *(3 marks)*

5 The sides of a triangle ABC lie on lines given as follows:

   side AB lies on $y = 3$,   side BC lies on $2x - 3y - 21 = 0$,   side AC lies on $3x + 2y - 12 = 0$

   a) Find the coordinates of the vertices of the triangle. *(5 marks)*

   b) Show that the triangle is right-angled. *(2 marks)*

   The point D has coordinates $(3, d)$.

   c) If point D lies outside the triangle, but not on it, show that either $d > 3$ or $d < 1.5$. *(3 marks)*

6   Given that the equation $3jx - jx^2 + 1 = 0$, where j is a constant, has no real roots,

    a)   Show that $9j^2 + 4j < 0$

         (3 marks)

    b)   Hence find the set of possible values of j

         (3 marks)

7   A curve has the equation $y = f(x)$, where $f(x) = x^3 - 3x + 2$.

    a)   Find $f'(x)$.

         (2 marks)

    b)   Show that $x^3 - 3x + 2$ factorises to $(x - 1)^2(x + 2)$.

         (2 marks)

    c)   Sketch the graphs of:

        (i) $y = f(x)$

         (3 marks)

        (ii) $y = f(x - 3)$

         (2 marks)

        (iii) $y = 2f(x)$

         (2 marks)

8   A function is defined by $f(x) = x^3 - 4x^2 - 7x + 10$.
    $(x - 1)$ is a factor of $f(x)$.
    Hence or otherwise solve the equation $x^3 - 4x^2 - 7x + 10 = 0$.

         (4 marks)

9   The diagram shows part of the graph with equation $y = x^3 - 2x^2 + 4$.

    a)   Find the equation of the tangent at $x = 1$.

         (3 marks)

    b)   Find the equation of the normal to the curve at $x = 2$.

         (3 marks)

    c)   Find the distance between where the tangent and the normal
        cross the x-axis.

         (2 marks)

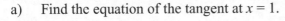

10 a)   A sequence is defined by the recurrence relation $x_{n+1} = 3x_n - 4$.
        Given that the first term is 6, find x_4.

         (2 marks)

    b)   The third term of an arithmetic progression is 9 and the seventh term is 33.

        (i)   Find the first term and the common difference.

         (3 marks)

        (ii) Find S_{12}, the sum of the first 12 terms in the series.

         (3 marks)

        (iii) Hence or otherwise find: $\sum_{1}^{12}(6n + 1)$

         (2 marks)

Algebraic Division

Algebraic division is one of those things that you have to learn when you do AS maths. You'll probably never use it again once you've done your exam, but hey ho... such is life.

Do **Algebraic Division** by means of **Subtraction**

$$(2x^3 - 3x^2 - 3x + 7) \div (x - 2) = \; ?$$

The trick with this is to see how many times you can <u>subtract</u> $(x - 2)$ from $(2x^3 - 3x^2 - 3x + 7)$.
The idea is to keep <u>subtracting</u> lumps of $(x - 2)$ until you've got rid of all the <u>powers of x</u>.

Do the subtracting in **Stages**

At each stage, always try to get rid of the <u>highest</u> power of x.
Then start again with whatever you've got left.

(1) Start with $2x^3 - 3x^2 - 3x + 7$, and <u>subtract</u> $2x^2$ lots of $(x - 2)$ to get rid of the x^3 term.

$(2x^3 - 3x^2 - 3x + 7) - 2x^2(x - 2)$
$(2x^3 - 3x^2 - 3x + 7) - 2x^3 + 4x^2$
$\qquad = x^2 - 3x + 7$

$2x^3 \div x = 2x^2$

This is what's left — so now you have to get rid of the $\underline{x^2}$ term.

(2) Now <u>start again</u> with $x^2 - 3x + 7$.
The highest power of x is the x^2 term.
So <u>subtract</u> x lots of $(x - 2)$ to get rid of that.

$(x^2 - 3x + 7) - x(x - 2)$
$(x^2 - 3x + 7) - x^2 + 2x$
$\qquad = -x + 7$

Now start again with this — and get rid of the $\underline{x}$ term.

(3) All that's left now is $-x + 7$.
Get rid of the $-x$ by <u>subtracting</u> -1 times $(x - 2)$.

$(-x + 7) - (-1(x - 2))$
$(-x + 7) + x - 2$
$\qquad = 5$

There are no more powers of x to get rid of — so <u>stop here</u>.
The <u>remainder's</u> 5.

Interpreting the results...

Time to work out exactly what all that <u>meant</u>...

Started with: $2x^3 - 3x^2 - 3x + 7$

Subtracted: $2x^2(x - 2) + x(x - 2) - 1(x - 2)$

$\qquad\qquad = (x - 2)(2x^2 + x - 1)$

Remainder: $= 5$

So... $2x^3 - 3x^2 - 3x + 7 = (x - 2)(2x^2 + x - 1) + 5$

...or to put that another way...

$$\frac{2x^3 - 3x^2 - 3x + 7}{x - 2} = 2x^2 + x - 1 \text{ with remainder 5.}$$

$2x^2 + x - 1$ is called the <u>quotient</u>.

Algebraic Division

$$(ax^3 + bx^2 + cx + d) \div (x - k) = \; ?$$

1) <u>SUBTRACT</u> a multiple of $(x - k)$ to get rid of the highest power of x.

2) <u>REPEAT</u> step 1 until you've got rid of all the powers of x.

3) <u>WORK OUT</u> how many lumps of $(x - k)$, you've subtracted, and the <u>REMAINDER</u>.

Algebraic division is a beautiful thing that we should all cherish...

Revising algebraic division isn't the most enjoyable way to spend an afternoon, it's true, but it's <u>in the specification</u>, and so you need to be <u>comfortable</u> with it. It involves the same process you use when you're doing long division with numbers — so if you're having trouble following the above, do $4863 \div 7$ really slowly. What you're doing at each stage is subtracting multiples of 7, and you do this until you can't take any more 7s away, which is when you get your remainder.

The Remainder and Factor Theorems

The Remainder Theorem and the Factor Theorem are easy, and possibly quite useful.

The **Remainder Theorem** is an easy way to work out **Remainders**

When you divide f(x) by (x − a), the remainder is f(a).

So in the example on the previous page, you could have worked out the remainder dead easily.

1) $f(x) = 2x^3 - 3x^2 - 3x + 7$.
2) You're dividing by $(x - 2)$, so $a = 2$.
3) So the remainder must be $f(2) = (2 \times 8) - (3 \times 4) - (3 \times 2) + 7 = 5$.

Careful now... when you're dividing by something like $(x + 7)$, a is negative — so here, a = −7.

If you want the remainder after dividing by something like (ax − b), there's an extension to the remainder theorem...

When you divide f(x) by (ax − b), the remainder is $f\left(\frac{b}{a}\right)$.

EXAMPLE Find the remainder when you divide $2x^3 - 3x^2 - 3x + 7$ by $2x - 1$.

$f(x) = 2x^3 - 3x^2 - 3x + 7$. You're dividing by $2x - 1$, so $a = 2$ and $b = 1$.

So the remainder must be: $f\left(\frac{1}{2}\right) = 2\left(\frac{1}{8}\right) - 3\left(\frac{1}{4}\right) - 3\left(\frac{1}{2}\right) + 7 = 5$

The **Factor Theorem** is just the Remainder Theorem with a **Zero Remainder**

If you get a remainder of zero when you divide f(x) by (x − a), then (x − a) must be a factor. That's the Factor Theorem.

If f(x) is a polynomial, and f(a) = 0, then (x − a) is a factor of f(x).

In other words: If you know the roots, you also know the factors — and vice versa.

EXAMPLE Show that $(2x + 1)$ is a factor of $f(x) = 2x^3 - 3x^2 + 4x + 3$.

The question's giving you a big hint here. Notice that $2x + 1 = 0$ when $x = -\frac{1}{2}$. So plug this value of x into f(x). If you show that $f(-\frac{1}{2}) = 0$, then the factor theorem says that $(x + \frac{1}{2})$ is a factor — which means that $2 \times (x + \frac{1}{2}) = (2x + 1)$ is also a factor.

$f(x) = 2x^3 - 3x^2 + 4x + 3$ and so $f\left(-\frac{1}{2}\right) = 2 \times \left(-\frac{1}{8}\right) - 3 \times \frac{1}{4} + 4 \times \left(-\frac{1}{2}\right) + 3 = 0$

So, by the factor theorem, $(x + \frac{1}{2})$ is a factor of f(x), and so $(2x + 1)$ is also a factor.

(x − 1) is a Factor if the coefficients **Add Up To 0**

This works for all polynomials — no exceptions. It could save a fair whack of time in the exam.

EXAMPLE Factorise the polynomial $f(x) = 6x^2 - 7x + 1$

The coefficients (6, −7 and 1) add up to 0. That means $f(1) = 0$, and so $(x − 1)$ is a factor. Easy.

Then just factorise it like any quadratic to get this: $f(x) = 6x^2 - 7x + 1 = (6x - 1)(x - 1)$

Factorising a **Cubic** given **No Factors**

If the question doesn't give you any factors, the best way to find a factor of a cubic is to guess — use trial and error.

First, add up the coefficients to check if (x−1) is a factor.
If that doesn't work, keep trying small numbers (find f(−1), f(2), f(−2), f(3), f(−3) and so on) until you find a number that gives you zero when you put it in the cubic. Call that number k. (x−k) is a factor of the cubic.
Then finish factorising the cubic using the method on page 19.

C2 Section 1 — Practice Questions

What a start that was. One thing that's definitely true about that opening section is that it was <u>short</u>. It was 2 pages long, to be precise. That's about as short as it could be without being <u>really silly</u>. So stop complaining that you're hard done by and have a crack at <u>these questions</u>...

Warm-up Questions

1) Write the following functions f(x) in the form f(x) = (x + 2)g(x) + remainder (where g(x) is a quadratic):
   a) f(x) = $3x^3 - 4x^2 - 5x - 6$,   b) f(x) = $x^3 + 2x^2 - 3x + 4$,   c) f(x) = $2x^3 + 6x - 3$

2) Find the remainder when the following are divided by: (i) (x + 1),   (ii) (x − 1),   (iii) (x − 2)
   a) f(x) = $6x^3 - x^2 - 3x - 12$,   b) f(x) = $x^4 + 2x^3 - x^2 + 3x + 4$,   c) f($x$) = $x^5 + 2x^2 - 3$

3) Find the remainder when f(x) = $x^4 - 3x^3 + 7x^2 - 12x + 14$ is divided by:
   a) $x + 2$ b) $2x + 4$ c) $x - 3$ d) $2x - 6$

4) Which of the following are factors of f(x) = $x^5 - 4x^4 + 3x^3 + 2x^2 - 2$?
   a) $x - 1$ b) $x + 1$ c) $x - 2$ d) $2x - 2$

5) Find the values of c and d so that $2x^4 + 3x^3 + 5x^2 + cx + d$ is exactly divisible by ($x - 2$)($x + 3$).

There now, that wasn't so bad, was it? Now take a <u>deep breath</u> and have a go at some questions that are a bit more like the ones you may have to face <u>in the exam</u>.

Exam Questions

1   f(x) = $2x^3 - 5x^2 - 4x + 3$

   a)   Find the remainder when f(x) is divided by

      (i)   ($x - 1$)

        (2 marks)

      (ii)   ($2x + 1$)

        (2 marks)

   b)   Show using the factor theorem that ($x + 1$) is a factor of f(x).

        (2 marks)

   c)   Factorise f(x) completely.

        (4 marks)

2   f(x) = ($4x^2 + 3x + 1$)($x - p$) + 5, where p is a constant.

   a)   State the value of f(p).

        (1 mark)

   b)   Find the value of p, given that when f(x) is divided by ($x + 1$), the remainder is −1.

        (2 marks)

   c)   Find the remainder when f(x) is divided by ($x - 1$).

        (1 mark)

Circles

I always say a <u>beautiful shape</u> deserves a <u>beautiful formula</u>, and here you've got one of my favourite double-acts...

Equation of a circle: $(x - a)^2 + (y - b)^2 = r^2$

The equation of a circle looks complicated, but it's all based on Pythagoras' theorem.
Take a look at the circle below, with centre (6, 4) and radius 3.

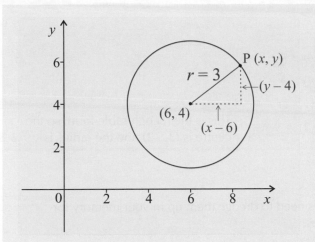

Joining a point P (x, y) on the circumference of the circle to its centre (6, 4), we can create a <u>right-angled triangle</u>.

Now let's see what happens if we use <u>Pythagoras' theorem</u>:

$$(x - 6)^2 + (y - 4)^2 = 3^2$$

or: $(x - 6)^2 + (y - 4)^2 = 9$

This is the equation for the circle. It's as easy as that.

In general, a circle with radius r and centre (a, b) has the equation: $(x - a)^2 + (y - b)^2 = r^2$

EXAMPLE:

i) What is the centre and radius of the circle with equation $(x - 2)^2 + (y + 3)^2 = 16$

ii) Write down the equation of the circle with centre (–4, 2) and radius 6.

SOLUTION:

i) Comparing $(x - 2)^2 + (y + 3)^2 = 16$ with the general form:

$$(x - a)^2 + (y - b)^2 = r^2$$

then $a = 2$, $b = -3$ and $r = 4$.

> **So the centre (a, b) is: (2, –3)**
>
> **and the radius (r) is: 4.**

And as if by magic, here it is.

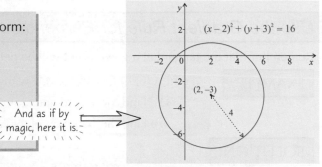

ii) The question says, 'Write down...', so you know you
don't need to do any working.
The centre of the circle is (–4, 2), so $a = -4$ and $b = 2$.
The radius is 6, so $r = 6$.
Using the general equation for a circle $(x - a)^2 + (y - b)^2 = r^2$
you can write: $(x + 4)^2 + (y - 2)^2 = 36$

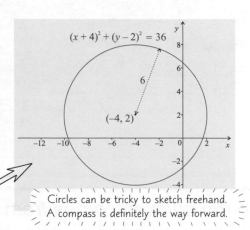

Circles can be tricky to sketch freehand.
A compass is definitely the way forward.

This is pretty much all you need to learn. Everything on the next page uses stuff you should know already.

Circles

Rearrange the equation into the **familiar form**

Sometimes you'll be given an equation for a circle that doesn't look much like $(x - a)^2 + (y - b)^2 = r^2$.
This is a bit of a pain, because it means you can't immediately tell what the **radius** is or where the **centre** is.
But all it takes is a bit of **rearranging**.

Let's take the equation: $x^2 + y^2 - 6x + 4y + 4 = 0$

You need to get it into the form $(x - a)^2 + (y - b)^2 = r^2$.

This is just like completing the square.

Have a look at C1 Section 2 for more on completing the square.

$x^2 + y^2 - 6x + 4y + 4 = 0$

$x^2 - 6x + y^2 + 4y + 4 = 0$

$(x - 3)^2 - 9 + (y + 2)^2 - 4 + 4 = 0$

$(x - 3)^2 + (y + 2)^2 = 9$ $\implies$ This is the recognisable form, so the centre is **(3, –2)** and the radius is $\sqrt{9} = 3$.

Don't forget the Properties of Circles

You will have seen the circle rules at GCSE. You'll sometimes need to dredge them up in your memory for these circle questions. Here's a reminder of a few useful ones.

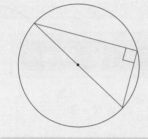

The angle in a semicircle is a right angle.

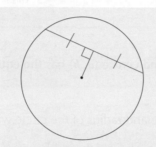

The perpendicular from the centre to a chord bisects the chord.

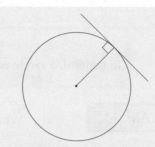

A radius and tangent to the same point will meet at right angles.

Use the Gradient Rule for Perpendicular Lines

Remember that the tangent at a given point will be perpendicular to the radius at that same point.

EXAMPLE: Point A (6, 4) lies on a circle with the equation $x^2 + y^2 - 4x - 2y - 20 = 0$.

    i) Find the centre and radius of the circle.

    ii) Find the equation of the tangent to the circle at A.

SOLUTION:

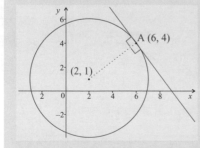

i) Rearrange the equation to show it as the sum of 2 squares:

$x^2 + y^2 - 4x - 2y - 20 = 0$

$x^2 - 4x + y^2 - 2y - 20 = 0$

$(x - 2)^2 - 4 + (y - 1)^2 - 1 - 20 = 0$

$(x - 2)^2 + (y - 1)^2 = 25$

This shows the centre is (2, 1) and the radius is 5.

ii) The tangent is at right angles to the radius at (6, 4).

Gradient of radius at (6, 4) = $\dfrac{4 - 1}{6 - 2} = \dfrac{3}{4}$

Gradient of tangent = $\dfrac{-1}{\frac{3}{4}} = -\dfrac{4}{3}$

Using $y - y_1 = m(x - x_1)$

$y - 4 = -\dfrac{4}{3}(x - 6)$

$3y - 12 = -4x + 24$

$3y + 4x - 36 = 0$

So the chicken comes from the egg, and the egg comes from the chicken...

Well folks, at least it makes a change from all those straight lines and quadratics.
I reckon if you know the **formula** and **what it means**, you should be absolutely **fine** with questions on circles.

Arc Length and Sector Area

Arc lengths and sector areas are easier than you'd think — once you've learnt two simple(ish) formulas.

Always work in *Radians* for *Arc Length* and *Sector Area Questions*

Remember — for arc length and sector area questions you've got to measure all the angles in <u>radians</u>.

The main thing is that you know how radians relate to <u>degrees</u>.
In short, 180 degrees = π radians. The table below shows you how to convert between the two units:

Converting angles	
<u>Radians to degrees:</u>	<u>Degrees to radians:</u>
Divide by π, multiply by 180.	**Divide by 180, multiply by π.**

Here's a table of some of the common angles you're going to need — in degrees and radians:

Degrees	0	30	45	60	90	120	180	270	360
Radians	0	$\frac{\pi}{6}$	$\frac{\pi}{4}$	$\frac{\pi}{3}$	$\frac{\pi}{2}$	$\frac{2\pi}{3}$	π	$\frac{3\pi}{2}$	2π

If you have <u>part of a circle</u> (like a section of pie chart), you can work out the <u>length of the curved side</u>, or the <u>area of the 'slice of pie'</u> — as long as you know the <u>angle</u> at the centre (θ) and the <u>length of the radius</u> (r). Read on...

You can find the *Length* of an *Arc* using a nice easy formula...

For a circle with a <u>radius of r</u>, where the angle θ is measured in <u>radians</u>, the <u>arc length of the sector S</u> is given by:

If you put $\theta = 2\pi$ in this formula (and so make the sector equal to the whole circle), you get that the distance all the way round the outside of the circle is $S = 2\pi r$.

This is just the normal circumference formula.

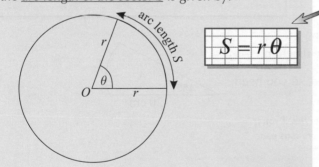

$$S = r\theta$$

...and the area of a *Sector* using a similar formula

For a circle with a <u>radius of r</u>, where the angle θ is measured in <u>radians</u>, you can work out A, the <u>area of the sector</u>, using:

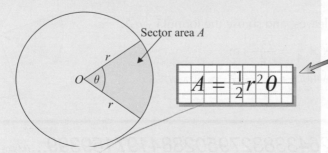

Sector area A

$$A = \frac{1}{2}r^2\theta$$

Again, if you put $\theta = 2\pi$ in the formula, you find that the area of the whole circle is $A = \frac{1}{2}r^2 \times 2\pi = \pi r^2$.

This is just the normal 'area of a circle' formula.

Arc Length and Sector Area

Questions on <u>trigonometry</u> quite often use the same angles — so it makes life easier if you know the sin, cos and tan of these commonly used angles. Or to put it another way, examiners expect you to know them — so learn them.

Draw Triangles to remember sin, cos and tan of the Important Angles

You should know the values of <u>sin</u>, <u>cos</u> and <u>tan</u> at 30°, 60° and 45°. But to help you remember, you can draw these two groovy triangles. It may seem a complicated way to learn a few numbers, but it does make it easier. Honest.

The idea is you draw the triangles below, putting in their angles and side lengths. Then you can use them to work out special trig values like <u>sin 45°</u> or <u>cos 60°</u> more accurately than any calculator (which only gives a few decimal places).

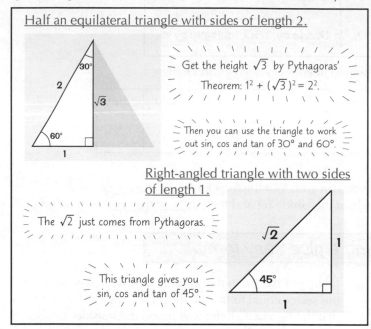

Half an equilateral triangle with sides of length 2.

Get the height $\sqrt{3}$ by Pythagoras' Theorem: $1^2 + (\sqrt{3})^2 = 2^2$.

Then you can use the triangle to work out sin, cos and tan of 30° and 60°.

Right-angled triangle with two sides of length 1.

The $\sqrt{2}$ just comes from Pythagoras.

This triangle gives you sin, cos and tan of 45°.

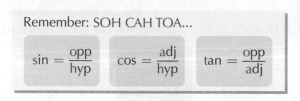

Remember: SOH CAH TOA...

$$\sin = \frac{opp}{hyp} \qquad \cos = \frac{adj}{hyp} \qquad \tan = \frac{opp}{adj}$$

Trig Values from Triangles

$\sin 30° = \frac{1}{2}$	$\sin 60° = \frac{\sqrt{3}}{2}$	$\sin 45° = \frac{1}{\sqrt{2}}$
$\cos 30° = \frac{\sqrt{3}}{2}$	$\cos 60° = \frac{1}{2}$	$\cos 45° = \frac{1}{\sqrt{2}}$
$\tan 30° = \frac{1}{\sqrt{3}}$	$\tan 60° = \sqrt{3}$	$\tan 45° = 1$

EXAMPLE Find the exact length L and area A in the diagram.

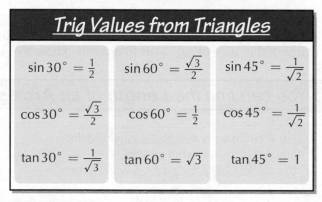

Right, first things first... it's an arc length and sector area, so you need the angle in radians.

$$45° = \frac{45 \times \pi}{180} = \frac{\pi}{4} \text{ radians}$$

Or you could just quote this if you've learnt the stuff on the previous page.

Now bung everything in your formulas:

$$L = r\theta = 20 \times \frac{\pi}{4} = 5\pi \text{ cm}$$

$$A = \frac{1}{2}r^2\theta = \frac{1}{2} \times 20^2 \times \frac{\pi}{4} = 50\pi \text{ cm}^2$$

EXAMPLE Find the area of the shaded part of the symbol.

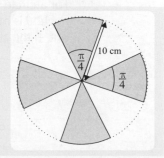

Instead, you could use the total angle of all the shaded sectors (π).

You need the area of the 'leaves' and so use the formula $\frac{1}{2}r^2\theta$.

Each leaf has area $\frac{1}{2} \times 10^2 \times \frac{\pi}{4} = \frac{25\pi}{2} \text{ cm}^2$

So the area of the whole symbol $= 4 \times \frac{25\pi}{2} = 50\pi \text{ cm}^2$

$\pi = 3.14159265358979323846264338327950288419716939...$ *(Make sure you know it)*

It's worth repeating, just to make sure — those formulas for arc length and sector area only work if the angle is in <u>radians</u>.

The Trig Formulas You Need to Know

There are some more trig formulas you <u>need to know</u> for the exam.
So here they are — learn them or you're seriously stuffed. Worse than an aubergine.

The **Sine Rule** and **Cosine Rule** work for **Any** triangle

Remember, these three formulas work for <u>ANY</u> triangle, not just right-angled ones.

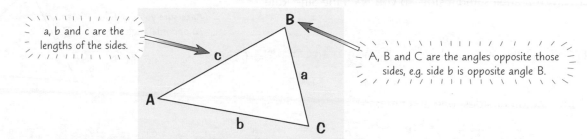

a, b and c are the lengths of the sides.

A, B and C are the angles opposite those sides, e.g. side b is opposite angle B.

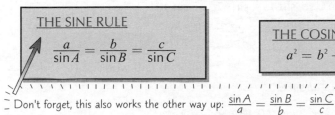

THE SINE RULE

$$\frac{a}{\sin A} = \frac{b}{\sin B} = \frac{c}{\sin C}$$

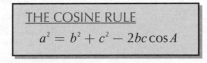

THE COSINE RULE

$$a^2 = b^2 + c^2 - 2bc \cos A$$

AREA OF ANY TRIANGLE

$$Area = \frac{1}{2}ab \sin C$$

Don't forget, this also works the other way up: $\frac{\sin A}{a} = \frac{\sin B}{b} = \frac{\sin C}{c}$

Sine Rule *or* Cosine Rule — *which one is it...*

To decide which of these two rules you need to use, look at what you <u>already</u> know about the triangle.

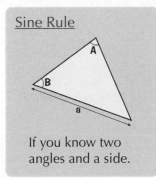

Sine Rule

If you know two angles and a side.

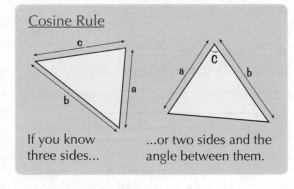

Cosine Rule

If you know three sides...

...or two sides and the angle between them.

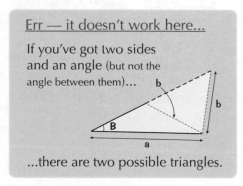

Err — it doesn't work here...

If you've got two sides and an angle (but not the angle between them)...

...there are two possible triangles.

The **Best** has been saved till last...

These two identities are really important. You'll need them <u>loads</u>.

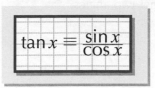

$$\tan x \equiv \frac{\sin x}{\cos x}$$

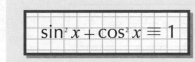

$$\sin^2 x + \cos^2 x \equiv 1$$

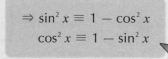

$$\Rightarrow \sin^2 x \equiv 1 - \cos^2 x$$
$$\cos^2 x \equiv 1 - \sin^2 x$$

Work out these two using $\sin^2 x + \cos^2 x \equiv 1$.

These two come up in exam questions <u>all the time</u>. Learn them.
Learnthemlearnthemlearnthemlearnthemlearnthemlear... okay, I'll stop now.

Tri angles — go on... you might like them.

Formulas and trigonometry go together even better than Richard and Judy. I can count 7 formulas on this page.
That's not many, so please, just make sure you know them. If you haven't learnt them I will cry for you. I will sob. ♦☹♦

Using the Sine and Cosine Rules

This page is about "solving" triangles, which just means finding all their <u>sides</u> and <u>angles</u> when you already know a few.

EXAMPLE   Solve △ABC, in which A = 40°, a = 27 m, B = 73°. Then find the area.

Draw a quick sketch first — don't worry if it's not deadly accurate, though.

You're given 2 angles and a side, so you need the Sine Rule.

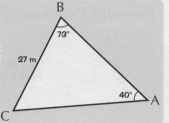

Make sure you put side a opposite angle A.

First of all, get the other angle: $\angle C = (180 - 40 - 73)° = 67°$

Then find the other sides, one at a time:

$$\frac{a}{\sin A} = \frac{b}{\sin B} \Rightarrow \frac{27}{\sin 40°} = \frac{b}{\sin 73°}$$
$$\Rightarrow b = \frac{\sin 73°}{\sin 40°} \times 27 = \underline{40.2\,m}$$

$$\frac{c}{\sin C} = \frac{a}{\sin A} \Rightarrow \frac{c}{\sin 67°} = \frac{27}{\sin 40°}$$
$$\Rightarrow c = \frac{\sin 67°}{\sin 40°} \times 27 = \underline{38.7\,m}$$

Now just use the formula to find its area.

$$\text{Area of } \triangle ABC = \tfrac{1}{2}ab\sin C$$
$$= \tfrac{1}{2} \times 27 \times 40.169 \times \sin 67°$$
$$= \underline{499.2\,m^2}$$

Use a more accurate value for b here, rather than the rounded value 40.2.

EXAMPLE   Find X, Y and z.

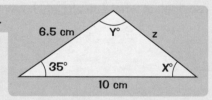

6.5 cm   Y°   z
35°   X°
10 cm

You've been given 2 sides and the angle between them, so you're first going to need the Cosine Rule to find side z.

$$a^2 = b^2 + c^2 - 2bc\cos A$$

$$z^2 = (6.5)^2 + 10^2 - 2(6.5)(10)\cos 35°$$
$$\Rightarrow z^2 = 142.25 - 130\cos 35°$$
$$\Rightarrow z^2 = 35.7602$$
$$\Rightarrow z = \underline{5.98\,cm}$$

In this case, angle A is 35°, and side a is actually z.

Now that you've got all the sides and one angle, you can use the Sine Rule to find the other two angles.

$$\frac{a}{\sin A} = \frac{b}{\sin B} = \frac{c}{\sin C}$$

Remember — if $\frac{a}{\sin A} = \frac{b}{\sin B}$ then $\frac{\sin A}{a} = \frac{\sin B}{b}$

$$\frac{\sin X}{6.5} = \frac{\sin 35°}{5.9800}$$
$$\Rightarrow \sin X = 0.6235$$
$$\Rightarrow X = \sin^{-1}0.6235$$
$$\Rightarrow X = \underline{38.6°}$$

$$\frac{\sin Y}{10} = \frac{\sin 35°}{5.9800}$$
$$\Rightarrow \sin Y = 0.9592$$
$$\Rightarrow Y = \sin^{-1}0.9592$$
$$\Rightarrow Y = \underline{73.6°} \text{ or } \underline{106.4°}$$

This is the answer you need. <u>Be careful</u>: your calculator only gives you values for $\sin^{-1}$ between −90° and 90°. See page 65.

Check your answers by adding up all the angles in the triangle. If they don't add up to 180°, you've gone wrong somewhere.

Graphs of Trig Functions

Before you leave this page, you should be able to close your eyes and picture these three graphs in your head, properly labelled and everything. If you can't, you need to learn them more. I'm not kidding.

sin x and cos x are always in the range –1 to 1

$\underline{\sin x}$ and $\underline{\cos x}$ are similar — they just bob up and down between –1 and 1.

sin x and cos x are both underline{periodic} (repeat themselves) with period 360°

$$\cos(x + 360°) = \cos x \qquad \sin(x + 360°) = \sin x$$

They bounce up and down from –1 to 1 — they can <u>never</u> have a value outside this range.

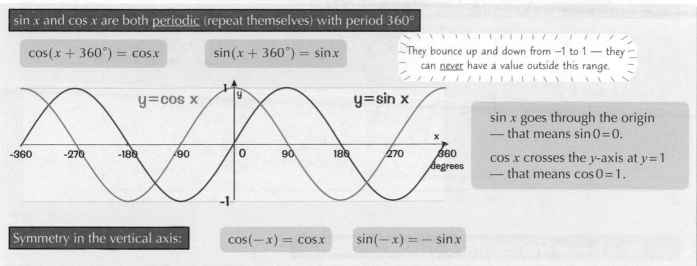

sin x goes through the origin — that means $\sin 0 = 0$.

cos x crosses the y-axis at $y = 1$ — that means $\cos 0 = 1$.

Symmetry in the vertical axis:

$$\cos(-x) = \cos x \qquad \sin(-x) = -\sin x$$

tan x can be Any Value at all

tan x is different from sin x or cos x.
It doesn't go gently up and down between –1 and 1 — it goes between $-\infty$ and $+\infty$.

TAN X IS ALSO <u>PERIODIC</u> — BUT WITH PERIOD 180°

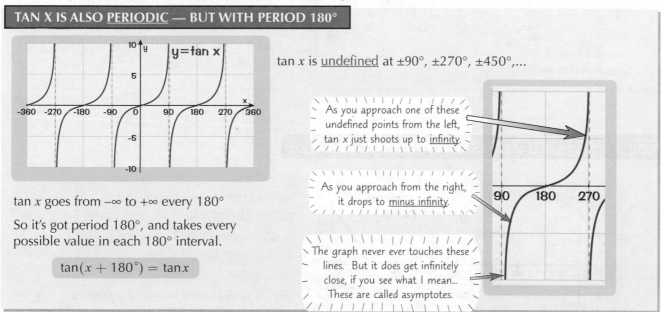

tan x is <u>undefined</u> at ±90°, ±270°, ±450°,...

As you approach one of these undefined points from the left, tan x just shoots up to <u>infinity</u>.

As you approach from the right, it drops to <u>minus infinity</u>.

The graph never ever touches these lines. But it does get infinitely close, if you see what I mean... These are called asymptotes.

tan x goes from $-\infty$ to $+\infty$ every 180°

So it's got period 180°, and takes every possible value in each 180° interval.

$$\tan(x + 180°) = \tan x$$

The easiest way to sketch any of these graphs is to plot the important points which happen every 90° (i.e. –180°, –90°, 0°, 90°, 180°, 270°, 360°...) and then just join the dots up.

Sin and cos can make your life worthwhile — give them a chance...

It's really really really really really important that you can draw the trig graphs on this page, and get all the labels right. Make sure you know what value sin, cos and tan have at the interesting points — i.e. 0°, 90°, 180°, 270°, 360°. It's easy to remember what the graphs look like, but you've got to know exactly <u>where</u> they're max, min, zero, etc.

Transformed Trig Graphs

Transformed trigonometric graphs look much the same as the bog-standard ones, just a little <u>different</u>. There are three main types of transformation.

There are 3 basic types of Transformed Trig Graph...

$y = n \sin x$ — a *Vertical Stretch* or *Squash*

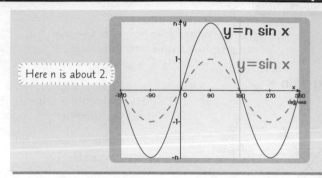

Here n is about 2.

If $n > 1$, the graph of $y = \sin x$ is <u>stretched vertically</u> by a factor of n.

If $0 < n < 1$, the graph is <u>squashed</u>.

And if $n < 0$, the graph is also <u>reflected</u> in the <u>x-axis</u>.

$y = \sin nx$ — a *Horizontal Squash* or *Stretch*

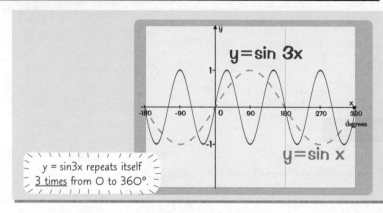

y = sin3x repeats itself <u>3 times</u> from O to 360°.

If $n > 1$, the graph of $y = \sin x$ is <u>squashed horizontally</u> by a factor of n.

If $0 < n < 1$, the graph is <u>stretched</u>.

And if $n < 0$, the graph is also <u>reflected</u> in the <u>y-axis</u>.

$y = \sin (x + c)$ — a *Translation* along the x-axis

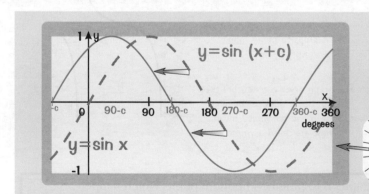

For $c > 0$, $\sin (x + c)$ is just $\sin x$ <u>shifted c to the left</u>.

Similarly, $\sin (x - c)$ is just $\sin x$ <u>shifted c to the right</u>.

For $y = \sin(x + c)$, the 'interesting' points are when $x + c = 0, 90°, 180°, 270°$, etc., i.e. when $x = -c, 90 - c, 180 - c, 270 - c,...$

Curling up on the sofa with 2cos x — that's my idea of cosiness ☺

One thing you've really got to be careful about is making sure you move or stretch the graphs in the <u>right</u> direction. In that last example, the graph would have moved to the right if "c" was negative. And it gets confusing with the horizontal and vertical stretching and squashing — in the first two examples above, $n > 1$ means a <u>vertical stretch</u> but a <u>horizontal squash</u>.

Solving Trig Equations in a Given Interval

I used to really hate trig stuff like this. But once I'd got the hang of it, I just couldn't get enough. I stopped going out, lost interest in the opposite sex — the CAST method became my life. Learn it, but be careful. It's addictive.

There are **Two Ways** to find Solutions in an **Interval**...

EXAMPLE Solve $\cos x = \frac{1}{2}$ for $-360° \leq x \leq 720°$.

Like I said — there are two ways to solve this kind of question. Just use the one you prefer...

You can draw a **graph**...

Your calculator gives you a solution of 60°. Then you have to work out what the others will be.

The other solutions are 60° either side of the graph's peaks.

1) Draw the graph of $y = \cos x$ for the range you're interested in...

2) Get the first solution from your calculator and mark this on the graph,

3) Use the symmetry of the graph to work out what the other solutions are:

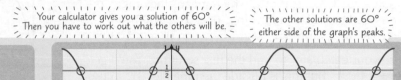

So the solutions are: $-300°, -60°, 60°, 300°, 420°$ and $660°$.

...or you can use the **CAST** diagram

CAST stands for COS, ALL, SIN, TAN — and the CAST diagram shows you where these functions are positive:

Between 90° and 180°, only SIN is positive.

Between 0 and 90°, ALL of sin, cos and tan are positive.

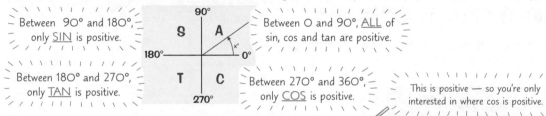

Between 180° and 270°, only TAN is positive.

Between 270° and 360°, only COS is positive.

This is positive — so you're only interested in where cos is positive.

First, to find all the values of x between $0°$ and $360°$ where $\cos x = \frac{1}{2}$ — you do this:

Put the first solution onto the CAST diagram.	Find the other angles between 0° and 360° that might be solutions.	Ditch the ones that are the wrong sign.

The angle from your calculator goes anticlockwise from the x-axis (unless it's negative — then it would go clockwise into the 4th quadrant).

The other possible solutions come from making the same angle from the horizontal axis in the other 3 quadrants.

$\cos x = \frac{1}{2}$, which is positive. The CAST diagram tells you cos is positive in the 4th quadrant — but not the 2nd or 3rd — so ditch those two angles.

So you've got solutions $60°$ and $300°$ in the range $0°$ to $360°$. But you need all the solutions in the range $-360°$ to $720°$. Get these by repeatedly adding or subtracting $360°$ onto each until you go out of range:

$$x = 60° \Rightarrow \underline{(adding}\ 360°)\ x = 420°,\ 780°\ (\text{too big})$$

$$\text{and}\ \underline{(subtracting}\ 360°)\ x = -300°,\ -660°\ (\text{too small})$$

$$x = 300° \Rightarrow \underline{(adding}\ 360°)\ x = 660°,\ 1020°\ (\text{too big})$$

$$\text{and}\ \underline{(subtracting}\ 360°)\ x = -60°,\ -420°\ (\text{too small})$$

So the solutions are: $x = -300°, -60°, 60°, 300°, 420°$ and $660°$.

And I feel that love is dead, I'm loving angles instead...

Suppose the first solution you get is negative, let's say $-d°$, then you'd measure it clockwise on the CAST diagram. So it'd be $d°$ in the 4th quadrant. Then you'd work out the other 3 possible solutions in exactly the same way, rejecting the ones which weren't the right sign. Got that? No? Got that? No? Got that? Yes? Good!

Solving Trig Equations in a Given Interval

Sometimes it's a bit more complicated. But only a bit.

Sometimes you end up with *sin kx = number*...

For these, it's definitely easier to draw the <u>graph</u> rather than use the CAST method —
that's one reason why being able to sketch trig graphs properly is so important.

> **EXAMPLE** Solve: $\sin 3x = -\frac{1}{\sqrt{2}}$ for $0° \leq x \leq 360°$.

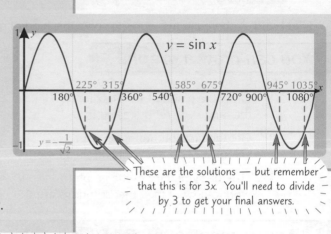

1) You've got $3x$ instead of x, which means the range
   you need to find solutions in is $0° \leq 3x \leq 1080°$.
   So draw the graph of $y = \sin x$ between $0°$ and $1080°$.

2) Use your calculator to find the first solution. You'll get
   $3x = -45°$ — but this is outside the range for $3x$, so use
   the pattern of the graph to find a solution in the range.
   As the sin curve repeats every $360°$, there'll be a solution
   at $360 - 45 = 315°$.

   These are the solutions — but remember that this is for $3x$. You'll need to divide by 3 to get your final answers.

3) Now use your graph to find the other 5 solutions.
   You can see that there's another solution at $180 + 45 = 225°$.
   Then add on $360°$ and $720°$ to both $225°$ and $315°$ to get:

   $$3x = 225°, 315°, 585°, 675°, 945° \text{ and } 1035°.$$

   It really is mega-important that you check these answers — it's dead easy to make a silly mistake. They should all be in the range $0° \leq x \leq 360°$.

   Divide by 3 to get the solutions for x:

   $$x = 75°, 105°, 195°, 225°, 315° \text{ and } 345°.$$

4) <u>Check</u> your answers by putting these values back into your calculator.

...or *sin (x + k) = number*

All the steps in this example are just the same as in the one above.

> **EXAMPLE** Solve $\sin(x + 60°) = \frac{3}{4}$ for $-360° \leq x \leq 360°$, giving your answers to 2 decimal places.

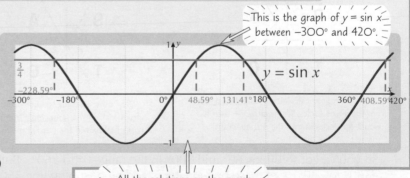

1) You've got $\sin(x + 60°)$ instead of $\sin x$ —
   so the range is $-300° \leq x + 60° \leq 420°$.

   This is the graph of $y = \sin x$ between $-300°$ and $420°$.

2) Use your calculator to get
   that first solution...

   $$\sin(x + 60°) = \frac{3}{4}$$
   $$\Rightarrow x + 60° = 48.59°$$

3) From the graph, you can see that the other
   solutions occur at $180 - 48.59$, $360 + 48.59$
   and $-180 - 48.59$. The solutions are:

   $$x + 60° = -228.58°, 48.59°, 131.41° \text{ and } 408.59°$$

   Subtract 60 to get the solutions for x:

   $$x = -11.41°, -288.59°, 71.41° \text{ and } 348.59°$$

   All the solutions on the graph are values of $x + 60°$. You need to subtract 60 to find the values of x — the solutions you want.

4) <u>Check</u> your answers by putting these values
   back into your calculator.

Live a life of sin (and cos and tan)...

Yep, the examples on this page are pretty fiddly. The most important bit is actually getting the sketch right for your new
range. If you don't, you're in big trouble. Then you've just got to carefully use the sketch to work out the other solutions.
It's tricky, but you'll feel better about yourself when you've got it mastered. Ah you will, you will, you will ...

Solving Trig Equations in a Given Interval

Now for something really exciting — trig identities. Mmm, well, maybe exciting was the wrong word. But they can be dead useful, so here goes...

For equations with **tan x** in, it often helps to use this...

$$\tan x \equiv \frac{\sin x}{\cos x}$$

This is a handy thing to know — and one the examiners love testing. Basically, if you've got a trig equation with a tan in it, together with a sin or a cos — chances are you'll be better off if you rewrite the tan using this formula.

EXAMPLE Solve: $3\sin x - \tan x = 0$, for $0 \le x \le 2\pi$.

It's got sin and tan in it — so writing $\tan x$ as $\frac{\sin x}{\cos x}$ is probably a good move:

$$3\sin x - \tan x = 0$$
$$\Rightarrow 3\sin x - \frac{\sin x}{\cos x} = 0$$

Get rid of the $\cos x$ on the bottom by multiplying the whole equation by $\cos x$.

$$\Rightarrow 3\sin x \cos x - \sin x = 0$$

Now — there's a common factor of $\sin x$. Take that outside a bracket.

$$\Rightarrow \sin x(3\cos x - 1) = 0$$

And now you're almost there. You've got two things multiplying together to make zero. That means either one or both of them is equal to zero themselves.

$$\Rightarrow \sin x = 0 \quad \text{or} \quad 3\cos x - 1 = 0$$

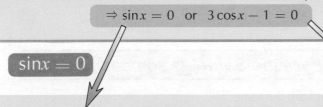

$\boxed{\sin x = 0}$

The first solution is... $\sin 0 = 0$

Now find the other points where $\sin x$ is zero in the interval $0 \le x \le 2\pi$.
(Remember the sin graph is zero every π radians.)

$$\Rightarrow x = 0, \pi, 2\pi \text{ radians}$$

$\boxed{3\cos x - 1 = 0}$

Rearrange... $\cos x = \frac{1}{3}$

So the first solution is...

$$\cos^{-1}\frac{1}{3} = 1.231$$

CAST (or the graph of $\cos x$) gives another solution in the 4th quadrant...

CAST gives any solutions in the interval $0 \le x \le 2\pi$.

So altogether you've got <u>five</u> possible solutions:

$$\Rightarrow x = 0, 1.231, \pi, 5.052, 2\pi \text{ radians}$$

And the two solutions from this part are:

$$\Rightarrow x = 1.231, 5.052 \text{ radians}$$

Trigonometry is the root of all evil...

What a page — you don't have fun like that every day, do you? No, trig equations are where it's at. This is a really useful trick, though — and can turn a nightmare of an equation into a bit of a pussycat. <u>Rewriting</u> stuff using <u>different</u> formulas is always worth trying if it feels like you're getting stuck — even if you're not sure why when you're doing it. You might have a flash of inspiration when you see the new version.

Solving Trig Equations in a Given Interval

Another trig identity — and it's a good 'un — examiners love it. And it's not difficult either.

And if you have a *sin² x* or a *cos² x*, think of this straight away...

$$\sin^2 x + \cos^2 x \equiv 1 \Rightarrow \begin{array}{l} \sin^2 x \equiv 1 - \cos^2 x \\ \cos^2 x \equiv 1 - \sin^2 x \end{array}$$

Use this identity to get rid of a $\sin^2$ or a $\cos^2$ that's making things awkward...

EXAMPLE   Solve: $2\sin^2 x + 5\cos x = 4$, for $0° \leq x \leq 360°$.

You can't do much while the equation's got both sin's and cos's in it. So replace the $\sin^2 x$ bit with $1 - \cos^2 x$.

$$2(1 - \cos^2 x) + 5\cos x = 4$$

Multiply out the bracket and rearrange it so that you've got zero on one side — and you get a quadratic in $\cos x$:

Now the only trig function is cos.

$$\Rightarrow 2 - 2\cos^2 x + 5\cos x = 4$$
$$\Rightarrow 2\cos^2 x - 5\cos x + 2 = 0$$

This is a quadratic in $\cos x$. It's easier to factorise this if you make the substitution $y = \cos x$.

$$2y^2 - 5y + 2 = 0$$
$$\Rightarrow (2y - 1)(y - 2) = 0$$
$$\Rightarrow (2\cos x - 1)(\cos x - 2) = 0$$

$2y^2 - 5y + 2 = (2y\ ?)(y\ ?)$
$= (2y - 1)(y - 2)$

Now one of the brackets must be 0. So you get 2 equations as usual:

You've already done this example on page 65.

$$2\cos x - 1 = 0 \quad \text{or} \quad \cos x - 2 = 0$$

$$\cos x = \tfrac{1}{2} \Rightarrow x = 60° \quad \text{or} \quad x = 300° \quad \text{and} \quad \cos x = 2$$

This is a bit weird. cos x is always between −1 and 1. So you don't get any solutions from this bracket.

So at the end of all that, the only solutions you get are $x = 60°$ and $x = 300°$. How boring.

Use the **Trig Identities** to prove something is the **Same** as something else

Another use for these trig identities is proving that two things are the same.

EXAMPLE   Show that $\dfrac{\cos^2 \theta}{1 + \sin \theta} \equiv 1 - \sin \theta$

The identity sign ≡ means that this is true for all θ, rather than just certain values.

Prove things like this by playing about with one side of the equation until you get the other side.

Left-hand side: $\dfrac{\cos^2 \theta}{1 + \sin \theta}$

The only thing I can think of doing here is replacing $\cos^2 \theta$ with $1 - \sin^2 \theta$. (Which is good because it works.)

$$\equiv \frac{1 - \sin^2 \theta}{1 + \sin \theta}$$

The next trick is the hardest to spot. Look at the top — does that remind you of anything?

The top line is a difference of two squares:

$$\equiv \frac{(1 + \sin \theta)(1 - \sin \theta)}{1 + \sin \theta}$$

$1 - a^2 = (1 + a)(1 - a)$
$\Rightarrow 1 - \sin^2 \theta = (1 + \sin \theta)(1 - \sin \theta)$

$$\equiv 1 - \sin \theta, \text{ the right-hand side}.$$

Trig identities — the path to a brighter future...

That was a pretty miserable section. But it's over. These trig identities aren't exactly a barrel of laughs, but they are a definite source of marks — you can bet your last penny they'll be in the exam. That substitution trick to get rid of a $\sin^2$ or a $\cos^2$ and end up with a quadratic in $\sin x$ or $\cos x$ is a real examiners' favourite. Those identities can be a bit daunting, but it's always worth having a few tricks in the back of your mind — always look for things that factorise, or fractions that can be cancelled down, or ways to use those trig identities. Ah, it's all good clean fun.

C2 Section 2 — Practice Questions

I know, I know — that was a long section. And, rather predictably, it's followed by lots of <u>practice questions</u>. <u>Brace yourself</u> for the warm-up — this is going to be a <u>heavy session</u>...

Warm-up Questions

1) Give the radius and the coordinates of the centre of the circles with the following equations:
   a) $x^2 + y^2 = 9$ b) $(x - 2)^2 + (y + 4)^2 = 4$ c) $x(x + 6) = y(8 - y)$

2) Write down the exact values of cos 30°, sin 30°, tan 30°, cos 45°, sin 45°, tan 45°, cos 60°, sin 60° and tan 60°.

3) Draw a triangle $\triangle XYZ$ with sides of length x, y and z.
   Write down the Sine and Cosine Rules for this triangle. Write down an expression for its area.

4) What is tan x in terms of cos x and sin x? What is $\cos^2 x$ in terms of $\sin^2 x$?

5) Solve :   a) $\triangle ABC$ in which A = 30°, C = 25°, $b = 6$ m, and find its area.
           b) $\triangle PQR$ in which $p = 3$ km, $q = 23$ km, R = 10°. (answers to 2 d.p.)

6) My pet triangle Freda has sides of length 10, 20 and 25.
   Find her angles (in degrees to 1 d.p.).

7) Find the 2 possible triangles $\triangle ABC$ which satisfy $b = 5$, $a = 3$, A = 35°. (This is tricky: Sketch it first, and see if you can work out how to make 2 different triangles satisfying the data given.)

8) Sketch the graphs for sin x, cos x and tan x.
   Make sure you label all the max/min/zero/undefined points.

9) Sketch the following graphs:
   a) $y = \frac{1}{2}\cos x$ (for $0° \le x \le 360°$)   b) $y = \sin(x + 30°)$ (for $0° \le x \le 360°$)   c) $y = \tan 3x$ (for $0° \le x \le 180°$)

10) a) Solve each of these equations for $0° \le \theta \le 360°$:
         (i) $\sin\theta = -\frac{\sqrt{3}}{2}$ (ii) $\tan\theta = -1$ (iii) $\cos\theta = -\frac{1}{\sqrt{2}}$
    b) Solve each of these equations for $-180° \le \theta \le 180°$ (giving your answer to 1 d.p.):
         (i) $\cos 4\theta = -\frac{2}{3}$ (ii) $\sin(\theta + 35°) = 0.3$ (iii) $\tan\left(\frac{1}{2}\theta\right) = 500$

11) Find all the solutions to $6\sin^2 x = \cos x + 5$ in the range $0° \le x \le 360°$ (answers to 1 d.p.).

12) Solve $3\tan x + 2\cos x = 0$ for $-90° \le x \le 90°$

13) Simplify: $(\sin y + \cos y)^2 + (\cos y - \sin y)^2$

14) Show that $\dfrac{\sin^4 x + \sin^2 x \cos^2 x}{\cos^2 x - 1} \equiv -1$

Well after that vigorous warm-up you should be more than ready to tackle this <u>mental marathon</u> of exam-style questions. Take your time to work out <u>what each question wants</u> — if it looks impossible, there's probably a way of simplifying it somehow. A picture paints a thousand words, so <u>sketch it out</u> if you're totally confused.

Exam Questions

1   C is a circle with the equation: $x^2 + y^2 - 2x - 10y + 21 = 0$.

     a)   Find the centre and radius of C.

(5 marks)

     The line joining $P(3, 6)$ and $Q(q, 4)$ is a diameter of C.

     b)   Show that $q = -1$.

(3 marks)

     c)   Find the equation of the tangent to C at Q, giving your answer in the form $ax + by + c = 0$, where a, b and c are integers.

(5 marks)

C2 Section 2 — Practice Questions

2   For an angle x, $3 \cos x = 2 \sin x$.

a)   Find $\tan x$.

(2 marks)

b)   Hence, or otherwise, find all the values of x, in the interval $0 \leq x \leq 360°$,
    for which $3 \cos x = 2 \sin x$, giving your answers to 1 d.p.

(2 marks)

3   A circle C is shown here.
    M is the centre of C, and J and K both lie on C.

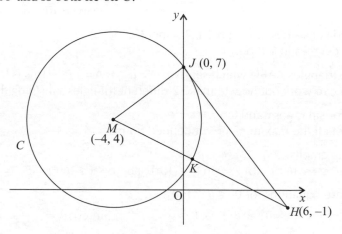

a)   Write down the equation of C in the form $(x - a)^2 + (y - b)^2 = r^2$.

(3 marks)

The line JH is a tangent to circle C at point J.

b)   Show that angle $JMH = 1.1071$ radians to 4 d.p.

(4 marks)

c)   Find the length of the shortest arc on C between J and K, giving your answer to 3 s.f.

(2 marks)

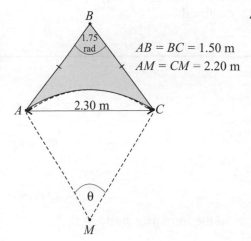

$AB = BC = 1.50$ m
$AM = CM = 2.20$ m

4   The shape ABC shown here is a concrete slab that forms part of a
    paved area around a circular pond.
    The curve AC is an arc of a circle with centre M and radius 2.20 m.

Find, to 3 significant figures:

a)   The size of angle θ in radians,

(2 marks)

b)   The perimeter of the slab in m,

(3 marks)

c)   The area of the slab in m².

(5 marks)

C2 Section 2 — Practice Questions

5   The diagram below shows the dimensions of a child's wooden toy. The toy is a prism with height 10 cm. Its cross-section is a sector of a circle with radius 20 cm and angle $\frac{\pi}{4}$ radians.

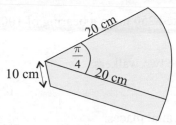

  a)   Show that the volume of the toy, $V = 500\pi$ cm^3.

(3 marks)

  b)   Show that the surface area of the toy, $S = (150\pi + 400)$ cm^2.

(5 marks)

6   a)   (i)   Sketch, for $0 \leq x \leq 360°$, the graph of $y = \cos(x + 60°)$.

(2 marks)

    (ii)   Write down all the values of x, for $0 \leq x \leq 360°$, where $\cos(x + 60°) = 0$.

(2 marks)

  b)   Sketch, for $0 \leq x \leq 180°$, the graph of $y = \sin 4x$.

(2 marks)

  c)   Solve, for $0 \leq x \leq 180°$, the equation: $\sin 4x = 0.5$, giving your answers in degrees.

(4 marks)

7   Solve, for $0 \leq x \leq 2\pi$:

  a) $\tan\left(x + \frac{\pi}{6}\right) = \sqrt{3}$.

(4 marks)

  b) $2\cos\left(x - \frac{\pi}{4}\right) = \sqrt{3}$

(4 marks)

  c) $\sin 2x = -\frac{1}{2}$

(6 marks)

8   a)   Show that the equation:

$$2(1 - \cos x) = 3\sin^2 x$$

    can be written as

$$3\cos^2 x - 2\cos x - 1 = 0$$

(2 marks)

  b)   Use this to solve the equation

$$2(1 - \cos x) = 3\sin^2 x$$

    for $0 \leq x \leq 360°$, giving your answers to 1 d.p.

(6 marks)

C2 Section 2 — Practice Questions

9   The diagram shows the locations of two walkers, X and Y, after walking in different directions from the same start position.

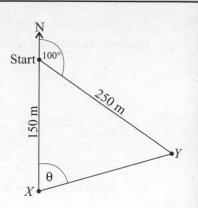

     X walked due south for 150 m. Y walked 250 m on a bearing of 100°.

     a)   Calculate the distance between the two walkers, in m to the nearest m.

           (2 marks)

     b)   Show that $\dfrac{\sin\theta}{\sin80°} = 0.93$ to 2 decimal places.

           (3 marks)

10   Find all the values of x, in the interval $0 \leq x \leq 2\pi$, for which:

$$2 - \sin x = 2 \cos^2 x$$

     giving your answers in terms of π.

           (6 marks)

11   Solve the following equations, for $-\pi \leq x \leq \pi$:

     a) $(1 + 2\cos x)(3 \tan^2 x - 1) = 0$

           (6 marks)

     b) $3 \cos^2 x = \sin^2 x$

           (4 marks)

12   The diagram shows a circle C, with centre P. M is the midpoint of AB, a chord.

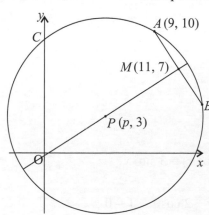

     a)   Show that $p = 5$.

           (5 marks)

     b)   Find the equation of circle C.

           (4 marks)

Logs

Don't be put off by your parents or grandparents telling you that logs are hard. <u>Logarithm</u> is just a fancy word for <u>power</u>, and once you know how to use them you can solve all sorts of equations.

You need to be able to **Switch** between **Different Notations**

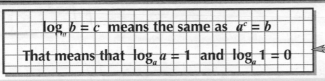

$\log_a b = c$ means the same as $a^c = b$

That means that $\log_a a = 1$ and $\log_a 1 = 0$

The little number 'a' after 'log' is called the <u>base</u>.
Logs can be to any base, but <u>base 10</u> is the most common.
The <u>button</u> marked '<u>log</u>' on your calculator uses base 10.

EXAMPLE   Index notation: $10^2 = 100$ log notation: $\log_{10} 100 = 2$

The <u>base</u> goes here but it's usually left out if it's 10.

So the <u>logarithm</u> of 100 to the <u>base 10</u> is 2, because 10 raised to the <u>power</u> of 2 is 100.

EXAMPLES   Write down the values of the following:

   a) $\log_2 8$ b) $\log_9 3$ c) $\log_5 5$

   a)   8 is 2 raised to the power of 3,
        so $2^3 = 8$ and $\log_2 8 = 3$

   b)   3 is the square root of 9, or $9^{1/2} = 3$,
        so $\log_9 3 = 1/2$

   c)   Anything to the power of 1 is itself,
        so $\log_5 5 = 1$

Write the following using log notation:

   a) $5^3 = 125$ b) $3^0 = 1$

You just need to make sure you get things in the right place.

   a)   3 is the power or <u>logarithm</u> that 5
        (the <u>base</u>) is raised to to get 125,
        so $\log_5 125 = 3$

   b)   You'll need to remember this one:
        $\log_3 1 = 0$

The **Laws of Logarithms** are **Unbelievably Useful**

Whenever you have to deal with <u>logs</u>, you'll end up using the <u>laws</u> below.
That means it's no bad idea to <u>learn them</u> by heart right now.

Laws of Logarithms

$$\log_a x + \log_a y = \log_a (xy)$$

$$\log_a x - \log_a y = \log_a \left(\frac{x}{y}\right)$$

$$\log_a x^k = k \log_a x$$

So $\log_a \frac{1}{x} = -\log_a x$

You've got to be able to change the <u>base</u> of a log too.

So $\log_7 4 = \dfrac{\log_{10} 4}{\log_{10} 7} = 0.7124$

To check: $7^{0.7124} = 4$

Change of Base

$$\log_a x = \frac{\log_b x}{\log_b a}$$

Use the **Laws** to **Manipulate Logs**

EXAMPLE   Write each expression in the form $\log_a n$, where n is a number.

     a) $\log_a 5 + \log_a 4$ b) $2 \log_a 6 - \log_a 9$

a) $\log_a x + \log_a y = \log_a (xy)$

You just have to <u>multiply</u>
the numbers together:

$$\log_a 5 + \log_a 4 = \log_a (5 \times 4)$$
$$= \log_a 20$$

b) $\log_a x^k = k \log_a x$

$$2 \log_a 6 = \log_a 6^2 = \log_a 36$$
$$\log_a 36 - \log_a 9 = \log_a (36 \div 9)$$
$$= \log_a 4$$

It's sometimes hard to see the wood for the trees — especially with logs...

Tricky, tricky, tricky... I think of $\log_a b$ as 'the <u>power</u> I have to raise a to if I want to end up with b' — that's all it is. And the log laws make a bit more sense if you think of 'log' as meaning 'power'. For example, you know that $2^a \times 2^b = 2^{a+b}$ — this just says that if you multiply the two numbers, you add the powers. Well, the first law of logs is saying the same thing. Any road, even if you don't really understand why they work, make sure you know the log laws like you know your own navel.

Exponentials and Logs

Okay, you've done the theory of logs. So now it's a bit of stuff about <u>exponentials</u> (the opposite of logs, kind of), and then it'll be time to get your calculator out for a bit of button pressing...

Graphs of a^x Never Reach Zero

All the graphs of $y = a^x$ (exponential graphs) where $a > 1$ have the <u>same basic shape</u>. The graphs for $a = 2$, $a = 3$ and $a = 4$ are shown on the right.

- All the a's are greater than 1 — so <u>y increases as x increases</u>.

- The <u>bigger</u> a is, the <u>quicker</u> the graphs increase.

- As x <u>decreases</u>, y <u>decreases</u> at a <u>smaller and smaller rate</u> — y will approach zero, but never actually get there.

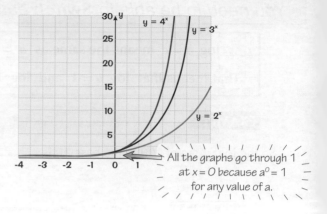

All the graphs go through 1 at $x = 0$ because $a^0 = 1$ for any value of a.

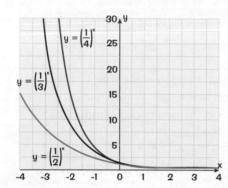

The graphs on the left are for $y = a^x$ where $a < 1$ (they're for $a = \frac{1}{2}$, $\frac{1}{3}$ and $\frac{1}{4}$).

- All the a's are less than 1 — meaning <u>y decreases as x increases</u>.

- As x <u>increases</u>, y <u>decreases</u> at a <u>smaller and smaller rate</u> — again, y will approach zero, but never actually get there.

*You can use **exponentials** and **logs** to **solve equations***

EXAMPLE   1)   Solve $2^{4x} = 3$ to 3 significant figures.

You want x on its own, so take logs of both sides (by writing 'log' in front of both sides):

$$\log 2^{4x} = \log 3$$

Now use one of the laws of logs: $\log x^k = k \log x$:

$$4x \log 2 = \log 3$$

You can now divide both sides by '4 log 2' to get x on its own:

$$x = \frac{\log 3}{4 \log 2}$$

But $\frac{\log 3}{4 \log 2}$ is just a number you can find using a calculator:

$$x = 0.396 \text{ (to 3 s.f.)}$$

I don't know about you, but I enjoyed that more than the biggest, fastest rollercoaster. You want another? OK then...

EXAMPLE   2)   Solve $7 \log_{10} x = 5$ to 3 significant figures.

You want x on its own, so begin by dividing both sides by 7:

$$\log_{10} x = \frac{5}{7}$$

You now need to take exponentials of both sides by doing '10 to the power of both sides' (since the log is to base 10):

$$10^{\log_{10} x} = 10^{\frac{5}{7}}$$

Logs and exponentials are inverse functions, so they cancel out:

$$x = 10^{\frac{5}{7}}$$

Again, $10^{\frac{5}{7}}$ is just a number you can find using a calculator:

$$x = 5.18 \text{ (to 3 s.f.)}$$

Exponentials and Logs

Use the **Calculator Log Button** Whenever You Can

EXAMPLE Use logarithms to solve the following for x, giving the answers to 4 s.f.

a) $10^{3x} = 4000$ b) $7^x = 55$ c) $\log_2 x = 5$

You've got the magic buttons on your calculator, but you'd better <u>follow the instructions</u> and show that you know how to use the <u>log rules</u> covered earlier.

a) $10^{3x} = 4000$ — there's an 'unknown' in the power, so <u>take logs of both sides</u>.
(In theory, it doesn't matter what <u>base</u> you use, but your calculator has a '$\log_{10}$' button, so base 10 is usually a good idea. But whatever base you use, <u>use the same one for both sides</u>.)

So taking logs to base 10 of both sides of the above equation gives:
$$\log 10^{3x} = \log 4000$$
$$\text{i.e. } 3x \log 10 = \log 4000$$

Since $\log_{10} 10 = 1 \Longrightarrow$ i.e. $3x = \log 4000$, so $x = 1.201$ (to 4 sig. fig.)

b) $7^x = 55$. Again, take logs of both sides, and use the log rules: $x \log_{10} 7 = \log_{10} 55$, so $x = \dfrac{\log_{10} 55}{\log_{10} 7} = 2.059$

c) $\log_2 x = 5$ — to get rid of a log, you 'take exponentials', meaning you do '2 (the base) to the power of each side'.

Think of 'taking logs' and 'taking exponentials' as opposite processes — one cancels the other out: $2^{\log_2 x} = 2^5$
$$\text{i.e. } x = 32$$

You might have to **Combine** the **Laws of Logs** to **Solve** equations

If the examiners are feeling particularly mean, they might make you use <u>more than one</u> law to solve an equation.

EXAMPLE Solve the equation $\log_3(2 - 3x) - 2\log_3 x = 2$.

First, combine the log terms into one term (you can do this because they both have the same base): $\log_3 \dfrac{2 - 3x}{x^2} = 2$

Remember that $2\log x = \log x^2$.

Then take exponentials of both sides: $3^{\log_3 \frac{2-3x}{x^2}} = 3^2 \Rightarrow \dfrac{2 - 3x}{x^2} = 9$

Ignore the negative solution because you can't take logs of a negative number.

Finally, rearrange the equation and solve for x: $2 - 3x = 9x^2 \Rightarrow 0 = 9x^2 + 3x - 2$
$$\Rightarrow 0 = (3x - 1)(3x + 2)$$

So $x = \dfrac{1}{3}$.

Exponential Growth and **Decay** Applies to **Real-life** Problems

Logs can even be used to solve real-life problems.

EXAMPLE The radioactivity of a substance decays by 20 per cent over a year. The initial level of radioactivity is 400. Find the time taken for the radioactivity to fall to 200 (the half-life).

$R = 400 \times 0.8^T$ where R is the <u>level of radioactivity</u> at time T years.
We need $R = 200$, so solve $200 = 400 \times 0.8^T$

The 0.8 comes from $1 - 20\%$ decay.

$0.8^T = \dfrac{200}{400} = 0.5 \Rightarrow T \log 0.8 = \log 0.5 \Rightarrow T = \dfrac{\log 0.5}{\log 0.8} = 3.106$ years

If in doubt, take the log of something — that usually works...

The thing about exponential growth is that it's really useful, as it happens in real life all over the place. Money in a bank account earns interest at <u>a certain percentage per year</u>, and so the balance rises <u>exponentially</u> (if you don't spend or save anything). Likewise, if you got 20% cleverer for every week you studied, that would also be exponential... and impressive.

C2 Section 3 — Practice Questions

Logs and exponentials are <u>surprisingly useful</u> things. As well as being in your exam they pop up all over the place in real life — <u>savings</u>, <u>radioactive decay</u>, growth of <u>bacteria</u> — all logarithmic. And now for something (marginally) different:

Warm-up Questions

1) Write down the values of the following:
   a) $\log_3 27$
   b) $\log_3 (1 \div 27)$
   c) $\log_3 18 - \log_3 2$

2) Simplify the following:
   a) $\log 3 + 2 \log 5$
   b) $\frac{1}{2} \log 36 - \log 3$
   c) $\log 2 - \frac{1}{4} \log 16$

3) Simplify $\log_b (x^2 - 1) - \log_b (x - 1)$

4) a) Copy and complete the table for the function $y = 4^x$:

x	−3	−2	−1	0	1	2	3
y							

   b) Using suitable scales, plot a graph of $y = 4^x$ for $-3 < x < 3$.
   c) Use the graph to solve the equation $4^x = 20$.

5) Solve these little jokers:
   a) $10^x = 240$
   b) $\log_{10} x = 5.3$
   c) $10^{2x+1} = 1500$
   d) $4^{(x-1)} = 200$

6) Find the smallest integer P such that $1.5^P > 1\,000\,000$.

Time for some practice at the <u>real thing</u>. On your marks... Get set... Go.

Exam Questions

1 a) Write the following expressions in the form $\log_a n$, where n is an integer:
   (i) $\log_a 20 - 2 \log_a 2$
   (3 marks)
   (ii) $\frac{1}{2} \log_a 16 + \frac{1}{3} \log_a 27$
   (3 marks)

   b) Find the value of:
   (i) $\log_2 64$
   (1 mark)
   (ii) $2 \log_3 9$
   (2 marks)

   c) Calculate the value of the following, giving your answer to 4 d.p.
   (i) $\log_6 25$
   (1 mark)
   (ii) $\log_3 10 + \log_3 2$
   (2 marks)

C2 Section 3 — Practice Questions

2 a) Solve the equation

$$2^x = 9$$

giving your answer to 2 decimal places.

(3 marks)

   b) Hence, or otherwise, solve the equation

$$2^{2x} - 13(2^x) + 36 = 0$$

giving each solution to an appropriate degree of accuracy.

(5 marks)

3 Solve the equation

$$\log_7 (y + 3) + \log_7 (2y + 1) = 1$$

where $y > 0$.

(5 marks)

4 a) Solve the equation:

$$\log_3 x = -\frac{1}{2}$$

leaving your answer as an exact value.

(3 marks)

   b) Find x, where

$$2 \log_3 x = -4$$

leaving your answer as an exact value.

(2 marks)

5 a) Find x, if

$$6^{(3x + 2)} = 9$$

giving your answer to 3 significant figures.

(3 marks)

   b) Find y, if

$$3^{(y^2 - 4)} = 7^{(y + 2)}$$

giving your answer to 3 significant figures.

(5 marks)

6 For the positive integers p and q,

$$\log_4 p - \log_4 q = \frac{1}{2}$$

   a) Show that $p = 2q$.

(3 marks)

   b) Solve the equation for p and q

$$\log_2 p + \log_2 q = 7$$

(5 marks)

Geometric Progressions

You have a geometric progression when each term in a sequence is found by multiplying the previous term by a (constant) number. Let me explain...

Geometric Progressions Multiply by a Constant each time

Geometric progressions work like this: the next term in the sequence is obtained by multiplying the previous one by a constant value. Couldn't be easier.

$$u_1 = a \qquad\qquad = a$$
$$u_2 = a \times r \qquad\quad = ar$$
$$u_3 = a \times r \times r \quad = ar^2$$
$$u_4 = a \times r \times r \times r = ar^3$$

The first term (u_1) is called 'a'.

The number you multiply by each time is called 'the common ratio', symbolised by 'r'.

Here's the formula describing any term in the geometric progression:

$$u_n = ar^{n-1}$$

EXAMPLE There is a chessboard with a 1p piece on the first square, 2p on the second square, 4p on the third, 8p on the fourth and so on until the board is full. Calculate how much money is on the board.

This is a geometric progression, where you get the next term in the sequence by multiplying the previous one by 2.

So a = 1 (because you start with 1p on the first square) and r = 2.

So $u_1 = 1$, $u_2 = 2$, $u_3 = 4$, $u_4 = 8$...

To be continued... (once we've gone over how to sum the terms of a geometric progression).

A Sequence becomes a Series when you Add the Terms to find the Total

S_n stands for the sum of the first n terms of the geometric progression.
In the example above, you're told to work out S_{64} (because there are 64 squares on a chessboard).

To work out the formula for the sum of a G.P. you use two series and subtract.

For a G.P.:	$S_n = a + ar + ar^2 + ar^3 + ... + ar^{n-1}$
Multiplying by r gives:	$rS_n = ar + ar^2 + ar^3 + ... + ar^{n-2} + ar^{n-1} + ar^n$
Subtracting gives:	$S_n - rS_n = a - ar^n$
Factorising:	$(1-r)S_n = a(1-r^n) \implies S_n = \dfrac{a(1-r^n)}{1-r}$

If the series were subtracted the other way around you'd get
$$S_n = \frac{a(r^n - 1)}{r - 1}.$$
Both versions are correct.

So, back to the chessboard example:

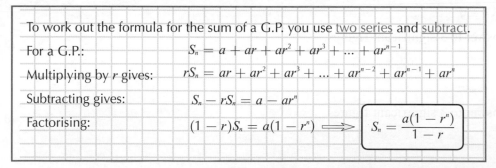

$$a = 1, \ r = 2, \ n = 64 \qquad S_{64} = \frac{1(1 - 2^{64})}{1 - 2}$$
$$S_{64} = 1.84 \times 10^{19} \text{ pence or } \pounds 1.84 \times 10^{17}$$

The whole is more than the sum of the parts — hmm, not in maths, it ain't...

You really need to understand the difference between arithmetic and geometric progressions — it's not hard, but it needs to be fixed firmly in your head. There are only a few formulas for sequences and series (the nth term of a sequence, the sum of the first n terms of a series), and these are in the formula book they give you — but make sure you know how to use them.

Geometric Progressions

Geometric progressions can either Grow or Shrink

In the chessboard example, each term was <u>bigger</u> than the previous one: 1, 2, 4, 8, 16, …
You can create a series where each term is <u>smaller</u> than the previous one by using a <u>small value of r</u>.

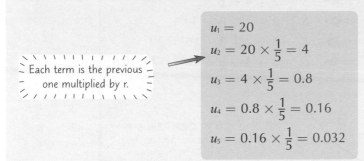

EXAMPLE If $a = 20$ and $r = \frac{1}{5}$, write down the first five terms of the sequence and the 20th term.

$u_1 = 20$

$u_2 = 20 \times \frac{1}{5} = 4$

$u_3 = 4 \times \frac{1}{5} = 0.8$

$u_4 = 0.8 \times \frac{1}{5} = 0.16$

$u_5 = 0.16 \times \frac{1}{5} = 0.032$

Each term is the previous one multiplied by r.

$u_{20} = ar^{19}$

$= 20 \times \left(\frac{1}{5}\right)^{19}$

$= 1.048576 \times 10^{-12}$

The sequence is <u>tending towards zero</u>, but won't ever get there.

In general, for each term to be <u>smaller</u> than the one before, you need $|r| < 1$.
A sequence with $|r| < 1$ is called <u>convergent</u>, since the terms converge to a limit.
Any other sequence (like the chessboard example on page 78) is called <u>divergent</u>.

$|r|$ means the modulus (or size) of r, <u>ignoring the sign</u> of the number. So $|r| < 1$ means that $-1 < r < 1$.

A Convergent Series has a Sum to Infinity

In other words, if you just <u>kept</u> adding terms to a <u>convergent series</u>, you'd get <u>closer and closer</u> to a certain number, but you'd never actually reach it.

If $|r| < 1$ and n is very, very <u>big</u>, then r^n will be very, very <u>small</u> — or to put it technically, $r^n \to 0$. (Try working out $(\frac{1}{2})^{100}$ on your calculator if you don't believe me.)

This means $(1 - r^n)$ is really, really close to 1.

So, as $n \to \infty$, $S_n \to \frac{a}{1-r}$.

It's easier to remember as $\boxed{S_\infty = \frac{a}{1-r}}$

S_∞ just means 'sum to infinity'.

EXAMPLE If $a = 2$ and $r = \frac{1}{2}$, find the sum to infinity of the geometric series.

$u_1 = 2$ $\Longrightarrow$ $S_1 = 2$

$u_2 = 2 \times \frac{1}{2} = 1$ $\Longrightarrow$ $S_2 = 2 + 1 = 3$

$u_3 = 1 \times \frac{1}{2} = \frac{1}{2}$ $\Longrightarrow$ $S_3 = 2 + 1 + \frac{1}{2} = 3\frac{1}{2}$

$u_4 = \frac{1}{2} \times \frac{1}{2} = \frac{1}{4}$ $\Longrightarrow$ $S_4 = 2 + 1 + \frac{1}{2} + \frac{1}{4} = 3\frac{3}{4}$

$u_5 = \frac{1}{4} \times \frac{1}{2} = \frac{1}{8}$ $\Longrightarrow$ $S_5 = 2 + 1 + \frac{1}{2} + \frac{1}{4} + \frac{1}{8} = 3\frac{7}{8}$

$u_6 = \frac{1}{8} \times \frac{1}{2} = \frac{1}{16}$ $\Longrightarrow$ $S_6 = 2 + 1 + \frac{1}{2} + \frac{1}{4} + \frac{1}{8} + \frac{1}{16} = 3\frac{15}{16}$

These values are getting <u>smaller</u> each time.

These values are getting closer (<u>converging</u>) to 4.

So, the sum to infinity is 4.

You can show this <u>graphically</u>:

The line on the graph is getting <u>closer and closer</u> to 4, but it'll never actually get there.

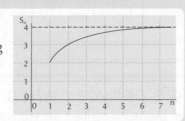

Of course, you could have saved yourself a lot of bother by using the <u>sum to infinity formula</u>:

$S_\infty = \frac{a}{1-r} = \frac{2}{1-\frac{1}{2}} = 4$

Geometric Progressions

A *Divergent* series *Doesn't* have a sum to infinity

EXAMPLE If $a = 2$ and $r = 2$, find the sum to infinity of the series.

$$u_1 = 2 \implies S_1 = 2$$
$$u_2 = 2 \times 2 = 4 \implies S_2 = 2 + 4 = 6$$
$$u_3 = 4 \times 2 = 8 \implies S_3 = 2 + 4 + 8 = 14$$
$$u_4 = 8 \times 2 = 16 \implies S_4 = 2 + 4 + 8 + 16 = 30$$
$$u_5 = 16 \times 2 = 32 \implies S_5 = 2 + 4 + 8 + 16 + 32 = 62$$

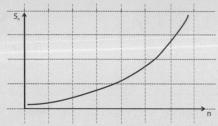

This is an exponential graph — see C2 Section 3.

As $n \to \infty$, $S_n \to \infty$ in a big way. So big, in fact, that eventually you <u>can't work it out</u> — so don't bother.

There is <u>no sum to infinity</u> for a <u>divergent</u> series.

EXAMPLE When a baby is born, £3000 is invested in an account with a fixed interest rate of 4% per year.

    a)   What will the account be worth at the start of the seventh year?

    b)   What age will the child be, to the nearest year, when the account has doubled in value?

a) $u_1 = a = 3000$

    $u_2 = 3000 + (4\% \text{ of } 3000)$ *This is the interest.*

      $= 3000 + (0.04 \times 3000)$

      $= 3000(1 + 0.04)$

      $= 3000 \times 1.04$ *So, r = 1.04*

    $u_3 = u_2 \times 1.04$

      $= (3000 \times 1.04) \times 1.04$

      $= 3000 \times (1.04)^2$

    $u_4 = 3000 \times (1.04)^3$

I've missed out some steps here — check that you understand what's happened.

    $u_7 = 3000 \times (1.04)^6$

      $= £3795.96$ (to the nearest penny)

b) You need to know when $u_n > 6000$ *double the original value.*

    From part a) you can tell that $u_n = 3000 \times (1.04)^{n-1}$

    So $3000 \times (1.04)^{n-1} > 6000$

           $(1.04)^{n-1} > 2$

To complete this you need to use logs (see C2 Section 3):

$$\log(1.04)^{n-1} > \log 2$$
$$(n - 1) \log(1.04) > \log 2$$
$$n - 1 > \frac{\log 2}{\log 1.04}$$
$$n - 1 > 17.67$$
$$n > 18.67 \quad \text{(to 2 d.p.)}$$

So u_{19} (the amount at the start of the 19th year) will be more than double the original amount — plenty of time to buy a Porsche for the 21st birthday.

So tell me — if my savings earn 4% per year, when will I be rich...

Now here's a funny thing — you can have a convergent geometric series if the common ratio is small enough. I find this odd — that I can keep adding things to a sum forever, but the sum never gets really really big.

Binomial Expansions

If you're feeling a bit stressed, just take a couple of minutes to relax before trying to get your head round this page — it's a bit of a stinker in places. Have a cup of tea and think about something else for a couple of minutes. Ready...

Writing **Binomial Expansions** is all about **Spotting Patterns**

Doing binomial expansions just involves <u>multiplying out</u> brackets. It would get nasty when you raise the brackets to <u>higher powers</u> — but once again I've got a <u>cunning plan</u>...

$$(1 + x)^0 = 1$$
$$(1 + x)^1 = 1 + x$$
$$(1 + x)^2 = 1 + 2x + x^2$$
$$(1 + x)^3 = 1 + 3x + 3x^2 + x^3$$
$$(1 + x)^4 = 1 + 4x + 6x^2 + 4x^3 + x^4$$

Anything to the power of O is 1.

$$(1 + x)^3 = (1 + x)(1 + x)^2$$
$$= (1 + x)(1 + 2x + x^2)$$
$$= 1 + 2x + x^2 + x + 2x^2 + x^3$$
$$= 1 + 3x + 3x^2 + x^3$$

A Frenchman named Pascal spotted the pattern in the coefficients and wrote them down in a <u>triangle</u>.
So it was called 'Pascal's Triangle' (imaginative, eh?).
The pattern's easy — each number is the <u>sum</u> of the two above it.

So, the next line will be: **1 5 10 10 5 1**
giving **$(1 + x)^5 = 1 + 5x + 10x^2 + 10x^3 + 5x^4 + x^5$**.

```
            1
         1     1
      1     2     1
   1     3     3   + 1
1     4     6   = 4     1
```

You **Don't** need to write out Pascal's Triangle for **Higher Powers**

There's a formula for the numbers in the triangle. The formula looks <u>horrible</u> (one of the worst in AS maths) so don't try to learn it letter by letter — look for the <u>patterns</u> in it instead. Here's an example:

EXAMPLE   Expand $(1 + x)^{20}$, giving the first four terms only.

So you can use this formula for any power, the power is called *n*. In this example *n* = 20.

$$(1 + x)^n = 1 + \frac{n}{1}x + \frac{n(n - 1)}{1 \times 2}x^2 + \boxed{\frac{n(n - 1)(n - 2)}{1 \times 2 \times 3}x^3} + \ldots\ldots\ldots + x^n$$

Here's a closer look at the term in the black box:

There are <u>three things</u> multiplied together on the top row. If n=20, this would be 20×19×18.

$$\frac{n(n - 1)(n - 2)}{1 \times 2 \times 3}x^3$$

<u>Start here</u>. The power of x is 3 and everything else here is based on 3.

There are <u>three integers</u> here multiplied together. 1×2×3 is written as 3! and called 3 <u>factorial</u>.

This means, if *n* = 20 and you were asked for '<u>the term in x^7</u>' you should write $\dfrac{20 \times 19 \times 18 \times 17 \times 16 \times 15 \times 14}{1 \times 2 \times 3 \times 4 \times 5 \times 6 \times 7}x^7$.

This can be <u>simplified</u> to $\dfrac{20!}{7!13!}x^7$ ← $20 \times 19 \times 18 \times 17 \times 16 \times 15 \times 14 = \dfrac{20!}{13!}$ because it's the numbers from 20 to 1 multiplied together, divided by the numbers from 13 to 1 multiplied together.

Believe it or not, there's an even <u>shorter</u> form: $\dfrac{20!}{7!13!}$ is written as $^{20}C_7$ or $\binom{20}{7}$ $^nC_r = \binom{n}{r} = \dfrac{n!}{r!(n - r)!}$

Your calculator will probably have an nCr button for working this out.
To calculate $\binom{20}{7}$ you'd put in 20 $\boxed{nCr}$ 7 =.

So, to finish the example, $(1 + x)^{20} = 1 + \dfrac{20}{1}x + \dfrac{20 \times 19}{1 \times 2}x^2 + \dfrac{20 \times 19 \times 18}{1 \times 2 \times 3}x^3 + \ldots$ $= 1 + 20x + 190x^2 + 1140x^3 + \ldots$

Binomial Expansions

It's slightly more complicated when the Coefficient of x isn't 1

EXAMPLE   What is the term in x^5 in the expansion of $(1 - 3x)^{12}$?

The term in x^5 will be as follows:

$$\frac{12 \times 11 \times 10 \times 9 \times 8}{1 \times 2 \times 3 \times 4 \times 5}(-3x)^5$$

Watch out — the –3 is included here with the x.

$$= \frac{12!}{5!7!}(-3)^5x^5 = -\frac{12!}{5!7!} \times 3^5x^5 = -192456x^5$$

Tip — the digits on the bottom of the fraction should always add up to the number on the top.

Note that $(-3)^{even}$ will always be <u>positive</u> and $(-3)^{odd}$ will always be <u>negative</u>.

Some Binomials contain More Complicated Expressions

The binomials so far have all had a <u>1</u> in the brackets — things get tricky when there's a <u>number other than 1</u>. Don't panic, though. The method is the same as before once you've done a bit of <u>factorising</u>.

EXAMPLE   What is the coefficient of x^4 in the expansion of $(2 + 5x)^7$?

Factorising $(2 + 5x)$ gives $2\left(1 + \frac{5}{2}x\right)$

So, $(2 + 5x)^7$ gives $2^7\left(1 + \frac{5}{2}x\right)^7$

It's really easy to forget the first bit (here it's 2^7) — you've been warned...

Here's the one you want.

$$(2 + 5x)^7 = 2^7\left(1 + \frac{5}{2}x\right)^7$$

$$= 2^7\left[1 + 7\left(\frac{5}{2}x\right) + \frac{7 \times 6}{1 \times 2}\left(\frac{5}{2}x\right)^2 + \frac{7 \times 6 \times 5}{1 \times 2 \times 3}\left(\frac{5}{2}x\right)^3 + \frac{7 \times 6 \times 5 \times 4}{1 \times 2 \times 3 \times 4}\left(\frac{5}{2}x\right)^4 + \ldots \right]$$

The coefficient of x^4 will be $2^7 \times \frac{7!}{4!3!}\left(\frac{5}{2}\right)^4 = 175000$

Don't forget the 2^7.

So, there's <u>no need</u> to work out all of the terms.
In fact, you could have gone <u>directly</u> to the term in x^4 by using the method on page 81.

> Note: the question asked for the <u>coefficient of x^4</u> in the expansion, so <u>don't include any x's</u> in your answer. If you'd been asked for the <u>term in x^4</u> in the expansion, then you <u>should</u> have included the x^4 in your answer.
>
> <u>Always</u> read the question very carefully.

Pascal was fine at maths but rubbish at music — he only played the triangle...

That ^nC_r button is pretty bloomin' useful (remember that it could be called something else on your calculator) — it saves a lot of button pressing and errors. Just make sure you put your numbers in the right way round. But that's all you need to know on binomial expansion, and on sequences and series in general — so it was reasonably short, if not exactly sweet.

C2 Section 4 — Practice Questions

What's that I hear you cry? You want revision questions — and lots of them. Well it just so happens I've got a few here. With sequences and series, get a <u>clear idea</u> in your head before you start, or you could go in completely the <u>wrong direction</u>. That would be bad.

Warm-up Questions

1) For the geometric progression 2, –6, 18, ..., find:

   a) the 10th term,

   b) the sum of the first 10 terms.

2) Find the sum of the first 12 terms of the following geometric series:

   a) $2 + 8 + 32 + ...$

   b) $30 + 15 + 7.5 + ...$

3) Find the common ratio for the following geometric series. State which ones are convergent and which are divergent.

   a) $1 + 2 + 4 + ...$

   b) $81 + 27 + 9 + ...$

   c) $1 + \frac{1}{3} + \frac{1}{9} + ...$

   d) $4 + 1 + \frac{1}{4} + ...$

4) For the geometric progression 24, 12, 6, ..., find:

   a) the common ratio,

   b) the seventh term,

   c) the sum of the first 10 terms,

   d) the sum to infinity.

5) A geometric progression begins 2, 6, ...

   Which term of the geometric progression equals 1458?

6) Write down the sixth row of Pascal's triangle. (Hint: it starts with a '1'.)

7) Give the first four terms in the expansion of $(1 + x)^{12}$.

8) What is the term in x^4 in the expansion of $(1 - 2x)^{16}$?

9) Find the coefficient of x^2 in the expansion of $(2 + 3x)^5$.

C2 Section 4 — Practice Questions

I like sequences, they're shiny and colourful and sparkly and... wait, what? ... Oh. Fiddlesticks.

Exam Questions

1   Find the coefficients of x, x^2, x^3 and x^4 in the binomial expansion of $(4 + 3x)^{10}$.

(4 marks)

2   A geometric series has the first term 12 and is defined by: $u_{n+1} = 12 \times 1.3^n$.

     a)   Is the series convergent or divergent?

(1 mark)

     b)   Find the values of the 3rd and 10th terms.

(2 marks)

3   In a geometric series, $a = 20$ and $r = \dfrac{3}{4}$.

     Find values for the following, giving your answers to 3 significant figures where necessary:

     a) S_∞

(2 marks)

     b) u_{15}

(2 marks)

     c)   The smallest value of n for which $S_n > 79.76$:

(5 marks)

4   To raise money for charity, Alex, Chris and Heather were sponsored £1 for each kilometre they ran over a 10-day period. They receive sponsorship proportionally for partial kilometres completed.

     Alex ran 3 km every day.

     Chris ran 2 km on day 1 and on each subsequent day ran 20% further than the day before.

     Heather ran 1 km on day 1 and on each subsequent day, she ran 50% further than the previous day.

     a)   How far did Heather run on day 5, to the nearest 10 metres?

(2 marks)

     b)   Show that day 10 is the first day that Chris runs further than 10 km.

(3 marks)

     c)   Find the total amount raised by the end of the 10 days, to the nearest penny.

(4 marks)

5   Two different geometric series have the same second term and sum to infinity:

         $u_2 = 5$ and $S_\infty = 36$.

     a)   Show that $36r^2 - 36r + 5 = 0$, where r represents the two possible ratios.

(4 marks)

     b)   Hence find the values of r, and the corresponding first terms, for both geometric series.

(4 marks)

Stationary Points

Differentiation is how you find gradients of curves. So you can use differentiation to find a <u>stationary point</u> (where a graph 'levels off') — that means finding where the <u>gradient</u> becomes <u>zero</u>.

Stationary Points are when the gradient is Zero

EXAMPLE   Find the stationary points on the curve $y = 2x^3 - 3x^2 - 12x + 5$, and work out the nature of each one.

A <u>stationary point</u> can be...

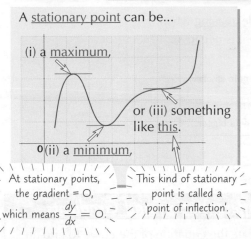

(i) a <u>maximum</u>,

or (iii) something like <u>this</u>.

0(ii) a <u>minimum</u>,

At stationary points, the gradient = O, which means $\frac{dy}{dx} = O$.

This kind of stationary point is called a 'point of inflection'.

You need to find where $\frac{dy}{dx} = f'(x) = 0$. So first, <u>differentiate</u> the function.

$$y = 2x^3 - 3x^2 - 12x + 5 \Rightarrow \frac{dy}{dx} = 6x^2 - 6x - 12$$

This is the expression for the gradient.

And then set this derivative equal to <u>zero</u>.

$f'(x)$ (pronounced "f-dash of x") is just another way to write a derivative.

$$6x^2 - 6x - 12 = 0$$
$$\Rightarrow x^2 - x - 2 = 0$$
$$\Rightarrow (x - 2)(x + 1) = 0$$
$$\Rightarrow x = 2 \text{ or } x = -1$$

So the graph has <u>two</u> stationary points, at $x = 2$ and $x = -1$.

The stationary points are actually at $(2, -15)$ and $(-1, 12)$.

Substitute the x values into the function to find the y-coordinates.

Decide if it's a Maximum or a Minimum by differentiating Again

Once you've found where the stationary points are, you have to decide whether each of them is a <u>maximum</u> or <u>minimum</u> — this is all a question means when it says, '...determine the nature of the turning points'.

A turning point is another name for a maximum or a minimum.

To decide whether a stationary point is a <u>maximum</u> or a <u>minimum</u> — just differentiate again to find $\frac{d^2y}{dx^2}$ (or $f''(x)$).

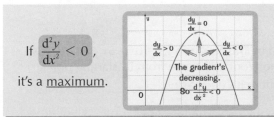

If $\frac{d^2y}{dx^2} < 0$, it's a <u>maximum</u>.

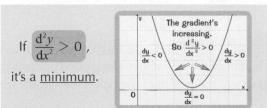

If $\frac{d^2y}{dx^2} > 0$, it's a <u>minimum</u>.

But if $\frac{d^2y}{dx^2} = O$, you can't tell what type of stationary point it is.

You've just found that $\frac{dy}{dx} = 6x^2 - 6x - 12$.

Stick in the x-coordinates of the stationary points.

So differentiating again gives $\frac{d^2y}{dx^2} = 12x - 6$.

At $x = -1$, $\frac{d^2y}{dx^2} = -18$, which is <u>negative</u> — so $x = -1$ is a <u>maximum</u>.

And at $x = 2$, $\frac{d^2y}{dx^2} = 18$, which is <u>positive</u> — so $x = 2$ is a <u>minimum</u>.

And since a cubic graph (where the coefficient of x^3 is <u>positive</u>) goes from <u>bottom-left</u> to <u>top-right</u>...

...you can draw a rough sketch of the graph, even though the roots would be hard to find.

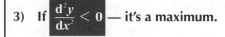

Stationary Points

1) **Find stationary points by solving** $\frac{dy}{dx} = 0$.

2) **Differentiate again to decide whether a point is a maximum or a minimum.**

3) **If** $\frac{d^2y}{dx^2} < 0$ **— it's a maximum.**

   If $\frac{d^2y}{dx^2} > 0$ **— it's a minimum.**

An anagram of differentiation is "Perfect Insomnia Cure"...

No joke, is it — this differentiation business — but it's a dead important topic in maths. It's so important to know how to find whether a stationary point is a max or a min — but it can get a bit confusing. Try remembering MINMAX — which is short for 'MINUS means a MAXIMUM'. Or make up some other clever way to remember what means what.

Increasing and Decreasing Functions

Differentiation is all about finding gradients. Which means that you can find out where a graph is going up...
...and where it's going down. Lovely.

Find out if a function is *Increasing* or *Decreasing*

You can use differentiation to work out exactly where a function is <u>increasing</u> or <u>decreasing</u> — and how quickly.

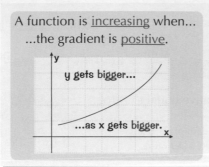

A function is <u>increasing</u> when...
...the gradient is <u>positive</u>.

y gets bigger...

...as x gets bigger.

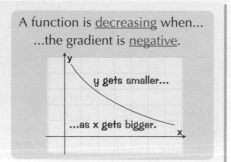

A function is <u>decreasing</u> when...
...the gradient is <u>negative</u>.

y gets smaller...

...as x gets bigger.

And there's more...

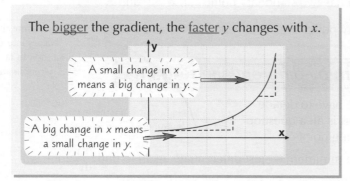

The <u>bigger</u> the gradient, the <u>faster</u> y changes with x.

A small change in x means a big change in y.

A big change in x means a small change in y.

Differentiation and Gradients

<u>Differentiate the equation</u> of the curve to find an expression for its gradient.

1) An increasing function has a <u>positive</u> gradient.

2) A decreasing function has a <u>negative</u> gradient.

EXAMPLE The path of a ball thrown through the air is described by the equation $y = 10x - 5x^2$, where y is the height of the ball above the ground and x is the horizontal distance from its starting point. Find where the height of the ball is increasing and where it's decreasing.

You have the equation for the path of the ball, and you need to know where y is increasing and where it's decreasing. That makes this a question about <u>gradients</u> — so <u>differentiate</u>.

$$y = 10x - 5x^2 \text{ so } \frac{dy}{dx} = 10 - 10x$$

This is an <u>increasing</u> function when: $10 - 10x > 0$, i.e. when $x < 1$, so the ball's height is increasing for $0 \le x < 1$.

And it's a <u>decreasing</u> function when: $10 - 10x < 0$, i.e. when $x > 1$, so its height decreases for $x > 1$ (until it lands).

Just to check: The gradient of the ball's path is given by $\frac{dy}{dx} = 10 - 10x$.

There's a turning point (i.e. the ball's flight levels out) when $x = 1$.

Differentiating again gives $\frac{d^2y}{dx^2} = -10$, which is <u>negative</u> — and so the turning point is a <u>maximum</u>.

This is what you'd expect — the ball goes <u>up then down</u>, not the other way round.

Decreasing function — also known as ironing...

Basically, you can tell whether a function is getting bigger or smaller by looking at the derivative. To make it more interesting, I wrote it as a nursery rhyme: the f(*duke of york*) = 10 000x, and when they were up the derivative was positive, and when they were down the derivative was negative, and when they were only halfway up at a stationary point the derivative was neither negative nor positive, it was zero. Catchy eh?

Curve Sketching

You'll even be asked to do some drawing in the exam... but don't get too excited — it's just drawing graphs... great.

Find where the curve crosses the Axes

Sketch the graph of $f(x) = \frac{x^2}{2} - 2\sqrt{x}$, for $x \geq 0$.

The curve crosses the _y-axis_ when $x = 0$ — so put $x = 0$ in the expression for y.

When $x = 0$, $f(x) = 0$ — and so the curve goes through the _origin_.

The curve crosses the _x-axis_ when $f(x) = 0$. So solve

$$\frac{x^2}{2} - 2\sqrt{x} = 0$$
$$\Rightarrow x^2 - 4x^{\frac{1}{2}} = 0$$
$$\Rightarrow x^{\frac{1}{2}}(x^{\frac{3}{2}} - 4) = 0 \quad \text{Factorising}$$
$$\Rightarrow x^{\frac{1}{2}} = 0 \Rightarrow x = 0$$
$$\text{or: } x^{\frac{3}{2}} = 4 \Rightarrow x = 4^{\frac{2}{3}} = \sqrt[3]{4^2} = \sqrt[3]{16} \approx 2.5$$

And so the curve crosses the x-axis when $x = 0$ (you knew this one already) and when $x \approx 2.5$.

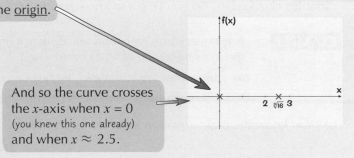

...Differentiate to find Gradient info...

Differentiating the function gives...

$$f(x) = \frac{1}{2}x^2 - 2x^{\frac{1}{2}}$$
$$\Rightarrow f'(x) = \frac{1}{2}(2x) - 2\left(\frac{1}{2}x^{-\frac{1}{2}}\right) = x - x^{-\frac{1}{2}} = x - \frac{1}{\sqrt{x}}$$

Using the derivative — you can find stationary points and tell when the graph goes 'uphill' and 'downhill'.

1) So there's a _stationary point_ when...

$$x - \frac{1}{\sqrt{x}} = 0$$
$$\Rightarrow x = \frac{1}{\sqrt{x}}$$
$$\Rightarrow x^{\frac{3}{2}} = 1 \Rightarrow x = 1$$

And at $x = 1$, $f(x) = \frac{1}{2} - 2 = -\frac{3}{2}$.

2) The gradient's _negative_ when...

$$x - \frac{1}{\sqrt{x}} < 0$$
$$\Rightarrow x < \frac{1}{\sqrt{x}}$$
$$\Rightarrow x^{\frac{3}{2}} < 1 \Rightarrow x < 1$$

So the function _decreases_ when $0 \leq x < 1$...

3) The gradient's _positive_ when...

$$x - \frac{1}{\sqrt{x}} > 0$$
$$\Rightarrow x > 1$$

...and _increases_ for $x > 1$.

This is often the quickest way to check if something's a max or a min.

You could check that $x = 1$ is a _minimum_ by differentiating again.

$$f''(x) = 1 - \left(-\frac{1}{2}x^{-\frac{3}{2}}\right) = 1 + \frac{1}{2\sqrt{x^3}}$$

This is _positive_ when $x = 1$, and so this is definitely a minimum.

...and find out what happens when x gets Big

You can also try and decide what happens as x gets very _big_ — in both the positive and negative directions.

Factorise f(x) by taking the _biggest_ power outside the brackets...

$$\frac{x^2}{2} - 2\sqrt{x} = x^2\left(\frac{1}{2} - 2x^{-\frac{3}{2}}\right) = x^2\left(\frac{1}{2} - \frac{2}{x^{\frac{3}{2}}}\right)$$

As x gets large, this bit disappears — and the bit in brackets gets closer to $\frac{1}{2}$.

And so as x gets larger, f(x) gets closer and closer to $\frac{1}{2}x^2$ — and this just keeps growing and growing.

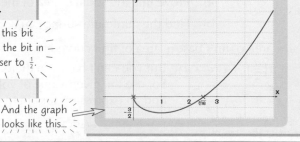

And the graph looks like this...

Curve sketching's important — but don't take my word for it...

Curve sketching — an underrated skill, in my opinion. As Shakespeare once wrote, 'Those who can do fab sketches of graphs and stuff are likely to get pretty good grades in maths exams, no word of a lie'. Well, he probably would've written something like that if he was into maths. And he would've written it because graphs are helpful when you're trying to work out what a question's all about — and once you know that, you can decide the best way forward. And if you don't believe me, remember the saying of the ancient Roman Emperor Julius Caesar, 'If in doubt, draw a graph'.

Real-life Problems

Differentiation isn't just mathematical daydreaming. It can be applied to <u>real-life</u> problems. For instance, you can use differentiation to find out the <u>maximum possible volume</u> of a box, given a limited amount of cardboard. Thrilling.

Finding *Maximum / Minimum Values* for *Volume* and *Area*

To find the maximum for a shape's volume, all you need is an equation for the volume <u>in terms of only one variable</u> — then just <u>differentiate as normal</u>. But examiners don't hand it to you on a plate — there's usually one too many variables chucked in. So you've got to know how to manipulate the information to get rid of that unwanted variable.

EXAMPLE

A jewellery box with a lid and dimensions $3x$ cm by x cm by y cm is made using a total of 450 cm² of wood.

a) Show that the volume of the box can be expressed as: $V = \dfrac{675x - 9x^3}{4}$.

b) Use calculus to find the maximum volume.

a) You know the basic equation for volume: $V = \text{width} \times \text{height} \times \text{depth}$
$$= x \times 3x \times y$$

But the question asks for volume in terms of x only — you don't want that pesky y in there. So you need to find y <u>in terms of x</u> and substitute that in. Use the given dimensions and surface area value to find y in terms of x:

① First, write an expression for the surface area:
$$2 \times [(3x \times x) + (3x \times y) + (x \times y)] = 450$$
$$\Rightarrow 6x^2 + 8xy = 450$$

Be careful when adding up the sides — here there's a lid so there are two of each side, but sometimes you'll get an open-topped shape.

② Then, rearrange to find an expression for y: $y = \dfrac{450 - 6x^2}{8x}$
$$y = \dfrac{225}{4x} - \dfrac{3x}{4}$$

③ Finally, substitute the new expression for y into the equation for V, so that it's all in terms of x...
$$V = x \times 3x \times y = 3x^2 \times \left(\dfrac{225}{4x} - \dfrac{3x}{4}\right)$$
$$V = \dfrac{675x - 9x^3}{4}$$
... and voila, the form the question asks for appears, as if by mathgic...

b) Now it's just differentiating as normal, hurrah...

① Differentiate and find x when $\dfrac{dV}{dx} = 0$: $\dfrac{dV}{dx} = \dfrac{675 - 27x^2}{4}$

$$\dfrac{675 - 27x^2}{4} = 0 \Rightarrow 675 = 27x^2 \Rightarrow 25 = x^2 \Rightarrow x = 5$$

The other solution, $x = -5$ isn't relevant in this context.

② Check $x = 5$ is a maximum: $\dfrac{d^2V}{dx^2} = \dfrac{-27x}{2} = \dfrac{-27 \times 5}{2} = -\dfrac{135}{2}$

$\dfrac{d^2V}{dx^2} < 0$ so yes, it's a maximum

③ Calculate V for $x = 5$: $V = \dfrac{(675 \times 5) - (9 \times 5^3)}{4} = 562.5 \text{ cm}^3$

Differentiation works for *Any Shape*

EXAMPLE

Ned uses a circular tin to bake his pies in. The tin is t cm high with a d cm diameter. The volume of the pie tin is 1000 cm³.

a) Prove that the surface area of the tin, $A = \dfrac{\pi}{4}d^2 + \dfrac{4000}{d}$.

b) Find the minimum surface area.

a) $A = $ area of tin's base + area of tin's curved face
$$\Rightarrow \pi\left(\dfrac{d}{2}\right)^2 + (\pi d \times t).$$ But you can't have t in there.
So, use the given value of volume to find an expression for t in terms of d:

This shape is open-topped, so only count the area of the circle once.

$$V = \pi\left(\dfrac{d}{2}\right)^2 t = 1000$$
$$\Rightarrow t = \dfrac{1000}{\pi\left(\dfrac{d}{2}\right)^2} = \dfrac{4000}{\pi d^2}$$

Substitute your expression for t into the equation for surface area:
$$A = \dfrac{\pi}{4}d^2 + \left(\pi d \times \dfrac{4000}{\pi d^2}\right) \Rightarrow A = \dfrac{\pi}{4}d^2 + \dfrac{4000}{d}$$

b) Differentiate and find the stationary point:
$$\dfrac{dA}{dd} = \dfrac{\pi}{2}d - \dfrac{4000}{d^2} \Rightarrow \dfrac{\pi}{2}d - \dfrac{4000}{d^2} = 0 \Rightarrow d^3 = \dfrac{8000}{\pi}$$
$$\Rightarrow d = \dfrac{20}{\sqrt[3]{\pi}}$$

Check it's a minimum:
$$\dfrac{d^2A}{dd^2} = \dfrac{\pi}{2} + \dfrac{8000}{d^3} = \dfrac{\pi}{2} + \dfrac{8000}{\left(\dfrac{8000}{\pi}\right)} = \dfrac{3\pi}{2}$$

Positive, so it is a minimum.

Calculate the area for that value of d:
$$A = \dfrac{\pi}{4}\left(\dfrac{20}{\sqrt[3]{\pi}}\right)^2 + \left(\dfrac{4000}{\left(\dfrac{20}{\sqrt[3]{\pi}}\right)}\right) = 439 \text{ cm}^2$$
(to 3 s.f.)

C2 Section 5 — Practice Questions

Well, I'll be... another section all over and done with. And there's some <u>mighty useful</u> stuff in that last section too. Now <u>drawing graphs</u> is something that it's always handy to be able to do, which makes differentiation a handy skill to have. The questions below will help you <u>get to grips with it all</u>. Try them and see how you get on.

Warm-up Questions

1) a) Write down what a <u>stationary point</u> is.

   b) Find the stationary points of the graph of $y = x^3 - 6x^2 - 63x + 21$.

2) How can you decide whether a stationary point is a <u>maximum</u> or a <u>minimum</u>?

3) Find the stationary points of the function $y = x^3 + \dfrac{3}{x}$.
   Decide whether each stationary point is a <u>minimum</u> or a <u>maximum</u>.

4) Find when these two functions are <u>increasing</u> and <u>decreasing</u>:
   a) $y = 6(x + 2)(x - 3)$
   b) $y = \dfrac{1}{x^2}$.

5) Sketch the graph of $y = x^3 - 4x$, clearly showing the <u>coordinates</u> of any <u>turning points</u>.

6) The height (h m) a firework can reach is related to the mass (m g) of fuel it carries as shown below:
   $$h = \frac{m^2}{10} - \frac{m^3}{800}$$
   Find the <u>mass of fuel</u> required to achieve the <u>maximum height</u> and state what the maximum height is.

If you get all the answers right, then well done... go <u>get yourself a pie</u>. But if you get any wrong, read the section again, work out where you went wrong, and then <u>try the questions again</u>.

Exam Questions

1 a) Find $\dfrac{dy}{dx}$ for the curve $y = 6 + \dfrac{4x^3 - 15x^2 + 12x}{6}$.

   (3 marks)

   b) Hence, find the coordinates of the stationary points on the curve.

   (5 marks)

   c) Determine the nature of each stationary point.

   (3 marks)

2 A steam train travels between Haverthwaite and Eskdale at a speed of x miles per hour and burns y units of coal, where y is given by: $2\sqrt{x} + \dfrac{27}{x}$, for $x > 2$.

   a) Find the speed that gives the minimum coal consumption.

   (5 marks)

   b) Find $\dfrac{d^2y}{dx^2}$, and hence show that this speed gives the minimum coal consumption.

   (2 marks)

   c) Calculate the minimum coal consumption.

   (1 mark)

C2 Section 5 — Practice Questions

3 a) Determine the coordinates of the stationary points for the curve $y = (x - 1)(3x^2 - 5x - 2)$.

(4 marks)

   b) Find whether each of these points is a maximum or minimum.

(3 marks)

   c) Sketch the curve of y.

(3 marks)

4 Ayesha is building a closed-back bookcase. She uses a total of 72 m² of wood (not including shelving) to make a bookcase that is x metres high, $\frac{x}{2}$ metres wide and d metres deep, as shown.

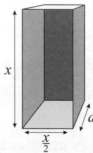

   a) Show that the full capacity of the bookcase is given by: $V = 12x - \frac{x^3}{12}$.

(4 marks)

   b) Find the value of x for which V is stationary. Leave your answer in surd form.

(4 marks)

   c) Show that this is a maximum point and hence calculate the maximum V.

(4 marks)

5 Function $f(x) = \frac{1}{2}x^4 - 3x$ has a single stationary point.

   a) Find the coordinates of the stationary point.

(3 marks)

   b) Determine the nature of the stationary point.

(2 marks)

   c) State the range of values of x for which $f(x)$ is:
       (i)   increasing,

(1 mark)

       (ii) decreasing.

(1 mark)

   d) Sketch the curve for the function $f(x) = \frac{1}{2}x^4 - 3x$

(2 marks)

Integration

Some integrals have <u>limits</u> (i.e. little numbers) next to the integral sign. You integrate them in exactly the same way — but you <u>don't</u> need a constant of integration. Much easier. And scrummier and yummier too.

A *Definite Integral* finds the *Area Under a Curve*

This definite integral tells you the <u>area</u> between the graph of $y = x^3$ and the x-axis between $x = -2$ and $x = 2$:

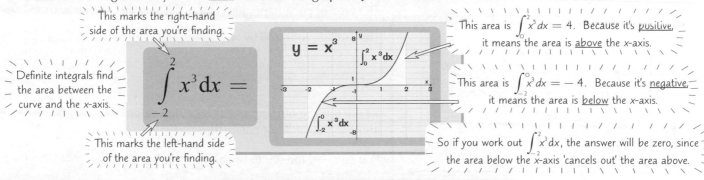

This marks the right-hand side of the area you're finding.

Definite integrals find the area between the curve and the x-axis.

$$\int_{-2}^{2} x^3\,\mathrm{d}x =$$

This marks the left-hand side of the area you're finding.

This area is $\int_0^2 x^3\,dx = 4$. Because it's <u>positive</u>, it means the area is <u>above</u> the x-axis.

This area is $\int_{-2}^{0} x^3\,dx = -4$. Because it's <u>negative</u>, it means the area is <u>below</u> the x-axis.

So if you work out $\int_{-2}^{2} x^3\,dx$, the answer will be zero, since the area below the x-axis 'cancels out' the area above.

Do the integration in the same way — then use the *Limits*

Finding a definite integral isn't really any harder than an indefinite one — there's just an <u>extra</u> stage you have to do. After you've integrated the function you have to work out the value of this new function by sticking in the <u>limits</u>.

EXAMPLE

Evaluate $\int_1^3 (x^2 + 2)dx$.

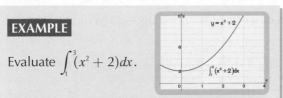

Definite Integrals
After you've integrated the function — put both the limits in and find the values. Then subtract what the bottom limit gave you from what the top limit gave you.

$2 = 2x^0$ — so increase the power (to 1) and divide by 1 to get 2x.

Find the integral in the normal way — then use the limits.

Put the integrated function in <u>square brackets</u> and rewrite the limits on the right-hand side.

$$\int_1^3 (x^2 + 2)dx = \left[\frac{x^3}{3} + 2x\right]_1^3$$
$$= \left(\frac{3^3}{3} + 6\right) - \left(\frac{1^3}{3} + 2\right)$$
$$= 15 - \frac{7}{3} = \frac{38}{3}$$

You don't need a constant of integration with a <u>definite</u> integral.

Integrate 'to *Infinity*' with the ∞ (infinity) sign

And you can integrate all the way to <u>infinity</u> as well. Just use the ∞ symbol as your upper limit. Or use −∞ as your lower limit if you want to integrate to '<u>minus infinity</u>'.

EXAMPLE   Find the area under the curve $y = \dfrac{15}{x^2} - \dfrac{30}{x^3}$ for $x \geq 2$.

For this, you need to integrate from $x = 2$ up to infinity (∞).

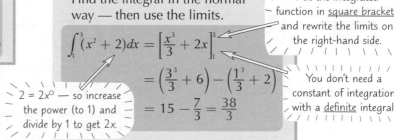

$$A = \int_2^\infty \left(\frac{15}{x^2} - \frac{30}{x^3}\right)dx = 15\int_2^\infty (x^{-2} - 2x^{-3})dx$$
$$= 15\left[\frac{x^{-1}}{-1} - \frac{2x^{-2}}{(-2)}\right]_2^\infty$$
$$= 15\left[-\frac{1}{x} + \frac{1}{x^2}\right]_2^\infty$$
$$= 15\left\{(-0 + 0) - \left(-\frac{1}{2} + \frac{1}{4}\right)\right\} = 15 \times \frac{1}{4} = \frac{15}{4}$$

Move <u>numbers</u> outside the integral sign or the square bracket as if you're <u>factorising</u> a normal bracket.

When you have to use the ∞ limit — remember: $\frac{1}{\infty} = \frac{1}{\infty^2} = \frac{1}{\infty^3} = 0$.

Curve continues forever in this direction.→

My hobbies? Well I'm really inte grating. Especially carrots.

It's still integration — but this time you're putting two numbers into an expression afterwards. So although this may not be the wild and crazy fun-packed time your teachers promised you when they were trying to persuade you to take AS maths, you've got to admit that a lot of this stuff is pretty similar — and if you can do one bit, you can use that to do quite a few other bits too. Maths is like that. But I admit it's probably not as much fun as a big banana-and-toffee cake.

The Trapezium Rule

Sometimes <u>integrals</u> can be just <u>too hard</u> to do using the normal methods — then you need to know other ways to solve them. That's where the <u>Trapezium Rule</u> comes in.

The **Trapezium Rule** is Used to Find the **Approximate Area** Under a Curve

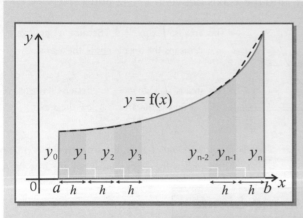

The area of each trapezium is $A = \frac{h}{2}(y_n + y_{n+1})$

The area represented by $\int_a^b y \, dx$ is approximately:

$$\int_a^b y \, dx \approx \frac{h}{2}[y_0 + 2(y_1 + y_2 + \ldots + y_{n-1}) + y_n]$$

where **n** is the number of strips or intervals and **h** is the width of each strip.

You can find the width of each strip using $h = \dfrac{(b-a)}{n}$

$y_0, y_1, y_2, \ldots, y_n$ are the heights of the sides of the trapeziums — you get these by putting the x-values into the curve.

So basically the formula for approximating $\int_a^b y \, dx$ works like this:

'Add the first and last heights $(y_0 + y_n)$ and add this to <u>twice</u> all the other heights added up — then multiply by $\frac{h}{2}$.'

EXAMPLE   Find an approximate value for $\int_0^2 \sqrt{4 - x^2} \, dx$ using 4 strips. Give your answer to 4 s.f.

Start by working out the width of each strip: $h = \dfrac{(b-a)}{n} = \dfrac{(2-0)}{4} = 0.5$

This means the x-values are $x_0 = 0$, $x_1 = 0.5$, $x_2 = 1$, $x_3 = 1.5$ and $x_4 = 2$ (the question specifies 4 strips, so $n = 4$).

Set up a table and work out the y-values or heights using the equation in the integral.

x	$y = \sqrt{4 - x^2}$
$x_0 = 0$	$y_0 = \sqrt{4 - 0^2} = 2$
$x_1 = 0.5$	$y_1 = \sqrt{4 - 0.5^2} = \sqrt{3.75} = 1.936491673$
$x_2 = 1.0$	$y_2 = \sqrt{4 - 1.0^2} = \sqrt{3} = 1.732050808$
$x_3 = 1.5$	$y_3 = \sqrt{4 - 1.5^2} = \sqrt{1.75} = 1.322875656$
$x_4 = 2.0$	$y_4 = \sqrt{4 - 2.0^2} = 0$

Now put all the y-values into the formula with h and n:

$$\int_a^b y \, dx \approx \frac{0.5}{2}[2 + 2(1.9365 + 1.7321 + 1.3229) + 0]$$
$$\approx 0.25[2 + 2 \times 4.9915]$$
$$\approx 2.996 \text{ to 4 s.f.}$$

Watch out — if they ask you to work out a question with 5 y-values (or 'ordinates') then this is the <u>same</u> as 4 strips. The x-values usually go up in <u>nice jumps</u> — if they don't then <u>check</u> your calculations carefully.

The Approximation might be an **Overestimate** or an **Underestimate**

It all depends on the shape of the curve...

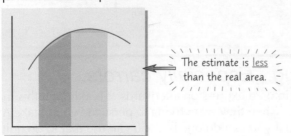

The estimate is <u>less</u> than the real area.

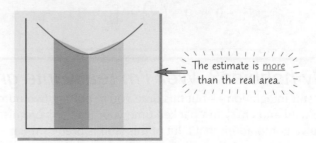

The estimate is <u>more</u> than the real area.

The Trapezium Rule

These are usually popular questions with examiners — as long as you're careful there are <u>plenty of marks</u> to be had.

The **Trapezium Rule** is in the **Formula Booklet**

...so don't try any heroics — always <u>look it up</u> and use it with these questions.

EXAMPLE Use the trapezium rule with 7 ordinates to find an approximation to $\int_1^{2.2} 2\log_{10}x \, dx$

Remember, <u>7 ordinates</u> means <u>6 strips</u> — so $n = 6$.

Calculate the width of the strips: $h = \dfrac{(b-a)}{n} = \dfrac{(2.2-1)}{6} = 0.2$

Set up a table and work out the y-values using $y = 2\log_{10}x$: $\longrightarrow$

x	$y = 2\log_{10}x$
$x_0 = 1.0$	$y_0 = 2\log_{10}1 = 0$
$x_1 = 1.2$	$y_1 = 2\log_{10}1.2 = 0.15836$
$x_2 = 1.4$	$y_2 = 0.29226$
$x_3 = 1.6$	$y_3 = 0.40824$
$x_4 = 1.8$	$y_4 = 0.51055$
$x_5 = 2.0$	$y_5 = 0.60206$
$x_6 = 2.2$	$y_6 = 0.68485$

$y_6 = 2\log_{10}b = 0.68485$

Putting all these values in the formula gives:

$$\int_a^b y \, dx \approx \frac{0.2}{2}[0 + 2(0.15836 + 0.29226 + 0.40824 + 0.51055 + 0.60206) + 0.68485]$$

$$\approx 0.1 \times [0.68485 + 2 \times 1.97147]$$

$$\approx 0.462779$$

$$\approx 0.463 \text{ to 3 d.p.}$$

EXAMPLE Use the trapezium rule with 8 intervals to find an approximation to $\int_0^\pi \sin x \, dx$

Whenever you get a calculus question using <u>trig functions</u>, you <u>have</u> to use <u>radians</u>. You'll probably be given a limit with π in, which is a pretty good reminder.

There are 8 intervals, so $n = 8$.

Keep your x-values in terms of π.

Calculate the width of the strips: $h = \dfrac{(b-a)}{n} = \dfrac{(\pi - 0)}{8} = \dfrac{\pi}{8}$

Set up a table and work out the y-values: $\longrightarrow$

So, putting all this in the formula gives:

x	$y = \sin x$
$x_0 = 0$	$y_0 = \sin 0 = 0$
$x_1 = \dfrac{\pi}{8}$	$y_1 = 0.38268$
$x_2 = \dfrac{\pi}{4}$	$y_2 = 0.70711$
$x_3 = \dfrac{3\pi}{8}$	$y_3 = 0.92388$
$x_4 = \dfrac{\pi}{2}$	$y_4 = 1$
$x_5 = \dfrac{5\pi}{8}$	$y_5 = 0.92388$
$x_6 = \dfrac{3\pi}{4}$	$y_6 = 0.70711$
$x_7 = \dfrac{7\pi}{8}$	$y_7 = 0.38268$
$x_8 = \pi$	$y_8 = 0$

$$\int_a^b y \, dx \approx \frac{1}{2}\cdot\frac{\pi}{8}[0 + 2(0.383 + 0.707 + 0.924 + 1 + 0.924 + 0.707 + 0.383) + 0]$$

$$\approx \frac{\pi}{16} \times [2 \times 5.028]$$

$$\approx 1.97 \text{ to 3 s.f.}$$

These values are quicker to work-out if you know that the graph is symmetrical.

Maths rhyming slang #3: Dribble and drool — Trapezium rule...

Take your time with trapezium rule questions — it's so easy to make a mistake with all those numbers flying around. Make a nice table showing all your ordinates (careful — this is always one more than the number of strips). Then add up y_1 to y_{n-1} and multiply the answer by 2. Add on y_0 and y_n. Finally, multiply what you've got so far by the width of a strip and divide by 2. It's a good idea to write down what you get after each stage, by the way — then if you press the wrong button (easily done) you'll be able to pick up from where you went wrong. They're not hard — just fiddly.

Areas Between Curves

With a bit of thought, you can use integration to find all kinds of areas — even ones that look quite tricky at first. The best way to work out what to do is draw a <u>picture</u>. Then it'll seem easier. I promise you it will.

Sometimes you have to **Add** integrals...

This looks pretty hard — until you draw a picture and see what it's all about.

EXAMPLE Find the area enclosed by the curve $y = x^2$, the line $y = 2 - x$ and the x-axis.

Find out where the graphs meet by <u>solving</u> $x^2 = 2 - x$ — they meet at $x = 1$ (they also meet at $x = -2$, but this isn't in A).

You have to find area A — but you'll need to <u>split</u> it into two smaller pieces.

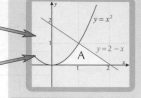

And it's pretty clear from the picture that you'll have to find the area in two lumps, A_1 and A_2.

The first area you need to find is A_1:

$$A_1 = \int_0^1 x^2 \, dx$$
$$= \left[\frac{x^3}{3} \right]_0^1 = \left(\frac{1}{3} - 0 \right) = \frac{1}{3}$$

The other area you need is A_2:

A_2 is just a triangle, with base length $2 - 1 = 1$ and height $= 1$. So the area of the triangle is $\frac{1}{2} \times b \times h = \frac{1}{2} \times 1 \times 1 = \frac{1}{2}$.

And the area the question actually asks for is $A_1 + A_2$. This is

$$A = A_1 + A_2$$
$$= \frac{1}{3} + \frac{1}{2} = \frac{5}{6}$$

You could also have integrated the line $y = 2 - x$ between $x = 1$ and $x = 2$, but finding the area of the triangle is easier.

...sometimes you have to **Subtract** them

Again, it's best to look at the <u>pictures</u> to work out exactly what you need to do.

EXAMPLE Find the area enclosed by the curves $y = x^2 + 1$ and $y = 9 - x^2$.

Solve $x^2 + 1 = 9 - x^2$ to find where the curves meet.
$x^2 + 1 = 9 - x^2 \Rightarrow 2x^2 = 8$
$\Rightarrow x^2 = 4$
$\Rightarrow x = \pm 2$

So you'll have to integrate between -2 and 2.

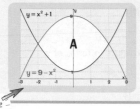

The area under the green curve A_1 is:

$$A_1 = \int_{-2}^{2} (9 - x^2) \, dx$$
$$= \left[9x - \frac{x^3}{3} \right]_{-2}^{2}$$
$$= \left(18 - \frac{2^3}{3} \right) - \left(-18 - \frac{(-2)^3}{3} \right)$$
$$= \left(18 - \frac{8}{3} \right) - \left(-18 - \left(\frac{-8}{3} \right) \right) = \frac{46}{3} - \left(-\frac{46}{3} \right) = \frac{92}{3}$$

The area under the red curve is:

$$A_2 = \int_{-2}^{2} (x^2 + 1) \, dx$$
$$= \left[\frac{x^3}{3} + x \right]_{-2}^{2}$$
$$= \left(\frac{2^3}{3} + 2 \right) - \left(\frac{(-2)^3}{3} + (-2) \right)$$
$$= \left(\frac{8}{3} + 2 \right) - \left(-\frac{8}{3} - 2 \right) = \frac{28}{3}$$

And the area you need is the difference between these:

$$A = A_1 - A_2$$
$$= \frac{92}{3} - \frac{28}{3} = \frac{64}{3}$$

Instead of integrating before subtracting — you could try 'subtracting the curves', and then integrating. This last area A is also:
$$A = \int_{-2}^{2} \{ (9 - x^2) - (x^2 + 1) \} \, dx$$

And so, our hero integrates the area between two curves, and saves the day...

That's the basic idea of finding the area enclosed by two curves and lines — draw a picture and then break the area down into <u>smaller, easier chunks</u>. Questions like this aren't hard — but they can sometimes take a long time. Great.

C2 Section 6 — Practice Questions

Penguins evolved with a layer of blubber under their skin to keep them warm.
They should've saved themselves the effort and <u>done these questions</u> instead — mmm, toasty warm...

1) How can you tell whether an integral is a <u>definite</u> one or an <u>indefinite</u> one?
(It's easy really — it just sounds difficult.)

2) What does a definite integral represent on a <u>graph</u>?

3) Evaluate the following <u>definite integrals</u>:
a) $\int_0^1 (4x^3 + 3x^2 + 2x + 1)\,dx$
b) $\int_1^2 \left(\frac{8}{x^5} + \frac{3}{\sqrt{x}}\right)dx$
c) $\int_1^6 \frac{3}{x^2}\,dx$.

4) Evaluate: a) $\int_{-3}^3 (9 - x^2)\,dx$
b) $\int_1^\infty \frac{3}{x^2}\,dx$.
<u>Sketch the areas</u> represented by these integrals.

5) Find <u>area A</u> in the diagram below:

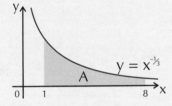

6) Use the <u>trapezium rule</u> with n intervals to estimate:
a) $\int_0^3 (9 - x^2)^{\frac{1}{2}}\,dx$ with $n = 3$
b) $\int_{0.2}^{1.2} x^{x^2}\,dx$ with $n = 5$

7) Use integration to find the <u>yellow area</u> in each of these graphs:

a)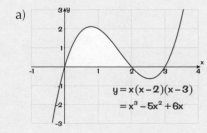
b)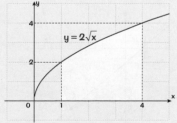
c)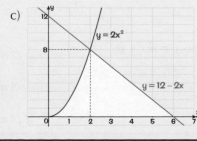
d)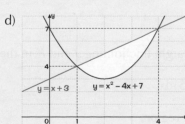

C2 Section 6 — Practice Questions

They think it's all over...

Exam Questions

1   Find the value of $\int_2^7 (2x - 6x^2 + \sqrt{x})\,dx$. Give your answer to 4 d.p.

(5 marks)

2   a)   Using the trapezium rule with n intervals, estimate the values of:

   (i) $\int_2^8 \left(\sqrt{3x^3} + \dfrac{2}{\sqrt{x}}\right)dx$, $n = 3$

   (4 marks)

   (ii) $\int_1^5 \left(\dfrac{x^3 - 2}{4}\right)dx$, $n = 4$

   (4 marks)

   b)   How could you change your application of the trapezium rule to get better approximations?   *(1 mark)*

3   Curve C, $y = (x - 3)^2 (x + 1)$, is sketched on the diagram below:

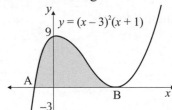

   Calculate the shaded area between point A, where C intersects the
   x-axis, and point B, where C touches the x-axis.

   (8 marks)

4   Complete the table and hence use the trapezium rule with 6 ordinates to estimate $\int_{1.5}^4 y\,dx$.

x	$x_0 = 1.5$	$x_1 =$	$x_2 =$	$x_3 =$	$x_4 = 3.5$	$x_5 = 4.0$
$y = 3x - \sqrt{2^x}$	$y_0 =$	$y_1 = 4$	$y_2 = 5.12156$	$y_3 =$	$y_4 =$	$y_5 = 8.0$

   (7 marks)

5   The diagram below shows the curve $y = (x + 1)(x - 5)$.
   Points J $(-1, 0)$ and K $(4, -5)$ lie on the curve.

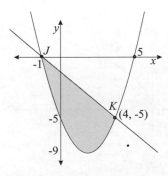

   a)   Find the equation of the straight line joining J and K in the form $y = mx + c$.

   (2 marks)

   b)   Calculate $\int_{-1}^4 (x + 1)(x - 5)\,dx$.

   (5 marks)

   c)   Find the area of the shaded region.

   (4 marks)

   ...it is now.

General Certificate of Education
Advanced Subsidiary (AS) and Advanced Level

Core Mathematics C2 — Practice Exam One

Time Allowed: 1 hour 30 min

Calculators may be used for this exam (except those with
facilities for symbolic algebra, differentiation or integration).

Give any non-exact numerical answers to an appropriate degree of accuracy.

There are 75 marks available for this paper.

1   a)   By sketching the graph of $y = \tan 2t$ for a suitable range of t, determine the number of
         solutions to the equation $\tan 2t = k$ in the range $0° \leq t < 360°$ where k is any number.

(3 marks)

    b)   Solve the equation $\sin 2t = \sqrt{2} \cos 2t$, giving all the solutions in the range $0° \leq t < 360°$.

(3 marks)

2   a)   Write down the value of $\log_3 3$.

(1 mark)

    b)   Given that $\log_a \chi = \log_a 4 + 3 \log_a 2$, show that $\chi = 32$.

(2 marks)

3   The binomial expansion of $(j + kx)^6$ is $j^6 + ax + bx^2 + cx^3 + ...$

    a)   Given that $c = 20{,}000$, show that $jk = 10$ (where both j and k are positive integers).

(3 marks)

    b)   Given that $a = 37500$, find the values of j and k.

(4 marks)

    c)   Find b.

(2 marks)

4   Given that the curve $y = 2x^3 + zx - 5$ is stationary at the point $(20, w)$:

    a)   Find the values of z and w.

(5 marks)

    b)   Calculate whether the curve is at a maximum or minimum at point $(20, w)$.

(3 marks)

5   a)   Sketch the curve $y = (x - 2)(x - 4)$ and the line $y = 2x - 4$ on the same set of axes,
         clearly marking the coordinates of the points of intersection.

(3 marks)

    b)   Evaluate the integral $\int_{2}^{4} (x - 2)(x - 4)\, dx$.

(3 marks)

    c)   Hence, or otherwise, show that the total area enclosed by the curve
         $y = (x - 2)(x - 4)$ and the line $y = 2x - 4$ is $\frac{32}{3}$.

(4 marks)

6   The diagram below shows a sector of a circle of radius r cm and angle 120°.
    The length of the arc of the sector is 40 cm.

    a)   Write 120° in radians.

    (1 mark)

    b)   Show that $r \approx 19.1$ cm.

    (2 marks)

    c)   Find the area of the sector to the nearest square centimetre.

    (2 marks)

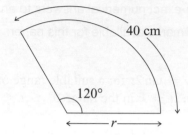

7   The diagram shows a circle. A (2, 1) and B (0, –5) lie on the circle
    and AB is a diameter. C (4, –1) is also on the circle.

    a)   Find the centre and radius of the circle.

    (4 marks)

    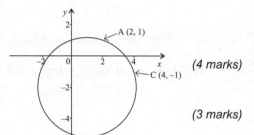

    b)   Show that the equation of the circle can be written in the form:
         $x^2 + y^2 - 2x + 4y - 5 = 0$

    (3 marks)

    c)   The tangent at A and the normal at C cross at D.
         Find the coordinates of D.

    (7 marks)

8   For the series with second term –2 and common ratio –½, find:

    a)   the 13th term,

    (3 marks)

    b)   the sum to infinity.

    (3 marks)

9   a)   Rewrite the following equation in the form f(x) = 0,
         where f(x) is of the form f(x) = $ax^3 + bx^2 + cx + d$:
         $$(x - 1)(x^2 + x + 1) = 2x^2 - 17$$

    (2 marks)

    b)   Show that $(x + 2)$ is a factor of f(x).

    (3 marks)

    c)   Hence factorise f(x) as the product of a linear factor and a quadratic factor.

    (3 marks)

    d)   By completing the square, or otherwise, show that f(x)=0 has only one root.

    (2 marks)

    e)   Divide the polynomial $x^3 - 2x^2 + 3x - 3$ by $(x - 1)$, showing both the quotient and remainder.

    (4 marks)

General Certificate of Education
Advanced Subsidiary (AS) and Advanced Level

Core Mathematics C2 — Practice Exam Two

Time Allowed: 1 hour 30 min

Calculators may be used for this exam (except those with
facilities for symbolic algebra, differentiation or integration).

Give any non-exact numerical answers to an appropriate degree of accuracy.

There are 75 marks available for this paper.

1 a) Sketch the graph of $y = \cos(x - 60°)$ for x between $0°$ and $360°$.

(2 marks)

   b) Show that the equation $2\sin^2(x - 60°) = 1 + \cos(x - 60°)$
       may be written as a quadratic in $\cos(x - 60°)$.

(4 marks)

   c) Hence solve this equation, giving all values of x such that $0° \leq x \leq 360°$.

(4 marks)

2 a) Write down the first four terms in the expansion of $(1 + ax)^{10}$, $a > 0$.

(2 marks)

   b) Find the coefficient of x^2 in the expansion of $(2 + 3x)^5$.

(2 marks)

   c) If the coefficients of x^2 in both expansions are equal, find the value of a.

(2 marks)

3 The diagram shows the graph of $y = 2^{x^2}$.

   a) Use the trapezium rule with 4 intervals to find an estimate for the
       area of the region bounded by the axes, the curve and the line $x = 2$.

(4 marks)

   b) State whether the estimate in a) is an overestimate or an
       underestimate, giving a reason for your answer.

(2 marks)

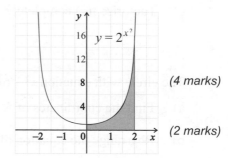

4 A geometric series $u_1 + u_2 + ... + u_n$ has 3rd term $\frac{5}{2}$ and 6th term $\frac{5}{16}$.

   a) Find the common ratio and the first term of the series.
       Hence give the formula for the nth term of the series.

(4 marks)

   b) Find $\sum_{i=1}^{10} u_i$. Give your answer as a fraction in its simplest terms.

(3 marks)

   c) Show that the sum to infinity of the series is 20.

(2 marks)

5 a) Find the missing length a in the triangle.

(2 marks)

   b) Find the angles θ and ϕ.

(3 marks)

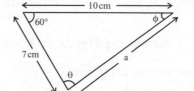

6   A new symmetrical mini-stage is to be built according to the design shown below. The design consists of a rectangle of length q metres and width $2r$ metres, two sectors of radius r and angle θ radians (shaded), and an isosceles triangle.

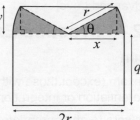

a)   (i)   Show that distance x is given by $x = r\cos\theta$.

(1 mark)

(ii)   Find a similar expression for distance y.

(1 mark)

b)   Find, in terms of r, q and θ, expressions for the perimeter P, and the area A, of the stage.

(4 marks)

c)   If the perimeter of the stage is to be 40 metres, and $\theta = \frac{\pi}{3}$,
show that A is given approximately by $A = 40r - 3.614r^2$.

(4 marks)

7   a)   Find the remainder when the function $f(x) = x^3 - 6x^2 - x + 30$ is divided by:

(i) $(x + 3)$

(2 marks)

(ii) $(4x - 1)$

(2 marks)

b)   Using the factor theorem, show that $(x - 3)$ is a factor of $f(x)$.

(2 marks)

c)   Factorise $f(x)$ completely.

(4 marks)

8   The circle with equation $x^2 - 6x + y^2 - 4y = 0$ crosses the y-axis at the origin and the point A.

a)   Find the coordinates of A.

(2 marks)

b)   Rearrange the equation of the circle in the form: $(x - a)^2 + (y - b)^2 = c$.

(3 marks)

c)   Write down the radius and the coordinates of the centre of the circle.

(2 marks)

d)   Find the equation of the tangent to the circle at A.

(4 marks)

9   a)   Sketch the curve $y = \frac{1}{x^2}$ for $x > 0$.

(1 mark)

b)   Show that $\displaystyle\int_{1}^{\infty} \frac{1}{x^2}\, dx = 1$.

(2 marks)

c)   Find $f(x)$, where $y = f(x)$ is the equation of the tangent
to the graph of $y = \frac{1}{x^2}$ at the point where $x = 1$.

(2 marks)

d)   Find $k < 1$ such that $\displaystyle\int_{0}^{k} f(x)\, dx = 1$, where $f(x)$ is the function found in part (c).
Give your answer using surds.

(3 marks)

Histograms

Histograms are glorified bar charts. The main difference is that you plot the <u>frequency density</u> rather than the frequency. Frequency density is easy to find — you just divide the <u>frequency</u> by the <u>width of a class</u>.

Using frequency density means it's a column's <u>area</u> (and <u>not</u> its height) that represents the <u>frequency</u>.

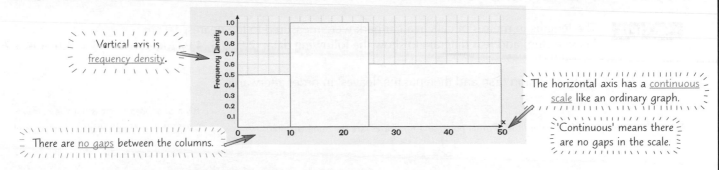

Vertical axis is <u>frequency density</u>.

There are <u>no gaps</u> between the columns.

The horizontal axis has a <u>continuous scale</u> like an ordinary graph.

'Continuous' means there are no gaps in the scale.

To Draw a **Histogram** it's best to Draw a **Table** First

Getting histograms right depends on finding the right <u>upper and lower boundaries</u> for each class.

EXAMPLE

Draw a histogram to represent the data below showing the masses of parcels (given to the nearest 100 g).

Mass of parcel (to nearest 100 g)	100 - 200	300 - 400	500 - 700	800 - 1100
Number of parcels	100	250	600	50

First draw a table showing the <u>upper and lower class boundaries</u>, plus the <u>frequency density</u>:

<u>Smallest</u> mass of parcel that will go <u>in that class</u>.

<u>Biggest</u> mass that will go <u>in that class</u>.

= <u>ucb − lcb</u>

Mass of parcel	Lower class boundary (lcb)	Upper class boundary (ucb)	Class width	Frequency	Frequency density = frequency ÷ class width
100 - 200	50	250	200	100	0.5
300 - 400	250	450	200	250	1.25
500 - 700	450	750	300	600	2
800 - 1100	750	1150	400	50	0.125

= 250 ÷ 200

Look — no gaps between a ucb and the next lcb.

= 1150 − 750

Now you can draw the histogram.

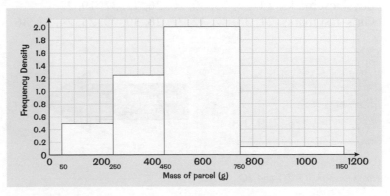

It's the <u>area</u> of each bar that shows the frequency — <u>not</u> the height.

Note: A class with a lower class boundary of 50 g and upper class boundary of 250 g can be written in different ways.

So you might see: "100 – 200 to nearest 100 g"
"50 ≤ mass < 250"
"50–", followed by "250–" for the next class and so on.

They all mean the same — just make sure you know how to spot the lower and upper class boundaries.

Stem and Leaf Diagrams

Stem and Leaf Diagrams *look nothing like stems or leaves*

They're just an easy way to represent your data. And they come in two flavours — plain and <u>back-to-back</u>.

EXAMPLE The lengths in metres of cars in a car park were measured to the nearest 10 cm.
Draw a stem and leaf diagram to show the following data: 2.9, 3.5, 4.0, 2.8, 4.1, 3.7, 3.1, 3.6, 3.8, 3.7

It's best to do a rough version first, and then put the 'leaves' in order afterwards.

It's a good idea to cross out the numbers
(in pencil) as you add them to your diagram.

My 'stems' are the numbers before the decimal point, and my 'leaves' are the numbers after.

```
2 | 9, 8
3 | 5, 7, 1, 6, 8, 7
4 | 0, 1
```

Put the digits after the
decimal point in order

```
2 | 8, 9
3 | 1, 5, 6, 7, 7, 8
4 | 0, 1
Key  2|9 means 2.9 m
```

Always give a key.

Digits after the decimal point — this row represents 4.0 m and 4.1 m.

EXAMPLE The heights of boys and girls in a year 11 class are given to the nearest cm
in the back-to-back stem and leaf diagram below. Write out the data in full.

First boy, 8|16|, has height 168 cm. The boys are read backwards.

Key 8|16|5 means
Boys 168 cm and girls 165 cm

First girl, |15|9, has height 159 cm.

Boys		Girls
	15	9
8	16	1, 5, 9
9, 8, 1	17	0, 2, 3, 5
5, 2	18	0
1	19	

<u>Boys</u>: 168, 171, 178, 179, 182, 185, 191

<u>Girls</u>: 159, 161, 165, 169, 170, 172, 173, 175, 180

EXAMPLE Construct a back-to-back stem and leaf diagram to represent the following data:
Boys' test marks: 34, 27, 15, 39, 20, 26, 32, 37, 19, 22
Girls' test marks: 21, 38, 37, 12, 27, 28, 39, 29, 25, 24, 31, 36

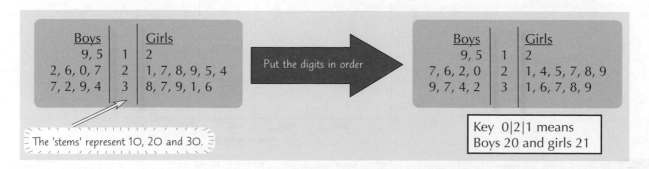

Boys		Girls
9, 5	1	2
2, 6, 0, 7	2	1, 7, 8, 9, 5, 4
7, 2, 9, 4	3	8, 7, 9, 1, 6

Put the digits in order

Boys		Girls
9, 5	1	2
7, 6, 2, 0	2	1, 4, 5, 7, 8, 9
9, 7, 4, 2	3	1, 6, 7, 8, 9

Key 0|2|1 means
Boys 20 and girls 21

The 'stems' represent 10, 20 and 30.

First things first: remember — there are lies, damned lies and statistics...

Histograms shouldn't really cause too many problems — this is quite a friendly topic really. The main things to remember are to work out the <u>lower and upper boundaries</u> of each class <u>properly</u>, and then make sure you use <u>frequency density</u> (rather than just the frequency). Stem and leaf diagrams — hah, they're easy, I do them in my sleep. Make sure you can too.

Location: Mean, Median and Mode

The mean, median and mode are measures of <u>location</u> (roughly speaking... where the <u>centre</u> of the data lies).

The **Definitions** are really GCSE stuff

You more than likely already know them. But if you don't, learn them now — you'll be needing them loads.

> **Mean** $= \bar{x} = \dfrac{\Sigma x}{n}$ or $\dfrac{\Sigma fx}{\Sigma f}$ The Σ (sigma) things just mean you add stuff up — so Σx means you add up all the values of x.
>
> where each x is a <u>data value</u>, f is the <u>frequency</u> of each x (the number of times it occurs), and n is the <u>total number</u> of data values.
>
> **Median** = <u>middle</u> data value when all the data values are placed <u>in order of size</u>.
>
> **Mode** = <u>most frequently occurring</u> data value.

There are two ways to find the <u>median</u> (but they amount to the same thing):   If ½(n+1) isn't a whole number, take the average of the terms either side.

<u>Either</u>: find the $\left(\dfrac{n+1}{2}\right)$th value in the ordered list. ◄

<u>Or</u>: (i)   if $\dfrac{n}{2}$ is a <u>whole number</u> (i.e. n is <u>even</u>), then the median is the <u>average of this term and the one above</u>.

     (ii)   if $\dfrac{n}{2}$ is <u>not a whole number</u> (i.e. n is <u>odd</u>), just <u>round the number up</u> to find the position of the median.

EXAMPLE   Find the mean, median and mode of the following list of data: 2, 3, 6, 2, 5, 9, 3, 8, 7, 2

Put in order first: 2, 2, 2, 3, 3, 5, 6, 7, 8, 9   **Mode = 2**

Mean $= \dfrac{2+2+2+3+3+5+6+7+8+9}{10} =$ **4.7**   **Median** = average of 5th and 6th values = **4**

Use a **Table** when there are a lot of **Numbers**

EXAMPLE   The number of letters received one day in 100 houses was recorded. Find the mean, median and mode of the number of letters.

Number of letters	Number of houses
0	11
1	25
2	27
3	21
4	9
5	7

The first thing to do is make a <u>table</u> like this one:

Number of letters x	Number of houses f		fx
0	11	(11)	0
1	25	(36)	25
2	27	(63)	54
3	21		63
4	9		36
5	7		35
totals	100		213

Multiply x by f to get this column.

The number of letters received by each house is a **discrete** quantity (e.g. 3 letters). There isn't a **continuous** set of possible values between getting 3 and 4 letters (e.g. 3.45 letters).

Put the <u>running total</u> in brackets — it's handy when you're finding the <u>median</u>. (But you can stop when you get past <u>halfway</u>.)

$\Sigma f = 100$ $\Sigma fx = 213$

① The <u>mean</u> is easy — just divide the <u>total</u> of the <u>fx-column</u> (sum of all the data values) by the total of the f-<u>column</u> (= n, the total number of data values).   **Mean** $= \dfrac{213}{100} =$ **2.13 letters**

② To find the <u>position</u> of the median, <u>add 1</u> to the total frequency (= $\Sigma f = n$) and then <u>divide by 2</u>. Here the median is in position: $(100 + 1) \div 2 =$ <u>50.5</u>.

So the median is <u>halfway between</u> the 50th and 51st data values.

Using your <u>running total</u> of f, you can see that the data values in positions 37 to 63 are all 2s. This means the data values at positions 50 and 51 are both 2 — so   **Median = 2 letters**

③ The <u>highest frequency</u> is for 2 letters — so   **Mode = 2 letters**

Location: Mean, Median and Mode

If the data's *Grouped* you'll have to *Estimate*

There are no precise readings here — each reading's been put into one of these groups.

If the data's grouped, you can only estimate the mean, median and mode.

EXAMPLE The height of a number of trees was recorded. The data collected is shown in this table:

Height of tree to nearest m	0 - 5	6 - 10	11 - 15	16 - 20
Number of trees	26	17	11	6

Find an estimate of the mean height of the trees.

Here, you assume that every reading in a class takes the mid-class value (which you find by adding the lower class boundary to the upper class boundary and dividing by 2). It's best to make another table...

Height of tree to nearest m	Mid-class value x	Number of trees f	fx
0 - 5	2.75	26 (26)	71.5
6 - 10	8	17 (43)	136
11 - 15	13	11	143
16 - 20	18	6	108
	Totals	60 (= Σf)	458.5(= Σfx)

Lower class boundary = 0.
Upper class boundary = 5.5.
So the mid-class value = (0 + 5.5) ÷ 2 = 2.75.

Estimated mean $= \dfrac{458.5}{60} = \textbf{7.64 m}$

Linear Interpolation Means Assuming Values are *Evenly Spread*

When you have grouped data, you can only estimate the median. To do this, you use (linear) interpolation.

The median position above is $(60 + 1) \div 2 = 30.5$, so the median is the 30.5th reading (halfway between the 30th and 31st). Your 'running total' tells you the median must be in the '6 - 10' class.

Now you have to assume that all the readings in this class are evenly spread.

There are 26 trees before class 6 - 10, so the 30.5th tree is the 4.5th value of this class.

Divide the class into 17 equally wide parts (as there are 17 readings) and assume there's a reading at the end of each part.

Then you want the '4.5th reading' (which is '4.5 × width of 1 part' along).

Width of class $\rightarrow \frac{5}{17}$; Number of readings $\frac{5}{17}$

5.5 ... 10.5

$5.5+(1\times\frac{5}{17})$ $5.5+(2\times\frac{5}{17})$

So the **estimated median** = lower class boundary + (4.5 × width of each 'part') $= 5.5 + \left[4.5 \times \frac{5}{17}\right] = \textbf{6.8 m}$ (to 1 d.p.)

The **modal class** is the class with the **highest frequency density**. In this example the modal class is **0 - 5 m**.

The Mean, Median and Mode are useful for *Different Kinds* of Data

These three different averages are useful for different kinds of data.

Mean:
- The mean's a good average because you use all your data in working it out.
- But it can be heavily affected by extreme values / outliers.
- And it can only be used with quantitative data (i.e. numbers).

See page 107 for more about outliers.

Median: The median is not affected by extreme values, so this is a good average to use when you have outliers.

Mode:
- The mode can be used even with non-numerical data.
- But some data sets can have more than one mode (and if every value in a data set occurs just once, then the mode isn't very helpful at all).

I can't deny it — these pages really are 'about average'...

If you have large amounts of grouped data ($n > 100$, say), it's usually okay to use the value in position $\frac{n}{2}$ (rather than $\frac{n+1}{2}$) as the median. With grouped data, you can only estimate the median anyway, and if you have a lot of data, that extra 'half a place' doesn't really make much difference. But if in any doubt, use the value in position $\frac{n+1}{2}$ — that'll always be okay.

Dispersion: Interquartile Range

'Dispersion' means how spread out your data is. You'll be happy to learn there are different ways to measure it.

The **Range** is a Measure of **Dispersion**...

The range is about the simplest measure of dispersion you could imagine.

> **Range** = highest value – lowest value

But the range is heavily affected by extreme values, so it isn't really the most useful way to measure dispersion.

Quartiles divide the data into **Four**

You've seen how the median divides a data set into two halves. Well, the quartiles divide the data into four parts — with 25% of the data less than the lower quartile, and 75% of the data less than the upper quartile.

There are various ways you can find the quartiles, and they sometimes give different results. But if you use the method below, you'll be fine.

The median is also known as Q_2.

1) To find the lower quartile (Q_1), first work out $\frac{n}{4}$.

    (i) if $\frac{n}{4}$ is a whole number, then the lower quartile is the average of this term and the one above.

    (ii) if $\frac{n}{4}$ is not a whole number, just round the number up to find the position of the lower quartile.

2) To find the upper quartile (Q_3), first work out $\frac{3n}{4}$.

    (i) if $\frac{3n}{4}$ is a whole number, then the upper quartile is the average of this term and the one above.

    (ii) if $\frac{3n}{4}$ is not a whole number, just round the number up to find the position of the upper quartile.

EXAMPLE Find the median and quartiles of the following data: 2, 5, 3, 11, 6, 8, 3, 8, 1, 6, 2, 23, 9, 11, 18, 19, 22, 7.

First put the list in order: 1, 2, 2, 3, 3, 5, 6, 6, 7, 8, 8, 9, 11, 11, 18, 19, 22, 23

You need to find Q_1, Q_2 and Q_3, so work out $\frac{n}{4} = \frac{18}{4}$, $\frac{n}{2} = \frac{18}{2}$, and $\frac{3n}{4} = \frac{54}{4}$.

1) $\frac{n}{4}$ is not a whole number (= 4.5), so round up and take the 5th term: $Q_1 = 3$

2) $\frac{n}{2}$ is a whole number (= 9), so find the average of the 9th and 10th terms: $Q_2 = \frac{7+8}{2} = 7.5$

3) $\frac{3n}{4}$ is not a whole number (= 13.5), so round up and take the 14th term: $Q_3 = 11$

If your data is grouped, you might need to use interpolation to find the quartiles. See page 104 for more info.

The **Interquartile Range** is Another Measure of **Dispersion**

> **Interquartile range** (IQR) = upper quartile (Q_3) – lower quartile (Q_1)

The IQR shows the range of the 'middle 50%' of the data.

EXAMPLE Find the interquartile range of the data in the previous example.

$Q_1 = 3$ and $Q_3 = 11$, so the interquartile range = $Q_3 – Q_1 = 11 – 3 = 8$

Percentiles divide the data into **100**

Percentiles divide the data into 100 — the median is the 50th percentile and Q_1 is the 25th percentile, etc.

For example, the position of the 11th percentile (P_{11}) is $\frac{11}{100} \times$ total frequency.

You find interpercentile ranges by subtracting two percentiles, e.g. the middle 60% of the readings = $P_{80} – P_{20}$.

Dispersion: Standard Deviation

Standard deviation and variance both measure how spread out the data is from the mean — the bigger the variance, the more spread out your readings are.

The Formulas look pretty Tricky

The formula is easier to use in this form.

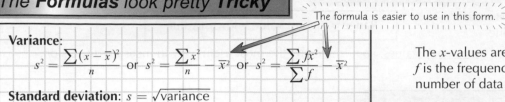

Variance:
$$s^2 = \frac{\sum(x - \bar{x})^2}{n} \quad \text{or} \quad s^2 = \frac{\sum x^2}{n} - \bar{x}^2 \quad \text{or} \quad s^2 = \frac{\sum fx^2}{\sum f} - \bar{x}^2$$

Standard deviation: $s = \sqrt{\text{variance}}$

The x-values are the data, $\bar{x}$ is the mean, f is the frequency of each x, and n is the number of data values.

EXAMPLE   Find the mean and standard deviation of the following numbers: 2, 3, 4, 4, 6, 11, 12

1) Find the <u>total</u> of the numbers first: $\sum x = 2 + 3 + 4 + 4 + 6 + 11 + 12 = 42$

2) Then the <u>mean</u> is easy: $\text{Mean} = \bar{x} = \frac{\sum x}{n} = \frac{42}{7} = 6$

3) Next find the <u>sum of the squares</u>: $\sum x^2 = 4 + 9 + 16 + 16 + 36 + 121 + 144 = 346$

4) Use this to find the <u>variance</u>: $\text{Variance}, s^2 = \frac{\sum x^2}{n} - \bar{x}^2 = \frac{346}{7} - 6^2 = 49.43 - 36 = 13.43$

5) And take the <u>square root</u> to find the standard deviation: $\text{Standard deviation} = \sqrt{13.43} = 3.66$ to 3 sig. fig.

Questions about Standard Deviation can look a bit Weird

They can ask questions about standard deviation in different ways. But you just need to use the same old formulas.

EXAMPLE   The mean of 10 boys' heights is 180 cm, and the standard deviation is 10 cm. The mean for 9 girls is 165 cm, and the standard deviation is 8 cm. Find the mean and standard deviation of the whole group of 19 girls and boys.

① Let the boys' heights be x and the girls' heights be y.

Write down the formula for the mean and put the numbers in for the boys: $\bar{x} = \frac{\sum x}{n} \Rightarrow 180 = \frac{\sum x}{10} \Rightarrow \sum x = 1800$

Do the same for the girls: $165 = \frac{\sum y}{9} \Rightarrow \sum y = 1485$

So the sum of the heights for the <u>boys and the girls</u> = $\sum x + \sum y = 1800 + 1485 = 3285$

And the <u>mean height</u> of the boys and the girls is: $\frac{3285}{19} = \underline{172.9 \text{ cm}}$ Round the fraction to 1 d.p. to give your answer. But if you need to use the mean in more calculations, use the <u>fraction</u> (or your <u>calculator's memory</u>) so you don't lose accuracy.

② Now for the variance — write down the formula for the boys first: $s_x^2 = \frac{\sum x^2}{n} - \bar{x}^2 \Rightarrow 10^2 = \frac{\sum x^2}{10} - 180^2 \Rightarrow \sum x^2 = 10 \times (100 + 32\,400) = 325\,000$

Do the same for the girls: $s_y^2 = \frac{\sum y^2}{n} - \bar{y}^2 \Rightarrow 8^2 = \frac{\sum y^2}{9} - 165^2 \Rightarrow \sum y^2 = 9 \times (64 + 27\,225) = 245\,601$

Okay, so the sum of the squares of the heights of the boys and the girls is: $\sum x^2 + \sum y^2 = 325\,000 + 245\,601 = 570\,601$

Which means the variance of all the heights is: $s^2 = \frac{570\,601}{19} - \left(\frac{3285}{19}\right)^2 = \underline{139.0 \text{ cm}^2}$ Don't use the <u>rounded</u> mean (172.9) — you'll lose accuracy.

And finally the standard deviation of the boys and the girls is: $s = \sqrt{139.0} = \underline{11.8 \text{ cm}}$

Phew.

People who enjoy this stuff are standard deviants...

The formula for the variance looks pretty scary, what with the s's and $\bar{x}$'s floating about. But it comes down to 'the mean of the squares minus the square of the mean'. That's how I remember it anyway — and my memory's rubbish.

Dispersion and Outliers

Use **Mid-Class Values** if your data's in a **Table**

With grouped data, assume every reading takes the <u>mid-class value</u>. Then use the <u>frequencies</u> to find $\sum fx$ and $\sum fx^2$.

EXAMPLE The heights of sunflowers in a garden were measured and recorded in the table below.
Estimate the mean height and the standard deviation.

Height of sunflower, h (cm)	$150 \le h < 170$	$170 \le h < 190$	$190 \le h < 210$	$210 \le h < 230$
Number of sunflowers	5	10	12	3

Draw up another table, and include columns for the <u>mid-class values x</u>, as well as <u>fx</u> and <u>fx^2</u>:

Height of sunflower (cm)	Mid-class value, x	x^2	f	fx	fx^2
$150 \le h < 170$	160	25600	5	800	128000
$170 \le h < 190$	180	32400	10	1800	324000
$190 \le h < 210$	200	40000	12	2400	480000
$210 \le h < 230$	220	48400	3	660	145200
		Totals	30 (= Σf)	5660 (= Σfx)	1077200 (= Σfx^2)

fx^2 means $f \times (x^2)$ — <u>not</u> $(fx)^2$.

Now you've got the totals in the table,
you can calculate the mean and variance:

$$\text{Mean} = \bar{x} = \frac{\sum fx}{\sum f} = \frac{5660}{30} = 189 \text{ cm to 3 sig. fig.}$$

$$\text{Variance} = s^2 = \frac{\sum fx^2}{\sum f} - \bar{x}^2 = \frac{1\,077\,200}{30} - \bar{x}^2 = 312 \text{ to 3 sig. fig.}$$

$$\text{Standard deviation} = \sqrt{s^2} = 17.7 \text{ cm to 3 sig. fig.}$$

Outliers fall **Outside Fences**

An <u>outlier</u> is a <u>freak</u> piece of data that lies a long way from the rest of the readings.
To decide whether a reading is an outlier you have to measure how far away from the rest of the data it is.

EXAMPLE A data value is considered to be an outlier if it is
more than 1.5 times the IQR above the upper quartile or
more than 1.5 times the IQR below the lower quartile.

There are various ways to decide if a reading is an outlier — the method you should use will always be described in the question.

The lower and upper quartiles of a data set are 70 and 100. Decide whether the data values 30 and 210 are outliers.

First you need the IQR: $Q_3 - Q_1 = 100 - 70 = 30$

Then it's a piece of cake to find where your <u>fences</u> are.

Lower fence first: $Q_1 - (1.5 \times IQR) = 70 - (1.5 \times 30) = 25$

And the upper fence: $Q_3 + (1.5 \times IQR) = 100 + (1.5 \times 30) = 145$

25 and 145 are called <u>fences</u>. Any reading lying <u>outside</u> the fences is considered an <u>outlier</u>.

30 is <u>inside the lower fence</u>, so it is <u>not</u> an outlier. 210 is <u>outside</u> the upper fence, so it <u>is</u> an outlier.

Outliers Affect what Measure of **Dispersion** is Best to Use

1) Outliers affect whether the <u>variance</u> and <u>standard deviation</u> are good measures of <u>dispersion</u>.
2) Outliers can make the variance (and standard deviation) <u>much</u> larger than it would otherwise be
   — which means these <u>freak</u> pieces of data are having more influence than they deserve.
3) If a data set contains outliers, then a better measure of dispersion is the <u>interquartile range</u>.

'Outlier' is the name I give to something that my theory can't explain...

Measures of <u>location</u> and <u>dispersion</u> are supposed to capture the essential characteristics of a data set in just one or two numbers. So don't choose an average that's heavily affected by freaky, far-flung outliers — it won't be much good.

Coding

Coding can make the Numbers much Easier

Coding means doing something to <u>every reading</u> (like <u>adding</u> or <u>multiplying</u> by a number) to make life easier.

Finding the mean of 1001, 1002 and 1006 looks hard(ish). But take 1000 off each number and finding the mean of what's left (1, 2 and 6) is much easier — it's <u>3</u>. So the mean of the original numbers must be <u>1003</u>. That's coding.

You usually change your original variable, x, to an easier one to work with, y (so here, if $x = 1001$, then $y = 1$).

Write down a formula connecting the two variables: e.g. $y = \dfrac{x - b}{a}$.

⟵ You can add/subtract a number, and multiply/divide by one as well — it all depends on what will make life easiest.

Then $\overline{y} = \dfrac{\overline{x} - b}{a}$ where $\overline{x}$ and $\overline{y}$ are the means of variables x and y.

Note that if you don't multiply or divide your readings by anything (i.e. if $a = 1$), then the spread isn't changed.

Also $s_y = \dfrac{s_x}{a}$ where s_x and s_y are the standard deviations of variables x and y. ⟸

EXAMPLE Find the mean and standard deviation of: 1 000 020, 1 000 040, 1 000 010 and 1 000 050.

The obvious thing to do is subtract a million from every reading to leave 20, 40, 10 and 50.

Then make life even simpler by dividing by 10 — giving 2, 4, 1 and 5.

① So use the coding: $y = \dfrac{x - 1000\,000}{10}$. Then $\overline{y} = \dfrac{\overline{x} - 1000\,000}{10}$ and $s_y = \dfrac{s_x}{10}$.

② Find the mean and standard deviation of the y values: $\overline{y} = \dfrac{2 + 4 + 1 + 5}{4} = \underline{3}$

$s_y = \sqrt{\dfrac{2^2 + 4^2 + 1^2 + 5^2}{4} - 3^2}$

$= \sqrt{\dfrac{46}{4} - 9} = \sqrt{2.5} = 1.58$ to 3 sig. fig.

③ Then use the formulas to find the mean and standard deviation of the original values:

$\overline{x} = 10\overline{y} + 1000\,000 = (10 \times 3) + 1000\,000 = \underline{1000\,030}$ $\qquad$ $s_x = 10s_y = 10 \times 1.58 = \underline{15.8}$

You can use coding with Summarised Data

This kind of question looks tricky at first — but use the same old formulas and it's a piece of cake.

EXAMPLE A set of 10 numbers (x-values) can be summarised as shown: Find the mean and standard deviation of the numbers. $\qquad \sum(x - 10) = 15$ and $\sum(x - 10)^2 = 100$

① Okay, the obvious first thing to try is: $y = x - 10 \implies$ That means: $\sum y = 15$ and $\sum y^2 = 100$

② Work out $\overline{y}$ and s_y^2 using the normal formulas: $\overline{y} = \dfrac{\sum y}{n} = \dfrac{15}{10} = 1.5$

$s_y^2 = \dfrac{\sum y^2}{n} - \overline{y}^2 = \dfrac{100}{10} - 1.5^2 = 10 - 2.25 = 7.75$

so $s_y = 2.78$ to 3 sig. fig.

③ Then finding the mean and standard deviation of the x-values is easy: $\overline{x} = \overline{y} + 10 = 1.5 + 10 = \underline{11.5}$

The spread of x is the same as the spread of y since you've only subtracted 10 from every number. $\implies$ $s_x = s_y = \underline{2.78}$ to 3 sig. fig.

I thought coding would be a little more... well, James Bond...

Coding data isn't hard — the only tricky thing can be to work out <u>how</u> best to code it, although there will usually be some pretty hefty clues in the question if you care to look. But remember that adding/subtracting a number from every reading won't change the spread (the variance or standard deviation), but multiplying/dividing readings by something will.

Skewness

Skewness tells you whether your data is <u>symmetrical</u> — or kind of <u>lopsided</u>.

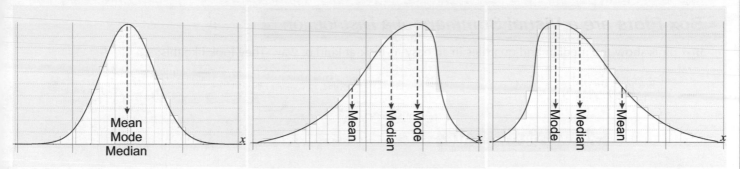

This is a typical <u>symmetrical</u> distribution.

Notice: <u>mean = median = mode</u>

A <u>negatively skewed</u> distribution has a <u>tail</u> on the <u>left</u>. Most data values are on the <u>higher side</u>.

A <u>positively skewed</u> distribution has a <u>tail</u> on the <u>right</u>. Most data values are on the <u>lower side</u>.

For <u>all</u> distributions: $\boxed{\text{mean} - \text{mode} = 3 \times (\text{mean} - \text{median})}$ — approximately.

Measure skewness using **Pearson's Coefficient of Skewness...**

A <u>coefficient of skewness</u> measures how 'all-up-one-end' your data is. There are a couple of important ones...

$$\text{Pearson's coefficient of skewness} = \frac{\text{mean} - \text{mode}}{\text{standard deviation}} = \frac{3(\text{mean} - \text{median})}{\text{standard deviation}}$$

This usually lies between −3 and +3.
So if Pearson's coefficient of skewness is −0.1, then the distribution is <u>slightly</u> negatively skewed.

...or the **Quartile Coefficient of Skewness**

Remember that Q_1 is the lower quartile, Q_3 is the upper quartile, and the median is Q_2.

These things are <u>box plots</u> — they're covered on the next page. So read that, and then come back here and notice how a box plot quickly tells you the skew of a data set.

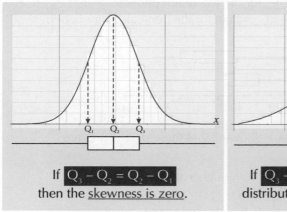

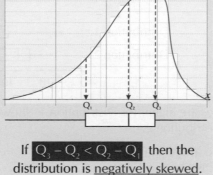

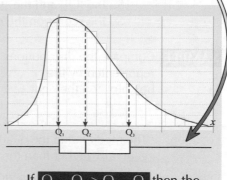

If $\boxed{Q_3 - Q_2 = Q_2 - Q_1}$ then the <u>skewness is zero</u>.

If $\boxed{Q_3 - Q_2 < Q_2 - Q_1}$ then the distribution is <u>negatively skewed</u>.

If $\boxed{Q_3 - Q_2 > Q_2 - Q_1}$ then the distribution is <u>positively skewed</u>.

$$\text{Quartile coefficient of skewness} = \frac{(Q_3 - Q_2) - (Q_2 - Q_1)}{Q_3 - Q_1} = \frac{Q_3 - 2Q_2 + Q_1}{Q_3 - Q_1}$$

I've worked out my wonky nose has a coefficient of skewness of 0.176 — cool...

Those definitions of positive and negative skew aren't the most obvious in the world — and it's easy to get them mixed up. Remember that <u>negative skew</u> involves a tail on the <u>left</u>, which means that a lot of your readings are on the <u>high</u> side. <u>Positive</u> skew is the opposite — a tail on the <u>right</u>, and a bunch of readings that are a little on the <u>low</u> side.

Comparing Distributions

To compare data sets, you need to know how to use all the formulas and what the results tell you.

Box Plots are a Visual Summary of a Distribution

Box plots show the median and quartiles in an easy-to-look-at kind of way. They look like this:

Box plots are sometimes called "box and whisker diagrams".

ALWAYS DRAW A SCALE

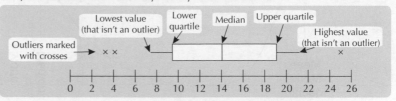

Outliers marked with crosses
Lowest value (that isn't an outlier)
Lower quartile
Median
Upper quartile
Highest value (that isn't an outlier)

Use Location, Dispersion and Skewness to Compare Distributions

EXAMPLE This table summarises the marks obtained in Maths 'calculator' and 'non-calculator' papers.

i) Calculate the Pearson's and Quartile coefficients of skewness for each paper. Comment on the results.

ii) Comment on the location and dispersion of the distributions.

Calculator Paper		Non-calculator paper
40	Lower quartile, Q_1	35
58	Median, Q_2	42
70	Upper quartile, Q_3	56
55	Mean	46.1
21.2	Standard deviation	17.8

i) The calculator paper scores have <u>negative</u> coefficients of skewness. This tells you that lots of scores were 'bunched up' towards the <u>higher</u> values.

Calculator Paper		Non-calculator Paper
$\dfrac{3 \times (55 - 58)}{21.2} = \dfrac{-9}{21.2} = -0.425$	Pearson's coefficient of skewness	$\dfrac{3 \times (46.1 - 42)}{17.8} = \dfrac{12.3}{17.8} = 0.691$
$\dfrac{70 - 2 \times 58 + 40}{70 - 40} = \dfrac{-6}{30} = -0.2$	Quartile coefficient of skewness	$\dfrac{56 - 2 \times 42 + 35}{56 - 35} = \dfrac{7}{21} = 0.333$

The non-calculator paper scores have <u>positive</u> coefficients of skewness. This tells you that <u>more scores</u> were nearer the <u>lower end</u> of the distribution than at the top.

ii) <u>Location:</u> The <u>mean</u>, the <u>median</u> and the <u>quartiles</u> are all higher for the calculator paper. This means that scores were <u>generally higher</u> on the calculator paper.

<u>Dispersion:</u> The <u>interquartile range</u> (IQR) for the calculator paper is $Q_3 - Q_1 = 70 - 40 = 30$. The <u>interquartile range</u> (IQR) for the non-calculator paper is $Q_3 - Q_1 = 56 - 35 = 21$. So the <u>IQR</u> and the <u>standard deviation</u> are both <u>higher</u> for the calculator paper. So the scores on the calculator paper are <u>more spread out</u> than for the non-calculator paper.

EXAMPLE Compare the distributions represented by the box plots on the right.

Distribution 1:
Distribution 2:

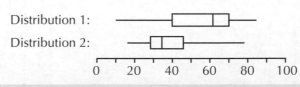

<u>Location:</u> The <u>median</u> and the <u>quartiles</u> are higher for Distribution 1, showing that these data values are <u>generally higher</u> than for Distribution 2.

<u>Dispersion:</u> The <u>interquartile range</u> (IQR) and the <u>range</u> for Distribution 1 are higher, showing that the values are more varied for Distribution 1 than for Distribution 2.

<u>Skewness:</u> Distribution 1 is <u>negatively skewed</u>, showing that there is a tail on the left of this distribution, meaning the <u>lower values</u> are <u>more spread out</u> / the <u>higher values</u> are more <u>tightly bunched</u>. Distribution 2 is <u>positively skewed</u>, showing that there is a tail on the right of this distribution, meaning the <u>higher values</u> are <u>more spread out</u> / the <u>lower values</u> are more <u>tightly bunched</u>.

That's the end of the Data section — hurrah for that...

On exam day, you could be asked to compare two distributions. Just work out any measures of <u>location</u>, <u>dispersion</u> and <u>skewness</u> you can. Then say which distribution has a <u>higher value</u> for each measure, and <u>what it means</u> — e.g. a higher variance means the values are <u>more spread out</u>, while a higher mean means the scores are <u>generally higher</u>. And so on.

S1 Section 1 — Practice Questions

It's important to make sure you know this stuff like the back of your hand — but unlike the back of your hand you won't have this book in the exam. So here are some practice questions to help you remember it all. We'll start off gently...

Warm-up Questions

```
12 | 8                    Key  12|8 means 12.8 cm
13 | 2, 5
14 | 3, 3, 6, 8
15 | 2, 9
16 | 1, 1, 2, 3
17 | 0, 2
```

1) The stem and leaf diagram on the right represents the lengths (in cm) of 15 bananas. Write down the original data as a list.

2) Twenty phone calls were made by a householder one evening. The lengths of the calls (in minutes to the nearest minute) are recorded below. Draw a histogram of the data.

Length of call	0 - 2	3 - 5	6 - 8	9 - 15
Number of calls	10	6	3	1

3) Calculate the mean, median and mode of the data in this table. $\implies$

x	0	1	2	3	4
f	5	4	4	2	1

4) The speeds of 60 cars travelling in a 40 mph speed limit area were measured to the nearest mph. The data is summarised in the table. Calculate estimates of the mean and median, and state the modal class.

Speed (mph)	30 - 34	35 - 39	40 - 44	45 - 50
Frequency	12	37	9	2

5) Find the mean and standard deviation of the following numbers: 11, 12, 14, 17, 21, 23, 27.

6) The scores in an IQ test for 50 people are recorded in the table below.

Score	100 - 106	107 - 113	114 - 120	121 - 127	128 - 134
Frequency	6	11	22	9	2

Calculate the mean and variance of the distribution.

7) For a set of data, $n = 100$, $\sum(x - 20) = 125$, and $\sum(x - 20)^2 = 221$. Find the mean and standard deviation of x.

8) The time taken (to the nearest minute) for a commuter to travel to work on 20 consecutive work days is recorded in the table. Use coding to find the mean and standard deviation of the times.

Time to nearest minute	30 - 33	34 - 37	38 - 41	42 - 45
Frequency	3	6	7	4

9) A data value is considered an outlier if it's more than 3 times the IQR above the upper quartile or more than 3 times the IQR below the lower quartile. If the lower and upper quartiles of a data set are 62 and 88, decide which of the following data are outliers:
   a) 161,   b) 176,   c) 0

10) Find the median and quartiles of the data below. Draw a box and whisker diagram, and comment on any skewness.
    Amount of pocket money (in £) received per week by twenty 15-year-olds:
    10, 5, 20, 50, 5, 1, 6, 5, 15, 20, 5, 7, 5, 10, 12, 4, 8, 6, 7, 30.

11) A set of data has a median of 132, a lower quartile prof 86 and an upper quartile of 150. Calculate the quartile coefficient of skewness, and draw a possible sketch of the distribution.

S1 Section 1 — Practice Questions

Well wasn't that lovely. I bet you're now ready and raring to test yourself with some proper exam-style questions. Here are some I made earlier, lucky you.

Exam Questions

1   The profits of 100 businesses are given in the table.

Profit, £x million.	Number of businesses
$4.5 \leqslant x < 5.0$	24
$5.0 \leqslant x < 5.5$	26
$5.5 \leqslant x < 6.0$	21
$6.0 \leqslant x < 6.5$	19
$6.5 \leqslant x < 8.0$	10

   a)   Represent the data in a histogram.

   [3 marks]

   b)   Comment on the distribution of the profits of the businesses.

   [2 marks]

2   A group of 19 people played a game. The scores, x, that the people achieved are summarised by:

$$\sum(x - 30) = 228 \text{ and } \sum(x - 30)^2 = 3040$$

   a)   Calculate the mean and the standard deviation of the 19 scores.

   [3 marks]

   b)   Show that $\sum x = 798$ and $\sum x^2 = 33\,820$.

   [3 marks]

   c)   Another student played the game. Her score was 32.
        Find the new mean and standard deviation of all 20 scores.

   [4 marks]

3   Two workers iron clothes. Each irons 10 items, and records the time it takes for each, to the nearest minute:

   Worker A: 3 5 2 7 10 4 5 5 4 12
   Worker B: 3 4 8 6 7 8 9 10 11 9

   a)   For worker A's times. Find:
        (i)   the median

   [1 mark]

        (ii) the lower and upper quartiles

   [2 marks]

   b)   On graph paper draw, using the same scale, two box plots
        to represent the times of each worker.

   [6 marks]

   c)   Make one statement comparing the two sets of data.

   [1 mark]

   d)   Which worker would be better to employ? Give a reason for your answer.

   [1 mark]

4   In a supermarket two types of chocolate drops were compared.
    The weights, a grams, of 20 chocolate drops of brand A are summarised by:
          $\Sigma a = 60.3 \text{ g}$ $\Sigma a^2 = 219 \text{ g}^2$
    The mean weight of 30 chocolate drops of brand B was 2.95 g, and the standard deviation was 1 g.

   a)   Find the mean weight of a brand A chocolate drop.

   [1 mark]

   b)   Find the standard deviation of the weight of the brand A chocolate drops.

   [3 marks]

   c)   Compare brands A and B.

   [2 marks]

   d)   Find the standard deviation of the weight of all 50 chocolate drops.

   [4 marks]

S1 Section 1 — Practice Questions

Despair not, valiant mathematician, for the end is nigh. Just a few more questions, then you can have a nap and a biscuit.

Exam Questions

5 The stem and leaf diagram shows the test marks for 30 male students and 16 female students.

Male students		Female students
8, 3, 3	4	
8, 7, 7, 7, 5, 3, 2	5	5, 6, 7
9, 7, 6, 6, 5, 5, 2, 2, 1, 1, 0	6	1, 2, 3, 3, 4, 5, 6, 7, 9
9, 9, 8, 5, 4, 3, 1, 0, 0	7	2, 4, 8, 9

Key 5|6|2 means Male student test mark 65 and
Female student test mark 62

a) Find the median test mark of the male students.

[1 mark]

b) Compare the distribution of the male and female marks.

[2 marks]

6 The table shows the number of hits received by people at a paint ball party.

No. of Hits	12	13	14	15	16	17	18	19	20	21	22	23	24	25
Frequency	2	4	6	7	6	4	4	2	1	1	0	0	0	1

a) Find the median and mode number of hits.

[3 marks]

b) An outlier is a data value which is greater than $1.5 \times (Q_3 - Q_1)$ above Q_3 or below Q_1.
Is 25 an outlier? Show your working.

[2 marks]

c) Sketch a box plot of the distribution and comment on any skewness.

[2 marks]

d) How would the shape of the distribution be affected if the value of 25 was removed?

[1 mark]

7 The histogram shows the nose-to-tail lengths of 50 lions in a game reserve.

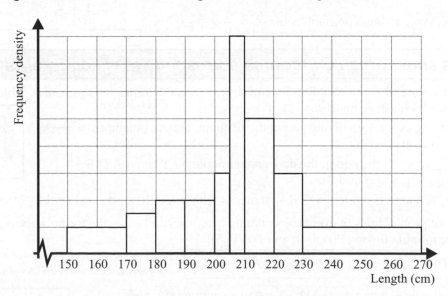

Use the histogram to find the number of lions who measured over 220 cm from nose to tail.

[5 marks]

Random Events and Venn Diagrams

Random events happen by chance. Probability is a measure of how likely they are. It can be a chancy business.

A Random Event has Various Outcomes

1) In a trial (or experiment) the things that can happen are called outcomes (so if I time how long it takes to eat my dinner, 63 seconds is a possible outcome).

2) Events are 'groups' of one or more outcomes (so an event might be 'it takes me less than a minute to eat my dinner every day one week').

3) When all outcomes are equally likely, you can work out the probability of an event by counting the outcomes:

$$P(\text{event}) = \frac{\text{Number of outcomes where event happens}}{\text{Total number of possible outcomes}}$$

EXAMPLE Suppose I've got a bag with 15 balls in — 5 red, 6 blue and 4 green.

If I take a ball out without looking, then any ball is equally likely — there are 15 possible outcomes.
Of these 15 outcomes, 5 are red, 6 are blue and 4 are green. And so...

$$P(\text{red ball}) = \frac{5}{15} = \frac{1}{3} \qquad P(\text{blue ball}) = \frac{6}{15} = \frac{2}{5} \qquad P(\text{green ball}) = \frac{4}{15}$$

You can find the probability of either a red or a green ball in a similar way...

$$P(\text{red or green ball}) = \frac{9}{15} = \frac{3}{5}$$

The Sample Space is the Set of All Possible Outcomes

Drawing the sample space (called S) helps you count the outcomes you're interested in.

EXAMPLE The classic probability machine is a dice. If you roll it twice, you can record all the possible outcomes in a 6 × 6 table (a possible diagram of the sample space).

There are 36 outcomes in total. You can find probabilities by counting the ones you're interested in (and using the above formula). For example:

(i) The probability of an odd number and then a '1'. There are 3 outcomes that make up this event, so the probability is: $\frac{3}{36} = \frac{1}{12}$

(ii) The probability of the total being 7. There are 6 outcomes that correspond to this event, giving a probability of: $\frac{6}{36} = \frac{1}{6}$

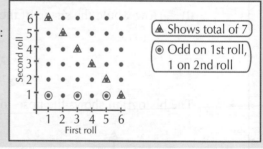

▲ Shows total of 7
◉ Odd on 1st roll, 1 on 2nd roll

Second roll / First roll

Venn Diagrams show which Outcomes correspond to which Events

Say you've got 2 events, A and B — a Venn diagram can show which outcomes satisfy event A, which satisfy B, which satisfy both, and which satisfy neither.

(i) All outcomes satisfying event A go in one part of the diagram, and all outcomes satisfying event B go in another bit.

(ii) If they satisfy 'both A and B', they go in the dark green middle bit, written A ∩ B (and called the intersection of A and B).

(iii) The whole of the green area is written A ∪ B — it means 'either A or B' (and is called the union of A and B).

Again, you can work out probabilities of events by counting outcomes and using the formula above.
You can also get a nice formula linking P(A ∩ B) and P(A ∪ B).

$$P(A \cup B) = P(A) + P(B) - P(A \cap B)$$

If you just add up the outcomes in A and B, you end up counting A ∩ B twice — that's why you have to subtract it.

EXAMPLE If you roll a dice, event A could be 'I get an even number', and B 'I get a number bigger than 4'. The Venn diagram would be:

$$P(A) = \frac{3}{6} = \frac{1}{2} \qquad P(B) = \frac{2}{6} = \frac{1}{3} \qquad P(A \cap B) = \frac{1}{6} \qquad P(A \cup B) = \frac{4}{6} = \frac{2}{3}$$

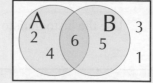

Here, I've just counted outcomes — but I could have used the formula.

Random Events and Venn Diagrams

EXAMPLE   A survey was carried out to find what pets people like.

The probability they like dogs is 0.6. The probability they like cats is 0.5. The probability they like gerbils is 0.4. The probability they like dogs and cats is 0.4. The probability they like cats and gerbils is 0.1, and the probability they like gerbils and dogs is 0.2. Finally, the probability they like all three kinds of animal is 0.1. You can draw all this in a Venn diagram. (Here I've used C for 'likes cats', D for 'likes dogs' and G for 'likes gerbils'.)

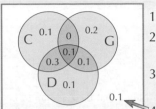

1) Stick in the middle one first — 'likes all 3 animals' (i.e. $C \cap D \cap G$).
2) Then do the 'likes 2 animals' probabilities by taking 0.1 from each of the given 'likes 2 animals' probabilities. (If they like 3 animals, they'll also be in the 'likes 2 animals' bits.)
3) Then do the 'likes 1 kind of animal' probabilities, by making sure the total probability in each circle adds up to the probability in the question.
4) Finally, subtract all the probabilities so far from 1 to find 'likes none of these animals'.

① From the Venn diagram, the probability that someone likes either dogs or cats is 0.7.

② The probability that someone likes gerbils but not dogs is 0.2.

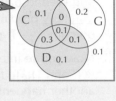

③ You can work out the probability that a dog-lover <u>also</u> like cats by ignoring everything outside the 'dogs' circle.

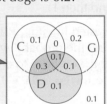

P(dog-lover also like cats)

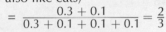

$$= \frac{0.3 + 0.1}{0.3 + 0.1 + 0.1 + 0.1} = \frac{2}{3}$$

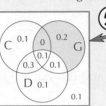

The **Complement** of 'Event A' is '**Not Event A**'

An event A will either happen or not happen. The event 'A doesn't happen' is called the <u>complement</u> of A (or <u>A'</u>). On a Venn diagram, it would look like this (because $A \cup A' = S$, the sample space): At least one of A and A' has to happen, so...

$$P(A) + P(A') = 1 \quad \text{or} \quad P(A') = 1 - P(A)$$

EXAMPLE   A teacher keeps socks loose in a box. One morning, he picks out a sock. He calculates that the probability of then picking out a matching sock is 0.56. What is the probability of him not picking a matching sock?

Call event A 'picks a matching sock'. Then A' is 'doesn't pick a matching sock'. Now A and A' are <u>complementary</u> events (and P(A) = 0.56), so $P(A) + P(A') = 1$, and therefore $P(A') = 1 - 0.56 = 0.44$

Mutually Exclusive Events Have **No Overlap**

If two events can't both happen at the same time (i.e. $P(A \cap B) = 0$) they're called <u>mutually exclusive</u> (or just '<u>exclusive</u>'). If A and B are exclusive, then the probability of A <u>or</u> B is: $P(A \cup B) = P(A) + P(B)$. ◄ Use the formula from page 114, but put $P(A \cap B) = 0$.
More generally,

> For n <u>exclusive</u> events (i.e. only one of them can happen at a time):
> $$P(A_1 \cup A_2 \cup ... \cup A_n) = P(A_1) + P(A_2) + ... + P(A_n)$$

EXAMPLE   Find the probability that a card pulled at random from a standard pack of cards (no jokers) is <u>either</u> a picture card (a Jack, Queen or King) <u>or</u> the 7, 8 or 9 of clubs.

Call <u>event A</u> — 'I get a picture card', and <u>event B</u> — 'I get the 7, 8 or 9 of clubs'.
Events A and B are <u>mutually exclusive</u> — they can't both happen. Also, $P(A) = \frac{12}{52} = \frac{3}{13}$ and $P(B) = \frac{3}{52}$.
So the probability of either A or B is: $P(A \cup B) = P(A) + P(B) = \frac{12}{52} + \frac{3}{52} = \frac{15}{52}$

Two heads are better than one — though only half as likely using two coins...

I must admit — I kind of like these pages. This stuff isn't too hard, and it's really useful for answering loads of questions. And one other good thing is that Venn diagrams look, well, nice somehow. But more importantly, when you're filling one in, the thing to remember is that you usually need to 'start from the inside and work out'.

Tree Diagrams

Tree Diagrams — they blossom from a tiny question-acorn into a beautiful tree of possibility. Inspiring <u>and</u> useful.

Tree Diagrams Show Probabilities for Two or More Events

Each 'chunk' of a tree diagram is a trial, and each branch of that chunk is a possible outcome. Multiplying probabilities along the branches gives you the probability of a <u>series</u> of outcomes.

EXAMPLE If Susan plays tennis one day, the probability that she'll play the next day is 0.2. If she doesn't play tennis, the probability that she'll play the next day is 0.6. She plays tennis on Monday. What is the probability she plays tennis:

(i) on both the Tuesday and Wednesday of that week?
(ii) on the Wednesday of the same week?

Let T mean 'plays tennis' (and then T' means 'doesn't play tennis').

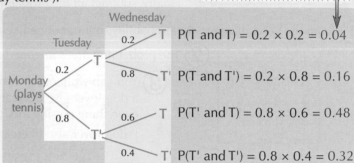

Notice that these add up to 1.

(i) Then the probability that she plays on Tuesday <u>and</u> Wednesday is $P(T \text{ and } T) = 0.2 \times 0.2 = \underline{0.04}$ (<u>multiply</u> probabilities since you need a <u>series</u> of outcomes — T and then T).

(ii) Now you're interested in <u>either</u> $P(T \text{ and } T)$ <u>or</u> $P(T' \text{ and } T)$. To find the probability of one event <u>or</u> another happening, you have to <u>add</u> probabilities: $P(\text{plays on Wednesday}) = 0.04 + 0.48 = \underline{0.52}$.

$P(T \text{ and } T) = 0.2 \times 0.2 = 0.04$

$P(T \text{ and } T') = 0.2 \times 0.8 = 0.16$

$P(T' \text{ and } T) = 0.8 \times 0.6 = 0.48$

$P(T' \text{ and } T') = 0.8 \times 0.4 = 0.32$

Sometimes a Branch is Missing

EXAMPLE A box of biscuits contains 5 chocolate biscuits and 1 lemon biscuit. George takes out 3 biscuits at random, one at a time, and eats them.

a) Find the probability that he eats 3 chocolate biscuits.
b) Find the probability that the last biscuit is chocolate.

Let C mean 'picks a chocolate biscuit' and L mean 'picks the lemon biscuit'.

After the lemon biscuit there are only chocolate biscuits left, so the tree diagram doesn't 'branch' after an 'L'.

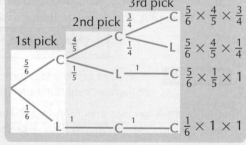

a) Three chocolate biscuits is shown by only one 'path' along the branches.
$$P(C \text{ and } C \text{ and } C) = \frac{5}{6} \times \frac{4}{5} \times \frac{3}{4} = \frac{60}{120} = \frac{1}{2}$$

b) The third biscuit being chocolate is shown by 3 'paths' along the branches — so you can add up the probabilities:
$$P(\text{third biscuit is chocolate}) = \left(\frac{5}{6} \times \frac{4}{5} \times \frac{3}{4}\right) + \left(\frac{5}{6} \times \frac{1}{5} \times 1\right) + \left(\frac{1}{6} \times 1 \times 1\right) = \frac{1}{2} + \frac{1}{6} + \frac{1}{6} = \frac{5}{6}$$

There's a quicker way to do this, since there's only one outcome where the chocolate <u>isn't</u> picked last:
$$P(\text{third biscuit is not chocolate}) = \frac{5}{6} \times \frac{4}{5} \times \frac{1}{4} = \frac{1}{6}, \text{ so } P(\text{third biscuit is chocolate}) = 1 - \frac{1}{6} = \frac{5}{6}$$

Working out the probability of the <u>complement</u> of the event you're interested in is sometimes easier.

Sampling with replacement — the probabilities stay the same

In the above example, each time George takes a biscuit he eats it before taking the next one (i.e. he doesn't replace it) — this is <u>sampling without replacement</u>. Suppose instead that each time he takes a biscuit he puts it back in the box before taking the next one — this is <u>sampling with replacement</u>. All this means is that the probability of choosing a particular item <u>remains the same</u> for each pick.

So part a) above becomes:

$$P(C \text{ and } C \text{ and } C) = \frac{5}{6} \times \frac{5}{6} \times \frac{5}{6} = \frac{125}{216} > \frac{1}{2}$$

So the probability that George picks 3 chocolate biscuits is slightly greater when sampling is done <u>with replacement</u>. This makes sense because now there are, on average, more chocolate biscuits available for his 2nd and 3rd picks, so he is more likely to choose one.

Conditional Probability

After the first set of branches, tree diagrams actually show <u>conditional probabilities</u>. Read on...

P(B|A) means **Probability of B**, given that **A has Already Happened**

Conditional probability means the probability of something, given that something else has already happened.
For example, P(B|A) means the probability of B, given that A has already happened. Back to tree diagrams...

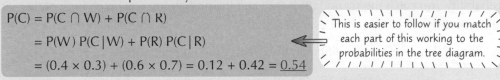

If you multiply probabilities along the branches, you get:

i.e. P(A and B) $\Longrightarrow$ $P(A \cap B) = P(A) \times P(B|A)$

You can rewrite this as:

$$P(B \mid A) = \frac{P(A \cap B)}{P(A)}$$

These are conditional probabilities, since
something (A or A') has already happened.

EXAMPLE Horace either walks (W) or runs (R) to the bus stop. If he walks he
catches (C) the bus with a probability of 0.3. If he runs he catches it with
a probability of 0.7. He walks to the bus stop with a probability of 0.4.
Find the probability that Horace catches the bus.

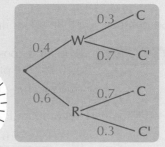

$P(C) = P(C \cap W) + P(C \cap R)$

$= P(W) P(C|W) + P(R) P(C|R)$

$= (0.4 \times 0.3) + (0.6 \times 0.7) = 0.12 + 0.42 = \underline{0.54}$

This is easier to follow if you match
each part of this working to the
probabilities in the tree diagram.

If **B is Conditional** on A then **A is Conditional** on B

If B depends on A then A depends on B — and it doesn't matter which event happens first.

EXAMPLE Horace turns up at school either late (L) or on time (L'). He is then either shouted at (S) or not (S').
The probability that he turns up late is 0.4. If he turns up late the probability that he is shouted at is 0.7.
If he turns up on time the probability that he is shouted at is 0.2.

If you hear Horace being shouted at, what is the probability that he turned up late?

1) The probability you want is P(L|S).   Get this the right way round — he's <u>already</u> being shouted at.

2) Use the conditional probability formula: $P(L \mid S) = \frac{P(L \cap S)}{P(S)}$

3) The best way to find P(L ∩ S) and P(S) is with a tree diagram.

   Be careful with questions like this — the information in the question tells you
   what you need to know to draw the tree diagram with L (or L') considered first.

   But you need P(L|S) — where S is considered first. So don't just rush in.

   $P(L \cap S) = 0.4 \times 0.7 = 0.28$
   $P(S) = P(L \cap S) + P(L' \cap S) = 0.28 + 0.12 = 0.40$

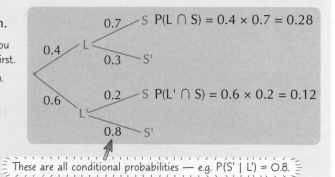

4) Put these in your conditional probability formula to get:

   $P(L \mid S) = \frac{0.28}{0.4} = 0.7$

These are all conditional probabilities — e.g. P(S' | L') = 0.8.

There's a **Formula** for Working this Out — but it's **Easier** to Use the **Tree Diagram**

This formula will be on the
formula sheet in your exam.

$$P(A \mid B) = \frac{P(A \cap B)}{P(B)} = \frac{P(B \mid A)P(A)}{P(B \mid A)P(A) + P(B \mid A')P(A')}$$

This is basically the
same working as with
the tree diagram above.

Here, this gives:

$$P(L \mid S) = \frac{P(L \cap S)}{P(S)} = \frac{P(S \mid L)P(L)}{P(S \mid L)P(L) + P(S \mid L')P(L')} = \frac{0.7 \times 0.4}{(0.7 \times 0.4) + (0.2 \times 0.6)} = \frac{0.28}{0.4} = 0.7$$

Independent Events

Independent Events Have No Effect on Each Other

If the probability of B happening doesn't depend on whether or not A has happened, then A and B are <u>independent</u>.
1) If A and B are independent, P(A | B) = P(A).
2) If you put this in the conditional probability formula, you get: $P(A \mid B) = P(A) = \dfrac{P(A \cap B)}{P(B)}$

Or, to put that another way:

> For independent events: $P(A \cap B) = P(A)P(B)$

EXAMPLE V and W are independent events, where P(V) = 0.2 and P(W) = 0.6.
  a) Find P(V ∩ W).   b) Find P(V ∪ W).

a) Just put the numbers into the formula for independent events: $P(V \cap W) = P(V)P(W) = 0.2 \times 0.6 = 0.12$

b) Using the formula on page 114: $P(V \cup W) = P(V) + P(W) - P(V \cap W) = 0.2 + 0.6 - 0.12 = 0.68$

Sometimes you'll be asked if two events are independent or not. Here's how you work it out...

EXAMPLE You are exposed to two infectious diseases — one after the other. The probability you catch the first (A) is 0.25, the probability you catch the second (B) is 0.5, and the probability you catch both of them is 0.2. Are catching the two diseases independent events?

You need to compare P(A | B) and P(A) — if they're different, the events <u>aren't independent</u>.

$P(A \mid B) = \dfrac{P(A \cap B)}{P(B)} = \dfrac{0.2}{0.5} = 0.4$ $P(A) = 0.25$ P(A | B) and P(A) are different, so they're <u>not independent</u>.

Take Your Time with Tough Probability Questions

EXAMPLE A and B are two events, with P(A) = 0.4, P(B | A) = 0.25, and P(A' ∩ B) = 0.2.
  a) Find: (i) P(A ∩ B),   (ii) P(A'),   (iii) P(B' | A),   (iv) P(B | A'),   (v) P(B),   (vi) P(A | B).
  b) Say whether or not A and B are independent.

a) i) $P(B \mid A) = \dfrac{P(A \cap B)}{P(A)} = 0.25$, so $P(A \cap B) = 0.25 \times P(A) = 0.25 \times 0.4 = 0.1$

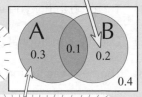

P(A' ∩ B)

A Venn diagram sometimes makes it easier to see what's going on. Fill it in as you find anything out.

  ii) $P(A') = 1 - P(A) = 1 - 0.4 = 0.6$

  iii) $P(B' \mid A) = 1 - P(B \mid A) = 1 - 0.25 = 0.75$

  iv) $P(B \mid A') = \dfrac{P(B \cap A')}{P(A')} = \dfrac{0.2}{0.6} = \dfrac{1}{3}$ ⟵ $P(B \cap A') = P(A' \cap B)$

From (i) P(A ∩ B) = 0.1, so you know this must be 0.3.

  v) $P(B) = P(B \mid A)P(A) + P(B \mid A')P(A') = (0.25 \times 0.4) + \left(\dfrac{1}{3} \times 0.6\right) = 0.3$ ⟵ Or use the Venn diagram.

  vi) $P(A \mid B) = \dfrac{P(A \cap B)}{P(B)} = \dfrac{0.1}{0.3} = \dfrac{1}{3}$

Or you could say that P(A ∩ B) = 0.1, while P(A)P(B) = 0.4 × 0.3 = 0.12 — they're different, which shows A and B are not independent.

b) If P(B | A) = P(B), then A and B are independent.
   But P(B | A) = 0.25, while P(B) = 0.3, so A and B are <u>not</u> independent.

Statisticians say: P(Having cake ∩ Eating it) = 0...

Probability questions can be tough. For tricky questions like the last one, try drawing a Venn diagram or a tree diagram, even if the question doesn't tell you to — they're really useful for getting your head round things and understanding what on earth is going on. And don't forget the tests for independent events — you're likely to get asked a question on those.

S1 Section 2 — Practice Questions

Gosh. A whole page of warm-up questions. By the time you've finished these, you'll be warmer than a wolf in woollen mittens. Oh, and you'll probably be <u>awesome at probability questions</u> too.

Warm-up Questions

1) A standard dice and a coin are thrown and the outcomes recorded.
   If a head is thrown, the score on the dice is doubled.
   If a tail is thrown, 4 is added to the score on the dice.

   a) Represent this by means of a sample space diagram.

   b) What is the probability that you score more than 5?

   c) If you throw a tail, what is the probability that you get an even score?

2) Half the students in a sixth-form college eat sausages for dinner and 20% eat chips.
   10% of those who eat chips also eat sausages. By use of a Venn diagram or otherwise, find:

   a) the percentage of students who eat both chips and sausages,

   b) the percentage of students who eat chips but not sausages,

   c) the percentage of students who eat either chips or sausages but not both.

3) Arabella rolls two standard dice and adds the two results together.

   a) What is the probability that she scores a prime number?

   b) What is the probability that she scores a square number?

   c) What is the probability that she scores a number that is
      either a prime number or a square number?

4) In a school orchestra (made up of pupils in either the upper or lower school),
   40% of the musicians are boys. Of the boys, 30% are in the upper school.
   Of the girls in the orchestra, 50% are in the upper school.

   a) Draw a tree diagram to show the various probabilities.

   b) Find the probability that a musician chosen at random is in the upper school.

5) Albert eats a limited choice of lunch. He eats either chicken or beef
   for his main course, and either chocolate pudding or ice cream for dessert.
   The probability that he eats chicken is 1/3, the probability that he eats
   ice cream given that he has chicken is 2/5, and the probability that he
   has ice cream given that he has beef is 3/4.

   a) Find the probability he has either chicken or ice cream — but not both.
   b) Find the probability that he eats ice cream.
   c) Find the probability that he had chicken given that you see him eating ice cream.

S1 Section 2 — Practice Questions

Boop. Boop. Boop. Exam simulation has begun. Repeat, exam simulation has begun. Please ensure your safety goggles are firmly attached. Emergency exits can be found on the right- and left-hand sides of the page.

Exam Questions

1 A soap company asked 120 people about the types of soap (from Brands A, B and C) they bought. Brand A was bought by 40 people, Brand B by 30 people and Brand C by 25. Both Brands A and B (and possibly C as well) were bought by 8 people, B and C (and maybe A) were bought by 10 people, and A and C (and maybe B) by 7 people. All three brands were bought by 3 people.

   a) Represent this information in a Venn diagram.

[5 marks]

   b) If a person is selected at random, find the probability that:

      (i) they buy at least one of the soaps.

[2 marks]

      (ii) they buy at least two of the soaps.

[2 marks]

      (iii) they buy soap B, given that they buy only one type of soap.

[3 marks]

2 A jar contains counters of 3 different colours. There are 3 red counters, 4 white counters and 5 green counters. Two random counters are removed from the jar one at a time. Once removed, the colour of the counter is noted. The first counter is not replaced before the second one is drawn.

   a) Draw a tree diagram to show the probabilities of the various outcomes.

[3 marks]

   b) Find the probability that the second counter is green.

[2 marks]

   c) Find the probability that both the counters are red.

[2 marks]

   d) Find the probability that the two counters are not both the same colour.

[3 marks]

3 Event J and Event K are independent events, where $P(J) = 0.7$ and $P(K) = 0.1$.

   a) Find:

      (i) $P(J \cap K)$

[1 mark]

      (ii) $P(J \cup K)$

[2 marks]

   b) If L is the event that neither J or K occurs, find $P(L|K')$

[3 marks]

4 For a particular biased dice, the event 'throw a 6' is called event B. $P(B) = 0.2$. This biased dice and a fair dice are rolled together. Find the probability that:

   a) the biased dice doesn't show a 6,

[1 mark]

   b) at least one of the dice shows a 6,

[2 marks]

   c) exactly one of the dice shows a 6, given that at least one of them shows a 6.

[3 marks]

Probability Distributions

This stuff isn't hard — but it can seem a bit weird at times.

Getting your head round this Basic Stuff will help a bit

This first bit isn't particularly interesting. But understanding the difference between X and x (bear with me) might make the later stuff a bit less confusing. Might.

1) X (upper case) is just the <u>name</u> of a <u>random variable</u>. So X could be 'score on a dice' — it's <u>just a name</u>.

2) A <u>random variable</u> doesn't have a <u>fixed</u> value. Like with a dice score — the value on any 'roll' is all down to chance.

3) x (lower case) is a <u>particular value</u> that X can take. So for one roll of a dice, x could be 1, 2, 3, 4, 5 or 6.

4) <u>Discrete</u> random variables only have a <u>certain number</u> of possible values. Often these values are whole numbers, but they don't have to be. Usually there are only a few possible values (e.g. the possible scores with one roll of a dice).

5) A <u>probability distribution</u> is a <u>table showing the possible values</u> of x, plus the <u>probability</u> for each one.

6) A <u>probability function</u> is a formula that generates the probabilities for different values of x.

All the Probabilities Add up to 1

For a discrete random variable X:

$$\sum_{\text{all } x} P(X = x) = 1$$

This says that if you add up the probabilities of all the possible values of X, you get 1.

EXAMPLE   The random variable X has probability function $P(X = x) = kx$ for $x = 1, 2, 3$. Find the value of k.

So X has three possible values ($x = 1, 2$ and 3), and the probability of each is kx (where you need to find k).

It's easier to understand with a table:

x	1	2	3
$P(X = x)$	$k \times 1 = k$	$k \times 2 = 2k$	$k \times 3 = 3k$

Now just use the formula: $\sum_{\text{all } x} P(X = x) = 1$ Here, this means: $k + 2k + 3k = 6k = 1$

i.e. $k = \dfrac{1}{6}$ Piece of cake.

The **mode** is the <u>most likely</u> value — so it's the value with the <u>biggest probability</u>.

EXAMPLE   The discrete random variable X has the probability distribution shown below.

x	0	1	2	3	4
$P(X = x)$	0.1	0.2	0.3	0.2	a

Find: (i) the value of a,   (ii) $P(2 \le X < 4)$,   (iii) the mode

(i) Use the formula $\sum_{\text{all } x} P(X = x) = 1$ again.

From the table: $0.1 + 0.2 + 0.3 + 0.2 + a = 1$
$0.8 + a = 1$
$\underline{a = 0.2}$

Careful with the inequality signs — you need to include $x = 2$ but not $x = 4$.

(ii) This is asking for the probability that 'X is greater than or equal to 2, but less than 4'. Easy — just add up the probabilities.

$P(2 \le X < 4) = P(X = 2) + P(X = 3) = 0.3 + 0.2 = \underline{0.5}$

(iii) The mode is the value of x with the biggest probability — so $\underline{\text{mode} = 2}$.

Probability Distributions

EXAMPLE An unbiased six-sided dice has faces marked 1, 1, 1, 2, 2, 3.
The dice is rolled twice. Let X be the random variable "sum of the two scores on the dice".
Show that $P(X = 4) = \frac{5}{18}$. Find the probability distribution of X.

① Make a table showing the 36 possible outcomes.
You can see from the table that 10 of these have the outcome $X = 4$

... so $P(X = 4) = \frac{10}{36} = \frac{5}{18}$

Score on roll 1

+	1	1	1	2	2	3
1	2	2	2	3	3	4
1	2	2	2	3	3	4
1	2	2	2	3	3	4
2	3	3	3	4	4	5
2	3	3	3	4	4	5
3	4	4	4	5	5	6

Score on roll 2

Don't forget to change the fractions into their simplest form.

② Use the table to work out the probabilities for the other outcomes and then fill in a table summarising the probability distribution. So...

... $\frac{9}{36}$ of the outcomes are a score of 2
... $\frac{12}{36}$ of the outcomes are a score of 3
... $\frac{4}{36}$ of the outcomes are a score of 5
... $\frac{1}{36}$ of the outcomes are a score of 6

x	2	3	4	5	6
$P(X = x)$	$\frac{1}{4}$	$\frac{1}{3}$	$\frac{5}{18}$	$\frac{1}{9}$	$\frac{1}{36}$

Do Complicated questions *Bit by bit*

EXAMPLE A game involves rolling two fair dice. If the sum of the scores is greater than 10 then the player wins 50p.
If the sum is between 8 and 10 (inclusive) then they win 20p. Otherwise they get nothing.
If X is the random variable "amount player wins", find the probability distribution of X.

There are 3 possible values for X (0, 20 and 50) and you need the probability of each.
To work these out, you need the probability of getting various totals on the dice.

① You need to know $P(8 \leq \text{score} \leq 10)$ — the probability that the score is between 8 and 10 inclusive (i.e. including 8 and 10) and $P(11 \leq \text{score} \leq 12)$ — the probability that the score is greater than 10.

This means working out: $P(\text{score} = 8)$, $P(\text{score} = 9)$, $P(\text{score} = 10)$, $P(\text{score} = 11)$ and $P(\text{score} = 12)$. Use a table...

②

Score on dice 1

+	1	2	3	4	5	6
1	2	3	4	5	6	7
2	3	4	5	6	7	8
3	4	5	6	7	8	9
4	5	6	7	8	9	10
5	6	7	8	9	10	11
6	7	8	9	10	11	12

Score on dice 2

There are 36 possible outcomes...
...5 of these have a total of 8 — so the probability of scoring 8 is $\frac{5}{36}$
...4 have a total of 9 — so the probability of scoring 9 is $\frac{4}{36}$,
...the probability of scoring 10 is $\frac{3}{36}$
...the probability of scoring 11 is $\frac{2}{36}$
...the probability of scoring 12 is $\frac{1}{36}$

③ To find the probabilities you need, you just add the right bits together:

$P(X = 20\text{p}) = P(8 \leq \text{score} \leq 10) = \frac{5}{36} + \frac{4}{36} + \frac{3}{36} = \frac{12}{36} = \frac{1}{3}$ $P(X = 50\text{p}) = P(11 \leq \text{score} \leq 12) = \frac{2}{36} + \frac{1}{36} = \frac{3}{36} = \frac{1}{12}$

To find $P(X = 0)$ just take the total of the two probabilities above from 1 (since $X = 0$ is the only other possibility).

$P(X = 0) = 1 - \left[\frac{12}{36} + \frac{3}{36}\right] = 1 - \frac{15}{36} = \frac{21}{36} = \frac{7}{12}$

④ Now just stick all this info in a table (and check that the probabilities all add up to 1):

x	0	20	50
$P(X = x)$	$\frac{7}{12}$	$\frac{1}{3}$	$\frac{1}{12}$

Useful quotes: All you need in life is ignorance and confidence, then success is sure[*]...

I said earlier that the 'counting the outcomes' approach was useful — well there you go. And if you remember how to do that, then you can work out a probability distribution. And if you can work out one of those, then you can often begin to unravel even fairly daunting-looking questions. But most of all, REMEMBER THAT ALL THE PROBABILITIES ADD UP TO 1.

* Mark Twain

The Cumulative Distribution Function

The probability function gives probabilities of underlined values of X. The cumulative distribution function tells you something else.

The Cumulative Distribution Function is a Running Total of Probabilities

The cumulative distribution function F(x) gives the probability that X will be less than or equal to a particular value.

$$F(x_0) = P(X \leq x_0) = \sum_{x \leq x_0} p(x) \quad \Longleftarrow \quad p(x) = P(X = x)$$

EXAMPLE The probability distribution of the discrete random variable H is shown in the table. Draw up a table to show the cumulative distribution function F(h).

h	0.1	0.2	0.3	0.4
P(H = h)	$\frac{1}{4}$	$\frac{1}{4}$	$\frac{1}{3}$	$\frac{1}{6}$

There are 4 values of h, so you have to find the probability that H is less than or equal to each of them in turn. It sounds trickier than it actually is — you only have to add up a few probabilities...

$F(0.1) = P(H \leq 0.1)$ — this is the same as $P(H = 0.1)$, since H can't be less than 0.1. So $F(0.1) = \frac{1}{4}$.

$F(0.2) = P(H \leq 0.2)$ — this is the probability that $H = 0.1$ or $H = 0.2$. So $F(0.2) = P(H = 0.1) + P(H = 0.2) = \frac{1}{4} + \frac{1}{4} = \frac{1}{2}$.

$F(0.3) = P(H \leq 0.3) = P(H = 0.1) + P(H = 0.2) + P(H = 0.3) = \frac{1}{4} + \frac{1}{4} + \frac{1}{3} = \frac{5}{6}$.

$F(0.4) = P(H \leq 0.4) = P(H = 0.1) + P(H = 0.2) + P(H = 0.3) + P(H = 0.4) = \frac{1}{4} + \frac{1}{4} + \frac{1}{3} + \frac{1}{6} = 1$.

$P(X \leq \text{largest value of } x)$ is always 1.

Finally, put these values in a table, and you're done...

h	0.1	0.2	0.3	0.4
F(h) = P(H ≤ h)	$\frac{1}{4}$	$\frac{1}{2}$	$\frac{5}{6}$	1

Sometimes they ask you to work backwards...

EXAMPLE The formula below gives the cumulative distribution function F(x) for a discrete random variable X. Find k, and the probability function.
$F(x) = kx$, for $x = 1, 2, 3$ and 4.

(1) First find k. You know that X has to be 4 or less — so $P(X \leq 4) = 1$.

Put $x = 4$ into the cumulative distribution function: $F(4) = P(X \leq 4) = 4k = 1$, so $k = \frac{1}{4}$.

(2) Now you can work out the probabilities of X being less than or equal to 1, 2, 3 and 4.

$F(1) = P(X \leq 1) = 1 \times k = \frac{1}{4}$, $F(2) = P(X \leq 2) = 2 \times k = \frac{1}{2}$, $F(3) = P(X \leq 3) = 3 \times k = \frac{3}{4}$, $F(4) = P(X \leq 4) = 1$

(3) This is the clever bit...

$P(X = 4) = P(X \leq 4) - P(X \leq 3) = 1 - \frac{3}{4} = \frac{1}{4}$

Think about it...
...if it's less than or equal to 4,
...but it's not less than or equal to 3,
...then it has to be 4.

$P(X = 3) = P(X \leq 3) - P(X \leq 2) = \frac{3}{4} - \frac{1}{2} = \frac{1}{4}$

$P(X = 2) = P(X \leq 2) - P(X \leq 1) = \frac{1}{2} - \frac{1}{4} = \frac{1}{4}$

$P(X = 1) = P(X \leq 1) = \frac{1}{4}$

Because x doesn't take any values less than 1.

(4) Finish it all off by making a table. The probability distribution of X is:

x	1	2	3	4
P(X = x)	$\frac{1}{4}$	$\frac{1}{4}$	$\frac{1}{4}$	$\frac{1}{4}$

So the probability function is: $P(X = x) = \frac{1}{4}$ for $x = 1, 2, 3, 4$

Fact: probability of this coming up in the exam is less than or equal to one...

So if you've got a probability distribution, you can easily work out the table for the cumulative distribution function. And from a cumulative distribution function, you can work out the probability distribution as long as you remember the wee trick on this page. Don't forget, all the stuff so far is for discrete variables — these can only take a certain number of values.

Expected Values, Mean & Variance

This is all about the mean and variance of <u>random variables</u> — <u>not</u> a load of data. It's a tricky concept, but bear with it.

Discrete Random Variables have an 'Expected Value' or 'Mean'

You can work out the <u>expected value</u> (or 'mean') $\underline{E(X)}$ for a discrete <u>random variable</u> X.
$E(X)$ is a kind of 'theoretical mean' — it's what you'd <u>expect</u> the mean of X to be if you
took <u>loads</u> of readings. In practice, the mean of your results is unlikely to match the
theoretical mean <u>exactly</u>, but it should be pretty near.

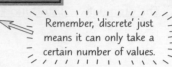 Remember, 'discrete' just means it can only take a certain number of values.

If the possible values of X are x_1, x_2, x_3,... then the expected value of X is:

$$\text{Mean} = \text{Expected Value } E(X) = \sum x_i P(X = x_i) = \sum x_i p_i \quad \longleftarrow \quad p_i = P(X = x_i)$$

EXAMPLE The probability distribution of X, the number of daughters in a family of 3 children, is shown in the table. Find the expected number of daughters.

x_i	0	1	2	3
p_i	$\frac{1}{8}$	$\frac{3}{8}$	$\frac{3}{8}$	$\frac{1}{8}$

$$\text{Mean} = \sum x_i p_i = \left[0 \times \tfrac{1}{8}\right] + \left[1 \times \tfrac{3}{8}\right] + \left[2 \times \tfrac{3}{8}\right] + \left[3 \times \tfrac{1}{8}\right] = 0 + \tfrac{3}{8} + \tfrac{6}{8} + \tfrac{3}{8} = \tfrac{12}{8} = 1.5$$

So the <u>expected</u> number of daughters is 1.5 — which sounds a bit weird.
But all it means is that if you check a <u>large number</u> of 3-child families, the <u>mean</u> will be close to 1.5.

The Variance measures how Spread Out the distribution is

You can also find the <u>variance</u> of a random variable. It's the 'expected variance' of a <u>large number</u> of readings.

$$\text{Var}(X) = E(X^2) - [E(X)]^2 = \sum x_i^2 p_i - \left[\sum x_i p_i\right]^2$$

This formula needs $E(X^2) = \sum x_i^2 p_i$ — take each possible value of x, square it, multiply it by its probability and then add up all the results.

EXAMPLE Work out the variance for the '3 daughters' example above:

First work out $E(X^2)$: $\quad E(X^2) = \sum x_i^2 p_i = \left[0^2 \times \tfrac{1}{8}\right] + \left[1^2 \times \tfrac{3}{8}\right] + \left[2^2 \times \tfrac{3}{8}\right] + \left[3^2 \times \tfrac{1}{8}\right]$

$$= 0 + \tfrac{3}{8} + \tfrac{12}{8} + \tfrac{9}{8} = \tfrac{24}{8} = \underline{3}$$

Now you take away the mean squared: $\text{Var}(X) = E(X^2) - [E(X)]^2 = 3 - 1.5^2 = 3 - 2.25 = \underline{0.75}$

EXAMPLE X has the probability function $P(X = x) = k(x + 1)$ for $x = 0, 1, 2, 3, 4$.
Find the mean and variance of X.

① First you need to find k — work out all the probabilities and make sure they add up to 1.

$P(X = 0) = k \times (0 + 1) = k$. Similarly, $P(X = 1) = 2k$, $P(X = 2) = 3k$, $P(X = 3) = 4k$, $P(X = 4) = 5k$.

So $k + 2k + 3k + 4k + 5k = 1$, i.e. $15k = 1$, and so $k = \dfrac{1}{15}$ $\longleftarrow$ Now you can work out p_1, p_2, p_3... where $p_1 = P(X = 1)$ etc.

② Now use the formulas — find the mean $E(X)$ first:

$$E(X) = \sum x_i p_i = \left[0 \times \tfrac{1}{15}\right] + \left[1 \times \tfrac{2}{15}\right] + \left[2 \times \tfrac{3}{15}\right] + \left[3 \times \tfrac{4}{15}\right] + \left[4 \times \tfrac{5}{15}\right] = \tfrac{40}{15} = \tfrac{8}{3}$$

For the variance you need $E(X^2)$:

$$E(X^2) = \sum x_i^2 p_i = \left[0^2 \times \tfrac{1}{15}\right] + \left[1^2 \times \tfrac{2}{15}\right] + \left[2^2 \times \tfrac{3}{15}\right] + \left[3^2 \times \tfrac{4}{15}\right] + \left[4^2 \times \tfrac{5}{15}\right] = \tfrac{130}{15} = \tfrac{26}{3}$$

And finally: $\quad \text{Var}(X) = E(X^2) - [E(X)]^2 = \tfrac{26}{3} - \left[\tfrac{8}{3}\right]^2 = \tfrac{14}{9}$

Expected Values, Mean and Variance

These formulas give the **Expected Value** and **Variance** for a **Function of X**

These little jokers are absolutely guaranteed to pop up somewhere in your S1 exam...

$$E(aX + b) = aE(X) + b \qquad Var(aX + b) = a^2Var(X)$$

Here a and b are any numbers.

EXAMPLE If $E(X) = 3$ and $Var(X) = 7$, find $E(2X + 5)$ and $Var(2X + 5)$.

Easy. $E(2X + 5) = 2E(X) + 5 = (2 \times 3) + 5 = 11$

$Var(2X + 5) = 2^2Var(X) = 4 \times 7 = 28$

EXAMPLE The discrete random variable X has the following probability distribution:

x	2	3	4	5	6
$P(X = x)$	0.1	0.2	0.3	0.2	k

Find: a) k, b) $E(X)$, c) $Var(X)$, d) $E(3X - 1)$, e) $Var(3X - 1)$

Slowly, slowly — one bit at a time...

a) Remember the probabilities add up to 1 — $0.1 + 0.2 + 0.3 + 0.2 + k = 1$, and so $k = 0.2$

b) Now you can use the formula to find $E(X)$: $E(X) = \sum x_ip_i = (2 \times 0.1) + (3 \times 0.2) + (4 \times 0.3) + (5 \times 0.2) + (6 \times 0.2) = 4.2$

c) Next work out $E(X^2)$: $E(X^2) = \sum x_i^2p_i = [2^2 \times 0.1] + [3^2 \times 0.2] + [4^2 \times 0.3] + [5^2 \times 0.2] + [6^2 \times 0.2] = 19.2$
and then the variance is easy: $Var(X) = E(X^2) - [E(X)]^2 = 19.2 - 4.2^2 = 1.56$

d) You'd expect the question to get harder but it doesn't: $E(3X - 1) = 3E(X) - 1 = 3 \times 4.2 - 1 = 11.6$

e) And finally: $Var(3X - 1) = 3^2Var(X) = 9 \times 1.56 = 14.04$

In a **Discrete Uniform Distribution** the Probabilities are **Equal**

('discrete' because there's only a certain number of possible outcomes.)

Sometimes you'll have a random variable where every value of X is equally likely — this is called a uniform distribution. For example, rolling a normal, unbiased dice gives you a discrete uniform distribution. The probability function of a discrete uniform distribution looks like this — here, there are only 4 possible values:

For a discrete uniform distribution X which can take consecutive whole number values $a, a + 1, a + 2,..., b$, you get these nice simple formulas for the mean and variance...

$$\text{Mean} = \frac{a + b}{2} \qquad \text{Variance} = \frac{(b - a + 1)^2 - 1}{12}$$

where a is the smallest value and b is the biggest.

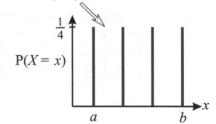

$P(X = x)$

EXAMPLE Find the mean and variance of the score on an unbiased, standard six-sided dice.

If X is the random variable 'score on a dice', then X has a discrete uniform distribution — like in this table:

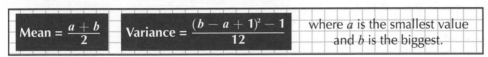

x	1	2	3	4	5	6
$P(X = x)$	$\frac{1}{6}$	$\frac{1}{6}$	$\frac{1}{6}$	$\frac{1}{6}$	$\frac{1}{6}$	$\frac{1}{6}$

The symmetry of the distribution should tell you where the mean is — it has to be halfway between 1 and 6.

The smallest value of x is 1 and the biggest is 6 — so $a = 1$ and $b = 6$. Now just stick the numbers in the formulas:

$$\text{Mean} = \frac{a + b}{2} = \frac{1 + 6}{2} = \frac{7}{2} = \underline{3.5} \qquad \text{Variance} = \frac{(b - a + 1)^2 - 1}{12} = \frac{(6 - 1 + 1)^2 - 1}{12} = \frac{35}{12} = \underline{2.92} \text{ to 3 sig. fig.}$$

Statisticians say: E(Bird in hand) = E(2 Birds in bush)...

The mean and variance here are theoretical values — don't get them confused with the mean and variance of a load of practical observations. This 'theoretical' variance has a similar formula to the variance formula on page 106, though — it's just "E(X-squared) minus E(X)-squared". And you can still take the square root of the variance to get the standard deviation.

S1 Section 3 — Practice Questions

Probability distribution, probability function, cumulative distribution function...
a lot of <u>fancy-looking names</u> for things that are actually quite straightforward...ish. Have a go at
these to make sure you know who's who in the <u>glitzy world of discrete random variables</u>.

Warm-up Questions

1) The <u>probability distribution</u> of Y is:

y	0	1	2	3
$P(Y = y)$	0.5	k	k	$3k$

a) Find the value of k.   b) Find $P(Y < 2)$.

2) The probability distribution for the random variable W is given in the table.
Draw up a table to show the <u>cumulative distribution function</u>.

w	0.2	0.3	0.4	0.5
$P(W = w)$	0.2	0.2	0.3	0.3

3) The cumulative distribution function for a random variable R is given in the table.
Calculate the <u>probability distribution</u> for R. Find $P(0 \leq R \leq 1)$.

r	0	1	2
$F(r) = P(R \leqslant r)$	0.1	0.5	1

4) The discrete random variable X has a <u>uniform distribution</u>, $P(X = x) = k$ for $x = 0, 1, 2, 3$ and 4.
Find the value of k, and then find the <u>mean</u> and <u>variance</u> of X.

5) A <u>discrete random variable</u> X has the probability distribution shown in the table,
where k is a constant.

x_i	1	2	3	4
p_i	$\frac{1}{6}$	$\frac{1}{2}$	k	$\frac{5}{24}$

a) Find k b) Find $E(X)$ and show that $\text{Var}(X) = 63/64$ c) Find $E(2X - 1)$ and $\text{Var}(2X - 1)$

6) A <u>discrete random variable</u> X has the probability distribution shown in the table.

x_i	1	2	3	4	5	6
p_i	0.1	0.2	0.25	0.2	0.1	0.15

a) Find $E(X)$ b) Find $\text{Var}(X)$

Exam Questions

1   In a game a player tosses three fair coins.
If three heads occur then the player gets 20p; if two heads occur then the player gets 10p.
For any other outcome, the player gets nothing.

(a) If X is the random variable 'amount received', tabulate the probability distribution of X.

[4 marks]

The player pays 10p to play one game.

(b) Use the probability distribution to find the probability that the player wins
(i.e. gets more money than they pay to play) in one game.

[2 marks]

S1 Section 3 — Practice Questions

There's nothing I enjoy more than pretending I'm in an exam. An eerie silence, sweaty palms, having to be escorted to the toilet by a responsible adult... and all these lovely maths questions too. Just like the real thing.

Exam Questions

2   A discrete random variable X can only take values 0, 1, 2 and 3.
Its probability distribution is shown below.

x	0	1	2	3
P($X = x$)	$2k$	$3k$	k	k

    a)   Find the value of k.

[1 mark]

    b)   Draw up a table to show the cumulative distribution function for X.

[4 marks]

    c)   Find P($X > 2$).

[1 mark]

3   The random variable X takes the following values with equal probability:
0, 1, 2, 3, 4, 5, 6, 7, 8, 9.

    a)   Write down the probability distribution of X.

[1 mark]

    b)   Find the mean and variance of X.

[3 marks]

    c)   Calculate the probability that X is less than the mean.

[2 marks]

4   A discrete random variable X has the probability function:
P($X = x$) $= ax$ for $x = 1, 2, 3$, where a is a constant.

    a)   Show $a = \dfrac{1}{6}$.

[1 mark]

    b)   Find E(X).

[2 marks]

    c)   If Var(X) $= \dfrac{5}{9}$ find E(X^2).

[2 marks]

    d)   Find E($3X + 4$) and Var($3X + 4$).

[3 marks]

5   The number of points awarded to each contestant in a talent competition is given by the discrete random variable X with the following probability distribution:

x	0	1	2	3
P($X = x$)	0.4	0.3	0.2	0.1

    a) Find E(X).

[2 marks]

    b) Find E($6X + 8$).

[2 marks]

    c) Show that Var(X) $= 1$

[4 marks]

    d) Find Var($5 - 3X$).

[2 marks]

Correlation

Correlation is all about how closely two quantities are <u>linked</u>. And it can involve a fairly hefty formula.

Draw a **Scatter Diagram** to see **Patterns** in Data

Sometimes variables are measured in <u>pairs</u> — maybe because you want to find out <u>how closely</u> they're <u>linked</u>.
These pairs of variables might be things like: — '<u>my age</u>' and '<u>length of my feet</u>', or
 — '<u>temperature</u>' and '<u>number of accidents on a stretch of road</u>'.

You can plot readings from a pair of variables on a <u>scatter diagram</u> — this'll tell you something about the data.

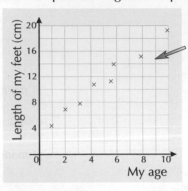

The variables 'my age' and 'length of my feet'
seem linked — all the points lie <u>close</u> to a <u>line</u>.
As I got older, my feet got bigger and bigger
(though I stopped measuring when I was 10).

It's a lot harder to see any connection between the
variables 'temperature' and 'number of accidents'
— the data seems <u>scattered</u> pretty much everywhere.

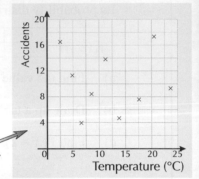

Correlation is a measure of **How Closely** variables are **Linked**

1) Sometimes, as one variable gets <u>bigger</u>, the other one also gets <u>bigger</u> — then the scatter diagram
might look like the one on the right. Here, a line of best fit would have a <u>positive gradient</u>.
The two variables are <u>positively correlated</u> (or there's a <u>positive correlation</u> between them).

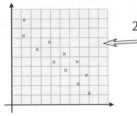

2) But if one variable gets <u>smaller</u> as the other one gets <u>bigger</u>,
then the scatter diagram might look like this one — and the
line of best fit would have a <u>negative gradient</u>.
The two variables are <u>negatively correlated</u> (or there's a
<u>negative correlation</u> between them).

3) And if the two variables <u>aren't</u> linked at all, you'd expect a <u>random</u>
scattering of points — it's hard to say where the line of best fit would be.
The variables <u>aren't correlated</u> (or there's <u>no correlation</u>).

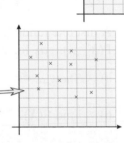

The **Product-Moment Correlation Coefficient (r)** measures Correlation

1) The <u>Product-Moment Correlation Coefficient</u> (<u>PMCC</u>, or <u>r</u>, for short) measures how close to a <u>straight line</u> the points
on a scatter graph lie.

2) The PMCC is always <u>between +1 and −1</u>.
If all your points lie <u>exactly</u> on a <u>straight line</u> with a <u>positive gradient</u> (perfect positive correlation), <u>r = +1</u>.
If all your points lie <u>exactly</u> on a <u>straight line</u> with a <u>negative gradient</u> (perfect negative correlation), <u>r = −1</u>.
(In reality, you'd never expect to get a PMCC of +1 or −1 — your scatter graph points might lie <u>pretty close</u> to a
straight line, but it's unlikely they'd all be <u>on</u> it.)

3) If r = 0 (or more likely, <u>pretty close</u> to 0), that would mean the variables <u>aren't correlated</u>.

4) The formula for the PMCC is a <u>real stinker</u>. But some calculators can work it out if you type in the pairs of readings,
which makes life easier. Otherwise, just take it nice and slow.

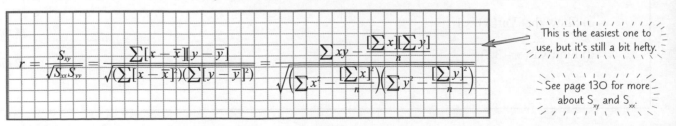

$$r = \frac{S_{xy}}{\sqrt{S_{xx}S_{yy}}} = \frac{\sum[x-\bar{x}][y-\bar{y}]}{\sqrt{(\sum[x-\bar{x}]^2)(\sum[y-\bar{y}]^2)}} = \frac{\sum xy - \frac{[\sum x][\sum y]}{n}}{\sqrt{\left(\sum x^2 - \frac{[\sum x]^2}{n}\right)\left(\sum y^2 - \frac{[\sum y]^2}{n}\right)}}$$

This is the easiest one to
use, but it's still a bit hefty.

See page 130 for more
about S_{xy} and S_{xx}.

Correlation

Don't rush questions on correlation. In fact, take your time and draw yourself a nice table.

EXAMPLE Illustrate the following data with a scatter diagram, and find the product-moment correlation coefficient (r) between the variables x and y. If $p = 4x - 3$ and $q = 9y + 17$, what is the PMCC between p and q?

x	1.6	2.0	2.1	2.1	2.5	2.8	2.9	3.3	3.4	3.8	4.1	4.4
y	11.4	11.8	11.5	12.2	12.5	12.0	12.9	13.4	12.8	13.4	14.2	14.3

1) The scatter diagram's the easy bit — just plot the points.

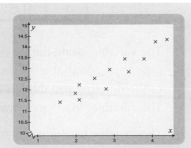

Now for the correlation coefficient. From the scatter diagram, the points lie pretty close to a straight line with a positive gradient — so if the correlation coefficient doesn't come out pretty close to +1, we'd need to worry...

2) There are 12 pairs of readings, so $n = 12$. That bit's easy — now you have to work out a load of sums. It's best to add a few extra rows to your table...

x	1.6	2	2.1	2.1	2.5	2.8	2.9	3.3	3.4	3.8	4.1	4.4	$35 = \Sigma x$
y	11.4	11.8	11.5	12.2	12.5	12	12.9	13.4	12.8	13.4	14.2	14.3	$152.4 = \Sigma y$
x^2	2.56	4	4.41	4.41	6.25	7.84	8.41	10.89	11.56	14.44	16.81	19.36	$110.94 = \Sigma x^2$
y^2	129.96	139.24	132.25	148.84	156.25	144	166.41	179.56	163.84	179.56	201.64	204.49	$1946.04 = \Sigma y^2$
xy	18.24	23.6	24.15	25.62	31.25	33.6	37.41	44.22	43.52	50.92	58.22	62.92	$453.67 = \Sigma xy$

Stick all these in the formula to get: $r = \dfrac{\left[453.67 - \dfrac{35 \times 152.4}{12} \right]}{\sqrt{\left[110.94 - \dfrac{35^2}{12} \right] \times \left[1946.04 - \dfrac{152.4^2}{12} \right]}} = \dfrac{9.17}{\sqrt{8.857 \times 10.56}} = \underline{0.948}$ (to 3 s.f.)

This is pretty close to 1, so there's a strong positive correlation between x and y.

3) Correlation coefficients aren't affected by linear transformations — you can multiply variables by a number, and add a number to them, and you won't change the PMCC between them.

So if p and q are given by $p = 4x - 3$ and $q = 9y + 17$, then the PMCC between p and q is also $\underline{0.948}$.

Don't make Sweeping Statements using Statistics

1) A high correlation coefficient doesn't necessarily mean that one factor causes the other.

EXAMPLE The number of televisions sold in Japan and the number of cars sold in America may well be correlated, but that doesn't mean that high TV sales in Japan cause high car sales in the US.

2) The PMCC is only a measure of a linear relationship between two variables (i.e. how close they'd be to a line if you plotted a scatter diagram).

EXAMPLE In the diagram on the right, the PMCC would be pretty low, but the two variables definitely look linked. It looks like the points lie on a parabola (the shape of an x^2 curve) — not a line.

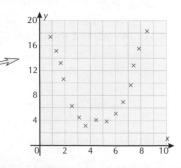

What's a statistician's favourite soap — Correlation Street... (Boom boom)

It's worth remembering that the PMCC assumes that both variables are normally distributed — chances are you won't get asked a question about that, but there's always the possibility that you might, so learn it.

Linear Regression

Linear regression is just fancy stats-speak for 'finding lines of best fit'. Not so scary now, eh...

Decide which is the Independent Variable and which is the Dependent

EXAMPLE The data below shows the load on a lorry, x (in tonnes), and the fuel efficiency, y (in km per litre).

x	5.1	5.6	5.9	6.3	6.8	7.4	7.8	8.5	9.1	9.8
y	9.6	9.5	8.6	8.0	7.8	6.8	6.7	6	5.4	5.4

1) The variable along the x-axis is the explanatory or independent variable — it's the variable you can control, or the one that you think is affecting the other. The variable 'load' goes along the x-axis here.

2) The variable up the y-axis is the response or dependent variable — it's the variable you think is being affected. In this example, this is the fuel efficiency.

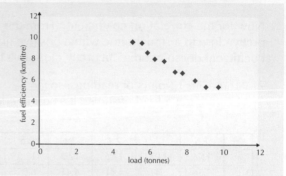

The Regression Line (Line of Best Fit) is in the form y = a + bx

To find the line of best fit for the above data you need to work out some sums. Then it's quite easy to work out the equation of the line. If your line of best fit is $y = a + bx$, this is what you do...

(1) First work out these four sums — a table is probably the best way: $\sum x$, $\sum y$, $\sum x^2$, $\sum xy$.

x	5.1	5.6	5.9	6.3	6.8	7.4	7.8	8.5	9.1	9.8	$72.3 = \sum x$
y	9.6	9.5	8.6	8	7.8	6.8	6.7	6	5.4	5.4	$73.8 = \sum y$
x^2	26.01	31.36	34.81	39.69	46.24	54.76	60.84	72.25	82.81	96.04	$544.81 = \sum x^2$
xy	48.96	53.2	50.74	50.4	53.04	50.32	52.26	51	49.14	52.92	$511.98 = \sum xy$

(2) Then work out S_{xy}, given by: $S_{xy} = \sum(x - \bar{x})(y - \bar{y}) = \sum xy - \frac{(\sum x)(\sum y)}{n}$

and S_{xx}, given by: $S_{xx} = \sum(x - \bar{x})^2 = \sum x^2 - \frac{(\sum x)^2}{n}$

These are the same as the terms used to work out the PMCC (see p. 128).

(3) The gradient (b) of your regression line is given by: $b = \frac{S_{xy}}{S_{xx}}$

(4) And the intercept (a) is given by: $a = \bar{y} - b\bar{x}$.

(5) Then the regression line is just: $y = a + bx$.

Loads of calculators will work out regression lines for you — but you still need to know this method, since they might give you just the sums from Step 1.

EXAMPLE Find the equation of the regression line of y on x for the data above.

The 'regression line of y on x' means that x is the independent variable, and y is the dependent variable.

1) Work out the sums: $\sum x = 72.3$, $\sum y = 73.8$, $\sum x^2 = 544.81$, $\sum xy = 511.98$.

2) Then work out S_{xy} and S_{xx}: $S_{xy} = 511.98 - \frac{72.3 \times 73.8}{10} = -21.594$, $S_{xx} = 544.81 - \frac{72.3^2}{10} = 22.081$

3) So the gradient of the regression line is: $b = \frac{-21.594}{22.081} = -0.978$ (to 3 sig. fig.)

Remember: $\bar{x} = \frac{\sum x}{n}$

4) And the intercept is: $a = \frac{\sum y}{n} - b\frac{\sum x}{n} = \frac{73.8}{10} - (-0.978) \times \frac{72.3}{10} = 14.451 = 14.5$ (to 3 sig. fig.)

5) This all means that your regression line is: $y = 14.5 - 0.978x$

The regression line always goes through the point $(\bar{x}, \bar{y})$.

This tells you: (i) for every extra tonne carried, you'd expect the lorry's fuel efficiency to fall by 0.978 km per litre, and (ii) with no load ($x = 0$), you'd expect the lorry to do 14.5 km per litre of fuel. Assuming the trend continues down to $x = 0$.

Linear Regression

Regression with **Coded Data** is a bit trickier

Examiners can be a bit sneaky so they might ask you to find a regression line for coded data. You could be asked to find the regression line of T on S, say, where $S = x + 15$ and $T = \frac{y}{10}$. Looks confusing, but the method's pretty straightforward. You know that $y = a + bx$ is the regression line for y on x, and you've already found values for a and b above. There's just a couple of extra steps.

1) Find expressions for x and y in terms of S and T. Here $x = S - 15$ and $y = 10T$. Substitute for x and y in $y = a + bx$ to find an expression for T. $\quad 10T = a + b(S - 15) \Rightarrow T = \frac{a - 15b}{10} + \frac{b}{10}S$

2) Now substitute your values for a and b into the expression $T = \frac{a - 15b}{10} + \frac{b}{10}S$ to find the regression line of T on S.

Residuals — the difference between **Practice** and **Theory**

A residual is the difference between an observed y-value and the y-value predicted by the regression line.

> Residual = Observed y-value − Estimated y-value

1) Residuals show the experimental error between the y-value that's observed and the y-value your regression line says it should be.

2) Residuals are shown by a vertical line from the actual point to the regression line.

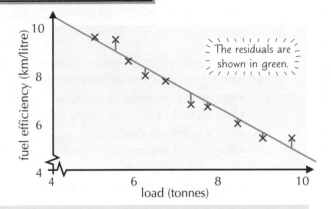

The residuals are shown in green.

EXAMPLE For the fuel efficiency example on the last page, calculate the residuals for: (i) $x = 5.6$, (ii) $x = 7.4$.

(i) When $x = 5.6$, the residual $= 9.5 - (-0.978 \times 5.6 + 14.451) = \underline{0.526}$ (to 3 sig. fig.)

(ii) When $x = 7.4$, the residual $= 6.8 - (-0.978 \times 7.4 + 14.451) = \underline{-0.414}$ (to 3 sig. fig.)

A positive residual means the regression line is too low for that value of x.
A negative residual means the regression line is too high.

> This kind of regression is called Least Squares Regression, because you're finding the equation of the line which minimises the sum of the squares of the residuals (i.e. $\sum e_k^2$ is as small as possible, where the e_k are the residuals).

Use Regression Lines **With Care**

You can use your regression line to predict values of y. But it's best not to do this for x-values outside the range of your original table of values.

EXAMPLE Use your regression equation to estimate the value of y when: (i) $x = 7.6$, (ii) $x = 12.6$

(i) When $x = 7.6$, $y = -0.978 \times 7.6 + 14.451 = \underline{7.02}$ (to 3 sig. fig.). This should be a pretty reliable guess, since $x = 7.6$ falls in the range of x we already have readings for — this is called interpolation.

(ii) When $x = 12.6$, $y = -0.978 \times 12.6 + 14.451 = \underline{2.13}$ (to 3 sig. fig.). This may well be unreliable since $x = 12.6$ is bigger than the biggest x-value we already have — this is called extrapolation.

99% of all statisticians make sweeping statements...

Be careful with that extrapolation business — it's like me saying that because I grew at an average rate of 10 cm a year for the first few years of my life, by the time I'm 50 I should be 5 metres tall. Residuals are always errors in the values of y — these equations for working out the regression line all assume that you can measure x perfectly all the time.

S1 Section 4 — Practice Questions

That was a short section, but chock-full of <u>fiddly terms</u> and <u>hefty equations</u>. The only way to learn all those details is by using them — so stretch your maths muscles and take a jog around this obstacle course of <u>practice questions</u>.

Warm-up Questions

1) The table below shows the results of some measurements concerning alcoholic cocktails. Here, x = total volume in ml, and y = percentage alcohol concentration by volume.

x	90	100	100	150	160	200	240	250	290	300
y	40	35	25	30	25	25	20	25	15	7

   a) Draw a scatter diagram representing this information.

   b) Calculate the product-moment correlation coefficient (PMCC) of these values.

   c) What does the PMCC tell you about these results?

2) For each pair of variables below, state which would be the dependent variable and which would be the independent variable.

   a) • the annual number of volleyball-related injuries
      • the annual number of sunny days

   b) • the annual number of rainy days
      • the annual number of Monopoly-related injuries

   c) • a person's disposable income
      • a person's spending on luxuries

   d) • the number of trips to the loo per day
      • the number of cups of tea drunk per day

   e) • the number of festival tickets sold
      • the number of pairs of Wellington boots bought

3) The radius in mm, r, and the weight in grams, w, of 10 randomly selected blueberry pancakes are given in the table below.

r	48.0	51.0	52.0	54.5	55.1	53.6	50.0	52.6	49.4	51.2
w	100	105	108	120	125	118	100	115	98	110

   a) Find: (i) $S_{rr} = \sum r^2 - \dfrac{\left(\sum r\right)^2}{n}$, (ii) $S_{rw} = \sum rw - \dfrac{\left(\sum r\right)\left(\sum w\right)}{n}$

   The regression line of w on r has equation $w = a + br$.

   b) Find b, the gradient of the regression line.

   c) Find a, the intercept of the regression line on the w-axis.

   d) Write down the equation of the regression line of w on r.

   e) Use your regression line to estimate the weight of a blueberry pancake of radius 60 mm.

   f) Comment on the reliability of your estimate, giving a reason for your answer.

4) Mmm, blueberry pancakes... must have another question about blueberry pancakes...

   The variables P and Q are defined as $P = r - 5$ and $Q = \dfrac{w}{8}$, where r and w are as used in Question 3.
   Find the regression line of Q on P.

S1 Section 4 — Practice Questions

Land ahoy, ye lily-livered yellow-bellies.
Just a quick heave-ho through these exam questions, then drop anchor and row ashore. Yaarrr, freedom...

Exam Questions

1   Values of two variables x and y are recorded in the table below.

x	1	2	3	4	5	6	7	8
y	0.50	0.70	0.10	0.82	0.50	0.36	0.16	0.80

a)   Represent this data on a scatter diagram.

[2 marks]

b)   Calculate the product-moment correlation coefficient (PMCC) between the two variables.

[4 marks]

c)   What does this value of the PMCC tell you about these variables?

[1 mark]

2   The following times (in seconds) were taken by eight different runners to complete distances of 20 metres and 60 metres.

Runner	A	B	C	D	E	F	G	H
20-metre time (x)	3.39	3.20	3.09	3.32	3.33	3.27	3.44	3.08
60-metre time (y)	8.78	7.73	8.28	8.25	8.91	8.59	8.90	8.05

a)   Plot a scatter diagram to represent the data.

[2 marks]

b)   Find the equation of the regression line of y on x, and plot it on your scatter diagram.

[8 marks]

c)   Use the equation of the regression line to estimate the value of y when:
(i) $x = 3.15$,   (ii) $x = 3.88$.
Comment on the reliability of your estimates.

[4 marks]

d)   Find the residuals for:
(i) $x = 3.32$ (ii) $x = 3.27$.

Illustrate them on your scatter diagram.

[4 marks]

3   A journalist at British Biking Monthly recorded the distance in miles, x, cycled by 20 different cyclists in the morning and the number of calories, y, eaten at lunch. The following summary statistics were provided:

$$S_{xx} = 310\ 880 \qquad S_{yy} = 788.95 \qquad S_{xy} = 12\ 666$$

a)   Use these values to calculate the product-moment correlation coefficient.

[2 marks]

b)   Give an interpretation of your answer to part (a).

[1 mark]

A Swedish cycling magazine calculated the product-moment correlation coefficient of the data after converting the distances to km.

c)   State the value of the product-moment correlation coefficient in this case.

[1 mark]

4   The equation of the line of regression for a set of data is $y = 211.599 + 9.602x$.

a)   Use the equation of the regression line to estimate the value of y when:
(i) $x = 12.5$ (ii) $x = 14.7$.

[2 marks]

b)   Calculate the residuals if the respective observed y-values were $y = 332.5$ and $y = 352.1$.

[2 marks]

Normal Distributions

The normal distribution is everywhere in statistics. Everywhere, I tell you. So learn this well...

The Normal Distribution is 'Bell-Shaped'

1) Loads of things in real life are most likely to fall 'somewhere in the middle', and are much less likely to take extremely high or extremely low values. In this kind of situation, you often get a normal distribution.

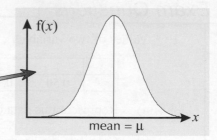

2) If you were to draw a graph showing how likely different values are, you'd end up with a graph that looks a bit like a bell. There's a peak in the middle at the mean (or expected value). And the graph is symmetrical — so values the same distance above and below the mean are equally likely.

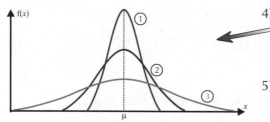

3) With a graph of a normal distribution, the probability of the random variable taking a value between two limits is the area under the graph between those limits. And since the total probability is 1, the total area under the graph must also be 1.

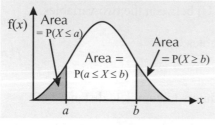

4) These three graphs all show normal distributions with the same mean (μ), but different variances (σ^2). Graph 1 has a small variance, and graph 3 has a larger variance — but the total area under all three curves is the same (= 1).

5) The most important normal distribution is the standard normal distribution, or Z — this has a mean of zero and a variance of 1.

Normal Distribution $N(\mu, \sigma^2)$

- If X is normally distributed with **mean** μ and **variance** σ^2, it's written $X \sim N(\mu, \sigma^2)$.
- The standard normal distribution Z has **mean** 0 and **variance** 1, i.e. $Z \sim N(0, 1)$.

Use *Tables* to Work out Probabilities of Z

Working out the area under a normal distribution curve is usually hard. But for Z, there are tables you can use (see p146). You look up a value of z and these tables (labelled $\Phi(z)$) tell you the probability that $Z \le z$.

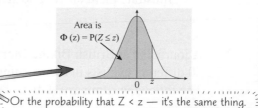

Area is
$\Phi(z) = P(Z \le z)$

Or the probability that $Z < z$ — it's the same thing.

EXAMPLE Find the probability that:
  a) $Z < 0.1$,   b) $Z \le 0.64$,   c) $Z > 0.23$,   d) $Z \ge -0.42$,   e) $Z \le -1.94$,   f) $0.12 < Z \le 0.82$

Tables only tell you the probability of Z being less than a particular value — use a sketch to work out anything else.

a) $P(Z < 0.1) = 0.5398$ ⟵ Just look up z = 0.1 in the tables.

b) $P(Z \le 0.64) = 0.7389$ ⟵ For continuous distributions like Z (i.e. with 'no gaps' between its possible values): $P(Z \le 0.64) = P(Z < 0.64)$.

c) $P(Z > 0.23) = 1 - P(Z \le 0.23) = 1 - 0.5910 = 0.4090$

d) $P(Z \ge -0.42) = P(Z \le 0.42) = 0.6628$ ⟵ Use the symmetry of the graph:

e) $P(Z \le -1.94) = P(Z \ge 1.94) = 1 - P(Z < 1.94) = 1 - 0.9738 = 0.0262$

f) $P(0.12 < Z \le 0.82) = P(Z \le 0.82) - P(Z \le 0.12)$
    $= 0.7939 - 0.5478 = 0.2461$

Again, draw a graph and use the symmetry:

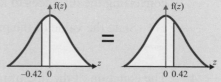

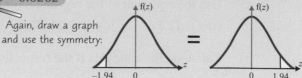

The Standard Normal Distribution, Z

You can use that big table of the normal distribution function ($\Phi(z)$) 'the other way round' as well — by starting with a probability and finding a value for z.

Use **Tables** to Find a **Value** of z if You're Given a **Probability**

EXAMPLE If $P(Z < z) = 0.9554$, then what is the value of z?

Using the table for $\Phi(z)$ (the normal cumulative distribution function), $z = 1.70$.

All the probabilities in the table of $\Phi(z)$ are greater than 0.5, but you can still use the tables with values less than this. You'll most likely need to subtract the probability from 1, and then use a sketch.

EXAMPLE If $P(Z < z) = 0.2611$, then what is the value of z?

① Subtract from 1 to get a probability greater than 0.5: $1 - 0.2611 = 0.7389$

② If $P(Z < z) = 0.7389$, then from the table, $z = 0.64$.

③ So if $P(Z < z) = 0.2611$, then, $z = -0.64$.

If $P(Z < z) = 0.2611$, then z must be negative.

The **Percentage Points** Table also Tells You z if You're Given a **Probability**

You use the percentage points table in a similar way (you start with a probability, and look up a value for z). But this time, the probability you start with is the probability that z is greater than a certain number.

EXAMPLE If $P(Z > z) = 0.15$, then what is the value of z?

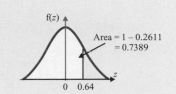
In the examples above, you started with the probability that z was less than a particular number.

Using the percentage points table, $z = 1.0364$.

You might need to use a bit of imagination and a sketch to get the most out of the percentage points table.

EXAMPLE Find z if: a) $P(Z < z) = 0.05$, b) $P(Z < z) = 0.9$, c) $P(Z \geq z) = 0.01$

a) Using the percentage points table, if $P(Z > z) = 0.05$, then $z = 1.6449$.
 So if $P(Z < z) = 0.05$, then z must equal -1.6449.

b) If $P(Z < z) = 0.9$, then $P(Z > z) = 0.1$.
 Using the percentage points table, $z = 1.2816$.

c) If $P(Z \geq z) = 0.01$, then $z = 2.3263$.

Again, you can treat the "greater than or equals sign" as a "greater than sign", since for a continuous distribution, $P(Z \geq z) = P(Z > z)$.

The medium of a random variable follows a paranormal distribution...

It's definitely worth sketching the graph when you're finding probabilities using a normal distribution — you're much less likely to make a daft mistake. So even if the question looks a simple one, draw a quick sketch — it's probably worth it.

Normal Distributions and Z-Tables

<u>All</u> normally-distributed variables can be transformed to Z — which is a marvellous thing.

Transform to Z by **Subtracting** μ, then **dividing by** σ

1) You can convert <u>any</u> normally-distributed variable to Z by:

    i)   <u>subtracting the mean</u>, and then

    ii)   <u>dividing by the standard deviation</u>.

> This means that if you subtract μ from any numbers in the question and then divide by σ — you can use your tables for Z

$$\text{If } X \sim N(\mu, \sigma^2), \text{ then } \frac{X - \mu}{\sigma} = Z, \text{ where } Z \sim N(0,1)$$

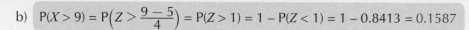

2) Once you've transformed a variable like this, you can use the <u>Z-tables</u>.

EXAMPLE   If $X \sim N(5, 16)$ find: a) $P(X < 7)$,   b) $P(X > 9)$,   c) $P(5 < X < 11)$

Subtract μ (= 5) from any numbers and divide by σ (= $\sqrt{16}$ = 4) — then you'll have a value for $Z \sim N(0, 1)$.

a) $P(X < 7) = P\left(Z < \frac{7 - 5}{4}\right) = P(Z < 0.5) = 0.6915$

> N(5, 16) means the <u>variance</u> is 16 — take the <u>square root</u> to find the <u>standard deviation</u>.

> Look up $P(Z < 0.5)$ in the big table on p.146.

b) $P(X > 9) = P\left(Z > \frac{9 - 5}{4}\right) = P(Z > 1) = 1 - P(Z < 1) = 1 - 0.8413 = 0.1587$

c) $P(5 < X < 11) = P\left(\frac{5 - 5}{4} < Z < \frac{11 - 5}{4}\right) = P(0 < Z < 1.5)$

     $= P(Z < 1.5) - P(Z < 0) = 0.9332 - 0.5 = 0.4332$

> Find the area to the left of 1.5 and subtract the area to the left of 0.

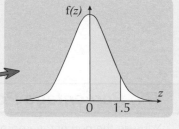

The **Z-Distribution** Can be Used in **Real-Life** Situations

EXAMPLE   The times taken by a group of people to complete an assault course are normally distributed with a mean of 600 seconds and a variance of 105 seconds. Find the probability that a randomly selected person took:
a) fewer than 575 seconds,   b) more than 620 seconds.

If X represents the time taken in seconds, then $X \sim N(600, 105)$.
It's a normal distribution — so your first thought should be to try and '<u>standardise</u>' it by converting it to Z.

a) <u>Subtract the mean</u> and <u>divide by the standard deviation</u>: $P(X < 575) = P\left(\frac{X - 600}{\sqrt{105}} < \frac{575 - 600}{\sqrt{105}}\right)$

                                      $= P(Z < -2.44)$

   So $P(Z < -2.44) = 1 - P(Z \le 2.44) = 1 - 0.9927 = 0.0073$.

b) Again, <u>subtract the mean</u> and <u>divide by the standard deviation</u>: $P(X > 620) = P\left(\frac{X - 600}{\sqrt{105}} > \frac{620 - 600}{\sqrt{105}}\right)$

                                        $= P(Z > 1.95)$

                                        $= 1 - P(Z \le 1.95)$

                                        $= 1 - 0.9744 = 0.0256$

Transform to Z and use Z-tables — I repeat: transform to Z and use Z-tables...

The basic idea is always the same — transform your normally-distributed variable to Z, and then use the Z-tables. Statisticians call this the "normal two-step". Well... some of them probably do, anyway. I admit, this stuff is all a bit weird and confusing at first. But as always, work through a few examples and it'll start to click. So get some practice.

Normal Distributions and Z-Tables

You might be given some <u>probabilities</u> and asked to find μ and σ. Just use the same old ideas...

Find μ and σ by First **Transforming** to Z

EXAMPLE $X \sim N(\mu, 2^2)$ and $P(X < 23) = 0.9015$. Find μ.

This is a normal distribution — so your first thought should be to convert it to Z.

(1) $P(X < 23) = P[\dfrac{X - \mu}{2} < \dfrac{23 - \mu}{2}] = P[Z < \dfrac{23 - \mu}{2}] = 0.9015$

Substitute Z for $\dfrac{X - \mu}{\sigma}$.

But 0.9015 isn't in the percentage points table, so you'll need to look it up in the 'big' table instead (the one for Φ(z)).

(2)   If $P(Z < z) = 0.9015$, then $z = 1.29$

(3)   So $\dfrac{23 - \mu}{2} = 1.29$ — now solve this to find $\mu = 23 - (2 \times 1.29) = 20.42$

EXAMPLE $X \sim N(53, \sigma^2)$ and $P(X < 50) = 0.2$. Find σ.

Again, this is a normal distribution — so you need to use that lovely <u>standardising equation</u> again.

(1) $P(X < 50) = P[Z < \dfrac{50 - 53}{\sigma}] = P[Z < -\dfrac{3}{\sigma}] = 0.2$, and so $P[Z > -\dfrac{3}{\sigma}] = 0.8$

Ideally, you'd look up 0.8 in the percentage points table to find $-\dfrac{3}{\sigma}$. Unfortunately, it isn't there, so you have to think a bit...

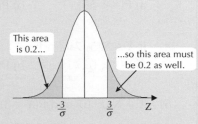

This area is 0.2...   ...so this area must be 0.2 as well.

(2) $P[Z > -\dfrac{3}{\sigma}]$ is 0.8, so from the symmetry of the graph, $P[Z > \dfrac{3}{\sigma}]$ must be 0.2 .

So look up 0.2 in the 'percentage points' table to find that an area of 0.2 is to the right of $z = 0.8416$.

This tells you that $\dfrac{3}{\sigma} = 0.8416$, or $\sigma = 3.56$ (to 3 sig. fig.)

If You Have to Find μ **and** σ, You'll Need to Solve **Simultaneous Equations**

EXAMPLE   The random variable $X \sim N(\mu, \sigma^2)$. If $P(X < 9) = 0.5596$ and $P(X > 14) = 0.0322$, then find μ and σ.

(1) $P(X < 9) = P[Z < \dfrac{9 - \mu}{\sigma}] = 0.5596$.

Using the table for Φ(z), this tells you that $\dfrac{9 - \mu}{\sigma} = 0.15$, or $9 - \mu = 0.15\sigma$.

(2) $P(X > 14) = P[Z > \dfrac{14 - \mu}{\sigma}] = 0.0322$, which means that $P[Z < \dfrac{14 - \mu}{\sigma}] = 1 - 0.0322 = 0.9678$.

Using the table for Φ(z), this tells you that $\dfrac{14 - \mu}{\sigma} = 1.85$, or $14 - \mu = 1.85\sigma$.

(3)   Subtract the equations: $(14 - \mu) - (9 - \mu) = 1.85\sigma - 0.15\sigma$, or $5 = 1.7\sigma$. This gives $\sigma = 5 \div 1.7 = 2.94$.
Now use one of the other equations to find μ: $\mu = 9 - (0.15 \times 2.94) = 8.56$.

The Norman Distribution — came to England in 1066...

It's always the same — you always need to do the <u>normal two-step</u> of subtracting the mean and dividing by the standard deviation. Just make sure you don't use the <u>variance</u> by mistake — remember, in $N(\mu, \sigma^2)$, the second number always shows the variance. I'm sure you wouldn't make a mistake... it's just that I'm such a worrier and I do so want you to do well.

S1 Section 5 — Practice Questions

Normal distributions... there's nothing terribly hard about them. They just get <u>fiddly and annoying</u>.
But you need to know how to answer normal distribution questions, so here are some to <u>practise</u> with.

Warm-up Questions

1) Find the probability that:
   a) $Z < 0.84$,
   b) $Z < 2.95$,
   c) $Z > 0.68$,
   d) $Z \geq 1.55$,
   e) $Z < -2.10$,
   f) $Z \leq -0.01$,
   g) $Z > 0.10$,
   h) $Z \leq 0.64$,
   i) $Z > 0.23$,
   j) $0.10 < Z \leq 0.50$,
   k) $-0.62 \leq Z < 1.10$,
   l) $-0.99 < Z \leq -0.74$.

2) Find the value of z if:
   a) $P(Z < z) = 0.9131$,
   b) $P(Z < z) = 0.5871$,
   c) $P(Z > z) = 0.0359$,
   d) $P(Z > z) = 0.01$,
   e) $P(Z \leq z) = 0.4013$,
   f) $P(Z \geq z) = 0.995$.

3) If $X \sim N(50, 16)$ find:
   a) $P(X < 55)$,
   b) $P(X < 42)$,
   c) $P(X > 56)$
   d) $P(47 < X < 57)$.

4) If $X \sim N(5, 7^2)$ find:
   a) $P(X < 0)$,
   b) $P(X < 1)$,
   c) $P(X > 7)$
   d) $P(2 < X < 4)$.

5) $X \sim N(\mu, 10)$ and $P(X < 8) = 0.8925$. Find μ.

6) $X \sim N(\mu, 8^2)$ and $P(X > 221) = 0.3085$. Find μ.

7) $X \sim N(11, \sigma^2)$ and $P(X < 13) = 0.6$. Find σ.

8) $X \sim N(108, \sigma^2)$ and $P(X \leq 110) = 0.9678$. Find σ.

9) The random variable $X \sim N(\mu, \sigma^2)$.
   If $P(X < 15.2) = 0.9783$ and $P(X > 14.8) = 0.1056$, then find μ and σ.

10) The mass of items produced by a factory is normally distributed with a mean of 55 grams and a standard deviation of 4.4 grams. Find the probability of a randomly chosen item having a mass of:
    a) less than 55 grams,
    b) less than 50 grams,
    c) more than 60 grams.

11) The mass of eggs laid by an ostrich is normally distributed with a mean of 1.4 kg and a standard deviation of 300 g. Find the probability of a randomly chosen egg from this bird having a mass of:
    a) less than 1 kg,
    b) more than 1.5 kg,
    c) between 1300 and 1600 g.

12) $X \sim N(80, 15)$
    a) If $P(X < a) = 0.99$, find a.
    b) If $P(|X - 80| < b) = 0.8$, find b.

S1 Section 5 — Practice Questions

The last questions for S1. Apart from the practice exams, I mean. Still... be positive.

1 The exam marks for 1000 candidates are normally distributed
with mean 50 marks and standard deviation 30 marks.

    a)   The pass mark is 41. Estimate the number of candidates who passed the exam.

        [3 marks]

    b)   Find the mark needed for a distinction if the top 10% of the candidates achieved a distinction.

        [3 marks]

2 The lifetimes of a particular type of battery are normally distributed with mean μ
and standard deviation σ. A student using these batteries finds that 40% last less
than 20 hours and 80% last less than 30 hours. Find μ and σ.

        [7 marks]

3 The random variable X has a normal distribution with mean 120 and standard deviation 25.

    a)   Find $P(X > 145)$.

        [3 marks]

    b)   Find the value of j such that $P(120 < X < j) = 0.4641$

        [4 marks]

4 The diameters of the pizza bases made at a restaurant are normally distributed.
The mean diameter is 12 inches, and 5% of the bases measure more than 13 inches.

    a)   Write down the median diameter of the pizza bases.

        [1 mark]

    b)   Find the standard deviation of the diameters of the pizza bases.

        [4 marks]

Any pizza base with a diameter of less than 10.8 inches is considered too small and is discarded.

    c)   If 100 pizza bases are made in an evening, approximately how many
would you expect to be discarded due to being too small?

        [3 marks]

Three pizza bases are selected at random.

    d)   Find the probability that at least one of these bases is too small.

        [3 marks]

5 A garden centre sells bags of compost. The volume of compost in the
bags is normally distributed with a mean of 50 litres.

    a)   If the standard deviation of the volume is 0.4 litres, find the probability that
a randomly selected bag will contain less than 49 litres of compost.

        [3 marks]

    b)   If 1000 of these bags of compost are bought, find the expected
number of bags containing more than 50.5 litres of compost.

        [5 marks]

A different garden centre sells bags of similar compost. The volume of compost, in litres, in these
bags is described by the random variable Y, where $Y \sim N(75, \sigma^2)$. It is found that 10% of the bags
from this garden centre contain less than 74 litres of compost.

    c)   Find σ.

        [3 marks]

General Certificate of Education
Advanced Subsidiary (AS) and Advanced Level

Statistics S1 — Practice Exam One

Time Allowed: 1 hour 30 min

Calculators may be used for this exam (except those with facilities for symbolic algebra, differentiation or integration).

Give any non-exact numerical answers to an appropriate degree of accuracy.

Statistical tables can be found on page 146.
Values used from these tables must be quoted in full.

There are 75 marks available for this paper.

1   The heights of giraffes in a game reserve were measured.
    This data is summarised in the box plot below.

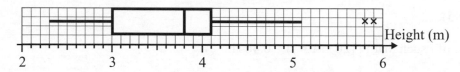

   a)   What height do only 25% of the giraffes exceed?

   (1 mark)

   b)   Two heights are marked with crosses. Explain why these two heights are marked in this way.

   (2 marks)

   The heights of giraffes living in a zoo in the same country range from 2.8 m to 5.8 m.
   The quartiles of their heights are 3.4 m, 4.1 m, and 5.2 m.

   c)   Draw a box plot below to represent the heights of the giraffes in the zoo.

   (4 marks)

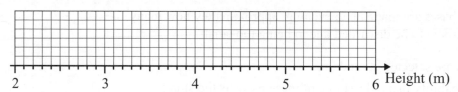

   d)   Compare the heights of these two groups of giraffes.

   (4 marks)

2   The events A and B are mutually exclusive.
    P(A) = 0.3 and P(B) = 0.4.

   a)   Write down $P(A \cap B)$.

   (1 mark)

   b)   Find $P(A \cup B)$.

   (2 marks)

   c)   Hence find the probability that neither event happens.

   (2 marks)

   d)   Write down $P(A \mid B)$. Explain your answer.

   (2 marks)

3   The probability distribution for the cash prizes, x pence, offered by a gambling machine is as follows:

x	0	10	20	50	100
$P(X = x)$	0.2	0.2	0.2	0.2	p

a)   Write down the value of p.

(1 mark)

b)   State the name of the distribution of X.

(1 mark)

c)   Find F(25).

(2 marks)

d)   Find $E(X)$ and $Var(X)$.

(4 marks)

e)   Find $E(3X - 4)$ and $Var(3X - 4)$.

(4 marks)

f)   The owner of the machine charges 40p per game. Comment on this cost.

(2 marks)

g)   Comment on whether the above distribution is really likely to be used in gambling machines.

(2 marks)

4   The duration in minutes, X, of a car wash is a little erratic, but it is normally distributed with a mean of 8 minutes and a variance of 1.2 minutes. Find:

a) $P(X < 7.5)$,

(3 marks)

b)   the probability that the duration deviates from the mean by more than 1 minute,

(4 marks)

c)   the duration in minutes, d, such that there is no more than a 1% probability that the car wash will take longer than this duration.

(4 marks)

5   Of 30 drivers interviewed, 9 have been involved in a car crash at some time.
Of those who have been involved in a crash, 5 wear glasses. The probability of wearing glasses, given that the driver has not had a car accident, is $\frac{1}{3}$.

a)   Represent this information in a tree diagram, giving all probabilities as fractions in their simplest form.

(3 marks)

b)   What is the probability that a person chosen at random wears glasses?

(3 marks)

c)   What is the probability that a glasses wearer has been a crash victim?

(3 marks)

6 A group of 10 friends play a round of minigolf.
   Their total score ($\sum x$) is 500 and $\sum x^2 = 25622$.

   a) Find the mean, μ, and the standard deviation, σ, for this data.

   (3 marks)

   b) Another friend wants to incorporate his score of 50. Without further calculation and giving reasons, explain the effect of adding this score on:

      (i) the mean,

      (2 marks)

      (ii) the standard deviation.

      (2 marks)

7 A teacher collects the following data showing students' marks in an examination (*M*) and the amount of revision undertaken in hours (*R*).

R	12	10	9	5	14	11	12	6
M	88	72	65	59	92	75	80	69

   ($\Sigma R = 79$, $\Sigma M = 600$, $\Sigma R^2 = 847$, $\Sigma M^2 = 45884$ and $\Sigma RM = 6143$.)

   a) Calculate S_{RR}, S_{MM} and S_{RM}.

   (3 marks)

   b) Calculate the product-moment correlation coefficient for *M* and *R*.

   (2 marks)

   c) Which of *M* and *R* is the explanatory variable? Explain your answer.

   (2 marks)

   d) The teacher believes she can fit a linear regression line $M = a + bR$ to the data.
      Give one reason to support her conclusion.

   (1 mark)

   e) Find the equation of the regression line of *M* on *R* in the form $M = a + bR$.

   (4 marks)

   f) The teacher must estimate the examination mark of a student who did not take the examination.
      The teacher knows this student to have done 4 hours of revision.
      Estimate the likely mark for this student.

   (1 mark)

   g) Comment on the reliability of your estimate in f).

   (1 mark)

General Certificate of Education
Advanced Subsidiary (AS) and Advanced Level

Statistics S1 — Practice Exam Two

Time Allowed: 1 hour 30 min

Calculators may be used for this exam (except those with
facilities for symbolic algebra, differentiation or integration).

Give any non-exact numerical answers to an appropriate degree of accuracy.

Statistical tables can be found on page 146.
Values used from these tables must be quoted in full.

There are 75 marks available for this paper.

1   A box of chocolates contains 20 chocolates, all of which are either hard or soft centred.
Some of the chocolates contain nuts. 13 chocolates have hard centres, of which 6 contain nuts.
There are 10 nutty chocolates in total.

   a)   Represent the data in a Venn diagram.

   (3 marks)

   b)   A chocolate is selected at random. Find the probability of:
   (i)   it having a soft centre.

   (1 mark)

   (ii)   it having a hard centre, given that it contains a nut.

   (2 marks)

   c)   If 3 chocolates are selected at random without replacement, find the probability that
   exactly one has a hard centre.

   (3 marks)

2   The discrete random variable X has the probability function shown below:

$$P(X = x) = \begin{cases} \dfrac{kx}{6} & \text{for } x = 1, 2, 3 \\[2mm] \dfrac{k(7-x)}{6} & \text{for } x = 4, 5, 6 \\[2mm] 0 & \text{otherwise} \end{cases}$$

   a)   Find the value of k.

   (3 marks)

   b)   Find F(3).

   (2 marks)

   c)   Show that $E(X) = 3.5$.

   (2 marks)

   d)   Given that $\text{Var}(X) = \dfrac{23}{12}$, find:

   (i) $E(2 - 3X)$ and $\text{Var}(2 - 3X)$,

   (4 marks)

   (ii) $E(X^2)$.

   (3 marks)

3   The sales figures of a gift shop for a 12-week period are shown below.

Week	1	2	3	4	5	6	7	8	9	10	11	12
Sales (£'000s)	5.5	4.2	5.8	9.1	3.8	4.6	6.4	6.2	4.9	5.9	6.0	4.1

(You may use $\Sigma x = 66.5$, and $\Sigma x^2 = 390.97$.)

a)   Find the mean and variance of the weekly sales.

(4 marks)

b)   Find the median and quartiles of the sales data.

(3 marks)

c)   (i)   Determine the quartile coefficient of skewness for the data, given by $\dfrac{Q_3 - 2Q_2 + Q_1}{Q_3 - Q_1}$.

(1 mark)

     (ii)   Describe the skewness of the sales data.

(2 marks)

This 12-week period included Christmas. The shop's manager wants to exclude any outliers from his analysis of the data to get a more realistic idea of how the shop is performing.

He decides to define outliers as readings satisfying either of the conditions below:

- below $Q_1 - 1.5 \times (Q_3 - Q_1)$ • above $Q_3 + 1.5 \times (Q_3 - Q_1)$.

d)   Identify any outliers in the data. Show your working.

(4 marks)

4   The temperature, $T\,°C$, reached by water in a kettle before the kettle switches off is normally distributed, with a mean of 93 °C. On 20% of occasions, the water reaches 95 °C before switching off.

a)   Write down the median temperature reached. Explain your answer.

(2 marks)

b)   Find the standard deviation of the switch-off temperature's distribution.

(4 marks)

c)   In a restaurant, tea made with water at less than 88 °C will attract complaints.
      Calculate the probability that a customer will complain if this kettle is used.

(2 marks)

d)   Find d such that $P(93 - d \leq T \leq 93 + d) = 0.99$.

(4 marks)

5   A construction company measures the length, y metres, of a cable when put under different amounts of tension (T, measured in kilonewtons, kN). The results of its tests are shown below.

T (kN)	1	2	3	5	8	10	15	20
y (metres)	3.05	3.1	3.13	3.15	3.27	3.4	3.5	3.6

a)   Draw a scatter graph to show these results.

(2 marks)

b)   Calculate S_{TT} and S_{Ty}.
(You may use $\Sigma T^2 = 828$ and $\Sigma Ty = 219.05$.)

(2 marks)

An engineer believes a linear regression line of the form $y = a + bT$ could be found to accurately describe the results.

c)   Find the equation of this regression line.

(4 marks)

d)   Explain what your values of a and b represent.

(2 marks)

e)   Use your regression line to predict the length of the cable when put under a tension of 30 kilonewtons.

(2 marks)

f)   Comment on the reliability of the estimate in e).

(1 mark)

6   The random variable X has the following probability distribution.

x	2	4	6	8	10
$P(X = x)$	0.1	0.2	p	q	0.2

The mean of the random variable X is 6.3.

a)   (i)   Write down two independent equations involving p and q.

(2 marks)

      (ii)   Solve your equations to find the values of p and q.

(3 marks)

b)   (i)   Find $E(X^2)$.

(2 marks)

      (ii)   Find $Var(X)$.

(2 marks)

c)   Draw up a table to show the cumulative distribution function, $F(x)$, for X.

(4 marks)

The normal distribution function

The cumulative distribution function $\Phi(z)$ is tabulated below. This is defined as $\Phi(z) = \frac{1}{\sqrt{2\pi}} \int_{-\infty}^{z} e^{-\frac{1}{2}t^2} dt$.

z	$\Phi(z)$	z	$\Phi(z)$	z	$\Phi(z)$	z	$\Phi(z)$	z	$\Phi(z)$
0.00	0.5000	0.50	0.6915	1.00	0.8413	1.50	0.9332	2.00	0.9772
0.01	0.5040	0.51	0.6950	1.01	0.8438	1.51	0.9345	2.02	0.9783
0.02	0.5080	0.52	0.6985	1.02	0.8461	1.52	0.9357	2.04	0.9793
0.03	0.5120	0.53	0.7019	1.03	0.8485	1.53	0.9370	2.06	0.9803
0.04	0.5160	0.54	0.7054	1.04	0.8508	1.54	0.9382	2.08	0.9812
0.05	0.5199	0.55	0.7088	1.05	0.8531	1.55	0.9394	2.10	0.9821
0.06	0.5239	0.56	0.7123	1.06	0.8554	1.56	0.9406	2.12	0.9830
0.07	0.5279	0.57	0.7157	1.07	0.8577	1.57	0.9418	2.14	0.9838
0.08	0.5319	0.58	0.7190	1.08	0.8599	1.58	0.9429	2.16	0.9846
0.09	0.5359	0.59	0.7224	1.09	0.8621	1.59	0.9441	2.18	0.9854
0.10	0.5398	0.60	0.7257	1.10	0.8643	1.60	0.9452	2.20	0.9861
0.11	0.5438	0.61	0.7291	1.11	0.8665	1.61	0.9463	2.22	0.9868
0.12	0.5478	0.62	0.7324	1.12	0.8686	1.62	0.9474	2.24	0.9875
0.13	0.5517	0.63	0.7357	1.13	0.8708	1.63	0.9484	2.26	0.9881
0.14	0.5557	0.64	0.7389	1.14	0.8729	1.64	0.9495	2.28	0.9887
0.15	0.5596	0.65	0.7422	1.15	0.8749	1.65	0.9505	2.30	0.9893
0.16	0.5636	0.66	0.7454	1.16	0.8770	1.66	0.9515	2.32	0.9898
0.17	0.5675	0.67	0.7486	1.17	0.8790	1.67	0.9525	2.34	0.9904
0.18	0.5714	0.68	0.7517	1.18	0.8810	1.68	0.9535	2.36	0.9909
0.19	0.5753	0.69	0.7549	1.19	0.8830	1.69	0.9545	2.38	0.9913
0.20	0.5793	0.70	0.7580	1.20	0.8849	1.70	0.9554	2.40	0.9918
0.21	0.5832	0.71	0.7611	1.21	0.8869	1.71	0.9564	2.42	0.9922
0.22	0.5871	0.72	0.7642	1.22	0.8888	1.72	0.9573	2.44	0.9927
0.23	0.5910	0.73	0.7673	1.23	0.8907	1.73	0.9582	2.46	0.9931
0.24	0.5948	0.74	0.7704	1.24	0.8925	1.74	0.9591	2.48	0.9934
0.25	0.5987	0.75	0.7734	1.25	0.8944	1.75	0.9599	2.50	0.9938
0.26	0.6026	0.76	0.7764	1.26	0.8962	1.76	0.9608	2.55	0.9946
0.27	0.6064	0.77	0.7794	1.27	0.8980	1.77	0.9616	2.60	0.9953
0.28	0.6103	0.78	0.7823	1.28	0.8997	1.78	0.9625	2.65	0.9960
0.29	0.6141	0.79	0.7852	1.29	0.9015	1.79	0.9633	2.70	0.9965
0.30	0.6179	0.80	0.7881	1.30	0.9032	1.80	0.9641	2.75	0.9970
0.31	0.6217	0.81	0.7910	1.31	0.9049	1.81	0.9649	2.80	0.9974
0.32	0.6255	0.82	0.7939	1.32	0.9066	1.82	0.9656	2.85	0.9978
0.33	0.6293	0.83	0.7967	1.33	0.9082	1.83	0.9664	2.90	0.9981
0.34	0.6331	0.84	0.7995	1.34	0.9099	1.84	0.9671	2.95	0.9984
0.35	0.6368	0.85	0.8023	1.35	0.9115	1.85	0.9678	3.00	0.9987
0.36	0.6406	0.86	0.8051	1.36	0.9131	1.86	0.9686	3.05	0.9989
0.37	0.6443	0.87	0.8078	1.37	0.9147	1.87	0.9693	3.10	0.9990
0.38	0.6480	0.88	0.8106	1.38	0.9162	1.88	0.9699	3.15	0.9992
0.39	0.6517	0.89	0.8133	1.39	0.9177	1.89	0.9706	3.20	0.9993
0.40	0.6554	0.90	0.8159	1.40	0.9192	1.90	0.9713	3.25	0.9994
0.41	0.6591	0.91	0.8186	1.41	0.9207	1.91	0.9719	3.30	0.9995
0.42	0.6628	0.92	0.8212	1.42	0.9222	1.92	0.9726	3.35	0.9996
0.43	0.6664	0.93	0.8238	1.43	0.9236	1.93	0.9732	3.40	0.9997
0.44	0.6700	0.94	0.8264	1.44	0.9251	1.94	0.9738	3.50	0.9998
0.45	0.6736	0.95	0.8289	1.45	0.9265	1.95	0.9744	3.60	0.9998
0.46	0.6772	0.96	0.8315	1.46	0.9279	1.96	0.9750	3.70	0.9999
0.47	0.6808	0.97	0.8340	1.47	0.9292	1.97	0.9756	3.80	0.9999
0.48	0.6844	0.98	0.8365	1.48	0.9306	1.98	0.9761	3.90	1.0000
0.49	0.6879	0.99	0.8389	1.49	0.9319	1.99	0.9767	4.00	1.0000
0.50	0.6915	1.00	0.8413	1.50	0.9332	2.00	0.9772		

Percentage points of the normal distribution

The z-values in the table are those which a random variable

$Z \sim N(0, 1)$ exceeds with probability p, i.e. $P(Z > z) = 1 - \Phi(z) = p$.

p	z	p	z
0.5000	0.0000	0.0500	1.6449
0.4000	0.2533	0.0250	1.9600
0.3000	0.5244	0.0100	2.3263
0.2000	0.8416	0.0050	2.5758
0.1500	1.0364	0.0010	3.0902
0.1000	1.2816	0.0005	3.2905

Constant Acceleration Equations

Welcome to the technicolour world of Mechanics 1. Fashions may change, but there will <u>always</u> be M1 questions that involve objects travelling in a <u>straight line</u>. It's just a case of picking the right equations to solve the problem.

There are **Five Constant Acceleration Equations**

Examiners call these "<u>uvast</u>" questions (pronounced ewe-vast, like a large sheep) because of the five variables involved:

u = <u>initial speed</u> (or <u>velocity</u>) in ms^{-1}

v = <u>final speed</u> (or <u>velocity</u>) in ms^{-1}

a = <u>acceleration</u> in ms^{-2}

s = <u>distance travelled</u> (or <u>displacement</u>) in m

t = <u>time</u> that passes in s (seconds)

The acceleration must be <u>constant</u>.

The constant acceleration equations are:

$$v = u + at$$
$$s = ut + \tfrac{1}{2}at^2$$
$$s = \tfrac{1}{2}(u + v)t$$
$$v^2 = u^2 + 2as$$
$$s = vt - \tfrac{1}{2}at^2$$

None of those equations are in the formula book, so you're going to have to <u>learn them</u>. Questions will usually give you <u>three variables</u> — your job is to choose the equation that will find you the missing <u>fourth variable</u>.

EXAMPLE

A jet ski travels in a straight line along a river. It passes under two bridges 200 m apart and is observed to be travelling at 5 ms^{-1} under the first bridge and at 9 ms^{-1} under the second bridge. Calculate its acceleration (assuming it is constant).

List the variables ("<u>uvast</u>"):

$u = 5$

$v = 9$ *You have to work out a.*

$a = a$

$s = 200$

$t = t$ *You're not told about the time taken.*

Choose the equation with u, v, s and a in it: $v^2 = u^2 + 2as$

Check you're using the right <u>units</u> — m, s, ms^{-1} and ms^{-2}.

<u>Substitute</u> values: $9^2 = 5^2 + (2 \times a \times 200)$

<u>Simplify</u>: $81 = 25 + 400a$

<u>Rearrange</u>: $400a = 81 - 25 = 56$

Then <u>solve</u>: $a = \dfrac{56}{400} = 0.14\,\text{ms}^{-2}$

Motion under Gravity just means taking $a = g$...

Don't be put off by questions involving objects <u>moving freely under gravity</u> — they're just telling you the <u>acceleration is g</u>.

Use the value of g given on the front of the paper or in the question. If you don't, you risk losing a mark because your answer won't match the examiners' answer.

EXAMPLE

A pebble is dropped into a well 18 m deep and moves freely under gravity until it hits the bottom. Calculate the time it takes to reach the bottom. (Take g = 9.8 ms^{-2}.)

First, list the variables:

$u = 0$ *Because the pebble was <u>dropped</u>, not thrown.*

$v = v$

$a = 9.8$

$s = 18$ *a = g = 9.8 ms^{-2}, because it's falling freely.*

$t = t$

You need the equation with u, a, s and t in it: $s = ut + \tfrac{1}{2}at^2$

Substitute values: $18 = (0 \times t) + (\tfrac{1}{2} \times 9.8 \times t^2)$

Simplify: $18 = 4.9t^2$

Rearrange to give t^2: $t^2 = \dfrac{18}{4.9} = 3.67...$

Solve by square-rooting: $t = \sqrt{3.67...} = 1.92$ s

Watch out for tricky questions like this — at first it <u>looks like</u> they've only given you <u>one variable</u>. You have to spot that the pebble was <u>dropped</u> (so it started with no velocity) and that it's <u>moving freely under gravity</u>.

Constant Acceleration Equations

...or $a = -g$

EXAMPLE

A ball is projected vertically upwards at 3 ms^{-1} from a point 1.5 m above the ground.
How long does it take to reach its maximum height?
How fast will the ball be travelling when it hits the ground? (Take g = 9.8 ms^{-2}.)

First, list the variables, taking up as the positive direction:

$u = 3$

$v = 0$ — *When projected objects reach the top of their motion, they stop momentarily.*

$a = -9.8$ — *Because g always acts downwards and up was taken as positive, a is negative.*

$s = s$

$t = ?$

s is negative as the ground is below the point of projection.

Use the equation with u, v, a and t in it: $v = u + at$

Substitute values: $0 = 3 + (-9.8 \times t)$

Simplify: $0 = 3 - 9.8t$

Rearrange and solve to find t: $t = \dfrac{3}{9.8} = 0.31$ s

To find the speed of the ball when it hits the ground consider the complete path of the ball.

$s = -1.5$ as it's the underlined displacement from the ball's original position, not total distance travelled.

Using $v^2 = u^2 + 2as$ where $u = 3$, $v = ?$, $a = -9.8$, $s = -1.5$:

$v^2 = u^2 + 2as = 3^2 + 2(-9.8 \times -1.5) = 38.4$, so $v = \sqrt{38.4} = 6.20$ ms^{-1} (to 3 s.f.)

Sometimes there's **More Than One Object Moving** at the **Same Time**

For these questions, t is often the same (or connected as in this example) because time ticks along for both objects at the same rate. The distance travelled might also be connected.

EXAMPLE

A car, A, travelling along a straight road at a constant 30 ms^{-1} passes point R at $t = 0$.
Exactly 2 seconds later, a second car, B, travelling at 25 ms^{-1}, moves in the same direction from point R. Car B accelerates at a constant 2 ms^{-2}. Show that the two cars are level when $t^2 - 9t - 46 = 0$, where t is the time taken by car A.

For each car, there are different "uvast" equations, so you write separate lists and separate equations.

CAR A

Constant speed so $a_A = 0$

$u_A = 30$ $\qquad$ $v_A = 30$

$a_A = 0$ $\qquad$ $s_A = s$

$t_A = t$

CAR B

$u_B = 25$ $\qquad$ $v_B = v$

$a_B = 2$ $\qquad$ $s_B = s$

$t_B = (t - 2)$

s is the same for both cars because they're level.

B starts moving 2 seconds after A passes point R.

The two cars are level, so choose an equation with s in it:

$$s = ut + \tfrac{1}{2}at^2$$

Substitute values: $s = 30t + (\tfrac{1}{2} \times 0 \times t^2)$

Simplify: $s = 30t$

Use the same equation for car B: $s = ut + \tfrac{1}{2}at^2$

Substitute values: $s = 25(t - 2) + (\tfrac{1}{2} \times 2 \times (t - 2)^2)$

Simplify: $s = 25t - 50 + (t - 2)(t - 2)$

$s = 25t - 50 + (t^2 - 4t + 4)$

$s = t^2 + 21t - 46$

The distance travelled by both cars is equal, so put the equations for s equal to each other:

$30t = t^2 + 21t - 46$

$t^2 - 9t - 46 = 0$ — *That's the result you were asked to find.*

Constant acceleration questions involve underlined modelling assumptions (simplifications to real life so you can use the equations):

1) <u>The object is a particle</u> — this just means it's very small and so isn't affected by air resistance as cars or stones would be in real life.

2) <u>Acceleration is constant</u> — without it, the equations couldn't be used.

As Socrates once said, "The unexamined life is not worth living"... *but what did he know...*

Make sure you: 1) Make a list of the uvast variables EVERY time you get one of these questions.

2) Look out for "hidden" values — e.g. "particle initially at rest..." means $u = 0$.

3) Choose and solve the equation that goes with the variables you've got.

Motion Graphs

You can use underline{displacement-time} (x/t), underline{velocity-time} (v/t) and underline{acceleration-time} (a/t) graphs to represent all sorts of motion.

Displacement-time Graphs: Height = Distance and Gradient = Velocity

The underline{steeper} the line, the underline{greater} the velocity. A underline{horizontal} line has a underline{zero gradient}, so that means the object isn't moving.

EXAMPLE A cyclist's journey is shown on this x/t graph. Describe the motion.

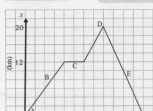

A: Starts from rest (when $t = 0$, $x = 0$)

B: Travels 12 km in 1 hour at a velocity of 12 kmh^{-1}

C: Rests for ½ hour ($v = 0$)

D: Cycles 8 km in ½ hour at a velocity of 16 kmh^{-1}

E: Returns to starting position, cycling 20 km in 1 hour at a velocity of -20 kmh^{-1} (i.e. 20 kmh^{-1} in the opposite direction)

EXAMPLE A girl jogs 2 km in 15 minutes and a boy runs 1.5 km in 6 min, rests for 1 min then walks the last 0.5 km in 8 min. Show the two journeys on an x/t graph.

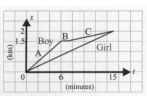

Girl: constant velocity, so there's just one straight line for her journey from (0, 0) to (15, 2)

Boy: three parts to the journey, so there are three straight lines: A - run, B - rest, C - walk

Velocity-time Graphs: Area = Distance and Gradient = Acceleration

The underline{area} under the graph can be calculated by underline{splitting} the area into rectangles, triangles or trapeziums. Work out the areas underline{separately}, then underline{add} them all up at the end.

EXAMPLE A train journey is shown on the v/t graph on the right. Find the distance travelled and the rate of deceleration as the train comes to a stop.

The time is given in minutes and the velocity as kilometres per hour, so divide the time in minutes by 60 to get the time in hours.

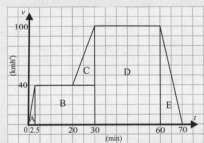

Area of A: $(2.5 \div 60 \times 40) \div 2 = 0.833...$

Area of B: $27.5 \div 60 \times 40 = 18.33...$

Area of C: $(10 \div 60 \times 60) \div 2 = 5$

Area of D: $30 \div 60 \times 100 = 50$

Area of E: $(10 \div 60 \times 100) \div 2 = 8.33...$

Total area = 82.5 so distance is 82.5 km

You might get a speed-time graph instead of a velocity-time graph — they're pretty much the same, except speeds are always positive, although you **can** have negative velocities.

The gradient of the graph at the end of the journey is -100 kmh$^{-1} \div (10 \div 60)$ hours = -600 kmh^{-2}
So the train decelerates at 600 kmh^{-2}.

Acceleration-time Graphs: Area = Velocity

EXAMPLE The acceleration of a parachutist who jumps from a plane is modelled by the a/t graph on the right. Describe the motion of the parachutist and find her velocity when she is no longer accelerating.

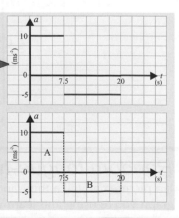

She falls with acceleration due to gravity of 10 ms^{-2} for 7.5 s. The parachute opens and the acceleration due to the air resistance of the parachute is 5 ms^{-2} acting upwards for 12.5 s. After 20 s, the acceleration is zero and so she falls with constant velocity. You need to find the area under the graph:

Area A: $10 \times 7.5 = 75$ ms^{-1} Area B: $5 \times 12.5 = 62.5$ ms^{-1}

Area B is underline{under} the horizontal axis, so underline{subtract} area B from area A:

Velocity = 75 ms^{-1} $-$ 62.5 ms^{-1} = 12.5 ms^{-1}

Motion Graphs

Graphs can be used to **Solve Complicated Problems**

As well as working out distance, velocity and acceleration from graphs, you can also solve more complicated problems. These might involve working out information <u>not shown directly on the graph</u>.

EXAMPLE

A jogger and a cyclist set off at the same time. The jogger runs with a constant velocity. The cyclist accelerates from rest, reaching a velocity of 5 ms⁻¹ after 6 s and then continues at this velocity. The cyclist overtakes the jogger after 15 s.

Draw a graph of the motion and find the velocity of the jogger.

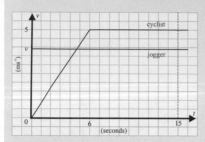

Call the velocity of the jogger v.

After 15 s the distance each has travelled is the same, so you can work out the area under the two graphs to get the distances:

Jogger: Area = distance = $15v$

Cyclist: Area = distance = $(6 \times 5) \div 2 + (9 \times 5) = 60$

> area of triangle + area of rectangle

So $15v = 60$
$v = 4 \text{ ms}^{-1}$

EXAMPLE

A man throws a pebble vertically upwards from ground level with a speed of u ms⁻¹. The pebble takes 2.6 s to return to ground level and reaches a maximum height of 9.4 m.

Draw a graph of the motion and find the value of u.

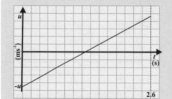 **OR**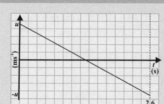

Using $s = \frac{1}{2}(u + v)t$ until ball reaches maximum height, where $v = 0$ ms⁻¹:

$9.4 = \frac{1}{2}(u + 0)1.3$

so $u = \frac{9.4}{0.65} = 14.5 \text{ ms}^{-1}$

> The ball reaches maximum height in half the time it takes to fall to the floor.

EXAMPLE

A bus is travelling at V ms⁻¹. When it reaches point A it accelerates uniformly for 4 s, reaching a speed of 21 ms⁻¹ as it passes point B. At point B, the driver brakes uniformly until the bus comes to a halt 7 s later. The rate of deceleration is twice the rate of acceleration.

Draw a speed-time graph of the motion and find the value of V.

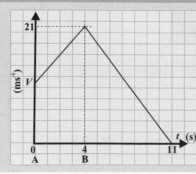

To find the deceleration use $a = \frac{(v - u)}{t}$:

$a = \frac{(0 - 21)}{7} = -3 \text{ ms}^{-2}$

> Just $v = u + at$ rearranged.

Rate of deceleration is twice the rate of acceleration, so rate of acceleration = 1.5 ms⁻².

Finding V using $u = v - at$:

$V = 21 - (1.5 \times 4)$

So, $V = 15 \text{ ms}^{-1}$

Random tongue-twister #1 — I wish to wash my Irish wristwatch...

If a picture can tell a thousand words then a graph can tell... um... a thousand and one. Make sure you know what type of graph you're using and learn what the gradient and the area under each type of graph tells you.

M1 Section 1 — Practice Questions

Time for the first set of M1 questions — just a few to get you up to speed gently.
A bit of advice — don't panic, take it step by step, and above all ~~don't get hurt~~ keep practising until it's second nature.

Warm-up Questions

Take $g = 9.8$ ms^{-2} in each of these questions.

1) A motorcyclist accelerates uniformly from 3 ms^{-1} to 9 ms^{-1} in 2 seconds.
   What is the distance travelled by the motorcyclist during this acceleration?

2) A runner starts from rest and accelerates at 0.5 ms^{-2} for 5 seconds. She maintains a constant
   velocity for 20 seconds then decelerates to a stop at 0.25 ms^{-2}.
   Draw a V/T graph to show the motion and find the distance the runner travelled.

3) The start of a journey is shown on the A/T graph to the right.
   Find the velocity when:　　a) $t = 3$　　b) $t = 5$　　c) $t = 6$

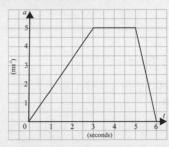

Hopefully those questions above were no trouble, so it's time to have a go at the kind of questions
you're likely to see in the exam. You'll find them below, in a different font and everything...

Exam Questions

Take $g = 9.8$ ms^{-2} in each of these questions.

1　A window cleaner of a block of flats accidentally drops his sandwich, which falls freely to the ground.
　The speed of the sandwich as it passes a high floor is u. After a further 1.2 seconds the sandwich is
　moving at a speed of 17 ms^{-1}. The distance between the consecutive floors of the building is h

　a)　Find the value of u.

(3 marks)

　b)　It takes the sandwich another 2.1 seconds to fall the remaining 14 floors to the ground.
　　　Find h.

(4 marks)

2　A train starts from rest at station A and travels with constant acceleration for 20 s. It then travels with
　constant speed V ms^{-1} for 2 minutes. It then decelerates with constant deceleration for 40 s before
　coming to rest at station B. The total distance between stations A and B is 2.1 km.

　a)　Sketch a speed-time graph for the motion of the train between stations A and B.

(3 marks)

　b)　Hence or otherwise find the value of V.

(3 marks)

　c)　Calculate the distance travelled by the train while decelerating.

(2 marks)

　d)　Sketch an acceleration-time graph for this motion.

(3 marks)

3　A rocket is projected vertically upwards from a point 8 m above the ground at a speed of u ms^{-1} and
　travels freely under gravity. The rocket hits the ground at 20 ms^{-1}. Find:

　a)　the value of u,

(3 marks)

　b)　how long it takes to hit the ground.

(3 marks)

Vectors

'Vector' might sound like a really dull Bond villain, but... well, it's not. Vectors have both size (or <u>magnitude</u>) and <u>direction</u>. If a measurement just has size but not direction, it's called a <u>scalar</u>.   - So, Vector, you expect me to talk?
- No, Mr Bond, I expect you to die! Ak ak ak!

Vectors have **Magnitude** and **Direction** — **Scalars** only have **Magnitude**

<u>Vectors</u>: velocity, displacement, acceleration, force. E.g. a train heading due east at 16 ms⁻¹.

<u>Scalars</u>: speed, distance. E.g. a car travelling at 4 ms⁻¹.

A really important thing to remember is that an object's speed and velocity <u>aren't always the same</u>:

EXAMPLE   A runner sprints 100 m along a track at a speed of 8 ms⁻¹ and then she jogs back 50 m at 4 ms⁻¹.

<u>Average</u> <u>Speed</u>

Speed = Distance ÷ Time

The runner takes $(100 \div 8) + (50 \div 4) = 25$ s to travel **150 m**.

So the average speed is $150 \div 25 = 6$ ms⁻¹

<u>Average</u> <u>Velocity</u>

Velocity = Change in displacement ÷ Time

In total, the runner ends up **50 m** away from her start point and it takes 25 s.

So the average velocity is $50 \div 25 = 2$ ms⁻¹ in the direction of the sprint.

She jogged back 50 m after she jogged forward 100 m.

The **Length** of the **Arrow** shows the **Magnitude of a Vector**

You can draw vectors as arrows where the <u>length</u> shows the <u>magnitude</u>:

① A train travelling due east at 16 ms⁻¹

16 ms⁻¹

② A train travelling due west at 8 ms⁻¹

8 ms⁻¹

The arrow is half the size since the vector has half the magnitude.

You can add vectors together by drawing the arrows <u>nose to tail</u>.
The single vector that goes from the start to the end of the vectors is called the <u>resultant</u> vector.

a + b

b

a

Resultant: **r = a + b**

b

a

r

b

a

r

b

a

a + b = b + a

You can also <u>multiply</u> a vector by a <u>scalar</u> (just a number): the <u>length changes</u> but the <u>direction stays the same</u>.

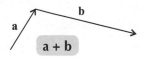

a

3a

Vectors can be described using **i + j** Units

You can <u>describe</u> vectors using the <u>unit vectors</u> **i + j**. They're called unit vectors because they each have a <u>magnitude of 1</u> and a <u>direction</u>. **i** is in the direction of the <u>x-axis</u> (horizontal or east), and **j** is in the direction of the <u>y-axis</u> (vertical or north).

E.g. $\overrightarrow{AB} = 5\mathbf{i} + 4\mathbf{j}$:

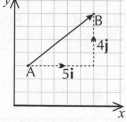

$\overrightarrow{AB}$ is just another way of writing a vector — it means 'the displacement of B from A'.

Vectors

Resolving means writing a vector as Component Vectors

Splitting a vector up like this means you can work things out with <u>one component at a time</u>.
When <u>adding</u> vectors to get a <u>resultant vector</u>, it's easier to <u>add</u> the <u>horizontal</u> and <u>vertical</u>
components <u>separately</u>. So you <u>split</u> the vector into components first — this is <u>resolving</u>.
If the vectors are in **i + j** notation <u>already</u>, you don't need to split them up — just add the **i** and **j**'s:

> **EXAMPLE** $\overrightarrow{AB} = 3\mathbf{i} + 2\mathbf{j}$ and $\overrightarrow{BC} = 5\mathbf{i} - 3\mathbf{j}$. Work out $\overrightarrow{AC}$.
>
> Add the horizontal and vertical components <u>separately</u>.
>
> $\overrightarrow{AC} = \overrightarrow{AB} + \overrightarrow{BC} = (3\mathbf{i} + 2\mathbf{j}) + (5\mathbf{i} - 3\mathbf{j}) = 8\mathbf{i} - \mathbf{j}$

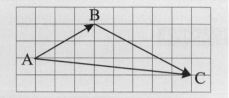

Use Trig and Pythagoras to Change a vector into Component Form

> **EXAMPLE** The speed of a ball is 5 ms⁻¹ at an angle of 30° to the horizontal.
> Find the horizontal and vertical components of the ball's velocity, **v**.
>
> First, draw a diagram and make a right-angle triangle:
>
>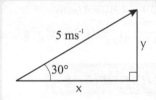
>
> Using trigonometry, we can find x and y:
>
> $\cos 30° = \dfrac{x}{5}$ so $x = 5\cos 30°$
>
> $\sin 30° = \dfrac{y}{5}$ so $y = 5\sin 30°$
>
> So $\mathbf{v} = (5\cos 30°\mathbf{i} + 5\sin 30°\mathbf{j})$ ms⁻¹

> **EXAMPLE** The acceleration of a body is given by the vector $\mathbf{a} = 6\mathbf{i} - 2\mathbf{j}$.
> Find the magnitude and direction of the acceleration.
>
> Start with a diagram again. Remember, the y-component "–2" means "down 2".
>
>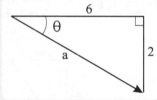
>
> Using Pythagoras' theorem, you can work out the magnitude of **a**:
>
> $a^2 = 6^2 + (-2)^2 = 40$
>
> so $a = \sqrt{40} = 6.32$ (to 3 s.f.)
>
> Use trigonometry to work out the angle:
>
> $\tan\theta = \dfrac{2}{6}$ so $\theta = \tan^{-1}\left(\dfrac{2}{6}\right) = 18.4°$
>
> So vector **a** has magnitude 6.32 and direction 18.4° below the horizontal.

> In general, a vector with magnitude **r** and direction θ can be written as **rcosθi + rsinθj**
>
> The vector $x\mathbf{i} + y\mathbf{j}$ has magnitude $\mathbf{r} = \sqrt{x^2 + y^2}$ and direction $\theta = \tan^{-1}\left(\dfrac{y}{x}\right)$

You can Resolve in any two Perpendicular Directions — not just x and y

> **EXAMPLE** Find the resultant of the forces shown in the diagram.
>
>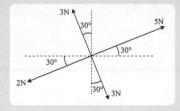
>
> Resolving in ＼ direction: 3 N – 3 N = 0
>
> Resolving in ／ direction: 5 N – 2 N = 3 N
>
> *The two forces of 3 N are acting in opposite directions, so you take one away from the other — balancing them out.*
>
> So the resultant force is 3 N in the direction of the 5 N force.

Vectors

Vectors are much more than just a pretty face (or arrow). Once you deal with all the waffle in the question, all sorts of problems involving <u>displacement</u>, <u>velocity</u>, <u>acceleration</u> and <u>forces</u> can be solved using vectors.

Draw a Diagram if there are Lots of Vectors floating around

In fact, draw a diagram when there are only a <u>couple</u> of vectors. But it's <u>vital</u> when there are lots of the little beggars.

EXAMPLE

A ship travels 100 km at a bearing of 025°, then 75 km at 140° before going 125 km at 215°. What is the displacement of the ship from its starting point?

> Remember that bearings are always measured <u>clockwise</u> starting from <u>north</u>.

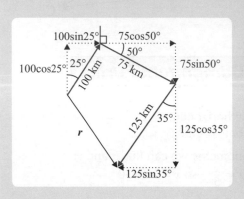

Resolve <u>East</u>: $100\sin25° + 75\cos50° - 125\sin35° = 18.8$ km

Resolve <u>North</u>: $100\cos25° - 75\sin50° - 125\cos35° = -69.2$ km

Magnitude of $\mathbf{r} = \sqrt{18.8^2 + (-69.2)^2} = 71.7$ km

direction $\theta = \tan^{-1}\left(\dfrac{69.2}{18.8}\right) = 74.8°$

Bearing is $90° + 74.8° = 164.8°$

So the displacement is 71.7 km on a bearing of 164.8°

The Direction part of a vector is Really Important

...and that means that you've got to make sure your <u>diagram</u> is <u>spot on</u>. These two problems look similar, and the final answers are pretty similar too. But <u>look closely</u> at the diagrams and you will see they are a bit <u>different</u>.

EXAMPLE

A canoe is paddled at 4 ms^{-1} in a direction perpendicular to the seashore. The sea current has a velocity of 1 ms^{-1} parallel to the shore. Find the resultant velocity $\mathbf{r}$ of the canoe.

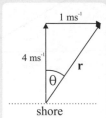

The resultant velocity $\mathbf{r}$ is the <u>hypotenuse</u> of the right-angled triangle.

Magnitude of $\mathbf{r} = \sqrt{4^2 + 1^2} = 4.1$ ms^{-1}

Direction: $\theta = \tan^{-1}\left(\dfrac{1}{4}\right) = 14.0°$ clockwise from the direction of paddling.

EXAMPLE

A canoe can be paddled at 4 ms^{-1} in still water. The sea current has a velocity of 1 ms^{-1} parallel to the shore. Find the angle at which the canoe must be paddled in order to travel in a direction perpendicular to the shore and the magnitude of the resultant velocity.

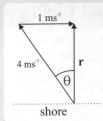

The resultant velocity $\mathbf{r}$ <u>isn't</u> the hypotenuse this time.

Magnitude of $\mathbf{r} = \sqrt{4^2 - 1^2} = 3.9$ ms^{-1}

Direction: $\theta = \sin^{-1}\left(\dfrac{1}{4}\right) = 14.5°$ anticlockwise from the perpendicular to the shore.

Bet you can't say 'perpendicular' 10 times fast...

Next time you're baffled by vectors, just start <u>resolving</u> and Bob's your mother's brother.

Forces and Modelling

Vector questions talk about forces all the time in M1, so you need to understand what each type of force is.
Then you can use that information to create a model to work from.

Hint: 'modelling' in maths doesn't have anything to do with plastic aeroplane kits... or catwalks.

Types of forces

WEIGHT (*W*)

Due to the particle's mass, <u>m</u> and the force of gravity, g: $W = mg$ — weight always acts <u>downwards</u>.

W = mg

THE NORMAL REACTION (*R* OR *N*)

The reaction from a surface. Reaction is always at <u>90° to the surface</u>.

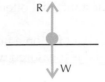

R

W

TENSION (*T*)

Force in a taut rope, wire or string.

T

W

FRICTION (*F*)

Due to the <u>roughness</u> between a body and a surface. Always acts <u>against</u> motion, or likely motion.

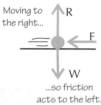

Moving to the right...

R

F

W

...so friction acts to the left.

THRUST

<u>Force in a rod</u> (e.g. the pole of an open umbrella).

Talk the Talk

Maths questions in M1 use a lot of words that you already know, but here they're used to mean something very <u>precise</u>. Learn these definitions so you don't get caught out:

<u>Particle</u>	the body is a point so its dimensions don't matter	<u>Rigid</u>	the body does not bend
<u>Light</u>	the body has no mass	<u>Thin</u>	the body has no thickness
<u>Static</u>	not moving	<u>Equilibrium</u>	nothing's moving
<u>Rough</u>	the surface will oppose motion with friction / drag	<u>Plane</u>	a flat surface
<u>Beam or Rod</u>	a long particle (e.g. a carpenter's plank)	<u>Inextensible</u>	the body can't be stretched
<u>Uniform</u>	the mass is evenly spread out throughout the body	<u>Smooth</u>	the surface doesn't have friction / drag opposing motion
<u>Non-uniform</u>	the mass is unevenly spread out		

Mathematical Modelling

You'll have to make lots of assumptions in M1. Doing this is called '<u>modelling</u>', and you do it to make a sticky real-life situation <u>simpler</u>.

EXAMPLE   **The ice hockey player**

<u>You might have to assume:</u>

- no friction between the skates and the ice
- no drag (air resistance)
- the skater generates a constant forward force S
- the skater is very small (a point mass)
- there is only one point of contact with the ice
- the weight acts downwards

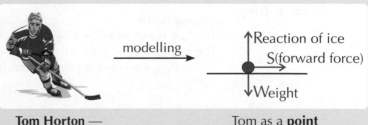

modelling →

Reaction of ice
S(forward force)
Weight

Tom Horton —
coffee shop owner

Tom as a **point mass** with forces

The simplified model you end up with can then be used as your vector diagram.

Forces and Modelling

Modelling is a Cycle

Having created a model you can later <u>improve</u> it by making more (or fewer) <u>assumptions</u>.
Solve the problem using the initial assumptions, <u>compare</u> it to real life, <u>evaluate</u> the model and then use that information to <u>change</u> the assumptions. Then keep going until you're <u>satisfied</u> with the model.

Always start by drawing a Simple Diagram of the Model

Here are a couple of old chestnuts that often turn up in M1 exams in one form or another.

EXAMPLE

The book on a table

A book is put flat on a table. One end of the table is slowly lifted and the angle to the horizontal is measured when the book starts to slide. What assumptions might you make?

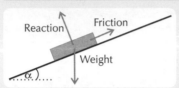

Assumptions:
The book is <u>rigid</u>, so it doesn't bend or open.
The book is a <u>particle</u>, so its dimensions don't matter.
There's <u>no</u> wind or other <u>external forces</u> involved.

EXAMPLE

The balance

A pencil is placed on a table and a ruler is put across the pencil so that the ruler balances. A 1p coin and a 10p coin are placed on the ruler either side of the pencil, so that the ruler still balances. Draw a model of the forces. What assumptions have you made?

Assumptions:
The coins are <u>point masses</u>.
The ruler is <u>rigid</u>.
The support acts at a <u>single point</u>.

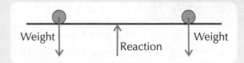

EXAMPLE

The sledge

A sledge is being steadily pulled by a small child on horizontal snow. Draw a force diagram for a model of the sledge. List your assumptions.

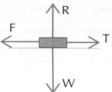

> It's quicker and easier to use just the first letter of the force in your diagram, e.g. F = friction.

Assumptions:
Friction is <u>too big</u> to be ignored (i.e. it's not ice).
The string is <u>horizontal</u> (it's a small child).
The sledge is a <u>small particle</u> (so its size doesn't matter).

EXAMPLE

The mass on a string

A ball is held by two strings, A and B, at angles α and β to the vertical. Draw a diagram to model this scenario. State your assumptions.

Assumptions:
The ball is modelled as a <u>particle</u> (its dimensions don't matter).
The strings are <u>light</u> (their mass can be ignored).
The strings are <u>inextensible</u> (they can't stretch).

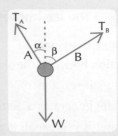

I used to be a model when I was younger, but then I fell apart...

Make sure you're completely familiar with the different forces and all the jargon that gets bandied about in mechanics. Keep your models as simple as possible — that will make answering the questions as simple as possible too.

Forces are Vectors

Forces have direction and magnitude, which makes them vectors — this means that all the stuff you learnt about vectors you'll need here. To help you out, we've given you some more vector examples all about forces...

Forces have **Components**

You've done a fair amount of <u>trigonometry</u> already, so this should be as straightforward as watching dry paint.

EXAMPLE A particle is acted on by a force of 15 N at 30° above the horizontal. Find the <u>horizontal</u> and <u>vertical components</u> of the force.

A bit of trigonometry is all that's required:

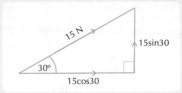

Force = 15cos30°**i** + 15sin30°**j**

$$= (13.0\,\mathbf{i} + 7.5\,\mathbf{j})\,\text{N}$$

(i.e. 13 N to the right and 7.5 N upwards)

Add Forces Nose to Tail to get the Resultant

The important bit when you're drawing a diagram to find the resultant is to make sure the <u>arrows</u> are the <u>right way round</u>. Repeat after me: nose to tail, nose to tail, nose to tail.

EXAMPLE A second horizontal force of 20 N to the right is also applied to the particle in the example above. Find the resultant of these forces.

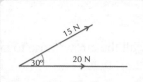

You need to put the arrows nose to tail:

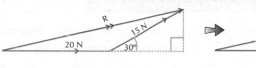 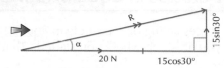

Using Pythagoras and trigonometry:

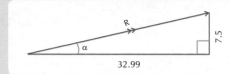

$$R = \sqrt{32.99^2 + 7.5^2} = 33.8\,\text{N}$$

$$\alpha = \tan^{-1}\left(\frac{7.5}{32.99}\right) = 12.8° \text{ above the horizontal}$$

EXAMPLE Find the magnitude and direction of the resultant of the forces shown acting on the particle.

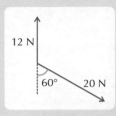 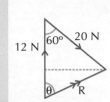 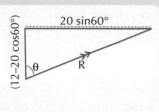

Hint: you could also use the cosine rule here to get R.

$$R = \sqrt{(12 - 20\cos 60°)^2 + (20\sin 60°)^2} = 17.4\,\text{N}$$

$$\theta = \tan^{-1}\left(\frac{20\sin 60°}{12 - 20\cos 60°}\right) = 83.4° \text{ to the vertical}$$

If a particle is released it will move in the direction of the resultant.

Forces are Vectors

Resolving *more than* Two Forces

A question could involve <u>more than two forces</u>. You still work it out the <u>same</u> though — <u>resolve, resolve, resolve</u>...

EXAMPLE Three forces of magnitudes 9 N, 12 N and 13 N act on a particle P in the directions shown in the diagram.
Find the magnitude and direction of the resultant of the three forces.

One of the forces is already aligned with the y-axis, so it makes sense to start by resolving the other forces relative to this.

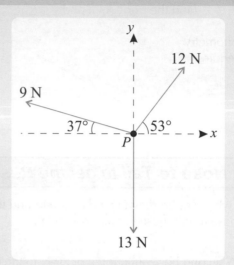

Along the *y*-axis:

9sin37° + 12sin53 – 13 = resultant
= 9(0.6) + 12(0.8) – 13 = 2 N

Along the *x*-axis:

12cos53° – 9cos37° = resultant
= 12(0.6) – 9(0.8) = 0 N

Overall: 2 N in the direction of the y-axis

Particles in *Equilibrium* Don't Move

Forces acting on a particle in <u>equilibrium</u> add up to zero force. That means when you draw all the arrows nose to tail, you finish up where you started. That's why diagrams showing equilibrium are called '<u>polygons of forces</u>'.

EXAMPLE Two perpendicular forces of magnitude 20 N act on a particle. A third force, P, acts at 45° to the horizontal, as shown. Given that the particle is in equilibrium, find the magnitude of P.

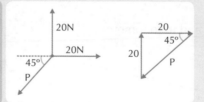

$P\cos 45° = 20$

$P = 28.3\,\text{N}$

EXAMPLE A force of 50 N acts on a particle at an angle of 20° to the vertical, as shown. Find the magnitude of the two other forces, T and S, if the particle is in equilibrium.

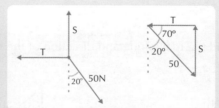

$S = 50\sin 70° = 47.0\,\text{N}$

$T = 50\cos 70° = 17.1\,\text{N}$

EXAMPLE Three forces act upon a particle. A force of magnitude 85 N acts horizontally, the force Q acts vertically, and the force P acts at 55° to the horizontal, as shown. The particle is in equilibrium. Find the magnitude of P and Q.

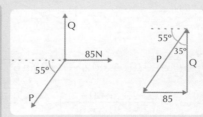

$\sin 35° = \dfrac{85}{P}$ so $P = 148\,\text{N}$

$\tan 35° = \dfrac{85}{Q}$ so $Q = 121\,\text{N}$

Forces are Vectors

An *Inclined Plane* is a *Sloping Surface*

EXAMPLE A particle of mass 0.1 kg is held at rest on a rough plane inclined at 20° to the horizontal by a friction force acting up the plane. Find the magnitude of this friction force and the normal reaction. ($g = 9.0 \text{ ms}^{-2}$.)

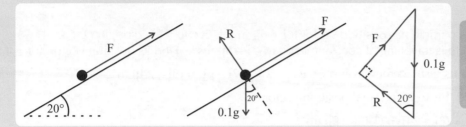

$$F = 0.1g\sin 20°$$
$$= 0.335 \text{ N}$$
$$R = 0.1g\cos 20°$$
$$= 0.921 \text{ N}$$

EXAMPLE A sledge of weight 1000 N is being held on a rough inclined plane at an angle of 35° by a force parallel to the slope of 700 N. Find the normal contact force N and the frictional force F acting on the sledge.

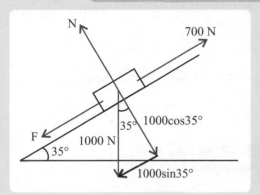

All the forces, except the weight, involved here act either <u>parallel</u> or <u>perpendicular</u> to the <u>slope</u> so it makes sense to resolve in these directions.

<u>Perpendicular</u> to the slope:

$N - 1000\cos 35° = 0$
So $N = 1000\cos 35°$ $= 819.2 \text{ N}$

<u>Parallel</u> to the slope:

$700 = 1000\sin 35° + F$

So $F = 700 - 1000\sin 35°$ $= 126.4 \text{ N}$

Masses on Strings *Produce Tension*

EXAMPLE A mass of 12 kg is held by two light strings, P and Q, acting at 40° and 20° to the vertical, as shown. Find the tension in each string. Take $g = 9.8 \text{ ms}^{-2}$. ⟹

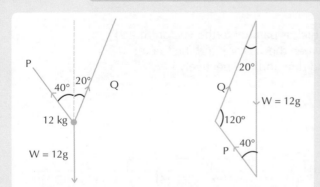

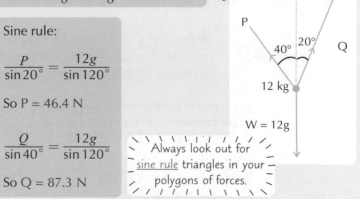

Sine rule:

$$\frac{P}{\sin 20°} = \frac{12g}{\sin 120°}$$

So $P = 46.4 \text{ N}$

$$\frac{Q}{\sin 40°} = \frac{12g}{\sin 120°}$$

So $Q = 87.3 \text{ N}$

Always look out for <u>sine rule</u> triangles in your polygons of forces.

Cliché #27 — "The more things change, the more they stay the same"...

As it makes their lives easier, examiners always stick the <u>same</u> kinds of questions into M1 exams. So, no need to panic — just keep practising the questions and then there'll be no surprises when it comes to the exam. Simple.

i + j Vectors

Bet you'd forgotten about vectors written with **i** + **j**. Well, they're back, and the only way to get your head around them is to practise, then practise a bit more. I'm sure you get the idea.

For particles travelling at **Constant Velocity**, s = vt

If a particle is travelling at a constant velocity, v, then its displacement, s, after time, t, can be found by $s = vt$. This is handy for questions involving <u>position vectors</u> — these are vectors which describe the position of something relative to an origin.

EXAMPLE
At $t = 0$ a particle has position vector $(6\mathbf{i} + 8\mathbf{j})$ m relative to a fixed origin O. The particle is travelling at constant velocity $(2\mathbf{i} - 6\mathbf{j})$ ms^{-1}. Find its position vector at $t = 4$ s.

First find its displacement using $s = vt$: $\quad s = 4(2\mathbf{i} - 6\mathbf{j}) = (8\mathbf{i} - 24\mathbf{j})$ m

Then add this to it's original position vector:

$(6\mathbf{i} + 8\mathbf{j}) + (8\mathbf{i} - 24\mathbf{j}) = (14\mathbf{i} - 16\mathbf{j})$ m

You might need to use the **Constant Acceleration Equations**

If a particle is accelerating, you'll need to use the constant acceleration equations from M1 section 1 (page 147).

EXAMPLE
Find the velocity and speed of a particle after 3 s if its initial velocity is $(6\mathbf{i} + 4\mathbf{j})$ ms^{-1} and acceleration is $(0.3\mathbf{i} + 0.5\mathbf{j})$ ms^{-2}.

Using $\mathbf{v} = \mathbf{u} + \mathbf{a}t$: $\quad \mathbf{v} = (6\mathbf{i} + 4\mathbf{j}) + 3(0.3\mathbf{i} + 0.5\mathbf{j}) = (6\mathbf{i} + 4\mathbf{j}) + (0.9\mathbf{i} + 1.5\mathbf{j}) = (6.9\mathbf{i} + 5.5\mathbf{j})$ ms^{-1}

Speed = magnitude of $\mathbf{v} = \sqrt{6.9^2 + 5.5^2} = 8.82$ ms^{-1}

EXAMPLE
A particle, P, has position vector $(3\mathbf{i} + \mathbf{j})$ m and velocity = $(2\mathbf{i} - 5\mathbf{j})$ ms^{-1} at $t = 0$. Given that P accelerates at a rate of $(-2\mathbf{i} + \mathbf{j})$ ms^{-2}, find its position vector at $t = 6$ s.

Using $\mathbf{s} = \mathbf{u}t + \frac{1}{2}\mathbf{a}t^2$:

$\mathbf{s} = 6(2\mathbf{i} - 5\mathbf{j}) + \frac{1}{2}(6^2)(-2\mathbf{i} + \mathbf{j}) = (12\mathbf{i} - 30\mathbf{j}) + (-36\mathbf{i} + 18\mathbf{j}) = (-24\mathbf{i} - 12\mathbf{j})$ m

New position vector = $(3\mathbf{i} + \mathbf{j}) + (-24\mathbf{i} - 12\mathbf{j}) = (-21\mathbf{i} - 11\mathbf{j})$ m

i + j vectors can be used to **Describe Forces**

Treat forces described in terms of **i** + **j** just the same as any other force or vector that you'll deal with in M1.

EXAMPLE
A force $(x\mathbf{i} + y\mathbf{j})$, acting in a direction which is parallel to the vector $(3\mathbf{i} - 2\mathbf{j})$, acts on a particle P. Find the angle between the vector **j** and the force. (The unit vectors i and j are due east and due north respectively.)

The force is parallel to $(3\mathbf{i} - 2\mathbf{j})$, so the angle between it and **j** is the same as the angle between **j** and $(3\mathbf{i} - 2\mathbf{j})$:

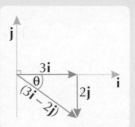

$\tan\theta = \dfrac{2}{3}$ and so $\theta = 33.7°$ below the horizontal

So, angle to **j** is $33.7 + 90 = 123.7° = 124°$ (to 3 s.f.)

An eye for an *i*, a tooth for a *j*...

Taking M1 exams might feel like punishment, unless you revise that is — then it'll feel like an invigorating walk in the park.

Friction

Friction tries to prevent motion, but don't let it prevent you getting marks in the exam — revise this page and it won't.

*Friction Tries to **Prevent Motion***

Push hard enough and a particle will move, even though there's friction opposing the motion. A <u>friction force</u>, F, has a <u>maximum value</u>. This depends on the <u>roughness</u> of the surface and the value of the <u>normal reaction</u> from the surface.

$$F \leq \mu R \quad \text{OR} \quad F \leq \mu N \quad \text{(where R and N both stand for normal reaction)}$$

μ has no units.
μ is pronounced as 'mu'.

μ is called the "<u>coefficient of friction</u>". The <u>rougher</u> the surface, the <u>bigger</u> μ gets.

EXAMPLE What range of values can a friction force take in resisting a horizontal force P acting on a particle Q, of mass 12 kg, resting on a rough horizontal plane which has a coefficient of friction of 0.4? (Take $g = 9.8$ ms^{-2}.)

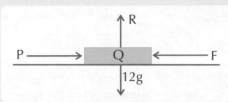

Resolving vertically: $R = 12g$

Use formula from above: $F \leq \mu R$

$$F \leq 0.4(12g)$$

$$F \leq 47.04 \text{ N}$$

So friction can take any value between 0 and 47.04 N, depending on how large P is.

If $P \leq 47.04$ N then Q remains in equilibrium. If $P = 47.04$ N then Q is <u>on the point of sliding</u> — i.e. friction is at its <u>limit</u>. If $P > 47.04$ then Q will start to move.

*Limiting Friction is When Friction is at **Maximum** (F = μR)*

EXAMPLE A particle of mass 6 kg is placed on a rough horizontal plane which has a coefficient of friction of 0.3. A horizontal force Q is applied to the particle. Describe what happens if Q is: a)   16 N
Take $g = 9.8$ ms^{-2} b)   20 N

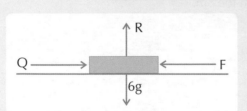

Resolving vertically: $R = 6g$

Using formula above: $F \leq \mu R$
$$F \leq 0.3(6g)$$
$$F \leq 17.64 \text{ N}$$

a)   Since Q < 17.64 it <u>won't move</u>.

b)   Since Q > 17.64 it'll <u>start moving</u>. No probs.

EXAMPLE A particle of mass 4 kg at rest on a rough horizontal plane is being pushed by a horizontal force of 30 N. Given that the particle is on the point of moving, find the coefficient of friction.

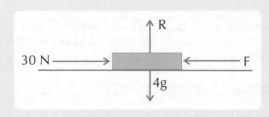

Resolving horizontally: $F = 30$
Resolving vertically: $R = 4g$
The particle's about to move, so friction is at its limit:

$$F = \mu R$$

$$30 = \mu(4g)$$

$$\mu = \frac{30}{4g} = 0.77$$

Sometimes friction really rubs me up the wrong way...

Friction can be a right nuisance, but without it we'd just slide all over the place, which would be worse (I imagine).

Moments

In this lifetime there are moments — moments of joy and of sorrow, and those moments where you have to answer questions on moments in exams.

Moments are *Clockwise* or *Anticlockwise*

A 'moment' is the turning effect a force has around a point.
The larger the force, and the greater the distance from the point, then the larger the moment.

Moment = Force × Perpendicular Distance

EXAMPLE A plank 2 m long is attached to a ship at one end, O, as shown. A bird lands on the other end of the plank, applying a force of 15 N. Model the plank as a light rod and find the moment applied by the bird.

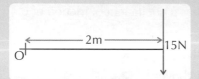

Moment $\quad = F \times d$
$\quad\quad\quad\quad = 15 \times 2$
$\quad\quad\quad\quad = 30 \text{ Nm}$

The units are just newtons × metres = Nm. Couldn't they have thought of a cleverer name?

EXAMPLE A force of 20 N is applied to a spanner, attached to a bolt at a point, O. The force is applied at an angle of 60° to the spanner head, as shown. Find the turning effect of the force upon the bolt.

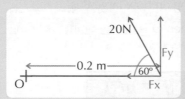

The 20 N force has components Fx and Fy.
Fx goes through O, so its moment is 0.

Resolve vertically: $Fy = 20\sin60°$

Moment $= 20\sin60° \times 0.2$
$\quad\quad\quad = 3.46 \text{ Nm (to 3 s.f.)}$

EXAMPLE A force of 25 N acts upon a point, O, via a light rod 5 m in length connected at an angle of 40°, as shown. What is the turning effect applied to O?

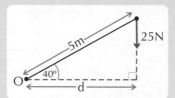

Remember it's got to be perpendicular distance, so you can't just plug in the 5m.

Here d is 5cos40°, so:

Moment $= 25 \times 5\cos40°$
$\quad\quad\quad = 95.8 \text{ Nm}$

In *Equilibrium* Moments Total *Zero*

...and that means that for a body in equilibrium the total moments either way must be equal:

Total Clockwise Moment = Total Anticlockwise Moment

EXAMPLE Two weights of 30 N and 45 N are placed on a light 8 m beam. The 30 N weight is at one end of the beam, as shown, whilst the other weight is a distance d from the midpoint M. The beam is in equilibrium held by a single wire with tension T attached at M. Find T and the distance d.

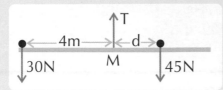

Resolving vertically: $30 + 45 = T = 75 \text{ N}$

Take moments about M:
Clockwise Moment $=$ Anticlockwise Moment
$\quad\quad\quad 45 \times d = 30 \times 4$
$\quad\quad\quad\quad d = \dfrac{120}{45} = 2\dfrac{2}{3} = 2.67 \text{ m}$

Moments

The **Weight** of a **Uniform Beam** Acts at its **Middle**

The weight of anything always acts at its centre of mass.
For a uniform beam, the centre of mass is always at the exact centre of the beam.

EXAMPLE

A 6 m long uniform beam AB of weight 40 N is supported at A by a vertical reaction R. AB is held horizontal by a vertical wire attached 1 m from the other end. A particle of weight of 30 N is placed 2 m from the support R.

Find the tension T in the wire and the force R.

Start by taking moments around the point of action of one of the unknown forces.

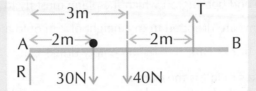

Take moments about A.

Clockwise Moment = Anticlockwise Moment

$(30 \times 2) + (40 \times 3) = T \times 5$

$T = 36$ N

Resolve vertically: $T + R = 30 + 40$

So: $R = 34$ N

EXAMPLE

A Christmas banner, AB, is attached to a ceiling by two pieces of tinsel. One piece of tinsel is attached to A, the other to the point C, where BC = 0.6 m. The banner has mass 8 kg and length 3.6 m and is held in a horizontal position.

Modelling the banner as a uniform rod held in equilibrium and the tinsel as light strings, find the tension in the tinsel at A and C. Take $g = 9.8$ ms^{-2}.

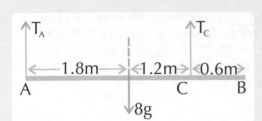

Take moments about A.

Clockwise Moment = Anticlockwise Moment

$8g \times 1.8 = T_C \times 3$

So, $T_C = \dfrac{141}{3} = 47$ N

Resolve vertically.

$T_A + T_C = 8g$

So, $T_A = 8g - 47 = 31.4$ N (to 3 s.f.)

The tinsel at A snaps and a downward force is applied at B to keep the banner horizontal. Find the magnitude of the force applied at B and the tension in the tinsel attached at C.

Take moments about C.

$8g \times 1.2 = F_B \times 0.6$

So, $F_B = \dfrac{94.1}{0.6} = 157$ N

Resolve vertically.

$T_C = 8g + 157$

So, $T_C = 235$ N

Significant moments in life — birthdays, exams, exam results...

Well, lots of stuff to learn in this section. I hope you got your head around it because up next is a load of questions to test your vector and statics know-how. Onwards and upwards folks, well, clockwise, anticlockwise, upwards and downwards...

M1 Section 2 — Practice Questions

Time to resolve the force applied to your revision along the question axis... That didn't sound as good as I hoped. Still, at least you can distract yourself from my terrible sense of humour with these excellent practice questions.

Warm-up Questions

Take $g = 9.8$ ms^{-2} in each of these questions.

1) Find the average velocity of a cyclist who cycles at 15 kmh^{-1} north for 15 minutes and then cycles south at 10 kmh^{-1} for 45 minutes.

2) Find $a + 2b - 3c$ where $a = 3i + 7j$; $b = -2i + 2j$; $c = i - 3j$.

3) A plane flies 40 miles due south, then 60 miles southeast before going 70 miles on a bearing of 020°. Find the distance and bearing on which the plane must fly to return to its starting point.

4) The following items are dropped from a height of 2 m onto a cushion:

   a)   a full 330 ml drinks can   b)   an empty drinks can   c)   a table tennis ball

   The time each takes to fall is measured.
   Draw a model of each situation and list any assumptions which you've made.

5) A car is travelling at 25 mph along a straight level road.
   Draw a model of the situation and list any assumptions which you've made.

6) Find the magnitudes and directions to the horizontal of the resultant force in each situation.

   a)   b)   c)

7) A mass of M kg is suspended by two light wires A and B, with angles 60° and 30° to the vertical respectively, as shown. The tension in A is 20 N. Find: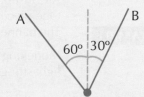

   a)   the tension in wire B

   b)   the value of M

8) 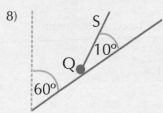   A particle, Q, of mass m kg, is in equilibrium on a smooth plane which makes an angle of 60° to the vertical. This is achieved by an attached string S, with tension 70 N, angled at 10° to the plane, as shown. Draw a force diagram and find both the mass of Q and the reaction on it from the surface.

9) A toy train of weight 25 N is pulled up a slope of 20°. The tension in the string is 25 N and a frictional force of 5 N acts on the train. Find the normal contact force and the resultant force acting on the train.

10)   a)   Describe the motion of a mass of 12 kg pushed by a force of 50 N parallel to the rough horizontal plane on which the mass is placed. The plane has coefficient of friction $\mu = \frac{1}{2}$.

       b)   What minimum force would be needed to move the mass in part a)?

11) A 60 kg uniform beam AE of length 14 m is in equilibrium, supported by two vertical ropes attached to B and D as shown.

    Find the tensions in the ropes to 1 d.p.

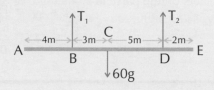

M1 Section 2 — Practice Questions

Now you can apply your vector and statics knowledge to some exam questions so that when you take the real thing there'll be no need for your pen to remain at equilibrium.

Exam Questions

Take $g = 9.8$ ms^{-2} in each of these questions.

1   The diagram shows two forces acting on a particle.
    Find the magnitude and direction of the resultant force.

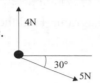

(4 marks)

2   A girl can swim at 2 ms^{-1} in still water. She is swimming across a river in a direction perpendicular to the riverbank. The river is flowing at 3 ms^{-1} and so it carries the girl downstream.
    Find the magnitude of the resultant velocity of the girl and the
    angle it makes with the riverbank.

(4 marks)

3   A box of mass 39 kg is at rest on a rough horizontal surface. The box is pushed with a force of 140 N from an angle of 20° above the horizontal. The box remains stationary.

a)   Draw a diagram to show the four forces acting on the box.

(2 marks)

b)   Calculate the magnitude of the normal reaction force and the frictional force on the box.

(4 marks)

4   Two forces, $A = (2\mathbf{i} - 11\mathbf{j})$ N and $B = (7\mathbf{i} + 5\mathbf{j})$ N, act on a particle:

a)   Work out the resultant of A and B.

(2 marks)

b)   Find the magnitude of the resultant force.

(2 marks)

5   A horizontal uniform beam with length x and weight 18 N is held in equilibrium by 2 strings.
    One string is attached to one end of the beam and the other at point A, 3 m from the first string.
    The tension in the string at point A is 12 N.
    Show that $x = 4$ m.

(3 marks)

6   A force of magnitude 7 N acts horizontally on a particle. Another force, of magnitude 4 N, acts on the particle at an angle of 30° to the horizontal force. The resultant of the two forces has a magnitude R at an angle α above the horizontal.
    Find:

a)   the force R.

(4 marks)

b)   the angle α.

(2 marks)

M1 Section 2 — Practice Questions

A couple more pages of questions to practise on here, so keep on truckin' (remembering that "truckin'" is a vector quantity with both magnitude *and* direction). The normal reaction to that joke is a sigh... on with the questions.

7 A mass, *M*, is attached to the end of a light inextensible rod fixed at point *X*. A force of magnitude 10 N is applied to *M* at 14° to the horizontal. *M* is held in equilibrium at an angle of 35° as shown in the diagram. Find the mass of *M*.

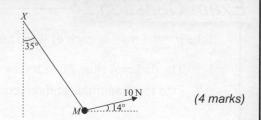

(4 marks)

8 Three forces of magnitudes 15 N, 12 N and *W* act on a particle as shown. Given that the particle is in equilibrium, find:

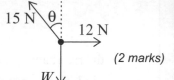

   a) the value of θ

(2 marks)

   b) the force *W*.

(2 marks)

The force *W* is now removed.

   c) State the magnitude and direction of the resultant of the two remaining forces.

(2 marks)

9 A sledge is held at rest on a smooth slope which makes an angle of 25° with the horizontal. The rope holding the sledge is at an angle of 20° above the slope. The normal reaction acting on the sledge due to contact with the surface is 80 N.

Find:

   a) The tension, *T*, in the rope.

(3 marks)

   b) The weight of the sledge.

(2 marks)

10 A particle is initially at position vector $(\mathbf{i} + 2\mathbf{j})$ m travelling with constant velocity $(3\mathbf{i} + \mathbf{j})$ ms^{-1}. After 8 s it reaches point *A*. A second particle has constant velocity $(-4\mathbf{i} + 2\mathbf{j})$ ms^{-1} and takes 5 s to travel from point *A* to point *B*.

Find the position vectors of points *A* and *B*.

(4 marks)

11 A 6 m long uniform beam of mass 20 kg is in equilibrium. One end is resting on a vertical pole, and the other end is held up by a vertical wire attached to that end so that the beam rests horizontally. There are two 10 kg weights attached to the beam, situated 2 m from either end.

   a) Draw a diagram of the beam including all the forces acting on it.

(2 marks)

Find, in terms of *g*:

   b) *T*, the tension in the wire.

(3 marks)

   c) *R*, the normal reaction at the pole.

(2 marks)

M1 Section 2 — Practice Questions

12   A particle of mass m kg is held in equilibrium by two light inextensible strings.
One string is attached at an angle of 50° to the horizontal, as shown, and has tension A N.
The other string is horizontal and has tension 58 N. Find:

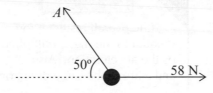

    a)   the magnitude of A.

                                 (3 marks)

    b)   the mass, m, of the particle.

                                   (3 marks)

13   At $t = 0$, a particle P is at position vector $(\mathbf{i} + 5\mathbf{j})$ m, relative to a fixed origin. P is moving with a constant velocity of $(7\mathbf{i} - 3\mathbf{j})$ ms^{-1}. After 4 s the velocity of P changes to $(a\mathbf{i} + b\mathbf{j})$ ms^{-1}. After a further 3.5 seconds P reaches position vector $(15\mathbf{i})$ m. Find:

    a)   the speed of P at $t = 0$

                                   (2 marks)

    b)   the bearing of P at $t = 0$

                                   (3 marks)

    c)   the values of a and b.

                                   (4 marks)

14   A 2 kg ring threaded on a rough horizontal rod is pulled by a rope having tension S and attached at 40° to the horizontal, as shown. The coefficient of friction between the rod and the ring is $\frac{3}{10}$. Given that the ring is about to slide, find the magnitude of S.

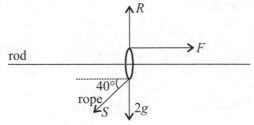

                                   (4 marks)

15   A uniform log XY rests horizontally on two rocks P and Q, where $XP = x$ m, $PQ = 0.4$ m, and 0.6 m $< x < 0.9$ m. The log has weight 80 N and length 1.8 m, and the reaction at Q is 50 N.

    a)   Draw a diagram to show the forces acting on the log.

                                   (2 marks)

    b)   Find the distance x.

                                   (3 marks)

    A large bird lands gently on the end of the log (Y), causing it to almost tip over about Q.

    c)   What assumptions would you make when modelling the log, rock and the bird?

                                   (3 marks)

    d)   Find the mass of the bird.

                                   (4 marks)

    e)   Find the magnitude of the normal reaction at Q.

                                   (2 marks)

Newton's Laws

That clever chap Isaac Newton established 3 laws involving motion. You need to know <u>all</u> of them.

Newton's Laws of Motion

Newton's First Law
A body will <u>stay at rest</u> or <u>maintain a constant velocity</u> — unless an extra force acts to <u>change</u> that motion.

Newton's Second Law

$$F_{net} = ma$$

F_{net} (the <u>overall resultant force</u>) is equal to the mass multiplied by the acceleration. Also, F_{net} and a are in the same direction.

Newton's Third Law
For <u>two bodies</u> in contact with each other, the force each applies to the other is <u>equal in magnitude</u> but <u>opposite in direction</u>.

<u>Hint</u>: $F_{net} = ma$ is sometimes just written as $F = ma$, but it means the same thing.

F = ma equations are sometimes known as 'equations of motion'.

Resolve Forces in Perpendicular Directions

EXAMPLE A mass of 4 kg at rest on a smooth horizontal plane is acted on by a horizontal force of 5 N. Find the acceleration of the mass and the normal reaction from the plane. Take $g = 9.8$ ms^{-2}.

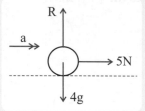

Resolve horizontally:

$F_{net} = ma$

Always write $F_{net} = ma$ first.

$5 = 4a$

$a = 1.25$ ms^{-2} in the direction of the horizontal force

Resolve vertically:

$F_{net} = ma$, so $R - 4g = 4 \times 0$

$R = 4g = 39.2$ N

EXAMPLE A particle of weight 30 N is being accelerated across a smooth horizontal plane by a force of 6 N acting at an angle of 25° to the horizontal, as shown. Given that the particle starts from rest, find:

a)   its speed after 4 seconds   b)   the magnitude of the normal reaction with the plane.

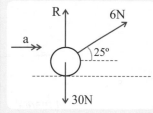

a) Resolve horizontally:

$F_{net} = ma$

$6\cos 25° = \dfrac{30}{g}a$ so $a = 1.776...$ ms^{-2}

$v = u + at$

$v = 0 + 1.776... \times 4 = 7.11$ ms^{-1} (to 3 s.f.)

b) Resolve vertically:

$F_{net} = ma$

$R + 6\sin 25° - 30 = \dfrac{30}{g} \times 0$

So $R = 30 - 6\sin 25° = 27.5$ N (to 3 s.f.)

You can apply F = ma to i + j Vectors too

EXAMPLE A particle of mass m kg is acted upon by two forces, $(6\mathbf{i} - \mathbf{j})$ N and $(2\mathbf{i} + 4\mathbf{j})$ N, resulting in an acceleration of 9 ms^{-2}. Find the value of m.

Resultant force, $R = (6\mathbf{i} - \mathbf{j}) + (2\mathbf{i} + 4\mathbf{j}) = (8\mathbf{i} + 3\mathbf{j})$ N

Magnitude of $R = |R| = \sqrt{8^2 + 3^2} = \sqrt{73} = 8.54$ N

$F = ma$, so $8.54 = 9m$
hence $m = 0.95$ kg

The force of $(2\mathbf{i} + 4\mathbf{j})$ N ceases to be applied to the particle. Calculate the new acceleration.

Magnitude $= \sqrt{6^2 + (-1)^2} = \sqrt{37} = 6.08$ N $a = \dfrac{F}{m} = \dfrac{6.08}{0.95} = 6.4$ ms^{-2}

Interesting Newton fact: Isaac Newton had a dog called Diamond...

Did you know that Isaac Newton and Stephen Hawking both held the same position at Cambridge University? And the dog fact about Newton is true — don't ask me how I know such things, just bask in my amazing knowledge of all things trivial.

Friction and Inclined Planes

Solving these problems involves careful use of $F_{net} = ma$, $F \le \mu R$ and the equations of motion.

Use F = ma in *Two Directions* for *Inclined Plane* questions

For inclined slope questions, it's much easier to resolve forces parallel and perpendicular to the plane's surface.

EXAMPLE
A mass of 600 g is propelled up the line of greatest slope of a smooth plane inclined at 30° to the horizontal. If its velocity is 3 ms⁻¹ after the propelling force has stopped, find the distance it travels before coming to rest and the magnitude of the normal reaction. Use $g = 9.8$ ms⁻².

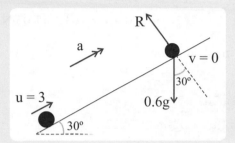

Resolve in ↗ direction:

$F_{net} = ma$

$-0.6g\sin30° = 0.6a$

Taking up the plane as +ve.

$a = -4.9$ ms⁻²

$v^2 = u^2 + 2as$

$0 = 3^2 + 2(-4.9)s$

So $s = 0.92$ m

Resolve in ↖ direction:

$F_{net} = ma$

$R - 0.6g\cos30° = 0.6 \times 0$

So $R = 5.09$ N

Remember that friction always acts in the opposite direction to the motion.

EXAMPLE
A small body of weight 20 N accelerates from rest and moves a distance of 5 m down a rough plane angled at 15° to the horizontal. Draw a force diagram and find the coefficient of friction between the body and the plane given that the motion takes 6 seconds. Take $g = 9.8$ ms⁻².

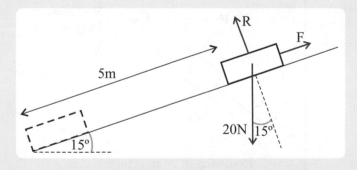

$u = 0, \quad s = 5, \quad t = 6, \quad a = ?$

Use one of the equations of motion: $s = ut + \frac{1}{2}at^2$

$5 = (0 \times 6) + (\frac{1}{2}a \times 6^2)$ so $a = 0.2778$ ms⁻²

Resolving in ↖ direction:

$F_{net} = ma$

$R - 20\cos15° = \frac{20}{g} \times 0$

So: $R = 20\cos15° = 19.32$ N

Resolving in ↙ direction:

$F_{net} = ma$

$20\sin15° - F = \frac{20}{g} \times 0.2778$

$F = 4.609$ N

It's sliding, so $F = \mu R$

$4.609 = \mu \times 19.32$

$\mu = 0.24$ (to 2 d.p.)

Friction and Inclined Planes

Friction opposes motion, so it also increases the tension in whatever's doing the pulling...

EXAMPLE   A mass of 3 kg is being pulled up a plane inclined at 20° to the horizontal by a rope parallel to the surface. Given that the mass is accelerating at 0.6 ms⁻² and that the coefficient of friction is 0.4, find the tension in the rope. Take g = 9.8 ms⁻².

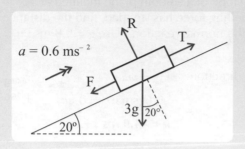

Resolving in ↖ direction:

$$F_{net} = ma$$
$$R - 3g\cos 20° = 3 \times 0$$

so: $R = 3g\cos 20° = 27.63$ N

The mass is sliding, so $F = \mu R$
$$= 0.4 \times 27.63 = 11.05 \text{ N}$$

Resolving in ↗ direction:

$$F_{net} = ma$$
$$T - F - 3g\sin 20° = 3 \times 0.6$$
$$T = 1.8 + 11.05 + 3g\sin 20° = 22.9 \text{ N}$$

Friction Opposes Limiting Motion

For a body <u>at rest</u> but on the point of moving <u>down</u> a plane, the friction force is <u>up</u> the plane. A body about to move <u>up</u> a plane is opposed by friction <u>down</u> the plane. Remember it well (it's about to come in handy).

EXAMPLE   A 4 kg box is placed on a 30° plane where $\mu = 0.4$. A force Q maintains equilibrium by acting up the plane parallel to the line of greatest slope. Find Q if the box is on the point of sliding:

a) up the plane   b) down the plane.

a)

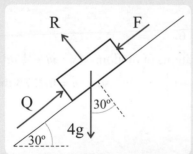

$$F_{net} = ma$$

Resolving in ↖ direction:
$$R - 4g\cos 30° = 0$$
$$R = 4g\cos 30°$$

$F = \mu R$
$$= 0.4 \times 4g\cos 30°$$
$$= 1.6g\cos 30°$$

Resolving in ↗ direction:
$$Q - 4g\sin 30° - F = 4 \times 0$$
$$Q = 4g\sin 30° + 1.6g\cos 30°$$
$$= 33.2 \text{ N}$$

b)

Resolving in ↖ direction:
$$R = 4g\cos 30°$$
$$F = 1.6g\cos 30°$$

Resolving in ↗ direction:
$$Q - 4g\sin 30° + F = 4 \times 0$$
$$Q = 4g\sin 30° - 1.6g\cos 30°$$
$$Q = 6.02 \text{ N}$$

So for equilibrium 6.02 N ≤ Q ≤ 33.2 N

Inclined planes — nothing to do with suggestible Boeing 737s...

The main thing to remember is that you can choose to resolve in any two directions as long as they're <u>perpendicular</u>. It makes sense to choose the directions that involve doing as little work as possible. Obviously.

Connected Particles

Like Laurel goes with Hardy and Posh goes with Becks, some particles are destined to be together...

Connected Particles act like One Mass

Particles connected together have the <u>same speeds</u> and <u>accelerations</u> as each other, unless the connection <u>fails</u>.
Train carriages moving together have the same acceleration.

> **EXAMPLE**
>
> A 30 tonne locomotive engine is pulling a single 10 tonne carriage as shown. They are accelerating at 0.3 ms^{-2} due to the force P generated by the engine. It's assumed that there are no forces resistant to motion. Find P and the tension in the coupling.

Here's the pretty picture:

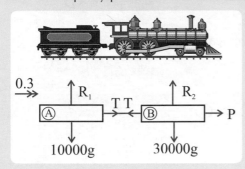

For A: $F_{net} = ma$
$T = 10\,000 \times 0.3$
$T = 3000$ N

For B: $F_{net} = ma$
$P - T = 30\,000 \times 0.3$
$P = 12000$ N

Pulleys (and 'Pegs') are always Smooth

In M1 questions, you can always assume that the <u>tension</u> in a string will be the <u>same</u> either side of a <u>smooth pulley</u>.

> **EXAMPLE**
>
> Masses of 3 kg and 5 kg are connected by an inextensible string and hang vertically either side of a smooth pulley. They are released from rest. Find their acceleration and the time it takes for each to move 40 cm. State any assumptions made in your model. Take $g = 9.8$ ms^{-2}.

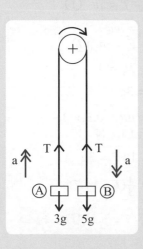

For A: $F_{net} = ma$
Resolving upwards: $T - 3g = 3a$ ①

For B: $F_{net} = ma$
Resolving downwards: $5g - T = 5a$
$T = 5g - 5a$ ②

Sub ② into ① : $(5g - 5a) - 3g = 3a$
$a = 2.45$ ms^{-2}

List variables: $u = 0$; $a = 2.45$; $s = 0.4$

Use an equation with u, a, s and t in it:

$s = ut + \frac{1}{2}at^2$

$0.4 = (0 \times t) + (\frac{1}{2} \times 2.45 \times t^2)$ So $t = \sqrt{\dfrac{0.8}{2.45}} = 0.57$ s

Assumptions: The 3 kg mass does not hit the pulley; there's no air resistance; the string is 'light' so the tension is the same for both A and B, and it doesn't break; the string is inextensible so the acceleration is the same for both masses.

Connected Particles

Use F = ma in the *Direction Each Particle Moves*

EXAMPLE A mass of 3 kg is placed on a smooth horizontal table. A light inextensible string connects it over a smooth peg to a 5 kg mass which hangs vertically as shown. Find the tension in the string if the system is released from rest. Take g = 9.8 ms⁻².

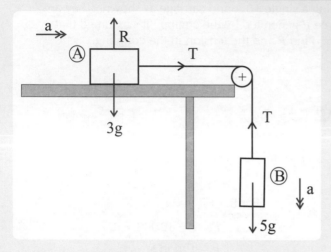

For A:
Resolve horizontally:

$$F_{net} = ma$$

$$T = 3a$$

$$a = \frac{T}{3} \quad \text{①}$$

For B:
Resolve vertically:

$$F_{net} = ma$$

$$5g - T = 5a \quad \text{②}$$

Sub ① into ②:

$$5g - T = 5 \times \frac{T}{3}$$

So $\frac{8}{3}T = 5g$

$$T = 18.4 \text{ N} \quad \text{(to 3 s.f.)}$$

Remember to use F ≤ μR on *Rough Planes*

More complicated pulley and peg questions have <u>friction</u> for you to enjoy too.

EXAMPLE The peg system of the example above is set up again. However, this time a friction force, F, acts on the 3 kg mass due to the table top now being rough, with coefficient of friction μ = 0.5. Find the new tension in the string when the particles are released from rest. Take g = 9.8 ms⁻².

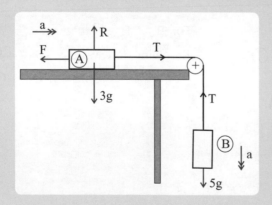

For B: Resolving vertically: $5g - T = 5a$ ①

For A: Resolving horizontally: $F_{net} = ma$
$$T - F = 3a \quad \text{②}$$

Resolving vertically: $R - 3g = 0$

$$\boldsymbol{R = 3g}$$

The particles are moving, so
$$F = \mu R = 0.5 \times 3g$$
$$= \textbf{14.7 N}$$

Sub this into ②: $T - 14.7 = 3a$

$$\boldsymbol{a = \frac{1}{3}(T - 14.7)}$$

Sub this into ①:

$$5g - T = 5 \times \frac{1}{3}(T - 14.7)$$

$$8T = 147 + 73.5$$

$$T = 27.6 \text{ N}$$

Useful if you're hanging over a Batman-style killer crocodile pit...

It makes things a lot easier when you know that connected particles act like one mass, and that in M1 pulleys can always be treated as smooth. Those examiners occasionally do try to make your life easier, honestly.

Connected Particles

Particles A and B of mass 4 kg and 10 kg respectively are connected by a light inextensible string over a smooth pulley as shown. A force of 15 N acts on A at an angle of 25° to a rough horizontal plane where $\mu = 0.7$. When B is released from rest it takes 2 s to fall d m to the ground. Find d. Take g = 9.8 ms^{-2}.

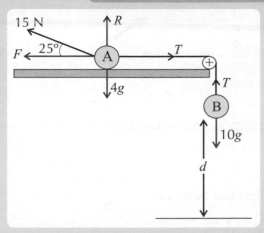

For A: Resolve vertically to find R:
$R = 4g - 15\sin25° = 32.9$ N
$F = \mu R = 0.7 \times 32.9 = 23.0$ N

$F_{net} = ma$
$T - 23.0 - 15\cos25° = 4a$
so $T = 4a + 36.6$ ①

For B: $F_{net} = ma$
$10g - T = 10a$
so $T = 98 - 10a$ ②

Substitute ① into ②:
$4a + 36.6 = 98 - 10a$
so $a = 4.4$ ms^{-2}

Using $s = ut + \frac{1}{2}at^2$: $s = d$, $u = 0$, $t = 2$, $a = 4.4$
so $d = (0 \times 2) + \frac{1}{2}(4.4 \times 2^2) = 8.8$ m

When B hits the ground A carries on moving along the plane. How long does it take A to stop after B hits the ground?

Speed of A when B hits the ground:
$v = u + at$: $u = 0$, $a = 4.4$, $t = 2$
$v = 0 + (4.4 \times 2) = 8.8$ ms^{-1}

Resolve to find stopping force on A:
$F = 23 + 15\cos25° = 36.6$ N

Deceleration, $a = \frac{F}{m} = \frac{36.6}{4} = 9.15$ ms^{-2}

Time taken to stop:
$t = \frac{(v - u)}{a}$: $v = 8.8$, $u = 0$, $a = 9.15$
so $t = \frac{8.8 - 0}{9.15} = 0.96$ s (to 2 d.p.)

Rough Inclined Plane questions need *Really Good* force diagrams

You know the routine... resolve forces parallel and perpendicular to the plane... yawn.

A 3 kg mass is held in equilibrium on a rough ($\mu = 0.4$) plane inclined at 30° to the horizontal. It is attached by a light, inextensible string to a mass of M kg hanging vertically beneath a smooth pulley, as shown. Find M if the 3 kg mass is on the point of sliding up the plane. Take g = 9.8 ms^{-2}.

For B: Resolving vertically: $F_{net} = ma$

$Mg - T = M \times 0$
$T = Mg$

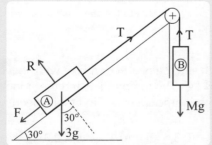

For A: Resolving in ↖ direction: $F_{net} = ma$

$R - 3g\cos30° = 3 \times 0$
$R = 3g\cos30°$

It's limiting friction, so: $F = \mu R = 0.4 \times 3g\cos30°$
$= 10.18$ N

For A: Resolving in ↗ direction:
$T - F - 3g\sin30° = 3 \times 0$
$Mg - 10.18 - 3g\sin30° = 0$
$M = 2.54$ kg (to 3 s.f.)

Connected particles — together forever... *isn't it beautiful?*

The key word here is <u>rough</u>. If a question mentions the surface is rough, then cogs should whirr and the word 'friction' should pop into your head. Take your time with force diagrams of rough inclined planes — I had a friend who rushed into drawing a diagram, and he ended up with a broken arm. But that was years later, now that I come to think of it.

Momentum

Momentum has Magnitude and Direction

Momentum is a measure of how much "umph" a <u>moving object</u> has, due to its <u>mass</u> and <u>velocity</u>.
Total momentum <u>before</u> a collision equals total momentum <u>after</u> a collision.
This idea is called "<u>Conservation of Momentum</u>".
Because it's a <u>vector</u>, the <u>sign</u> of the velocity in momentum is important.

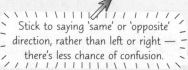

Momentum = Mass × Velocity

The unit of momentum is kgms⁻¹ or Ns

EXAMPLE Particles A and B, each of mass 5 kg, move in a straight line with velocities 6 ms⁻¹ and 2 ms⁻¹ respectively. After collision mass A continues in the same direction with velocity 4.2 ms⁻¹. Find the velocity of B after impact.

Before

Momentum A + Momentum B = Momentum A + Momentum B

$(5 \times 6) + (5 \times 2) = (5 \times 4.2) + (5 \times v)$

$40 = 21 + 5v$

After

So $v = 3.8$ ms⁻¹ in the same direction as before

Draw '<u>before</u>' and '<u>after</u>' diagrams to help you see what's going on.

Stick to saying 'same' or 'opposite' direction, rather than left or right — there's less chance of confusion.

EXAMPLE Particles A and B of mass 6 kg and 3 kg are moving towards each other at speeds of 2 ms⁻¹ and 1 ms⁻¹ respectively. Given that B rebounds with speed 3 ms⁻¹ in the opposite direction to its initial velocity, find the velocity of A after the collision.

Before   **After**

$(6 \times 2) + (3 \times -1) = (6 \times v) + (3 \times 3)$

$9 = 6v + 9$

$v = 0$

Masses Joined Together have the Same Velocity

Particles that <u>stick together</u> after impact are said to "<u>coalesce</u>". After that you can treat them as just <u>one object</u>.

EXAMPLE Two particles of mass 40 g and M kg move towards each other with speeds of 6 ms⁻¹ and 3 ms⁻¹ respectively. Given that the particles coalesce after impact and move with a speed of 2 ms⁻¹ in the same direction as that of the 40 g particle's initial velocity, find M.

Before   **After**

$(0.04 \times 6) + (M \times -3) = [(M + 0.04) \times 2]$

$0.24 - 3M = 2M + 0.08$

$5M = 0.16$

$M = 0.032$ kg

Don't forget to convert all masses to the same units.

EXAMPLE A lump of ice of mass 0.1 kg is slid across the smooth surface of a frozen lake with speed 4 ms⁻¹. It collides with a stationary stone of mass 0.3 kg. The lump of ice and the stone then move in opposite directions to each other with the same speeds. Find their speed.

Before   **After**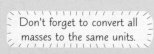

$(0.1 \times 4) + (0.3 \times 0) = (0.1 \times -v) + 0.3v$

$0.4 = -0.1v + 0.3v$

$v = 2$ ms⁻¹

Ever heard of Hercules?

Well, he carried out 12 tasks. Nothing to do with momentum, but if you're feeling sorry for yourself for doing M1, think on.

Impulse

An impulse changes the momentum of a particle in the direction of motion.

Impulse is **Change in Momentum**

To work out the impulse that's acted on an object, just <u>subtract</u> the object's <u>initial</u> momentum from its <u>final</u> momentum. Impulse is measured in <u>newton seconds</u> (<u>Ns</u>).

$$\text{Impulse} = mv - mu$$

EXAMPLE A body of mass 500 g is travelling in a straight line. Find the magnitude of the impulse needed to increase its speed from 2 ms^{-1} to 5 ms^{-1}.

Impulse = Change in momentum

$= mv - mu$

$= (0.5 \times 5) - (0.5 \times 2)$

$= 1.5$ Ns

This is called the impulse-momentum principle. Ooooh, aaaaaah.

Momentum = mass × velocity

EXAMPLE A 20 g ball is dropped 1 m onto the ground. Immediately after rebounding the ball has a speed of 2 ms^{-1}. Find the impulse given to the ball by the ground. How high does the ball rebound? Take $g = 9.8$ ms^{-2}.

First you need to work out the ball's speed as it reaches the ground:

List the variables you're given: $u = 0$
$s = 1$
$a = 9.8$
$v = ?$

The ball was <u>dropped</u>, so it started from $u = 0$.

Acceleration due to gravity.

Choose an equation containing u, s, a and v: $v^2 = u^2 + 2as$

$v^2 = 0^2 + (2 \times 9.8 \times 1)$

$v = 4.43$ ms^{-1}

The sign is really important. Make sure that <u>down</u> is <u>positive</u> in this part of the question.

Now work out the impulse as the ball hits the ground and rebounds:

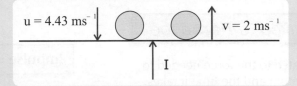

Impulse $= mv - mu$

$= (0.02 \times 2) - (0.02 \times -4.43)$

$= 0.129$ Ns (to 3 s.f.)

Finally you need to use a new equation of motion to find s (the greatest height the ball reaches after the bounce):

List the variables: $u = 2$
$v = 0$
$a = -9.8$
$s = ?$

$v = 0$ at the ball's greatest height.

a is negative because the ball is <u>decelerating</u>.

$v^2 = u^2 + 2as$

$0^2 = 2^2 + (2 \times -9.8 \times s)$

$s = 0.204$ m (to 3 s.f.)

Impulse

Impulses always Balance in Collisions

During impact between particles A and B, the impulse that A gives to B is the same as the impulse that B gives to A, but in the opposite direction.

EXAMPLE

A mass of 2 kg moving at 2 ms^{-1} collides with a mass of 3 kg which is moving in the same direction at 1 ms^{-1}. The 2 kg mass continues to move in the same direction at 1 ms^{-1} after impact. Find the impulse given by the 2 kg mass to the other mass.

Using "conservation of momentum":

$(2 \times 2) + (3 \times 1) = (2 \times 1) + 3v$

So $v = 1\frac{2}{3}$ ms^{-1}

Impulse (on B) = $mv - mu$ (for B)

$= (3 \times 1\frac{2}{3}) - (3 \times 1)$

$= 2$ Ns

The impulse B gives to A is $(2 \times 1) - (2 \times 2) = -2$ Ns. Aside from the different direction, you can see it's the same — so you didn't actually need to find v for this question.

EXAMPLE

Two snooker balls A and B have a mass of 0.6 kg and 0.9 kg respectively. The balls are initially at rest on a snooker table. Ball A is given an impulse of magnitude 4.5 Ns towards ball B. Modelling the snooker table as a smooth horizontal plane, find the speed of ball A before it collides with B.

Impulse = $mv - mu = 4.5 = (0.6 \times v) - (0.6 \times 0) = 0.6v$

Therefore $v = \dfrac{4.5}{0.6} = 7.5$ ms^{-1}

The balls collide and move away in the direction A was travelling before the collision. Find the speed of ball A after the collision, given that the speed of ball B is 4 ms^{-1}.

Using conservation of momentum:

$(0.6 \times 7.5) + (0.9 \times 0) = 0.6v + (0.9 \times 4)$

So, $v = 1.5$ ms^{-1}

Impulse is linked to Force too

Impulse is also related to the force needed to change the momentum and the time it takes.

Impulse = Force × Time

EXAMPLE

A 0.9 tonne car increases its speed from 30 kmh^{-1} to 40 kmh^{-1}. Given that the maximum additional constant forward force the car's engine can produce is 1 kN, find the shortest time it will take to achieve this change in speed.

Impulse = $mv - mu$

$= (900 \times \dfrac{40\,000}{3600}) - (900 \times \dfrac{30\,000}{3600})$

$= 2500$ Ns

Now use Impulse = Force × Time:

$2500 = 1000 \times t$

$t = 2.5$ s

You should understand all of M1 by now — *if not, turn to page 147 and have another go...*

Impulse is change in momentum — remember that and you'll be laughing. Anyway, I know all about impulse. Those orange nylon flares looked great in the shop window, but I really should have tried them on before I bought them.

M1 Section 3 — Practice Questions

Find the coefficient of friction between a student's pen and a sheet of paper. Model the pen as a rod and the paper as a rough plane... or else you could just answer the questions below, which would be a better use of your time.

Warm-up Questions

Take g = 9.8 ms⁻² in each of these questions.

1) A horizontal force of 2 N acts on a 1.5 kg particle initially at rest on a smooth horizontal plane. Find the speed of the particle 3 seconds later.

2) Two forces act on a particle of mass 8 kg which is initially at rest on a smooth horizontal plane. The two forces are $(24\mathbf{i} + 18\mathbf{j})$ N and $(6\mathbf{i} + 22\mathbf{j})$ N (with $\mathbf{i}$ and $\mathbf{j}$ being perpendicular unit vectors in the plane). Find the magnitude and direction of the resulting acceleration of the particle and its displacement after 3 seconds.

3) A horizontal force P acting on a 2 kg mass generates an acceleration of 0.3 ms⁻². Given that the mass is in contact with a rough horizontal plane which resists motion with a force of 1 N, find P. Then find the coefficient of friction, μ, to 2 d.p.

4) A brick of mass 1.2 kg is sliding down a rough plane which is inclined at 25° to the horizontal. Given that its acceleration is 0.3 ms⁻², find the coefficient of friction between the brick and the plane. What assumptions have you made?

5) An army recruit of weight 600 N steps off a tower and accelerates down a "death slide" wire as shown. The recruit hangs from a light rope held between her hands and looped over the wire. The coefficient of friction between the rope and wire is 0.5. Given that the wire is 20 m long and makes an angle of 30° to the horizontal throughout its length, find how fast the recruit is travelling when she reaches the end of the wire.

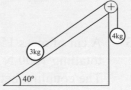

6) A 2 tonne tractor experiences a resistance force of 1000 N whilst driving along a straight horizontal road. If the tractor engine provides a forward force of 1500 N and it's pulling a 1 tonne trailer, find the resistance force acting on the trailer, and the tension in the coupling between the tractor and trailer, if they are moving with constant speed.

7) Two particles are connected by a light inextensible string, and hang in a vertical plane either side of a smooth pulley. When released from rest the particles accelerate at 1.2 ms⁻². If the heavier mass is 4 kg, find the weight of the other.

8) Two particles of mass 3 kg and 4 kg are connected by a light, inextensible, string passing over a smooth pulley as shown. The 3 kg mass is on a smooth slope angled at 40° to the horizontal. Find the acceleration of the system if released from rest, and find the tension in the string. What force acting on the 3 kg mass parallel to the plane would be needed to maintain equilibrium?

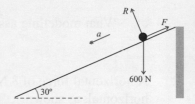

9) Each diagram represents the motion of two particles moving in a straight line. Find the missing mass or velocity (all masses are in kg and all velocities are in ms⁻¹).

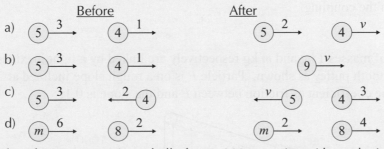

10) An impulse of 2 Ns acts against a ball of mass 300 g moving with a velocity of 5 ms⁻¹. Find the ball's new velocity.

11) A particle of mass 450 g is dropped 2 m onto a floor. It rebounds to two thirds of its original height. Find the impulse given to the particle by the ground.

M1 Section 3 — Practice Questions

Right, those warm-up questions should have given you the momentum to get straight into these exam questions.
May the *ma* be with you...

Exam Questions

Take g = 9.8 ms⁻² in each of these questions.

1   A crane moves a mass of 300 kg, A, suspended by two light cables AB and AC attached to a horizontal
     movable beam BC. The mass is moved in the direction of the line of the supporting beam BC during
     which time the cables maintain a constant angle of 40° to the horizontal, as shown.

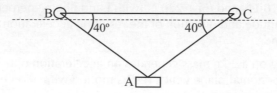

a)   The mass is initially moving with constant speed. Find the tension in each cable.

(4 marks)

b)   The crane then moves the mass with a constant acceleration of 0.4 ms⁻².
     Find the tension in each cable.

(6 marks)

c)   What modelling assumptions have you made in part b)?

(2 marks)

2   A horizontal force of 8 N just stops a mass of 7 kg from sliding down a plane inclined at 15° to the
     horizontal, as shown.

8 N

15°

a)   Calculate the coefficient of friction between the mass and the plane to 2 d.p.

(5 marks)

b)   The 8 N force is now removed. Find how long the mass takes to slide a distance
     of 3 m down the line of greatest slope.

(7 marks)

3   A car of mass 1500 kg is pulling a caravan of mass 500 kg. They experience resistance forces
     totalling 1000 N and 200 N respectively. The forward force generated by the car's engine is 2500 N.
     The coupling between the two does not break.

a)   Find the acceleration of the car and caravan.

(3 marks)

b)   Find the tension in the coupling.

(2 marks)

4   Two particles *P* and *Q* of masses 1 kg and *m* kg respectively are linked by a light inextensible
     string passing over a smooth pulley as shown. Particle *P* is on a rough slope inclined at 20° to
     the horizontal, where the coefficient of friction between *P* and the slope is 0.1.

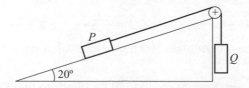

a)   Given that *P* is about to slide down the plane, find the mass of *Q*.

(5 marks)

b)   Describe the motion of the system if the mass of *Q* is 1 kg.

(5 marks)

M1 Section 3 — Practice Questions

5   Two particles of mass 0.8 kg and 1.2 kg are travelling in the same direction along a straight line with speeds of 4 ms^{-1} and 2 ms^{-1} respectively until they collide. After the collision the 0.8 kg mass has a velocity of 2.5 ms^{-1} in the same direction. The 1.2 kg mass then continues with its new velocity until it collides with a mass of m kg travelling with a speed of 4 ms^{-1} in the opposite direction to it.

Given that both particles are brought to rest by this collision, find the mass m.

(4 marks)

6   A coal wagon of mass 4 tonnes is rolling along a straight rail track at 2.5 ms^{-1}. It collides with a stationary wagon of mass 1 tonne. During the collision the wagons become coupled and move together along the track.

   a)   Find their speed after the collision.

(2 marks)

   b)   Find the impulse given to the more massive wagon.

(2 marks)

   c)   State two assumptions made in your model.

(2 marks)

7   Two forces, $(x\mathbf{i} + y\mathbf{j})$ N and $(5\mathbf{i} + \mathbf{j})$ N, act on a particle P of mass 2.5 kg.
The resultant of the two forces is $(8\mathbf{i} - 3\mathbf{j})$ N.
Find:

   a)   the values of x and y.

(2 marks)

   b)   the acceleration of P.

(3 marks)

8   Two particles A and B are connected by a light inextensible string which passes over a smooth fixed pulley as shown. A has a mass of 7 kg and B has a mass of 3 kg. The particles are released from rest with the string taut, and A falls freely until it strikes the ground travelling at a speed of 5.9 ms^{-1}. A does not rebound after hitting the floor.

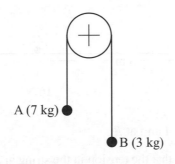

A (7 kg)

B (3 kg)

   a)   Find the time taken for A to hit the ground.

(4 marks)

   b)   How far will B have travelled when A hits the ground?

(2 marks)

   c)   Find the time (in s) from when A hits the ground until the string becomes taut again.

(4 marks)

General Certificate of Education
Advanced Subsidiary (AS) and Advanced Level

Mechanics M1 — Practice Exam One

Time Allowed: 1 hour 30 min

Calculators may be used for this exam (except those with
facilities for symbolic algebra, differentiation or integration).

Whenever a numerical value of g is required, take $g = 9.8$ ms^{-2}.

Give any non-exact numerical answers to an appropriate degree of accuracy.

There are 75 marks available for this paper.

1   A motorcyclist is travelling at 15 ms^{-1}. As he passes point A on a straight section of road,
he accelerates uniformly for 4 s until he passes point B at 40 ms^{-1}. He then immediately
decelerates at 2.8 ms^{-2} so that when he passes point C he is travelling at 26 ms^{-1}.
Find:

    a)   his acceleration between A and B,

(2 marks)

    b)   the time to travel from B to C,

(2 marks)

    c)   the distance from A to C.

(3 marks)

2   A uniform rod AB of length 4 m and weight 20 N hangs in equilibrium in a horizontal position supported
by two vertical inextensible strings attached at A and B. A bead of weight 16 N rests on the rod at a point C,
which is 3 m from A. The bead can be modelled as a particle.

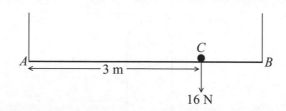

    a)   Find the tensions in the strings at A and at B.

(3 marks)

The bead is now moved along the rod so that the tension in the string at B is twice the tension in the string at A.

    b)   Find the distance of the bead from A.

(3 marks)

    c)   Explain how you have used the information that the rod is uniform.

(1 mark)

3   A particle of mass 0.5 kg moves under the action of two forces, $\mathbf{F}_1$ and $\mathbf{F}_2$,
    where $\mathbf{F}_1 = (3\mathbf{i} + 2\mathbf{j})$ N and $\mathbf{F}_2 = (2\mathbf{i} - \mathbf{j})$ N.

    a)   Find, in degrees correct to one decimal place, the angle which the resultant force
         makes with the direction of $\mathbf{i}$.

         (3 marks)

    b)   Find, correct to three significant figures, the magnitude of the acceleration
         of the particle.

         (4 marks)

4   Two railway trucks A and B, with masses 3.5 tonnes and 1.5 tonnes respectively, are travelling towards
    each other on straight horizontal rails. Both trucks are travelling at 3 ms^{-1} when they collide.
    In the collision the trucks couple together and then move as a single body on the rails.
    Modelling the trucks as particles, find:

    a)   the speed and direction of the combined trucks immediately after the collision,

         (4 marks)

    b)   the magnitude of the impulse exerted by A on B in the collision.

         (3 marks)

As soon as the trucks have collided a constant braking force is applied, which brings them to rest in 18 m.

    c)   Find the magnitude of the braking force.

         (4 marks)

5   A cyclist starts from rest and accelerates at 1.5 ms^{-2} for 8 s along a straight horizontal road and then
    continues at a constant speed. A motorist sets off from the same point as the cyclist at the same time and
    accelerates at a constant rate for 6 s, reaching a maximum speed of V ms^{-1}. The car then decelerates at
    a constant rate until it comes to rest after a further 18 s. At the instant when the car comes to rest, it is
    overtaken by the cycle.

    a)   Calculate the greatest speed attained by the cyclist.

         (2 marks)

    b)   Sketch a speed-time graph for the motion of the cyclist.

         (2 marks)

    c)   Sketch a speed-time graph for the motion of the car.

         (2 marks)

    d)   Calculate the distance travelled by the cyclist before overtaking the car.

         (3 marks)

    e)   Calculate the value of V.

         (3 marks)

6   A helicopter P sets off from its base O on the coast at 12:00 and flies at constant height at a constant velocity of $(45\mathbf{i} + 108\mathbf{j})$ kmh⁻¹. Ten minutes later a second helicopter Q sets off from O and flies at the same constant height at a constant velocity of $(-70\mathbf{i} + 90\mathbf{j})$ kmh⁻¹.

At 12:40 the pilot of helicopter P lets the base know that he has engine failure and is ditching in the sea. The base immediately informs the pilot of helicopter Q, who changes course to go to the aid of helicopter P. Find:

   a)   the speed of helicopter P before engine failure,

(2 marks)

   b)   the position vector of helicopter Q relative to O at 12:40,

(2 marks)

   c)   the position vector of helicopter P relative to helicopter Q at 12:40.

(3 marks)

Helicopter Q flies at 160 kmh⁻¹ directly to the last known position of helicopter P.

   d)   Give the new direction of helicopter Q as a bearing, correct to the nearest degree.

(3 marks)

Assuming that helicopter Q sets off towards helicopter P at 12:40,

   e)   find, to the nearest minute, its estimated time of arrival at the last known position of helicopter P.

(4 marks)

7   A particle A of mass 3 kg is placed on a rough plane inclined at an angle of $\tan^{-1}\dfrac{3}{4}$ to the horizontal and is attached by a light inextensible string to a second particle B of mass 4 kg.
The string passes over a smooth pulley at the top of the inclined plane so that particle B hangs freely. When the system is released, particle B moves downwards vertically with an acceleration of 1.4 ms⁻² and A moves up the plane. Take $g = 9.8$ ms⁻².

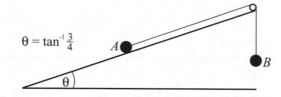

   a)   Draw a diagram showing clearly *all* the forces acting on the particles.

(1 mark)

Find:

   b)   the normal reaction between particle A and the plane, to 2 decimal places,

(2 marks)

   c)   the tension in the string,

(2 marks)

   d)   the friction force acting on particle A,

(3 marks)

   e)   the coefficient of friction between particle A and the plane.

(2 marks)

Two seconds after the particles are released, the string breaks. A then continues up the plane until it comes instantaneously to rest. At no point does A reach the pulley. Find:

   f)   how far particle B moves from the start of the motion until the string breaks,

(2 marks)

   g)   how far particle A moves from the instant the string breaks until it comes to rest.

(5 marks)

General Certificate of Education
Advanced Subsidiary (AS) and Advanced Level

Mechanics M1 — Practice Exam Two

Time Allowed: 1 hour 30 min

Calculators may be used for this exam (except those with
facilities for symbolic algebra, differentiation or integration).

Whenever a numerical value of g is required, take $g = 9.8$ ms^{-2} unless otherwise stated.

Give any non-exact numerical answers to an appropriate degree of accuracy.

There are 75 marks available for this paper.

1 Three forces act at a point, O, as shown below.

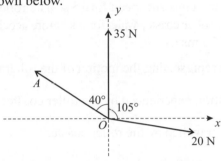

a) Given that there is no net force in the direction of the x-axis, find A.

(2 marks)

b) Find the magnitude and direction of the resultant force, R.

(3 marks)

2 A rocket with mass 1 400 000 kg is launched vertically upwards by engines providing a force of 34 000 000 N.

a) Calculate the expected acceleration of the rocket assuming no other forces
are acting on the rocket.

(2 marks)

b) The actual acceleration is measured at 12 ms^{-2}.
Find the magnitude of the total resistive force, R, acting on the rocket.

(2 marks)

c) Find the time taken to reach a height of 50 km.

(3 marks)

d) At 50 km, the engines stop firing and the rocket moves freely under gravity.
Find the maximum height reached by the rocket (use $g = 9.6$ ms^{-2}).

(4 marks)

3   A small ring of mass *m* is threaded onto a rough inextensible rope held taut and horizontal. The ring is held in limiting equilibrium by a string pulling upwards at an angle of 51.3° to the horizontal. The coefficient of friction between the ring and the rope is 0.6 and the magnitude of the frictional force is 1.5 N.

a)   Sketch a diagram showing the ring and the forces acting upon it.

(2 marks)

b)   Find the tension in the string, *T*, and the mass of the ring, *m*.

(6 marks)

4   A roller coaster moves from rest along a track with constant acceleration. The roller coaster takes 2 s to reach a speed of 6 ms^{-1}, then travels at a constant speed for 15 s. The roller coaster decelerates to rest in 1 s as it reaches the top of the hill. The roller coaster waits for 5 s before accelerating down a vertical slope for 4 s, reaching a maximum speed of 30 ms^{-1}.

a)   Draw a speed-time graph representing the motion of the roller coaster.

(2 marks)

b)   Find the greatest acceleration experienced by the roller coaster.

(3 marks)

c)   What is the total distance travelled by the roller coaster?

(3 marks)

5   A skateboarder balances a skateboard on a rail. The skateboard has mass 4 kg and can be modelled as a uniform horizontal rod, *AB*, 0.8 m in length. The skateboard is in contact with the rail at point *C*, where *AC* = 0.5 m. The skateboarder has mass 80 kg and places their feet at points *X* and *Y*. The skateboard is in equilibrium when the force applied at *Y* is three times the force applied at *X* and *CY* is half the magnitude of *CB*.

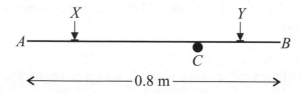

a)   Find the distance *AX*.

(5 marks)

b)   Find the magnitude of the normal reaction, *R*, at the rail.

(2 marks)

An object of mass 3 kg is placed on the skateboard at point *A*. The skateboarder moves their foot such that *AX* is increased by 13 cm. The skateboard remains horizontal and in equilibrium.

c)   Find the magnitude of the forces applied at *X* and *Y*.

(5 marks)

6   A particle, A (m = 3 kg), collides directly with a stationary particle, P (m = 0.4 kg).
     Immediately before the collision A has speed 7 ms^{-1}. After the collision P has speed 9 ms^{-1}. Find:

   a)   the speed of A immediately after the collision.

                                                                                    (3 marks)

   b)   the magnitude of the impulse exerted on P by A.

                                                                                    (3 marks)

   P then collides with another particle, Q (m = 0.1 kg), which has speed 25 ms^{-1} immediately
   after the collision. The magnitude of the impulse received by Q is 2 Ns.

   c)   What was the speed of Q before the collision?

                                                                                    (3 marks)

7   A pirate ship is travelling at a constant velocity of $(6\mathbf{i} + 9\mathbf{j})$ kmh^{-1}.

   a)   Given that the unit vectors $\mathbf{i}$ and $\mathbf{j}$ are due east and north respectively, find the bearing on
        which the ship is moving.

                                                                                    (2 marks)

   The pirate ship is heading for a port 81 km away along its current bearing. At 1350 hours the pirate ship is
   at a point O. At the same time a naval ship is at position vector $(-18\mathbf{i} + 6\mathbf{j})$ km relative to O.

   b)   At what time should the pirate ship arrive at the port?

                                                                                    (3 marks)

   The naval ship travels at a constant velocity and intercepts the pirate ship at 1950 hours.

   c)   Find the velocity of the naval ship.

                                                                                    (4 marks)

8

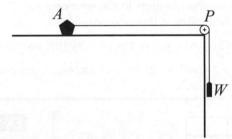

   A particle, A, is attached to a weight, W, by a light inextensible string which passes over a smooth pulley, P,
   as shown. When the system is released from rest, with the string taut, A and W experience an acceleration of
   4 ms^{-2}. A moves across a rough horizontal plane and W falls vertically. The mass of W is 1.5 times the mass of A.

   a)   Given that the mass of A is 0.2 kg, find the coefficient of friction, μ, between A and
        the horizontal plane.

                                                                                    (5 marks)

   W falls for h m until it impacts the ground and does not rebound. A continues to move until
   it reaches P with speed 3 ms^{-1}. The initial distance between A and P is $\frac{7}{4}h$.

   b)   Find the time taken for W to impact the ground.

                                                                                    (7 marks)

   c)   How did you use the information that the string is inextensible?

                                                                                    (1 mark)

Algorithms

Welcome to the wonderful world of Decision Maths. And what a good decision it was too. This page is on algorithms, which aren't as scary as they sound. You've probably come across algorithms before, though you might not know it.

Algorithms are sets of *Instructions*

An underline{algorithm} is just a fancy mathematical name for a underline{set of instructions} for underline{solving} a problem. You come across lots of algorithms in everyday life — underline{recipes}, underline{directions} and underline{assembly instructions} are all examples of algorithms.

1) Algorithms start with an underline{input} (e.g. in a recipe, the input is the raw ingredients). You carry out the algorithm on the input, following the instructions underline{in order}.

2) Algorithms have an underline{end result} — something that you underline{achieve} by carrying out the algorithm (e.g. a cake).

3) This means that algorithms will underline{stop} when you've reached a underline{solution}, or produced your underline{finished product}. They'll often have a underline{stopping condition} — an underline{instruction} that tells you to stop when you've reached a certain point.

4) Algorithms are often written so that underline{computers} could follow the instructions. underline{Computer programming} is an important use of decision maths.

In Maths, the **End Result** is the **Solution**

Algorithms can be used to solve underline{mathematical problems} too.

1) The underline{input} in a mathematical algorithm is the underline{number} (or numbers) you start with. Your underline{end result} is the underline{final number} you end up with — this'll be the underline{solution} to the original problem.

2) It's a good idea to underline{write down} the numbers each instruction produces in a underline{table} — sometimes the algorithm will underline{tell you} when to do this. This table is called a underline{trace table}.

The **Russian Peasant** algorithm **Multiplies** two numbers

The Russian Peasant algorithm is a well-known algorithm that multiplies two numbers together. And I've no idea what Russian peasants have to do with it.

1) Write down the two numbers that you are multiplying in a table. Call them x and y.

2) Divide x by 2 and write down the result underneath x, ignoring any halves (e.g. if $x = 11$, when you divide it by 2 you write down 5, not 5.5).

3) Multiply y by 2 and write down the result underneath y.

4) Repeat steps 2) - 3) for the numbers in the new row. Keep going until the number in the x-column is 1.

5) Work down your table and cross out every row that has an even value for x.

6) Add up the remaining numbers in the y-column (i.e. the ones that haven't been crossed out). This is the solution xy.

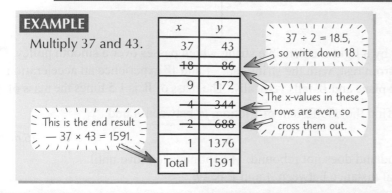

EXAMPLE

Multiply 37 and 43.

x	y
37	43
~~18~~	~~86~~
9	172
~~4~~	~~344~~
~~2~~	~~688~~
1	1376
Total	1591

37 ÷ 2 = 18.5, so write down 18.

The x-values in these rows are even, so cross them out.

This is the end result — 37 × 43 = 1591.

EXAMPLE

Multiply 21 and 52.

x	y
21	52
~~10~~	~~104~~
5	208
~~2~~	~~416~~
1	832
Total	1092

This is the end result — 21 × 52 = 1092.

Feel the beat of the rithm of the night...

For your homework tonight, I would like you to make me a cake please. My favourite's carrot cake, with walnuts in it and cream cheese icing on the top. Mmmm... I will settle for chocolate fudge cake, Victoria sponge cake, or Madeira cake.

Flow Charts

Right, now you know what an algorithm is, it's time to have some fun. Well, I say fun, but really it's just putting algorithms into flow charts and stuff like that.

Algorithms can be written as Flow Charts

Instead of giving instructions in <u>words</u> (like in the Russian Peasant example on the previous page), some algorithms are written as <u>flow charts</u>. There are three different types of <u>boxes</u> which are used for different things:

Start / Stop   Instruction   Decision

The boxes are connected with <u>arrows</u> to guide you through the flow chart. 'Decision' boxes will ask a question, and for each one you have a <u>choice</u> of arrows, one arrow for '<u>yes</u>' and one for '<u>no</u>', which will take you to another box. Sometimes flow charts will include a loop which takes you back to an earlier stage in the chart. Loops are a way of <u>repeating steps</u> until the algorithm is <u>finished</u>.

Put the Results from the flow chart into a Trace Table

It can sometimes be a bit tricky keeping track of the <u>results</u> of a flow chart, especially if you have to go round a <u>loop</u> lots of times. It's a good idea to put your results into a <u>trace table</u> — it's much easier to see the <u>solutions</u> that way.

EXAMPLE

This flow chart works out the factors of a number, a.

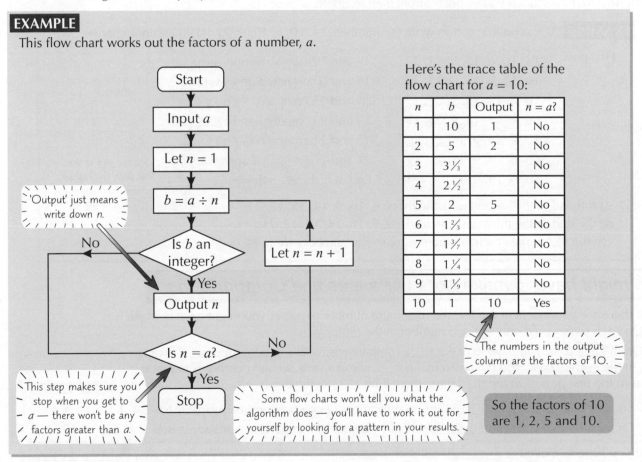

Here's the trace table of the flow chart for $a = 10$:

n	b	Output	$n = a$?
1	10	1	No
2	5	2	No
3	$3\frac{1}{3}$		No
4	$2\frac{1}{2}$		No
5	2	5	No
6	$1\frac{2}{3}$		No
7	$1\frac{3}{7}$		No
8	$1\frac{1}{4}$		No
9	$1\frac{1}{9}$		No
10	1	10	Yes

The numbers in the output column are the factors of 10.

So the factors of 10 are 1, 2, 5 and 10.

'Output' just means write down n.

This step makes sure you stop when you get to a — there won't be any factors greater than a.

Some flow charts won't tell you what the algorithm does — you'll have to work it out for yourself by looking for a pattern in your results.

Things seem to be flowing nicely...

Don't worry — you won't be expected to write algorithms in your exam, but you need to be able to use them. Some algorithms can look a bit confusing if there are lots of steps and decisions, but if you work through them slowly step by step they aren't too bad. Have a go at the one on this page for some different values of a — try $a = 12$, 17 and 18.

Sorting

You've probably been able to sort things into alphabetical or numerical order since you were knee-high to a hamster, but in D1 you need to know how to sort things using an algorithm. At least it'll be easy to check your answer.

A **Bubble Sort** compares **Pairs** of numbers

The <u>bubble sort</u> is an <u>algorithm</u> that <u>sorts numbers</u> (or <u>letters</u>). It's pretty easy to do, but it can be a bit fiddly, so <u>take care</u>.

The Bubble Sort

1) Look at the <u>first two numbers</u> in your list. If they're in the right order, you don't have to do anything with them. If they're the wrong way round, <u>swap</u> them. It might help you to <u>make note</u> of which numbers you swap each time.

2) Move on to the <u>next</u> pair of numbers (the first will be one of the two you've just compared) and <u>repeat step 1</u>. Keep going through the list until you get to the <u>last two numbers</u>. This set of comparisons is called a <u>pass</u>.

3) When you've finished the first pass, go back to the beginning of the list and <u>start again</u>. You won't have to compare the <u>last pair</u> of numbers, as the last number is now <u>in place</u>. Each pass has <u>one less comparison</u> than the one before it. When there are <u>no swaps</u> in a pass, the list is in <u>order</u>.

If there are n numbers in your list, there will be $n-1$ comparisons in the first pass.

Stop when there are **No More Swaps**

It's called the bubble sort because the highest numbers <u>rise</u> to the end of the list like <u>bubbles</u> (apparently). It'll all be a lot easier once you've been through an example...

EXAMPLE Use a bubble sort to write the numbers 14, 10, 6, 15, 9, 21, 17 in ascending order.

First pass:

<u>14, 10</u>, 6, 15, 9, 21, 17	14 and 10 compared and swapped
10, <u>14, 6</u>, 15, 9, 21, 17	14 and 6 compared and swapped
10, 6, <u>14, 15</u>, 9, 21, 17	14 and 15 compared — no swap
10, 6, 14, <u>15, 9</u>, 21, 17	15 and 9 compared and swapped
10, 6, 14, 9, <u>15, 21</u>, 17	15 and 21 compared — no swap
10, 6, 14, 9, 15, <u>21, 17</u>	21 and 17 compared and swapped
10, 6, 14, 9, 15, 17, 21	End of first pass. ←

At the end of the first pass, the highest number has moved to the end of the list.

At the end of the second pass the list is: 6, 10, 9, 14, 15, 17, 21.
At the end of the third pass the list is: 6, 9, 10, 14, 15, 17, 21.
On the fourth pass there are no swaps, so the numbers are in <u>ascending order</u>.

You might have to make **Lots** of Passes and **Comparisons**

1) If there are n numbers in the list, the <u>maximum</u> number of <u>passes</u> you might have to make is n. On each pass, you'd only get <u>one</u> number in the <u>right place</u>.

2) On the <u>first</u> pass, you have to make $n - 1$ comparisons, with a <u>maximum</u> of $n - 1$ swaps. On the <u>second</u> pass, you'll have to make $n - 2$ comparisons, as one number is in place from the first pass. On the <u>third</u> pass, there'll be $n - 3$ comparisons etc.

So for a bubble sort with <u>7 numbers</u>, max. number of comparisons (or swaps) is $6 + 5 + 4 + 3 + 2 + 1 = 21$
Or for a bubble sort with <u>50 numbers</u>, max. number of comparisons (or swaps) is $\frac{1}{2} \times 49 \times 50 = 1225$

3) If you have to make the <u>maximum</u> number of <u>swaps</u>, it means that the original list was in <u>reverse order</u>.

4) You can also use the algorithm to put numbers in <u>descending</u> order — on each comparison, just put the <u>higher</u> number <u>first</u> instead.

For big lists, use the formula $S_k = \frac{1}{2}k(k + 1)$ for the sum of the <u>first k whole numbers</u> (though if $n = 50$, you want the sum of the first 49 numbers).

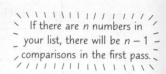

Double, double, toil and trouble, fire burn and cauldron bubble...

Bubble sorts are fairly easy, but they aren't the most efficient way of sorting numbers. If you had a list of 100 numbers, you might have to do 100 passes with 4950 comparisons (or swaps). It'd take ages. If only there were a quicker way...

Sorting

... of course there is, and it's imaginatively called the 'quick sort'. With a name like that, you'd hope it was pretty quick.

A **Quick Sort** breaks the list into **Smaller Lists**

The quick sort algorithm works by choosing a pivot (see below) which breaks down the list into two smaller lists, which are then broken down in the same way until the numbers are in order.

The Quick Sort

1) Choose a pivot. Move any numbers that are <u>less</u> than the pivot to a new list on the <u>left</u> of it and the numbers that are <u>greater</u> to a new list on the <u>right</u>. Don't change the <u>order</u> of the numbers though.

2) <u>Repeat step 1)</u> for each of the smaller lists you've just made. You'll need to choose <u>new pivots</u> for the new lists.

3) When <u>every number</u> has been chosen as a pivot, you can <u>stop</u>, as the list is in order.

Sometimes the smaller lists will only have one number in them — you don't need to do anything with these as they're already in order.

The pivot should be the 'middle' item in the list. In a list of n items:

- If n is <u>odd</u>, the middle item is the $\frac{1}{2}(n + 1)$th item.
  For $n = 7$, the pivot is the $\frac{1}{2}(7 + 1) = $ 4th item in the list.

- If n is <u>even</u>, the middle item is the $\frac{1}{2}(n + 2)$th item.
  For $n = 8$, the pivot is the $\frac{1}{2}(8 + 2) = $ 5th item in the list.

The 'middle' item is only in the <u>middle</u> if n is <u>odd</u>. For <u>even</u> n, the 'middle' item is the one to the <u>right</u> of the middle.

EXAMPLES Find the pivot for the list 7, 12, 9, 16, 24.

> There are 5 items in the list, which means the pivot is the $\frac{1}{2}(5 + 1) = $ 3rd item in the list. So the pivot is 9.

Find the pivot for the list 4, 23, 5, 28, 17, 32.

> There are 6 items in the list, which means the pivot is the $\frac{1}{2}(6 + 2) = $ 4th item in the list. So the pivot is 28.

Keep track of the **Pivots** you use

It's a good idea to circle or underline the pivots you're using at each step of the quick sort — it helps you keep track of where you're up to. In the example below, the numbers are written in grey when they're in the correct place.

EXAMPLE Sort the numbers 54, 36, 29, 56, 45, 39, 32, 27 into ascending order using a quick sort.

There are 8 items in the list, which means the pivot is the $\frac{1}{2}(8 + 2) = $ 5th item in the list. So the pivot is 45.

Now make new lists by moving numbers < 45 to the left, and numbers > 45 to the right. Don't reorder the numbers, just write them down in the order they appear in the original list on the correct side of the pivot.

These are all the numbers < 45 (in the same order as in the original list). →
$$\underbrace{36, 29, 39, 32, 27}_{l_1}, 45, \underbrace{54, 56}_{l_2}$$
← *These are all the numbers > 45 (in the same order as in the original list).*

The list has been divided into smaller lists, l_1 and l_2. There are 5 items in l_1, so the pivot is the 3rd item (39). There are only 2 items in l_2, so the pivot is the 2nd item (56). Rearranging around the new pivots gives:

$$\underbrace{36, 29, 32, 27}_{l_3}, \underline{39}, 45, \underbrace{54}_{l_4}, 56$$
← *The pivots are in the right place now.*

The list l_4 only has one item in it, so 54 is in the correct place (it's a pivot in a list of its own). There are now 4 items in l_3, so the pivot is the 3rd item (32). Rearranging again using the new pivot gives:

$$\underbrace{29, 27}_{l_5}, \underline{32}, \underbrace{36}_{l_6}, 39, 45, 54, 56$$

The list l_6 only has one item in it, so 36 is now in the correct place. There are just 2 items in l_5, so the pivot is the 2nd item (27). Rearranging the list using the new pivot gives:

$$\underline{27}, \underbrace{29}_{l_7}, 32, 36, 39, 45, 54, 56$$

As there's only one item left in the final list (l_7), the list is now in the right order.

Quick — let's get out of here...

A quick sort is usually quicker than a bubble sort, but you need to know both methods — you could be asked for either (or both) in the exam. Of course, you may be thinking that it would be much easier to sort them just by looking at them, but the point of these algorithms is that you could program them into a computer and sort huge lists of numbers really quickly.

Searching

My preferred method of searching involves at least three helicopters, a metal detector and a highly-trained sniffer dog. In case you don't have access to those things, this page gives you a handy algorithm for searching a list.

A **Binary Search** looks for items in an **Ordered List**

1) The binary search algorithm works by <u>dividing</u> the list in <u>half</u> over and over again until it finds the item. At each stage, you can <u>get rid of</u> half the list to narrow the search.

2) If the item you're looking for <u>isn't</u> in the list, you can use the binary search method to <u>show</u> that it isn't there.

3) The list has to be <u>in order</u> (e.g. a list of names in <u>alphabetical order</u>) for the algorithm to work. You can <u>number</u> the items to show their <u>position</u> in the list.

4) You need to be able to find the '<u>middle</u>' item in a list — use the method on the previous page for finding the pivot.

Find the **Middle Item** and **Compare**

This is how to use the binary search algorithm:

> 1) Find the <u>middle item</u> in the list (see p. 189).
>
> 2) If the middle item is the item you're looking for, then you can <u>stop</u> — you've found it.
>
> 3) If not, <u>compare</u> the item you're looking for to the middle item. If it comes <u>before</u> the middle item in the list, you can <u>throw away</u> the <u>second half</u> of the list (<u>including</u> the middle item).
>
> 4) If the item you're looking for comes <u>after</u> the middle item, you can <u>throw away</u> the <u>first half</u> of the list (<u>including</u> the middle item).
>
> 5) You'll be left with a list that's <u>half</u> (or just under half) the size of the original list. <u>Repeat steps 1) – 4)</u> on this smaller list to get an even smaller one, and keep going until you find the item you're looking for.

If the list has shrunk to just <u>one</u> item, and this <u>isn't</u> the item you're looking for, this shows that the item <u>isn't in the list</u>.

EXAMPLE Use the binary search algorithm to locate the name Jackman in the following list:

1 — Armitage
2 — Cooke
3 — Garcia
4 — Horner
5 — Jackman
6 — Marsden
7 — Wales

The middle item is the $\frac{1}{2}(7 + 1)$ = 4th item = Horner.

Jackman ≠ Horner, and Jackman comes after Horner, so throw away the first half of the list (including Horner). This leaves a list that looks like this:

5 — Jackman
6 — Marsden
7 — Wales

The middle item is the $\frac{1}{2}(3 + 1)$ = 2nd item. As this list starts with 5, the middle item is the 6th item = Marsden.

Jackman ≠ Marsden, and Jackman comes before Marsden, so throw away the second half of the list (including Marsden). This leaves a list that looks like this:

5 — Jackman

There is now only one name left, and it's the name you're looking for, so the search is complete. Jackman is the 5th name in the list.

If you were looking for a name that wasn't in the list (e.g. Lewis), you'd follow the same steps, but when you were left with just Jackman, it would show that Lewis wasn't in the list.

If you're left with a list that doesn't start with item 1, you might find it easier to find the middle item by using the formula $\frac{1}{2}(a + l)$ and rounding up where necessary, where a is the position of the first item and l is the position of the last item. So for the example above, the middle item of the second list is the $\frac{1}{2}(5 + 7)$ = 6th term.

Send out a search party — I've lost my mojo...

The main thing to remember about the binary search algorithm is that it only works on ordered lists. If you try it on an unordered list, it'll be like looking for your car in a theme park car park on the busiest day of the year. With your eyes shut.

Packing

Packing to go on holiday can be a pain. You've got so much to fit in your suitcase, and there's always the risk of your shampoo leaking. Well, you'll be pleased to know that there are a couple of algorithms that are about packing (admittedly not holiday packing, but I'm sure you could adjust them to make them work).

These algorithms are called **Bin Packing Algorithms**

1) In bin packing problems, you have a set of items that you need to fit into the minimum number of bins.

2) One of the most common examples is fitting boxes of different heights on top of each other into bins of a given height. You need to arrange the boxes to use the fewest bins possible.

3) Other examples include things like cutting specified lengths of wood from planks of a fixed length (you want to minimise the number of planks used), or loading items of different weights into lorries that have a maximum weight capacity (again, you want to use the smallest number of lorries possible).

4) An optimal solution is one that uses the least possible number of bins. There's often more than one possible optimal solution.

> Because the optimal solution has the least possible no. of bins, it also has the least wasted space.

The **Lower Bound** gives a **Minimum** for the **Number of Bins** you'll need

1) To work out the lower bound, you add up the height / weights etc. of the items, and divide the total by the capacity of the bins. For the wooden planks example above, you'd add up the lengths of the different pieces of wood and divide by the length of the planks (the planks are all the same length).

2) Always round up your answer — if you get a lower bound of 2.25, you'd need a minimum of 3 bins (you wouldn't fit the items in 2 bins).

3) Just because you've worked out a lower bound, it doesn't mean you can definitely fit the items into this number of bins — you'll need at least this, but possibly more...

4) ...in other words, the optimal solution will not necessarily match the lower bound — it might need more bins.

5) But... if your solution does match the lower bound, you know it's definitely optimal.

> **EXAMPLE** Five boxes of heights 20 cm, 43 cm, 35 cm, 29 cm and 38 cm are to be packed into bins of height 60 cm. Calculate a lower bound for the number of bins.
>
> $20 + 43 + 35 + 29 + 38 = 165$ cm
>
> $165 \div 60 = 2.75$, so the lower bound is 3 bins.

The **First-Fit** algorithm puts the items in the **First Bin** they'll go in

The first-fit algorithm is quick and easy, but it probably won't give you an optimal solution. Here's how it works:

1) Take the first item in the list and put it in the first bin.

2) Move on to the next item, and put it in the first bin it'll fit into. It might fit in the first bin, or you might have to move on to another bin.

3) Repeat step 2) until all the items are in a bin. For each item, try the first bin before you move on to the next.

> **EXAMPLE** The ad breaks in a TV programme can be no longer than 150 seconds. Use the first-fit algorithm to sort the adverts into the breaks, saying how many breaks are needed and how much time is wasted.
>
> A: 90s   B: 75s   C: 30s   D: 65s   E: 120s   F: 45s   G: 60s
>
> | | | |
> |---|---|---|
> | Ad break 1: | A: 90s, C: 30s | space left: ~~60s~~ 30s |
> | Ad break 2: | B: 75s, D: 65s | space left: ~~75s~~ 10s |
> | Ad break 3: | E: 120s | space left: 30s |
> | Ad break 4: | F: 45s, G: 60s | space left: ~~105s~~ 45s |
>
> B won't fit in break 1, as there's only 60s left after A, so it goes in break 2. However, C < 60s, so it fits in break 1.
>
> So the adverts can be fitted into 4 ad breaks, with $30 + 10 + 30 + 45 = 115$ s wasted.

I've bin packing my suitcase for my holiday...

Sorry, sorry, that was terrible. Try not to confuse first-fit with first-past-the-post — one's a bin packing algorithm, the other's a voting system. If you get them mixed up, you might end up trying to see how many politicians you can fit in a box. Hmm...

Packing

If you decide to carry out your own bin packing problem with bits of wood, do be careful with the saw. Maths is a lot harder if you don't have all 10 fingers.

The **First-Fit Decreasing** algorithm needs the items in **Descending Order**

1) The <u>first-fit decreasing</u> algorithm is very <u>similar</u> to the <u>first-fit</u> algorithm except you need to put the items in <u>descending order</u> first.

2) You can do this by using one of the <u>sorting algorithms</u> on pages 188-189.

3) Once you've got your ordered list, you just carry out the <u>first-fit algorithm</u> from the previous page.

4) The first-fit decreasing algorithm usually gives you a <u>better solution</u> than the first-fit algorithm, but it still might <u>not</u> be <u>optimal</u>.

EXAMPLE Ribbon comes in rolls of length 5 m. For the lengths of ribbon given below, use the first-fit decreasing algorithm to work out how the lengths can be cut from the rolls. You should also say how many rolls are needed and how much ribbon is wasted. All lengths are in metres.

2.5   1.9   2.9   3.1   2.7   2.2   1.8   2.0

First, use a sorting algorithm to put the lengths in order. The new list is:

3.1   2.9   2.7   2.5   2.2   2.0   1.9   1.8

Now use the first-fit algorithm to sort the lengths into rolls:

Roll 1:   3.1, 1.9   length left: ~~1.9~~ 0
Roll 2:   2.9, 2.0   length left: ~~2.1~~ 0.1
Roll 3:   2.7, 2.2   length left: ~~2.3~~ 0.1
Roll 4:   2.5, 1.8   length left: ~~2.5~~ 0.7

The ribbon can be cut from 4 rolls, with 0 + 0.1 + 0.1 + 0.7 = 0.9 m wasted.

If you solved this problem using the first-fit algorithm, you'd use 5 rolls and waste 5.9 m of ribbon.

The **Full-Bin** packing algorithm usually wastes the **Least Space**

The <u>full-bin packing algorithm</u> needs a bit more <u>work</u> than the other two, but it's more likely to produce an <u>optimal solution</u>. However, it can be quite <u>hard</u> to do if you've got a lot of items.

1) In the full-bin algorithm, first you need to <u>look</u> at the items and find items that will <u>add up</u> to give a <u>full bin</u>. You just have to do this <u>by eye</u>, so it can get a bit tricky.

2) Once you've <u>filled</u> as many bins as you can, you use the <u>first-fit algorithm</u> (see previous page) to put the rest of the items in the <u>remaining spaces</u> in the bins.

EXAMPLE Boxes of the same length and width need to be packed in bins of height 2.5 m. Use the full-bin packing algorithm to pack the boxes, and say how many bins are used, how much space is wasted and whether the solution is optimal. The heights of the boxes (in metres) are:

0.7   1.1   1.2   2.3   0.8   1.4   0.9   1.0   2.5

Just by looking at the heights, you can see that 0.7 + 0.8 + 1.0 = 2.5, 1.1 + 1.4 = 2.5 and 2.5 = 2.5, so you can fill 3 bins straight away. Then use the first-fit algorithm to pack the remaining boxes.

Bin 1:   0.7, 0.8, 1.0   space left: 0
Bin 2:   1.1, 1.4   space left: 0
Bin 3:   2.5   space left: 0
Bin 4:   1.2, 0.9   space left: ~~1.3~~ 0.4
Bin 5:   2.3   space left: 0.2

The boxes can be packed into 5 bins, with 0.4 + 0.2 = 0.6 m wasted space.

Using the method on the previous page, the lower bound for this problem is 11.9 ÷ 2.5 = 4.76 = 5 bins. As this solution uses 5 bins, it is optimal. Also, the amount of space wasted is less than one bin.

Don't forget to pack your pyjamas...

So, the first-fit algorithm is the quickest but doesn't give the best solution, the first-fit decreasing algorithm needs a little more work but usually gives a better solution, and the full-bin algorithm is a bit harder but is more likely to give an optimal solution.

D1 Section 1 — Practice Questions

That's the first section of D1 done and dusted — and I don't think it was that bad. Once you get your head round the fact that 'algorithm' is just a fancy word for 'set of instructions', there's not much to it. Time now for a few questions to check you know your stuff. Here are some straightforward ones to start off with.

Warm-up Questions

1) For each of the following sets of instructions, identify the input and output.
   a) a recipe for vegetable soup,
   b) directions from Leicester Square to the Albert Hall,
   c) flat-pack instructions for building a TV cabinet.

2) Use the Russian Peasant algorithm to multiply 17 and 56.

3) What are diamond-shaped boxes in flow diagrams used for?

4) Use the flow chart on p.187 to work out the factors of 16.

5) Use a bubble sort to write the numbers 72, 57, 64, 54, 68, 71 in ascending order. How many passes do you need to make?

6) If you had to put a list of 10 numbers in order using a bubble sort, what is the maximum number of comparisons you'd need to make?

7) Find the pivot for the list 104, 97, 111, 108, 95, 91.

8) Sort the numbers 0.8, 1.2, 0.7, 0.5, 0.4, 1.0, 0.1 into ascending order using a quick sort. Write down the pivots you use at each step.

9) Use the binary search algorithm to locate 'Brooklyn' in the list below:
   1) Bronx 2) Brooklyn 3) Manhattan 4) Queens 5) Staten Island.

10) Six items of weights 5 kg, 11 kg, 8 kg, 9 kg, 12 kg, 7 kg need to be packed into boxes that can hold a maximum weight of 15 kg.
   a) Find a lower bound for the number of boxes needed.
   b) Pack the items into the boxes using the first-fit algorithm.
   c) Pack the items into the boxes using the first-fit decreasing algorithm.
   d) Pack the items into the boxes using the full-bin packing algorithm.
   For parts b)-d), say how many boxes are needed, how much space is wasted and whether the solution is optimal (if you can say).

If those warm-up questions have left you hungry for more, here are some tasty exam-style questions for you to sink your teeth into. They'll need a bit more work than the warm-up questions, but it'll be worth it, I promise.

Exam Questions

1   77   83   96   105   78   89   112   80   98   94

(a) Use a quick sort to arrange the list of numbers above into ascending order. You must clearly show the pivots you use at each stage.

(5 marks)

(b) A list of six numbers is to be sorted into ascending order using a bubble sort.

(i) Which number(s) will definitely be in the correct position after the first pass?

(1 mark)

(ii) Write down the maximum number of passes and the maximum number of swaps needed to sort a list of six numbers into ascending order.

(2 marks)

D1 Section 1 — Practice Questions

If you thought I'd leave you hanging with just <u>one exam question</u>, you're very much mistaken.
I'd <u>never</u> be that mean to you — and to prove it, here's a page full of <u>beautiful exam questions</u>.

2   Consider the following algorithm:

        Step 1:   Input A, B with $A < B$
        Step 2:   Input $N = 1$
        Step 3:   Calculate $C = A \div N$
        Step 4:   Calculate $D = B \div N$
        Step 5:   If both C and D are integers, output N
        Step 6:   If $N = A$, then stop. Otherwise let $N = N + 1$ and go back to Step 3.

(a)   Carry out the algorithm with $A = 8$ and $B = 12$. Record your results.

(3 marks)

(b)   (i)   What does this algorithm produce?

(1 mark)

        (ii)   Using your answer to part (i) or otherwise, write down the output that would be produced
            if you applied the algorithm to $A = 19$ and $B = 25$, and explain your answer. You do not need
            to carry out the algorithm again.

(2 marks)

3   A joiner has planks of wood that are 3 m long. He needs to cut pieces of wood from the planks in the
    following lengths:
        1.2 m   2.3 m   0.6 m   0.8 m   1.5 m   1.0 m   0.9 m   2.5 m

(a)   Calculate a lower bound for the number of planks of wood he will need to use.

(2 marks)

(b)   Use the first-fit bin packing algorithm to fit the lengths of wood onto the planks.
        State how many planks are needed and how much wood is wasted.

(3 marks)

(c)   (i)   Use the full-bin packing algorithm to fit the lengths of wood onto the planks.
            Again, state how many planks are needed and how much wood is wasted.

(3 marks)

        (ii)   Is this solution optimal? Explain your answer.

(1 mark)

4   Mark   Adam   Dan   James   Stella   Helen   Robert

(a)   Use a quick sort to list the above names in alphabetical order. Show clearly the pivots you use.

(4 marks)

(b)   (i)   Use the binary search algorithm on the list from part (a) to locate the name 'Adam'.

(3 marks)

        (ii)   Use the binary search algorithm again to try to locate the name 'Laura' in the list.

(3 marks)

Graphs

You probably reckon you're an old pro at graphs. But the graphs coming up are rather different. For a start, there's not a scrap of squared paper in sight. Fret not — soon they'll be as innocuous to you as a bar chart.

Graphs have **Points** Connected by **Lines**

Here's the definition of a graph:

> A **graph** is made up of **points** (called vertices or nodes) joined by **lines** (called edges or arcs).

1) Graphs can be used to <u>model</u> or <u>solve</u> real-life problems. Here are a few examples:

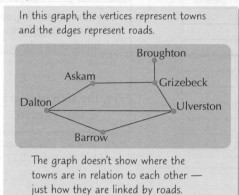

In this graph, the vertices represent towns and the edges represent roads.

The graph doesn't show where the towns are in relation to each other — just how they are linked by roads.

This graph shows the jobs a group of students would prefer to do at the end-of-term barbecue.

Andy — lighting barbecue
Ben — flipping burgers
Caley — making fruit punch
Ed — squirting ketchup
Fred — clearing up

This is a bipartite graph — it has two sets of vertices. The edges only join vertices in the opposite set. So you could never join Andy and Caley, or "lighting barbecue" and "flipping burgers".

2) <u>Weighted graphs</u>, or <u>networks</u>, have a number associated with each edge (called the edge's <u>weight</u>).

Weights often give you <u>lengths</u> — like in this network showing points in a nature reserve and the footpaths joining them. They sometimes give you costs, or times too.

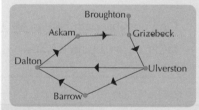

3) Sometimes edges have <u>directions</u>, e.g. to show one-way streets. If they do, they're called <u>directed edges</u> and the graph is a <u>digraph</u>.

The edges on this digraph show the bus routes between the towns. A bus goes from Dalton to Askam, but not from Askam to Dalton.

There's no direction on the edge connecting Broughton and Grizebeck, so the buses run in both directions.

Subgraphs are Just **Bits of Another Graph**

1) These are absolute doddles. If you take a graph, and rub a few bits out, then you're left with a <u>subgraph</u>.

2) Here's the posh definition to learn: A **subgraph** of graph *G* is a graph where all the vertices and edges belong to *G*.

EXAMPLE

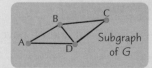

Graph G

Subgraph of G

Subgraph of G

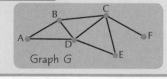

Subgraph of G

All the Vertices are **Directly Connected** in a **Complete Graph**

1) Each vertex in a <u>complete graph</u> is joined <u>directly</u> to every other vertex by an edge.

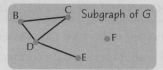

2) In a <u>complete bipartite graph</u>, each vertex is joined directly to each vertex in the <u>opposite set</u>.

A subgiraph — just the neck or a leg...

A page jam-packed with definitions. You'll need them very soon, so learn them and then scribble them out from memory. Then check back to the page to see what you missed. Graphs are all over the place — the London Underground map is one.

Graphs

More terminology coming up. Make sure you totally get it, or confusion will be the next thing coming up.

The **Degree** of a **Vertex** is the **Number of Lines** coming off it

1) Here's the formal definition:

> The **degree** or **valency** of a vertex is the number of edges connected to it.

EXAMPLE Calculate the degree of each vertex in the graph below.

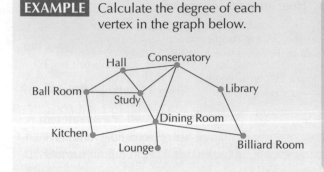

Vertex	Degree
Ball Room	3
Billiard Room	2
Conservatory	4
Dining Room	5
Hall	3
Kitchen	2
Library	2
Lounge	1
Study	4

Rules of Degrees

The sum of the degrees is always <u>double</u> the number of edges.
(It's a count of how many edge ends there are.)

So, the sum of degrees is always even.

Here, there are 13 edges and the sum of the degrees is 26 (2 × 13).

2) A vertex with an <u>odd degree</u> is <u>odd</u>, and one with an <u>even degree</u> is — wait for it — <u>even</u>. So the Billiard Room, Conservatory, Kitchen, Library and Study are all even, and the rest are odd.

A **Path** is just a Route in the Graph

A path is a <u>sequence of edges</u> that flow on, end to end. The only thing is, you <u>can't</u> go through a vertex more than once.

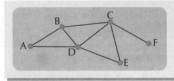

One possible path here is <u>ABDECF</u>. Another is <u>CBAD</u>.

DCECF <u>isn't</u> a path because you go through vertex C more than once.

A **Cycle** is a Path that **Brings you Back** to Your Starting Point

1) A <u>cycle</u> (or circuit) is a <u>closed path</u>. The <u>end</u> vertex is the same as the <u>start</u> vertex.
2) So on the graph above, <u>ABDA</u> is a cycle. Other cycles are <u>ABCDA</u> and <u>CEDBC</u>.

You have to follow the rules for paths still — so you can't go through a vertex twice. So ABDCEDA isn't a cycle because it goes through vertex D twice.

Graphs Can be **Connected** or Not Connected

1) Two vertices are connected if there's a path between them — it doesn't have to be direct.
2) A graph is connected if all its vertices are connected.

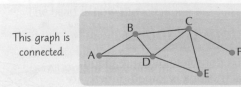

This graph is connected.

This graph is NOT connected — you can't get from some vertices to others. E.g. there's no path between B and C, or between E and A.

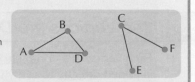

And your leg bone's connected to your foot bone...

I didn't put the bit about the <u>sum of degrees</u> always being <u>even</u> in just because it's fascinating. (Although it certainly is.) It actually comes up quite often in exam questions. They don't ask you straight out though. They'll phrase it in a cunning way. E.g. Why can't you have a graph consisting of three vertices with odd degrees? Just stay calm and it'll be OK.

Graphs

Only trees and matrices to go. And not the leafy sorts of trees.

Trees are Graphs that have No Cycles

They also have to be <u>connected graphs</u> (see page 196).

Both graphs here are connected — but <u>only</u> the
first is a <u>tree</u> (the graphical type of tree that is).

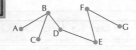

This is a tree — there are no cycles.

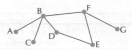

This one <u>isn't</u> a tree —
there's a cycle (BDEFB.)

Spanning Trees are Subgraphs that are Also Trees

1) They can't be just any old subgraph though — they have to include <u>all the vertices</u>.

2) So if you're asked to draw a <u>spanning tree of a graph</u>, you can only delete <u>edges</u> from the original graph.

3) The number of <u>edges</u> in a spanning tree is always <u>one less</u> than the number of <u>vertices</u>.

4) Like most things in this section, a few diagrams speak a thousand words...

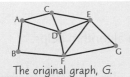

The original graph, G.

A spanning tree of G.
There's no cycle, and it
contains all the vertices
from G.

Another spanning
tree of G. There are
plenty more that
could be drawn too.

Adjacency Matrices show the Number of Links between Vertices

1) To draw an <u>adjacency matrix</u> from a graph, go through each space in the matrix and count the <u>number of direct connections</u> from the vertex at the left of the <u>row</u> to the vertex at the top of the <u>column</u>.

EXAMPLE Represent this graph with an adjacency matrix.

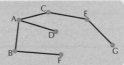

There's a loop from A to A.
You can go in either direction,
so it counts as 2 links.

There's no direct link
from D to A.

There's 1 direct link
from A to E.

There are 2 direct links
from D to E.

	A	B	C	D	E
A	2	0	1	0	1
B	0	0	1	0	0
C	1	1	0	1	0
D	0	0	1	0	2
E	1	0	0	2	0

Notice that the matrix is
symmetrical along this diagonal.

2) You might have to draw a <u>graph</u> from an adjacency matrix too. Mark the <u>vertices</u> first. Then go through each space in the matrix and <u>draw edges in</u>. But remember, one edge between A and B is the <u>same</u> as one between B and A — don't draw two.

Distance Matrices show the Weights between Vertices

1) To draw a distance <u>matrix</u> from a weighted digraph (p. 195), go through each space in the matrix and write down the <u>weight</u> between the two vertices. You only include <u>direct links</u> — don't start adding weights together.

2) Be really careful with <u>directed edges</u>. A weight on a directed edge only goes in <u>one</u> space of the matrix, as below.

EXAMPLE Represent this graph with a distance matrix.

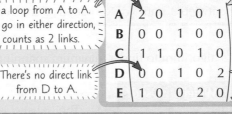

These are the "from"
vertices down the side.

There's an edge of weight
7 from C to A.

There's an edge of
6 from E to A.

	TO				
FROM	A	B	C	D	E
A	–	–	–	9	6
B	–	–	2	–	–
C	7	2	–	–	–
D	9	–	4	–	12
E	6	–	–	10	–

There's no edge from
A to C, so put a
dash, NOT a 0.

There's an edge of
12 from D to E...

... but only one of 10 from E to D.

3) If you're asked to draw a <u>digraph</u> from a distance matrix, mark the <u>vertices</u>, then go through the matrix, adding <u>edges</u> and <u>weights</u> in. If a weight only appears <u>once</u> in the matrix, the edge must be directed, so add an arrow.

Now what would a tree do with a cycle anyway...

The best thing you can do now is cover up the graphs above and practise drawing them from the matrices. Then draw the matrices from the graphs. And watch out for those blasted directed edges — they ruin the aesthetically pleasing symmetry.

Minimum Spanning Trees

The stuff you've seen so far in this section might seem like it's been dreamed up by bored mathematicians to provide you with useless facts to learn. But this minimum spanning tree stuff is actually rather handy in the real world.

A **Minimum Spanning Tree** is the **Shortest** way to Connect All the Points

See page 197 if you've forgotten what spanning trees are. And remember — an <u>arc</u> is just another name for an edge.

A **minimum spanning tree** (MST) is a spanning tree where the total length of the arcs is as **small as possible**.

An MST is also known as a minimum connector.

<u>Minimum spanning trees</u> come in handy for cable or pipe-laying companies. If they need to connect several buildings in a town, say, they'd want to find the <u>cheapest path</u> — this may be the <u>shortest</u> route, or have the <u>easiest</u> ground to dig up.

Kruskal's Algorithm Finds Minimum Spanning Trees

Being absolutely certain that you've got the minimum spanning tree is tricky, so using an algorithm helps.

> ### KRUSKAL'S ALGORITHM
>
> 1) List the arcs in <u>ascending order of weight</u>.
>
> 2) Pick the arc of <u>least weight</u> — this starts the tree.
>
> 3) Look at the <u>next arc</u> in your list.
>   - if it forms a cycle, <u>DON'T</u> use it and go on to the next arc.
>   - if it <u>doesn't</u> form a cycle, add it to the tree.
>
> 4) Repeat step 3 until you've joined <u>all</u> the vertices.

This is a 'greedy algorithm'. You make the choice that seems best at each stage, without worrying about later choices.

If there are n vertices in a network, there will always be $(n - 1)$ arcs in an MST.

EXAMPLE Use Kruskal's algorithm to find a minimum spanning tree for this network.

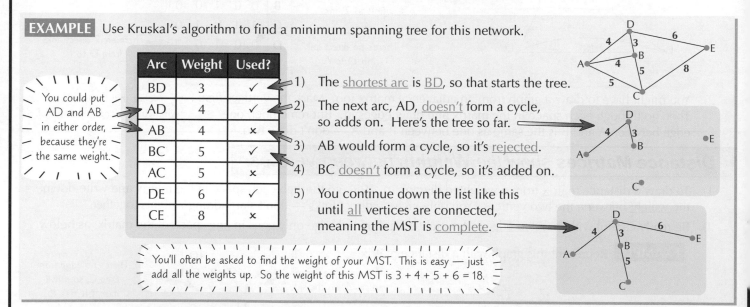

Arc	Weight	Used?
BD	3	✓
AD	4	✓
AB	4	✗
BC	5	✓
AC	5	✗
DE	6	✓
CE	8	✗

You could put AD and AB in either order, because they're the same weight.

1) The <u>shortest arc</u> is <u>BD</u>, so that starts the tree.

2) The next arc, AD, <u>doesn't</u> form a cycle, so adds on. Here's the tree so far.

3) AB would form a cycle, so it's <u>rejected</u>.

4) BC <u>doesn't</u> form a cycle, so it's added on.

5) You continue down the list like this until <u>all</u> vertices are connected, meaning the MST is <u>complete</u>.

You'll often be asked to find the weight of your MST. This is easy — just add all the weights up. So the weight of this MST is 3 + 4 + 5 + 6 = 18.

There are often <u>a few different</u> MSTs that can be found for a network — and you might be asked to find them.

In the list above, <u>AB</u> could have been used instead of AD, or <u>AC</u> instead of BC. All the different combinations of these arcs give <u>three more MSTs</u>:

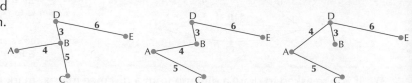

MST = Mathematical Stress and Tension...

All the minimum spanning trees for a network have the same total weight — if they didn't, they wouldn't all be minimums. Kruskal's algorithm works fine, but you can't do it straight from matrix forms of graphs, which is what computers often use. And if you've got a whopping network, this might be important. This is where Prim's algorithm is tops — see the next page...

Minimum Spanning Trees

Prim's algorithm does exactly the same job as Kruskal's algorithm. I'd love to say you only need to learn the one you like best, but that'd be a fib. You've got to learn them both, of course.

Prim's Algorithm *Finds Minimum Spanning Trees Too*

PRIM'S ALGORITHM

1) Pick a vertex, <u>any vertex</u> — this <u>starts</u> the tree.

2) Choose the arc of <u>least weight</u> that'll join a vertex <u>in the tree</u> to one <u>not yet</u> in the tree.

3) Repeat step 2 until you've joined <u>all</u> the vertices.

> If there's more than one to pick from, just choose randomly.

EXAMPLE Use Prim's algorithm to find a minimum spanning tree for the network on the right.

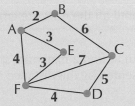

1) I'm going to randomly pick <u>E</u> to start the tree.

   There are <u>two arcs</u> that join E to vertices not yet in the tree. EA and EF. They're both the same weight (3), so I'll randomly pick <u>EA</u>.

2) There are three arcs that join a vertex <u>in</u> the tree (A or E) to a vertex <u>outside</u> the tree: AB (2), AF (4), EF (3).

   3) <u>AB</u> has the <u>least weight</u>, so that's the one to add.

4) Now the choice is from arcs AF (4), EF (3) or BC (6) — they join a vertex in the tree (A, B or E) to one not yet in the tree.

   5) <u>EF</u> has the <u>least weight</u>, so it's added on next.

> You <u>don't</u> have to check for cycles like you did with Kruskal's algorithm. Connecting to a vertex <u>outside</u> the tree will never make a cycle (one less thing to worry about).

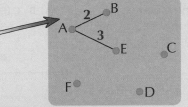

6) Nearly there now — the next choice is from BC (6), FC (7) or FD (4).

   7) <u>FD</u> has <u>least weight</u>, so that's the one to add.

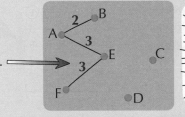

8) Finally, C could be joined by BC (6), FC (7) or DC (5).

   9) <u>DC</u> has the <u>least weight</u>, so that's the final arc.

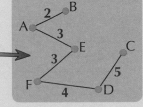

Wasn't Mr Prim one of the Mr Men?...

Practice, practice, practice is definitely the key for this algorithm too. There are only three steps to memorise, so that's not too tricky. You've just got to apply them accurately — so check you've considered all the possible arcs. Exam questions usually ask you to show your working, or to state the order that you add the arcs in, so winging it with a different method won't do.

Minimum Spanning Trees

The easiest way of putting a graph into a computer is to use a matrix. And the reason why Prim's algorithm is so useful is that it can be used on a <u>distance matrix</u>. It looks like a horrible mess of crossing out and circling randomly, but stay with me, and it'll all come clear.

Prim's Algorithm *Can be Used on* Distance Matrices

<u>PRIM'S ALGORITHM</u>

1) Pick <u>any vertex</u> to start the tree.

2) Cross out the <u>row</u> for the new vertex and circle the <u>column</u> for it.

3) Look for the <u>smallest weight</u> that's in <u>ANY circled column</u> AND <u>isn't</u> yet crossed out. Circle it. This is the <u>next arc</u> to add to the tree. The row it's in gives you the <u>new vertex</u>.

4) <u>Repeat</u> steps 2 and 3 until all the rows are crossed out.

EXAMPLE Use Prim's algorithm to find a minimum spanning tree for the graph represented by this distance matrix.

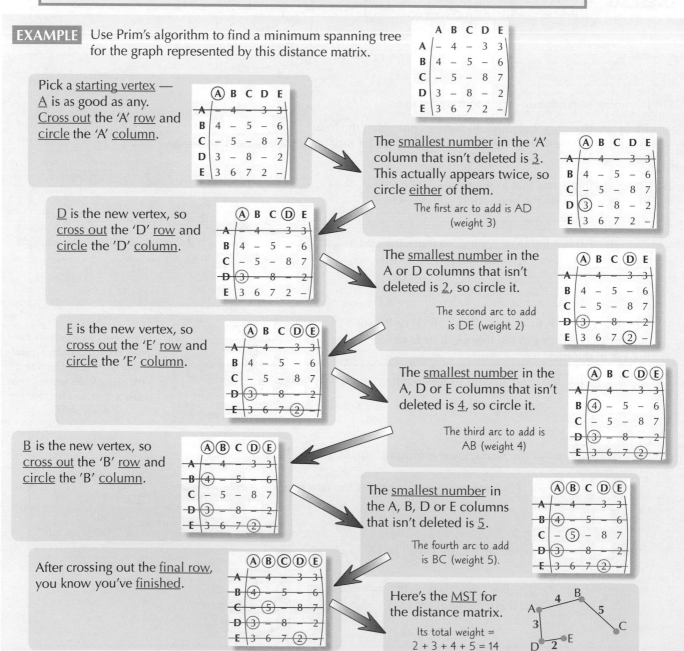

Pick a <u>starting vertex</u> — <u>A</u> is as good as any. <u>Cross out</u> the 'A' <u>row</u> and <u>circle</u> the 'A' <u>column</u>.

The <u>smallest number</u> in the 'A' column that isn't deleted is <u>3</u>. This actually appears twice, so circle <u>either</u> of them.

The first arc to add is AD (weight 3)

<u>D</u> is the new vertex, so <u>cross out</u> the 'D' <u>row</u> and <u>circle</u> the 'D' <u>column</u>.

The <u>smallest number</u> in the A or D columns that isn't deleted is <u>2</u>, so circle it.

The second arc to add is DE (weight 2)

<u>E</u> is the new vertex, so <u>cross out</u> the 'E' <u>row</u> and <u>circle</u> the 'E' <u>column</u>.

The <u>smallest number</u> in the A, D or E columns that isn't deleted is <u>4</u>, so circle it.

The third arc to add is AB (weight 4)

<u>B</u> is the new vertex, so <u>cross out</u> the 'B' <u>row</u> and <u>circle</u> the 'B' <u>column</u>.

The <u>smallest number</u> in the A, B, D or E columns that isn't deleted is <u>5</u>.

The fourth arc to add is BC (weight 5).

After crossing out the <u>final row</u>, you know you've <u>finished</u>.

Here's the <u>MST</u> for the distance matrix.

Its total weight = 2 + 3 + 4 + 5 = 14

This method also calculates the birthdate of your one true love...

When you're doing this yourself, you don't have to draw out the matrix a zillion times like I've done. Phew, I hear you say. I just wanted you to see all my steps. It took me ages, so do admire them all, and then have a practice for yourself.

Dijkstra's Algorithm

This is another of those algorithms that look really, really complicated. But when you've <u>learnt the steps</u>, you can string them together pretty rapidly. This one does a different job from the last two, so don't just skim the first bit.

Dijkstra's Algorithm Finds the Shortest Path between Two Vertices

1) If you're driving between <u>two cities</u> with a complicated road network between them, it's good to be able to work out which route is <u>quickest</u> (just like satnavs do).

2) <u>Dijkstra's algorithm</u> is a foolproof way to do this. Basically, you <u>label each vertex</u> with the length of the shortest path found so far from the starting point. If you find a <u>shorter path</u>, then you <u>change the label</u>. You keep doing this until you're sure that you've got the shortest distance to it.

<u>DIJKSTRA'S ALGORITHM</u>

1) Give the <u>Start vertex</u> the <u>final value '0'</u>.

> Once you've given a vertex a final label, you can't change it.

2) Find all the vertices <u>directly connected</u> to the vertex you've just given a final value to. Give each of these vertices a <u>working value</u>.

$$\text{Working value} = \frac{\text{Final value at}}{\text{previous vertex}} + \frac{\text{weight of arc}}{\text{between previous vertex and this one.}}$$

If one of these vertices already has a working value, replace it <u>ONLY</u> if the new working value is <u>lower</u>.

3) Look at the <u>working values</u> of vertices that <u>don't</u> have a final value yet. Pick the <u>smallest</u> and make this the <u>final value</u> of that vertex.

> If two vertices have the same smallest working value, pick either.

4) Now repeat steps 2 and 3 until the <u>End vertex</u> has a <u>final value</u> (this is the shortest path length).

5) Trace the route <u>backwards</u> (from the End vertex to the Start vertex). An arc is on the path if:

$$\text{Weight of arc} = \text{Difference in final values of the arc's vertices}$$

EXAMPLE: Use Dijkstra's algorithm to find the shortest route between A and G.

In the exam, you'll be given a version of the graph with <u>boxes</u> like this to complete. Here's what goes in each box:

Vertex	Order of labelling	Final value
	Working values	

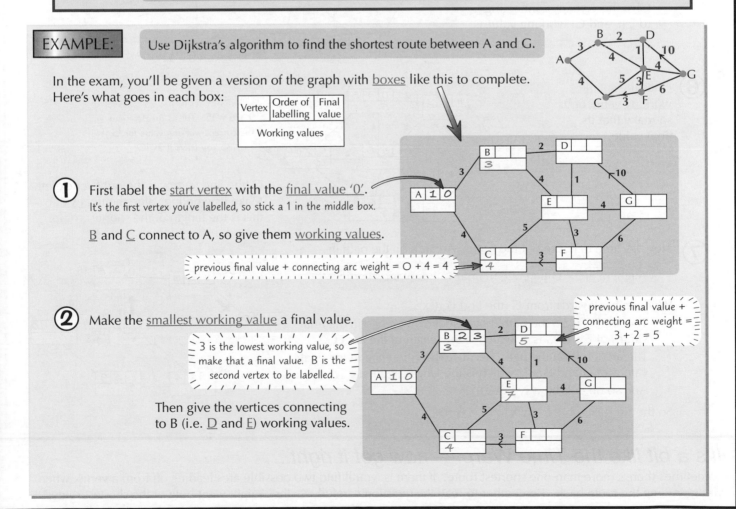

① First label the <u>start vertex</u> with the <u>final value '0'</u>.
It's the first vertex you've labelled, so stick a 1 in the middle box.

<u>B</u> and <u>C</u> connect to A, so give them <u>working values</u>.

> previous final value + connecting arc weight = 0 + 4 = 4

② Make the <u>smallest working value</u> a final value.

> 3 is the lowest working value, so make that a final value. B is the second vertex to be labelled.

> previous final value + connecting arc weight = 3 + 2 = 5

Then give the vertices connecting to B (i.e. <u>D</u> and <u>E</u>) working values.

Dijkstra's Algorithm

③ Again, make the <u>smallest working value</u> a final value.

4 is the lowest working value, so make that a final value. C is the third vertex to be labelled.

Then give all vertices without final values connecting to C a working value. It's only <u>E</u>, because F is connected by a <u>directed edge</u> that only goes from <u>F to C</u>, not from C to F.

4 + 5 = 9, but this is greater than the current working value for E, so you DON'T replace it.

④ You're probably getting the idea now. Make the smallest working value a <u>final value</u>. In this case it's <u>5</u> (making D the <u>4th</u> vertex to be labelled).

Give all vertices without final values connecting to D a working value. It's only E in this case. (G is connected by a <u>directed edge</u> running in the opposite direction. And of course B <u>already</u> has a final value.)

5 + 1 = 6 (you're coming from vertex D this time). This is smaller than the current working value for E, so you DO replace it.

⑤ <u>E</u> has the <u>smallest working value</u> (6), (in fact the only working value) so make that its <u>final value</u>.

Give the vertices connecting to <u>E</u> (F and G) working values.

6 + 4 = 10

6 + 3 = 9

⑥ <u>F</u> has the <u>smallest working value</u> of 9, so make that its final value.

9 + 6 = 15. This is greater than the current working value for G, so you leave it as 10.

G (the <u>End vertex</u>) is the only vertex left. Make its working value the <u>final value</u> — this is the length of the <u>shortest route</u>.

⑦ Now it's time to figure out the <u>route</u>. An arc's on the path if:

> Weight of arc = Difference in final values of arc's vertices

Working backwards from G (the <u>End vertex</u>):

- The arc <u>EG</u> is on the path, because the <u>difference</u> in the final values of E and G is <u>4</u>, which is the length of the arc <u>EG</u>.
- The arc <u>DE</u> is on the path, because the <u>difference</u> in the final values of D and E is <u>1</u>, which is the length of the arc <u>DE</u>.
- And so on, all the way back to <u>A</u>.

So the <u>shortest route</u> from A to G is <u>ABDEG</u>.

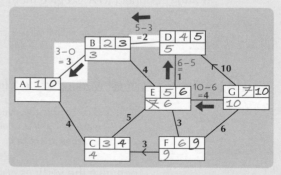

It's a bit like the Time Warp — now get it right...

Sometimes there's more than one shortest route. If there is, you'll find two possible arcs leading off from a vertex when you're tracing the route back. Remember to read the question carefully — they might want both of the shortest routes.

D1 Section 2 — Practice Questions

Right, now for this algorithm. 1) <u>Try</u> the questions. 2) <u>Check</u> your answers. 3) Reread the page on any you got <u>wrong</u>. 4) Repeat steps 1-3 until you get them all <u>right</u>.

Warm-up Questions

1) Explain what the following are; a) network, b) digraph, c) tree, d) spanning tree

   Questions 2–7 are about the graph on the right.

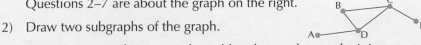

2) Draw two subgraphs of the graph.

3) How many arcs do you need to add to the graph to make it into a complete graph?

4) Describe a possible path in the graph.

5) Describe a possible cycle in the graph.

6) The graph is currently connected. Delete some edges so that it isn't connected any more.

7) List the degree of each vertex.
   Explain the link between the number of edges and the sum of the degrees.

8) Here's an adjacency matrix. Draw the graph it represents.

$$\begin{array}{c} \\ A \\ B \\ C \\ D \end{array} \begin{array}{c} A\ B\ C\ D \\ \left(\begin{array}{cccc} 0 & 0 & 1 & 1 \\ 0 & 2 & 0 & 1 \\ 1 & 0 & 0 & 1 \\ 1 & 1 & 1 & 0 \end{array}\right) \end{array}$$

9) Using Dijkstra's algorithm on the graph on page 201: (a) Find the shortest route from A to F.
   (b) Delete edge BD. Now find the shortest route from A to G.

Now for some questions just like you'll get in the exam. They really are the best sort of practice you can do.

Exam Questions

1 Figure 1 shows the potential connections for a sprinkler system between greenhouses at a plant nursery. The numbers on each arc represent the cost in pounds of each connection.

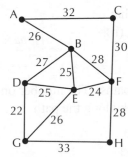

Figure 1

(a) Use Kruskal's algorithm to find a minimum spanning tree for the network in Figure 1. List the edges in the order that you consider them and state whether you are adding them to your minimum spanning tree.

(3 marks)

(b) State the minimum cost of connecting the sprinkler system.

(1 mark)

(c) Draw the minimum spanning tree obtained in a).

(2 marks)

(d) If Prim's algorithm had been used to find the minimum spanning tree, starting from E, find which edge would have been the final edge added. Show your working.

(2 marks)

(e) State two advantages of Prim's algorithm over Kruskal's algorithm for finding a minimum spanning tree.

(2 marks)

D1 Section 2 — Practice Questions

Keep going. You can do it. No need for gas and air yet.

2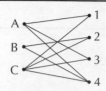

Figure 2   Figure 3

(a) Name the type of graph drawn in Figure 2.

(1 mark)

(b) State the number of edges that would need to be added to Figure 2 to make the graph complete.

(1 mark)

(c) State the number of edges that would need to be added to Figure 3 to make the graph connected.

(1 mark)

(d) What is the sum of the orders of the vertices in Figure 3?

(1 mark)

(e) Explain why it is impossible to add edges to Figure 3 so that all vertices have an odd order.

(2 marks)

3 The table shows the lengths, in miles, of the roads between five towns.

(a) Use Prim's algorithm, starting from A, to find a minimum spanning tree for this table. Write down the arcs in the order that they are selected.

(3 marks)

(b) Draw your tree and state its total weight.

(2 marks)

(c) State the number of other spanning trees that are the same length as your answer in part (b).

(1 mark)

	A	B	C	D	E
A	–	14	22	21	18
B	14	–	19	21	20
C	22	19	–	21	15
D	21	21	21	–	24
E	18	20	15	24	–

4 The diagram below shows a network of forest paths. The number on each edge represents the time, in minutes, required to walk along the path.

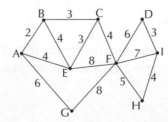

(a) Write down the number of edges in a minimum spanning tree of the network shown.

(1 mark)

(b) Use Dijkstra's algorithm to find the fastest route from A to I. State how long the route will take.

(6 marks)

(c) A new path, requiring x minutes to walk along, is to be made between G and H. The new path reduces the time required to walk between A and I. Find and solve an inequality for x.

(2 marks)

Traversable Graphs

You know that puzzle where you have to draw the shape on the right <u>without</u> taking your pencil off the paper? Well, you're about to find out the logic behind it and exactly which points you can <u>start</u> drawing from.

Graphs can be **Eulerian**...

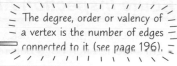

The degree, order or valency of a vertex is the number of edges connected to it (see page 196).

If <u>all</u> the vertices in a graph have an <u>even degree</u>, the graph is Eulerian.

1) These three graphs are all <u>Eulerian</u>. Every vertex is <u>even</u>.

   The numbers show the degrees.

 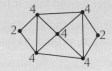

2) Eulerian graphs are <u>traversable</u>. This means it's <u>always possible</u> for you to start at <u>any point</u>, draw along each edge <u>exactly once</u> without taking your pen off the paper, and end up back at your <u>starting position</u>. Not every route works, but there'll definitely be some that do.

3) Or to look at it another way, if the graph represents <u>roads</u>, it's possible to walk down each of them <u>exactly once</u> before getting back to your <u>starting point</u>.

   EXAMPLE   Find a route that traverses the graph on the right.

   The graph is Eulerian. So you can start at any point.
   A possible route is: AGDBCDEGBAFEA.
   Another route is: EDCBDGAFEABGE.

   No matter which route you take, it'll always involve passing through the same number of vertices (13 in this case).

...Semi-Eulerian...

If <u>exactly two vertices</u> have an <u>odd degree</u>, and the rest are even, the graph is semi-Eulerian.

1) These three graphs are all <u>semi-Eulerian</u>. There are exactly <u>two odd</u> vertices.

2) Semi-Eulerian graphs are <u>semi-traversable</u>. This means it's possible to go along every edge on the graph <u>exactly once</u>, but <u>ONLY</u> if you start at one odd vertex and end up at the <u>other</u> odd vertex.

   EXAMPLE   Find a route that traverses the graph on the right.

   This graph is semi-Eulerian, so you have to start and end at the odd vertices (A and D).
   A possible route is: ABDCA around the square, then ABDCA around the circle,
                          then across the diagonal to D.

...or **Neither**

Remember — the sum of degrees is always even (see page 196). This means there'll always be an even number of odd vertices in a graph. So you'll never have 3 or 5 odd vertices, but you might have 4 or 6.

1) If a graph has <u>more than two odd vertices</u>, you <u>can't</u> traverse it.

2) There's <u>no route</u> that travels along each edge exactly once. You have to go along some of them <u>twice</u>.

   Some lovely non-Eulerian graphs. Try to find a route through them that goes along each edge exactly once. No taking your pen off the paper. See — you can't, can you?

Eul 'er now and it might stop 'er squeaking...

Right. So, Eulerian = traverse from any point and get back to starting point. Semi-Eulerian = traverse from one odd vertex to the other odd vertex. Non-Eulerian = You can't traverse. You have to go over some edges twice. And that's about it.

Route Inspection Problems

The <u>route inspection problem</u> is also called the <u>Chinese postman problem</u>. (It's named after a Chinese guy called Kwan Mei-Ko who discovered it in the 1960s. He wasn't a postman though — he was a mathematician.)

The **Route Inspection Algorithm** Finds the **Shortest Route** Covering **All Edges**

1) Route inspection problems ask you to find the <u>shortest route</u> through a network that goes along <u>each edge</u> before returning to the <u>starting point</u>.

2) It's the route that, say, a railway engineer would take if he had to <u>inspect</u> all the tracks.

3) In <u>postman</u> terms, the postman wants to find the <u>shortest route</u> that allows him to deliver letters to <u>every street</u> in a city, and brings him back to his <u>starting point</u> for a cup of tea.

4) And you'll never guess what. There are <u>algorithms</u> for solving these types of problems.

5) The first step is always to consider whether the graph is <u>Eulerian</u>, <u>semi-Eulerian</u>, or <u>neither</u>.

See the previous page.

A pre-1960s postman still manages to raise a smile, despite walking further than he has to each day.

Eulerian Graphs are Most **Straightforward**

Remember — in an Eulerian graph, all vertices have an even degree.

> **Length of inspection route in an Eulerian graph = Weight of network**

1) Eulerian graphs are <u>traversable</u>. So whatever point you start from, you can travel along each edge <u>exactly once</u>, and end up back at your <u>start point</u>.

2) Because you've gone down <u>each</u> edge <u>once</u>, you find the length of the route by just <u>adding up</u> all the <u>edge weights</u>.

EXAMPLE Find an inspection route for the network on the right.
Your route must start and finish at A. State the length of the route.

The graph is Eulerian — the vertices all have even degrees (they're all 4). So the graph is traversable, and a possible route is: <u>ABCDABCDA</u> (once round the quadrilateral, then once around the circle).

Length of the route = sum of weights = 4 + 6 + 5 + 2 + 5 + 7 + 6 + 3 = <u>38</u>

Semi-Eulerian Graphs are a **Bit Trickier**

In a semi-Eulerian graph, exactly two vertices have an odd degree, remember.

In semi-Eulerian networks, you have to repeat <u>the shortest path</u> between the two <u>odd vertices</u> in an inspection route. You can think of it as adding arcs to make the network <u>Eulerian</u> so that it can be traversed.

This formula lets you find the length of the inspection route:

> **Length of inspection route in a semi-Eulerian graph** = **Weight of network** + **weight of the shortest path between the two odd vertices**

EXAMPLE Find an inspection route for the network on the right.
Your route must start and finish at B. State the length of the route.

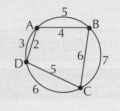

The graph is semi-Eulerian — the odd vertices are A and D. The shortest path between them is <u>AED</u>, of length **10** (5 + 5). So extra edges <u>AE</u> and <u>ED</u> are added.
A possible path is <u>BCDEFABDEAEB</u>.

Rather than use the formula, you could just find the length of your path by adding all the weights.

The weight of the network = 3 + 6 + 4 + 5 + 4 + 3 + 5 + 4 + 8 = <u>42</u>

This is worked out just by looking at the different possibilities, i.e. 'by inspection'.

Length of inspection route = Weight of network + weight of shortest path between odd vertices
= 42 + 10 = **52**

You only have to do this if you want to <u>start and end</u> at the <u>same vertex</u>. If you can pick and choose, just start at one odd vertex and end at the other. Then the length of the route will <u>equal</u> the network's weight (see the previous page).

Of course, during postal strikes, the postman stays put...

In the exam, you'll either be able to figure out the distance between the odd vertices by just looking, or you'll have worked it out earlier in the question using Dijkstra's algorithm (p.201). Right, now to find out what to do with non-Eulerian graphs...

Route Inspection Problems

The previous page told you how to find the shortest inspection route for Eulerian and semi-Eulerian networks. Now it's time to find out how to do it with <u>non-Eulerian networks</u>. It's a teensy bit more complicated — but I guess that's why it comes up more often in exam questions.

Learn the **Route Inspection Algorithm** for **Non-Eulerian** Networks

1) Non-Eulerian networks have <u>more than</u> two odd vertices.

2) But the exam board promises <u>not</u> to give you a network with more than four odd vertices, and networks <u>never</u> have odd numbers of odd vertices (see p.205). So you <u>only</u> have to worry about doing this stuff with <u>four odd vertices</u>.

> FINDING THE SHORTEST INSPECTION ROUTE FOR A NON-EULERIAN NETWORK:
> With this method you can start at <u>any vertex</u>, and you'll end up at the <u>same one</u>.
> <u>Pair</u> the four odd vertices in all the possible ways, find the pairing that gives you the <u>smallest total</u>, then <u>repeat the paths</u> between these pairs of vertices.

This'll become clearer with an example:

> **EXAMPLE** Find an inspection route for the network below. Your route must start and finish at E. State the length of the route.

1) Pick out the vertices with <u>odd degrees</u>.

They're marked in pink on this network — A, B, C, D ⟹

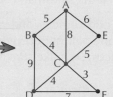

2) <u>Pair</u> the vertices in <u>all the ways possible</u>.

There are 3 ways to pair 4 vertices: AB + CD
                                         AC + BD
                                         AD + BC

3) Work out the <u>minimum total distance</u> for each pairing.

E.g. for AB + CD, you find the smallest distance between AB, and the smallest distance between CD, and add them together: AB + CD = 5 + 4 = 9

> There are loads of paths from A to D. By inspection, ACD is shortest (8 + 4 = 12), so use that one.

AC + BD = 8 + 8 = 16
AD + BC = 12 + 4 = 16

4) Pick the pairing with the <u>smallest total distance</u>.

With a length of 9, it's <u>AB + CD</u> that has the smallest total. AB and CD will be the paths you repeat in the inspection route.

5) Add <u>extra edges</u> along each path in your pair.

Add edges to repeat the path between <u>A and B</u> and to repeat the path between <u>C and D</u>.

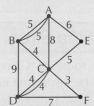

6) The graph is now <u>Eulerian</u> — so you can find an <u>inspection route</u> through it.

A possible route starting and finishing at E = E A B D C A B C F D C E. ⟸ I've marked the pink paths for you.

> Alternatively, you could just add up all the edges on your path. ⟹

7) Add the <u>lengths of the new edges</u> to the <u>weight of the network</u> to get the length of the inspection route.

The length of the route = weight of network + length of extra edges

= 51 + 9 = <u>60</u>

> Weight of network = 5 + 6 + 8 + 4 + 5 + 9 + 4 + 7 + 3 = 51

He ain't heavy — so he's the best pairing possible...

Even if it's screamingly obvious which pairing is going to have the smallest value, always show the examiner that you've considered all the possible pairings and didn't just get lucky this once. Oh, and the most direct path mightn't be the shortest, so do check carefully, or it'll cock the whole thing up, and you'll shed marks like a German Shepherd sheds fur.

Route Inspection Problems

Those examiners don't want you thinking they're the least bit cuddly. After asking you to find an inspection route for a network starting and finishing at <u>a certain point</u>, they'll often tell you to find the minimum inspection route if you can <u>start and finish wherever you like</u>.

Starting and Finishing at **Different Odd Vertices** Always **Shortens** the Route

In a network with <u>four odd vertices</u>, the shortest route which goes down each path at least once always involves <u>starting at one odd vertex</u> and <u>ending at a different odd vertex</u>.

> FINDING THE SHORTEST INSPECTION ROUTE STARTING AND ENDING AT DIFFERENT POINTS OF YOUR CHOICE:
>
> You <u>start at one odd vertex</u> and <u>end at another odd vertex</u>, so you <u>only</u> have to <u>repeat the path</u> between <u>one pair</u> of vertices. You want this path to be <u>as short as possible</u>.

> **EXAMPLE** A feather duster salesman wants to travel along each street in a housing estate. He can start his journey at any point, and end it at any point. The graph represents the streets in the estate, and the numbers represent the lengths of each street in hundreds of metres.
>
> (a) State the vertices that the salesman could start at to minimise his journey.
>
> > There are four odd vertices, and you're going to <u>start and end</u> at <u>two of them</u>, so this leaves one pair of odd vertices. You have to <u>repeat the path</u> between these two vertices, so make sure it's the <u>shortest possible</u>.
> >
> > > The odd vertices are <u>B, C, D, E</u>.
> > > The distance between each possible pair is:
> > > BC = 5, DE = 4, BD = 8, CE = 9, BE = 8, CD = 5
> >
> > > The distance between <u>D and E</u> is shortest, at only 4. So that's the path you need to repeat. So you start at <u>either</u> of vertices <u>B or C</u>, and end at the other.
>
> (b) Find the length of his journey.
>
> > You just have to repeat the path between D and E.
> > So, total length of journey = weight of network + path between D and E
> > $$= 54 + 4 = \underline{58}$$
> >
> > Always <u>check the units</u> — the weights represent <u>hundreds of metres</u>, so the journey is actually <u>5800 metres</u> long, or <u>5.8 km</u>.

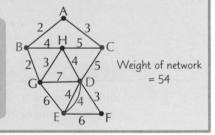

Weight of network = 54

Walking on **Both Sides** of the Street Makes it an **Eulerian Network**

1) Sometimes, the usual 'inspection route' or 'Chinese postman' problem is <u>changed</u> so that the person has to go down <u>each path twice</u>. Perhaps they'll be inspecting each pavement, or delivering leaflets to both sides of the streets.

2) This actually makes it <u>easier</u> to solve. It effectively <u>doubles</u> the edges at each vertex, making all the vertices effectively <u>even</u>. The network is now <u>Eulerian</u>, so you can <u>traverse</u> it, and the length of the inspection route will just be <u>double the weight</u> of the network.

> **EXAMPLE** The feather duster salesman decides to go down each street twice, once on each side. What is the length of his new route?
>
> Weight of original network = 54.
> Weight of doubled network = 54 × 2 = 108
> Length of new route = <u>10 800 metres</u>, or <u>10.8 km</u>.

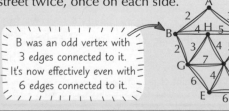

B was an odd vertex with 3 edges connected to it. It's now effectively even with 6 edges connected to it.

How's your network? — dunno, the fish just sort of get caught in it...

Remember — always read the question carefully to see what it's asking. Don't just scan it, say "Ah, Chinese postman" and jump in there. The examiners might have added a subtle twist. But to be kind, they'll often just tell you the network's weight.

D1 Section 3 — Practice Questions

Right. Time to see if you have the information from this section stuck in your head...

Warm-up Questions

1) Say whether each of these graphs is <u>Eulerian</u>, <u>semi-Eulerian</u> or <u>neither</u>.

(a) 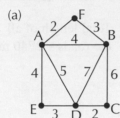 (b) 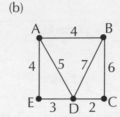 (c) 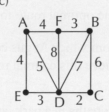 (d)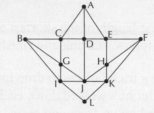

2) Identify the <u>odd</u> vertices in the graph on the right.
   Write down all the possible ways of <u>pairing</u> the odd vertices.

3) Find the length of the shortest "<u>Chinese postman</u>" route
   for each of these networks. Start and end at vertex A.

(a)   (b)    (c)

4) Repeat question 3. But this time you can start and finish at <u>any</u> vertices.
   State <u>which</u> vertices you're starting and finishing at.

That's all the main skills brushed up on. Now it's time to see if you can apply them to some exam-style questions.

Exam Questions

1   A machinist is embroidering logos on sportsbags. The two logos are shown below.

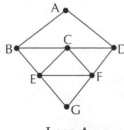 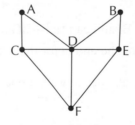

Logo A   Logo B

(a) Say whether each logo consists of an Eulerian graph, a semi-Eulerian graph or neither. *(2 marks)*

(b) The sewing machine needle is positioned at any starting point, and sews a route without
    stitching any line more than once. It can be lifted and moved to a new starting point.

   (i)   For each logo, how many times must the needle be lifted? *(2 marks)*

   (ii) For logo A, state an efficient starting vertex. *(1 mark)*

(c) Extra arcs are added to each logo to make them Eulerian.
    State the minimum number of arcs that must be added to each logo.

   (2 marks)

D1 Section 3 — Practice Questions

These are your typical "Chinese postman" problems — of course they contain subtle variations. Just like in real exams.

2   The diagram on the right shows all the streets in a town, and their lengths in metres.

Angus is considering moving to the town and wants to walk down each street at least once.
He parks his car at K.

(a)   Explain why it is not possible to walk down each street only once and return to the starting point.

(1 mark)

(b)   Find the length of the shortest route Angus could follow, starting and finishing at K.

(6 marks)

(c)   Angus's friend offers to drop him off at any point, and after he's walked down each street at least once, to pick him up from any point.

(i)   Find the length of the optimal route for Angus.

(2 marks)

(ii)   State the vertices from which Angus could start in order to achieve this optimal route.

(1 mark)

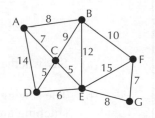

Total length of all the roads = 2740 m

3   The diagram below shows the paths in a park, and the time taken to walk them in minutes.

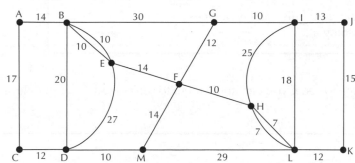

Alice the park keeper needs to walk down each path to check for storm-damaged trees.
She parks her car at F.

Time required to walk along all paths
= 336 minutes

(a)   Find the time for the shortest route Alice could follow, starting and finishing at F.

(6 marks)

(b)   If Alice starts at point B, and can finish at any point:

(i)   What point should she end at for the optimal route? Show your working.

(2 marks)

(ii)   How long will it take her to walk along all the paths now?

(1 mark)

4   The diagram on the right shows the distances between towns in miles.
The total road distance is 106 miles.

Jamie is inspecting the hedgerows along the roads.
He needs to go along each road, starting and finishing at A.

(a)   Find the length of the optimal 'Chinese postman' route for Jamie.

(6 marks)

(b)   There are ice-cream shops at points C and E.
If Jamie follows his optimal route, how many times will he pass an ice-cream shop?

(2 marks)

(c)   Jamie decides it would be better if he went along each road twice.
What is the length of his new optimal route?

(2 marks)

Activity Networks

A complicated project like building a house involves <u>lots of different activities</u>. Some activities can't be <u>started</u> until others are <u>finished</u>, e.g. you can't put the roof on until the walls are built, and you can't build the walls until the foundations are in place. If you want to get the house built as quickly and cheaply as possible, you have to do a lot of <u>planning</u> (there's no point in the decorator turning up on the same day as the bricklayer and hanging round for 3 months). Luckily, there are a few tricks you can use to help you...

Precedence Tables Show Which Activities Need Doing **Before** Others

<u>Precedence tables</u> show what activities must be <u>finished</u> before others are <u>started</u>.
They're not as exciting as an episode of Futurama, but they're pretty easy to draw.

| EXAMPLE | Below is a list of activities involved in rustling up an apple pie. On the right is a precedence table of the information. |

Activities	
A	Mix pastry
B	Roll out pastry
C	Grease tin
D	Line tin with pastry
E	Chop apples
F	Stew apples
G	Fill pastry with filling
H	Add pastry lid

The activities are usually given letters so you can refer to them easily.

Activity	Immediately preceding activities
A	—
B	A
C	—
D	B, C
E	—
F	E
G	D, F
H	G

A (mixing pastry) doesn't depend on anything being done first.

You need to finish A before starting B (mix the pastry before rolling it out).

D depends on both B and C being finished. A must be finished too, but you've already said that B can't start until A is done, so you don't have to show it again.

Activity Networks Show the Information **More Clearly**

1) Precedence diagrams are OK, but they're not great. Putting the information in an <u>activity network</u> makes it easier to understand, and it looks more impressive too.

2) In activity networks, 'arcs' represent the <u>activities</u>, and 'nodes' represent the <u>completion of activities</u> or '<u>events</u>'.

3) The nodes are <u>numbered</u> as they're added to the network. The first one is numbered <u>zero</u> and is called the <u>source node</u>. The final node is called the <u>sink node</u>.

4) Constructing activity networks can be quite awkward — here's one for the <u>apple pie example</u> above:

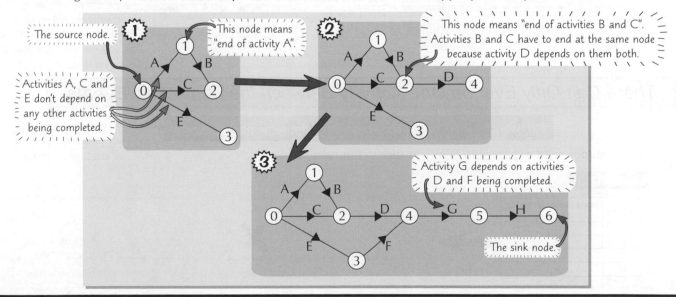

The source node.

This node means "end of activity A".

Activities A, C and E don't depend on any other activities being completed.

This node means "end of activities B and C". Activities B and C have to end at the same node because activity D depends on them both.

Activity G depends on activities D and F being completed.

The sink node.

Arcs are always straight lines — not a general rule for life...

It's very tricky getting the layout of an activity network right first time round, and you're likely to have to do some rubbing out — so draw your activity network in pencil. Getting the layout right does get easier the more you practise though.

Activity Networks

Some situations can only be shown on an activity network using curiously named "dummies". You'll see...

Dummies Help Show the Order that Activities Must be Done in

1) Dummy activities aren't real activities. They just show that one real activity depends on another real activity when it can't be shown clearly otherwise.

2) The information in this precedence table can't be shown on an activity network without a dummy:

EXAMPLE 1:
Show the information in this table in an activity network.

Activity	Immediately preceding activities
A	—
B	A
C	—
D	A, C

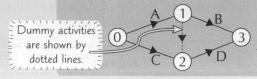

Dummy activities are shown by dotted lines.

The dummy activity shows that activity A must be done before activity D.

Activity D must come after the completion of activities A AND C. B doesn't depend on C being completed though.

EXAMPLE 2:
Show the information in the table in an activity network. (It's a tricky one that needs two dummies.)

Activity	Immediately preceding activities
A	—
B	—
C	—
D	A
E	B, C
F	C
G	D
H	D, E, F

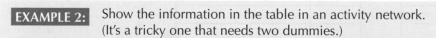

The first bit's straightforward. Activities A, B and C don't depend on any other activities. Activity D depends only on A.

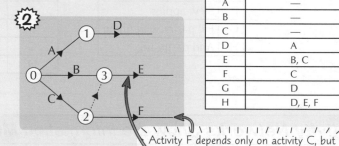

Activity F depends only on activity C, but activity E depends on BOTH B and C. You need a dummy to show this.

Activity H depends on both activity E and activity F — so you have to make E and F's arcs end at a shared node.

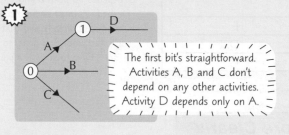

Activity G depends only on D, but activity H depends on D, E and F. You show this with a dummy.

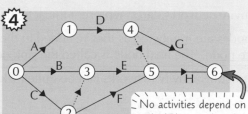

No activities depend on G, so make the 'G' arc end at the sink node. You only ever have one final node.

There Can Only Ever be One Activity Between the Same Two Events

No two activities can be shown between the **same pair of events**.

In other words each activity must be shown between a different pair of nodes.

You sometimes need a dummy to help you stick to this rule:

EXAMPLE 3:

Activity	Immediately preceding activities
A	—
B	A
C	A
D	B, C

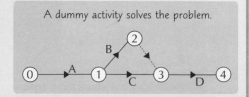

You can't do it like this. Activities B and C are between the same pair of nodes.

A dummy activity solves the problem.

Shop window dummies, crash test dummies, baby dummies — now these...
Dummies are handy tools. But that's all they are, they don't exist in the real world. Just like monsters are a tool for scaring small children. Oh, shouldn't you do that? Ooops. Well, try drawing precedence tables from the networks above anyway.

Activity Networks with Early and Late Times

When planning real projects, you normally have a <u>deadline</u> to work to. So <u>how long</u> each activity takes is important.

Each Activity has a **Duration**

1) The <u>duration</u> of an activity is <u>how long</u> it takes to complete.

> The durations are given in brackets.
> They can have the units hours, days, weeks, etc.

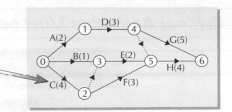

2) Each node or 'event' has an <u>early event time</u> and a <u>late event time</u>:

> An event is the completion of an activity or activities (see page 211).

EARLY EVENT TIME — the **earliest time** you can reach an event. It depends on the durations of the **preceding activities**.	**LATE EVENT TIME** — the **latest time** you can get to an event **without** increasing the duration of the entire project.

3) The convention is to use <u>boxes</u> at each node to show the early and late event times.

3	← The early event time goes in the top box.
6	← The late event time goes in the bottom box.

Work Out the **Early Event Times** Starting from the **Source**...

You work out an <u>early event time</u> by adding the <u>activity duration</u> to the <u>previous early event time</u>. If there's a <u>choice of paths</u> to a node you use the <u>biggest number</u>. Have a look at the example below (the times are in hours).

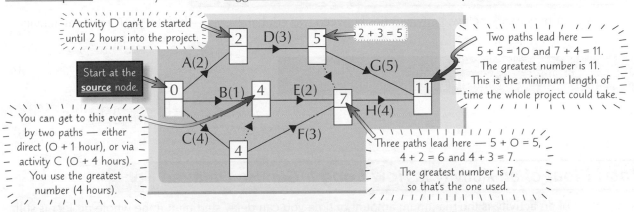

> Activity D can't be started until 2 hours into the project.

> Start at the **source** node.

> You can get to this event by two paths — either direct (0 + 1 hour), or via activity C (0 + 4 hours). You use the greatest number (4 hours).

> 2 + 3 = 5

> Two paths lead here — 5 + 5 = 10 and 7 + 4 = 11. The greatest number is 11. This is the minimum length of time the whole project could take.

> Three paths lead here — 5 + 0 = 5, 4 + 2 = 6 and 4 + 3 = 7. The greatest number is 7, so that's the one used.

...Then Work Out the **Late Event Times** Starting from the **Sink**

To work out <u>late event times</u>, you start at the <u>sink node</u>, and work back towards the source. You subtract each <u>activity duration</u> from its <u>late event time</u>. If there's a choice of paths, you use the <u>smallest number</u> each time.

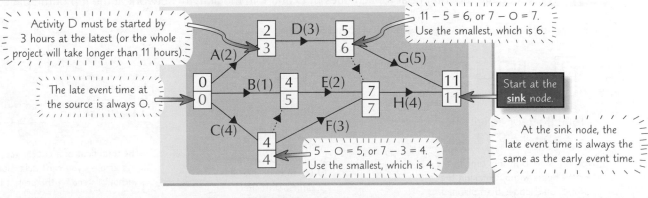

> Activity D must be started by 3 hours at the latest (or the whole project will take longer than 11 hours).

> The late event time at the source is always 0.

> 11 − 5 = 6, or 7 − 0 = 7. Use the smallest, which is 6.

> G(5)

> Start at the **sink** node.

> At the sink node, the late event time is always the same as the early event time.

> 5 − 0 = 5, or 7 − 3 = 4. Use the smallest, which is 4.

Help — I'm sinking. Too late... Glugg glugg glugg...

It's good to check at the end that your early event times are all less than or equal to your late event times. Remember, use the biggest numbers to work out early event times. Try imagining that your teammates are running along different length paths to the node. You haven't won until you all arrive there. Then remember that working out late event times is the opposite.

Critical Paths

All the activities for a project have to be done, but with some of them you can take your time and <u>still meet</u> the deadline. With other activities you <u>can't</u> — these are the <u>critical</u> ones.

Critical Activities Must be Completed *Within* their Allotted Time

If the duration of a **CRITICAL ACTIVITY** increases, the duration of the whole project increases by the same amount.

A **CRITICAL PATH** runs from the source node to the sink node and is made up of **critical activities**.

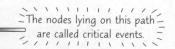
The nodes lying on this path are called critical events.

1) All the nodes on a <u>critical path</u> have the same <u>early and late event times</u>.

2) This means they <u>MUST</u> be started at a particular time — there's no 'slack'.

3) <u>Adding up</u> the <u>durations</u> of the activities on the critical path gives you the duration of the <u>whole project</u> or the '<u>critical time</u>'.

4) You can have an activity network with <u>more than one</u> critical path. The critical paths will all have the same durations.

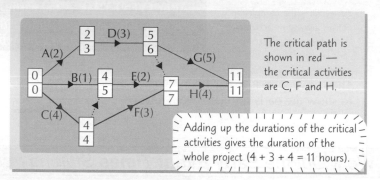

The critical path is shown in red — the critical activities are C, F and H.

Adding up the durations of the critical activities gives the duration of the whole project (4 + 3 + 4 = 11 hours).

Activities *Between* Critical Events *Aren't* Always Critical

1) In the really simple network below, <u>activity B</u> lies between <u>two critical events</u> — but it's <u>NOT</u> a critical activity.

2) An activity is <u>only critical</u> if:

$$\text{Activity Duration} = \text{Event time of critical node after} - \text{Event time of critical node before}$$

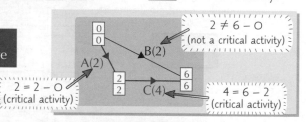

$2 \neq 6 - 0$
(not a critical activity)

$2 = 2 - 0$
(critical activity)

$4 = 6 - 2$
(critical activity)

The *Total Float* of an Activity is *How Long* It Can Be *Delayed* For

1) The <u>total float</u> of an activity is the <u>maximum amount of time</u> you can <u>delay starting it</u> if the whole project is still going to be <u>completed on time</u>.

2) There's a lovely formula for it: Total float of an activity = latest finish time – duration – earliest start time

EXAMPLE: Calculate the total float for activities B, D and F in the activity network at the top of the page.

I've copied the chunks of the activity network you need for each part.

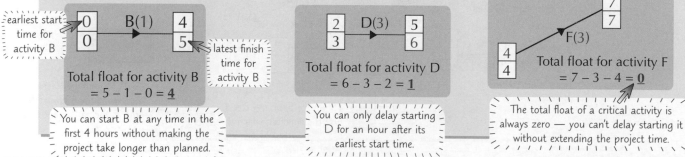

earliest start time for activity B

latest finish time for activity B

Total float for activity B
= 5 – 1 – 0 = **4**

You can start B at any time in the first 4 hours without making the project take longer than planned.

Total float for activity D
= 6 – 3 – 2 = **1**

You can only delay starting D for an hour after its earliest start time.

Total float for activity F
= 7 – 3 – 4 = **0**

The total float of a critical activity is always zero — you can't delay starting it without extending the project time.

Critical path: You're walking on me all wrong...

Make sure you're really hot at finding critical paths and identifying critical activities. Remember, there may be more than one — just use the same rules for finding each of them. There's practice at spotting multiple critical paths on p. 218-219.

Gantt Charts

The Gantt chart, Henry Gantt's greatest triumph. I bet his mum was proud.

Gantt Charts Show Possible Start and Finish Times For Activities

1) <u>Gantt charts</u> are also called <u>cascade charts</u>. They let you show the <u>possible time periods</u> that each activity can happen in for the project to be completed on time.

2) Here's how to plot a single activity on a Gantt chart:

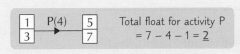

Total float for activity P = 7 − 4 − 1 = <u>2</u>

The earliest P can start is 1 hour into the project.

P has a duration of 4 hours. If it starts at 1 hour, it'll finish at 5 hours.

The latest P can finish is at 7 hours.

The total float is 2 hours.

3) This example shows a whole activity network plotted on a Gantt chart:

EXAMPLE: Display this information on a Gantt chart. The times are all shown in hours.

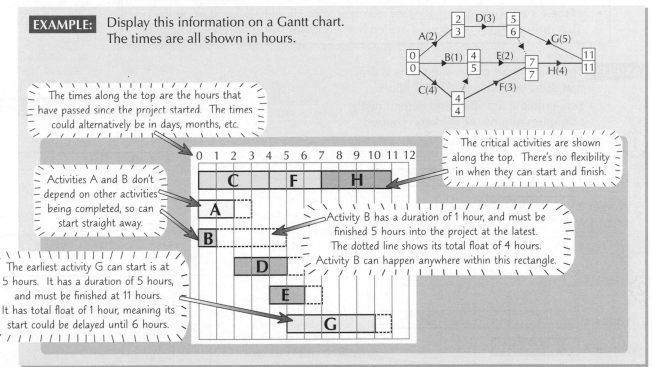

The times along the top are the hours that have passed since the project started. The times could alternatively be in days, months, etc.

Activities A and B don't depend on other activities being completed, so can start straight away.

The critical activities are shown along the top. There's no flexibility in when they can start and finish.

Activity B has a duration of 1 hour, and must be finished 5 hours into the project at the latest. The dotted line shows its total float of 4 hours. Activity B can happen anywhere within this rectangle.

The earliest activity G can start is at 5 hours. It has a duration of 5 hours, and must be finished at 11 hours. It has total float of 1 hour, meaning its start could be delayed until 6 hours.

4) The solid activity rectangles can slide <u>right</u> — as long as they stay within the <u>dotted rectangles</u>. But sliding them might affect <u>other activities</u>. E.g. if A slides forward an hour, then D must too. This means G has to slide later too.

You Have to Interpret Gantt Charts Too

1) You'll sometimes be asked what activities will be happening at a particular time. Some activities will <u>definitely</u> be happening at this time, whereas other activities only <u>might</u> be happening.

2) Don't forget — the times along the top indicate the number of days or hours that have <u>elapsed</u>. So day 4 is the space <u>before</u> 4 on the scale.

EXAMPLE: Which activities will definitely be happening and which might be happening at noon on day 4 of this 12 day project?

Only activity <u>D</u> will <u>definitely</u> be happening on day 4. Activities <u>B</u> and <u>C</u> <u>might</u> be happening.

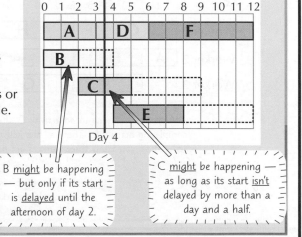

B <u>might</u> be happening — but only if its start is <u>delayed</u> until the afternoon of day 2.

C <u>might</u> be happening — as long as its start <u>isn't</u> delayed by more than a day and a half.

I Gantt help it — I didn't write the Exam specification...

Well, actually, it's not so bad, this stuff. If I had written the spec, I reckon I'd have put it on too. To test whether you really know it, copy out the activity network above but without the times. Then try to fill them in by just looking at the Gantt chart.

Scheduling

Gantt charts are useful for working out <u>how many people</u> are needed to complete a project on time. An employer wouldn't want to have people standing about, because they're all going to want to be <u>paid</u>.

Scheduling Diagrams Show How Many Workers are Needed

1) Scheduling diagrams show <u>which activities</u> are assigned to each worker, and <u>how many workers</u> are needed to complete the project by the deadline.

2) There are some <u>rules</u> to follow when you're scheduling workers:

- Assume that <u>each activity</u> requires <u>one worker</u>.

- Once a worker starts an activity, they have to <u>carry on</u> with it until it's finished. They <u>can't</u> break off from the activity to get another one started.

- Assume that once a worker finishes one activity, they're ready to start on another <u>immediately</u> — workers in Decision Maths don't need a lunch break.

- If there's a <u>choice of activities</u> for a worker, they should always start on the one that must be <u>finished soonest</u> (which has the lowest latest finishing time).

EXAMPLE: The Gantt chart on the right is for a project that must be completed in 12 hours. Schedule the activities and determine the minimum number of workers required.

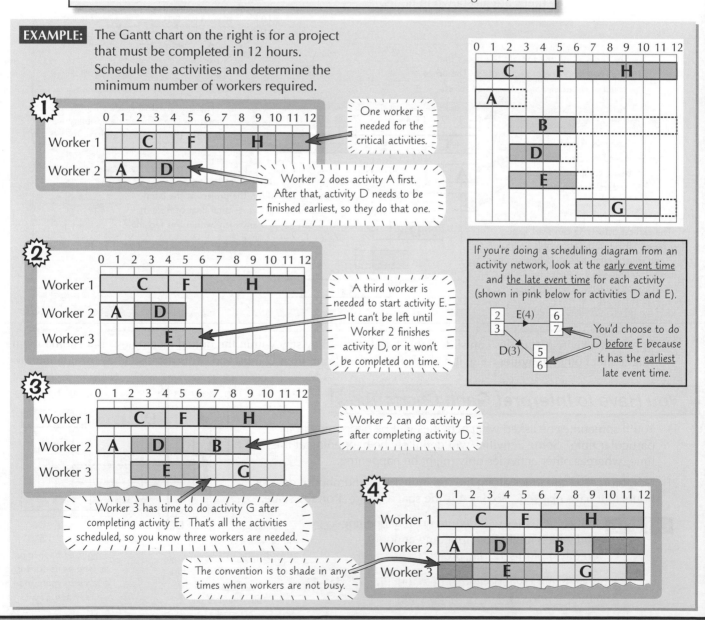

One worker is needed for the critical activities.

Worker 2 does activity A first. After that, activity D needs to be finished earliest, so they do that one.

A third worker is needed to start activity E. It can't be left until Worker 2 finishes activity D, or it won't be completed on time.

If you're doing a scheduling diagram from an activity network, look at the <u>early event time</u> and <u>the late event time</u> for each activity (shown in pink below for activities D and E).

You'd choose to do D <u>before</u> E because it has the <u>earliest</u> late event time.

Worker 2 can do activity B after completing activity D.

Worker 3 has time to do activity G after completing activity E. That's all the activities scheduled, so you know three workers are needed.

The convention is to shade in any times when workers are not busy.

I left my job because of something the boss said — it was "You're fired"...

Of course, in real life, there'll be complications. You might need a worker with specialist skills to do particular activities, and things will happen to scupper your scheduling diagrams, like people being off sick, or machines breaking. But hey, who cares about that — this is maths. Your workers can work nonstop for hours. They don't even need loo breaks.

Scheduling

There are a <u>few more things</u> about scheduling that you need to know.
Hang on in there — after this it's only the questions to go.

The **Lower Bound** for the Number of Workers is the **Absolute Minimum**

The <u>lower bound</u> for the number of workers is the <u>fewest workers</u> that could <u>possibly</u>
complete the project within the critical time (the length of the critical path).

> The lower bound for the number of workers is: The smallest integer $\geq \dfrac{\text{sum of all activity durations}}{\text{critical time of project}}$

> **EXAMPLE:** Calculate the lower bound for the number of workers needed to complete
> the project on the right within the critical time. The times are all in days.
> The sum of all the durations of activities is 38 days.
>
> Critical time of project = 15 days
>
> $$\frac{\text{sum of all activity durations}}{\text{critical time of project}} = \frac{38}{15} = 2.53$$
>
> The smallest integer greater than or equal to 2.53 is 3.
> So the lower bound is <u>3</u>.

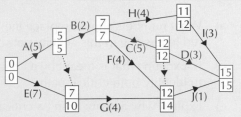

But the Lower Bound **Might Not be Enough**

1) The <u>lower bound</u> doesn't take into account the <u>overlap</u> and <u>orders</u> of
   activities. So there's a fair chance that the lower bound number of
   workers <u>won't</u> be enough to complete the project within the <u>critical time</u>.

2) The Gantt chart for the activity network above shows that on day 11,
   <u>four</u> activities <u>must</u> be happening — three workers <u>isn't enough</u>.

3) Constructing a <u>scheduling</u>
   <u>diagram</u> from the Gantt chart,
   shows that you do need
   <u>four workers</u> to complete
   the project in <u>15 days</u>.

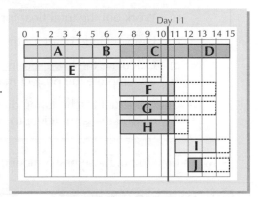

Without **Enough** Workers, the Project will take **Longer**

1) If you don't have enough workers, the project will take <u>longer than its critical time</u>.

2) If you're creating a scheduling diagram for <u>fewer</u> than the minimum number of workers,
   it's best to do it from the <u>activity network</u> so you don't miss any precedences.

> **EXAMPLE:** Only two workers are available to complete the project represented in the activity network above.
> Schedule the activities in the minimum number of days.
>
>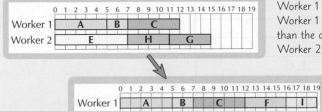
>
> Worker 1 starts activity A, Worker 2 starts activity E.
> Worker 1 finishes first and does activity B (it has to be completed sooner
> than the other activities, and the activities it depends on (just A) are finished).
> Worker 2 finishes activity E and goes on to activity H (using same logic as before).
>
> This system carries on until all the activities are scheduled.
> The whole project takes 19 days.

You're getting through in leaps and lower bounds...

The lower bound assumes that the tasks can all be slotted together neatly. But when some activities depend on others,
it doesn't usually work that way. Remember to round up to an integer — because you can't have part of a worker.

D1 Section 4 — Practice Questions

Oooh... precedence tables, activity networks, Gantt charts, scheduling diagrams.
To make sure you know how to construct and use each of them, try these juicy little questions.

Warm-up Questions

1) Activities A and B don't depend on any other activities. Activity C depends on activity A, activity D depends on activity B, and activity E depends on both activities C and D.

   a) Draw a precedence table to represent this information.

   b) Draw an activity network for the project.

2) Draw an activity network for the precedence table on the right. You'll need two dummies.

Activity	Immediate preceding activities
A	—
B	—
C	A, B
D	A, B
E	C
F	C
G	D
H	E, G
I	E, F, G

3) a) What numbers should replace the red letters *a*–*g* in this activity network?

   b) i) Explain what is meant by a critical path.
      ii) Identify both critical paths.

   c) Explain why activity E isn't a critical activity.

   d) State the critical time for the project. (The times given are in hours.)

   e) Work out the total float for activities B, E and I.

   f) Draw a Gantt chart for the project.

   g) Calculate the lower bound for the number of workers required to complete the project in the critical time.

   h) Construct a scheduling diagram to show that the lower bound number of workers is sufficient to complete the project in the critical time.

Now it's time to put those newly-honed skills to the test with some exam-style questions.

Exam Questions

1 a) This precedence table contains information about a project.
        Draw an activity network to show the information. Use exactly two dummies.

(5 marks)

Activity	Immediate preceding activities
A	—
B	—
C	A
D	B, C
E	B
F	D
G	E
H	F, G
I	F, G
J	H, I

   b) Explain why each dummy is needed.

(2 marks)

D1 Section 4 — Practice Questions

Just <u>two more questions</u> to have a bash at. There's nothing too devilish about them...

2   The network in Figure 1 shows the activities involved in a process. The number in brackets on each arc gives the time, in days, taken to complete the activity.

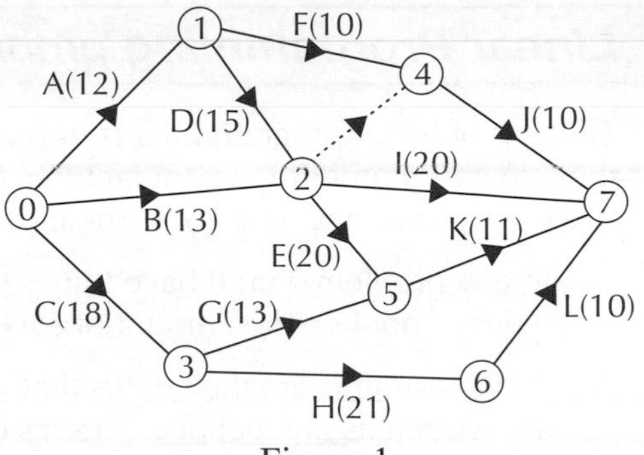

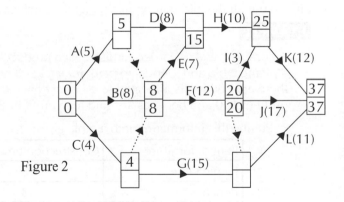

Figure 1

a)   Calculate the early time and the late time for each event. Show them on a copy of the diagram.   *(4 marks)*

b)   Determine the critical activities and the length of the critical path.   *(3 marks)*

c)   Calculate the total float on each of activities F and G. Show your working.   *(3 marks)*

d)   Draw a cascade (Gantt) chart for the process.   *(4 marks)*

e)   Calculate a lower bound for the number of workers needed to complete the project in the minimum time. You must show your working.   *(2 marks)*

f)   Schedule the activities, using as few workers as possible, so that the project is completed in 58 days.   *(3 marks)*

g)   Comment on your answer to part f) regarding the lower bound you calculated in part e).   *(1 mark)*

3   Figure 2 shows an activity network for a project. The duration of each activity is shown in days. The sum of all the activity times is 112 days.

Figure 2

a)   Some of the early and late event times are shown. Complete the diagram by calculating the missing early and late event times.   *(3 marks)*

b)   There are two critical paths for this network. State them both.   *(2 marks)*

c)   Find a lower bound for the number of workers required to complete the project in the critical time. Show your working.   *(2 marks)*

d)   Which activities must be happening on day 23? Explain how you know this.   *(2 marks)*

e)   Schedule the activities so that the project is completed in the minimum number of days. You must use as few workers as possible.   *(3 marks)*

f)   The supervisor realises at midday on day 14 that activity D has not yet been started. Determine if the project can still be finished on time. Explain your answer.   *(2 marks)*

Linear Programs

Linear programming is a way of <u>solving problems</u> that have lots of inequalities, often to do with <u>money</u> or <u>business</u>. So if you're a budding entrepreneur, pay attention — this section could help you make your <u>first million</u>. Or just pass D1.

Linear Programming problems use Inequalities

The <u>aim</u> of linear programming is to produce an <u>optimal solution</u> to a <u>problem</u>, e.g. to find the solution that gives the <u>maximum profit</u> to a manufacturer, based on <u>conditions</u> that would affect it, such as <u>limited time</u> or <u>materials</u>. Before you start having a go at linear programming problems, there are a few <u>terms</u> you need to know.

1) In any problem, you'll have things that are being <u>produced</u> (or <u>bought</u> or <u>sold</u> etc.) — e.g. jars of jam or different types of books. The <u>amount</u> of each thing is represented by x, y, z etc — these are called the <u>decision variables</u>.

2) The <u>constraints</u> are the <u>factors</u> that <u>limit</u> the problem, e.g. a limited amount of workers available. The constraints are written as <u>inequalities</u> in terms of the <u>decision variables</u>. Most problems will have <u>non-negativity constraints</u>. This just means that the decision variables <u>can't</u> be <u>negative</u>. It makes sense really — you can't have −1 books.

3) The <u>objective function</u> is what you're trying to <u>maximise</u> or <u>minimise</u> (e.g. maximise <u>profit</u> or minimise <u>cost</u>). It's usually in the form of an <u>equation</u> written in terms of the <u>decision variables</u>.

4) A <u>feasible solution</u> is a solution that <u>satisfies</u> all the <u>constraints</u>. It'll give you a <u>value</u> for each of the <u>decision variables</u>. On a <u>graph</u>, the <u>set</u> of feasible solutions lie in the <u>feasible region</u> (see p.221).

5) You're aiming to <u>optimise</u> the objective function — that's finding a solution within the feasible region that maximises (or minimises) the <u>objective function</u>. This is the <u>optimal solution</u>, and there can be <u>more than one</u>.

Put the Information you're given in a Table

Linear programming questions can look a bit confusing because you're given a lot of <u>information</u> in one go. But if you put all the information in a <u>table</u>, it's much easier to work out the <u>inequalities</u> you need.

EXAMPLE

A company makes garden furniture, and produces both picnic tables and benches. It takes 5 hours to make a picnic table and 2 hours to paint it. It takes 3 hours to make a bench and 1 hour to paint it. In a week, there are 100 hours allocated to construction and 50 hours allocated to painting. Picnic tables are sold for a profit of £30 and benches are sold for a profit of £10. The company wants to maximise their weekly profit.

Putting this information into a table gives:

Item of furniture	Construction time (hours)	Painting time (hours)	Profit (£)
Picnic table	5	2	30
Bench	3	1	10
Total time available:	100	50	

Now use the table to identify all the different parts of the problem and come up with the inequalities:

- The <u>decision variables</u> are the <u>number of picnic tables</u> and the <u>number of benches</u>, so let x = number of picnic tables and y = number of benches.

- The <u>constraints</u> are the <u>number of hours available</u> for <u>each stage</u> of manufacture. Making a picnic table takes 5 hours, so x tables will take $5x$ hours. Making a bench takes 3 hours, so y benches will take $3y$ hours. There are a total of 100 hours available. From this, you get the inequality $5x + 3y \leq 100$. Using the same method for the painting hours produces the inequality $2x + y \leq 50$. You also need $x, y \geq 0$. ◄——— *Don't forget the non-negativity constraints.*

- The <u>objective function</u> to be <u>maximised</u> is <u>profit</u>. Each picnic table makes a profit of £30, so x tables make a profit of £$30x$. Each bench makes a profit of £10, so y benches make a profit of £$10y$. Let P be the profit, then the aim is to maximise $P = 30x + 10y$.

I'd like −3 picnic tables please...

In fancy examiner speak, the example above could be written like this: 'maximise $P = 30x + 10y$ subject to the constraints $5x + 3y \leq 100$, $2x + y \leq 50$ and $x, y \geq 0$'. Watch out for constraints such as 'there have to be at least twice as many benches as picnic tables — this would be written as $2x \leq y$. You might have to think about this one to get your head round it.

Feasible Regions

If you're a fan of <u>drawing graphs</u>, you'll love this page. It's a bit like the <u>graphical inequality problems</u> you came across at GCSE. Even if you're not that keen on graphs, or if the mere thought of them brings you out in a <u>rash</u>, don't worry — they're only <u>straight line graphs</u>.

Drawing **Graphs** can help solve **Linear Programming Problems**

<u>Plotting the constraints</u> on a <u>graph</u> is probably the easiest way to <u>solve</u> a linear programming problem it helps you see the <u>feasible solutions</u> clearly. Get your ruler and graph paper ready.

1) Draw each of the <u>constraints</u> as a <u>line</u> on the graph. All you have to do is <u>change</u> the <u>inequality sign</u> to an <u>equals sign</u> and plot the line. If you find it easier, <u>rearrange</u> the equation into the form <u>$y = mx + c$</u>.

2) Then you have to <u>decide</u> which bit of the graph you <u>want</u> — whether the solution will be <u>above</u> or <u>below</u> the line. This will depend on the <u>inequality sign</u> — <u>rearrange</u> the inequality into the form $y = mx + c$, then think about which sign you'd use. For <u>$y \leq mx + c$</u> (or <), you want the bit <u>underneath</u> the line, and if it's <u>$y \geq mx + c$</u> (or >) then you want the bit <u>above</u> the line. If you're not sure, put the <u>coordinates</u> of a point in one region (e.g. the origin) into the equation and see if it <u>satisfies</u> the inequality.

3) Once you've decided which bit you want, <u>shade</u> the region you <u>don't want</u>. This way, when you put all the constraints on the graph, the <u>unshaded region</u> (the bit you want) is easy to see.

4) If the inequality sign is < or >, use a <u>dotted line</u> — this means you <u>don't</u> include the line in the region. If the inequality sign is ≤ or ≥ then use a <u>normal</u> line, so the line is <u>included</u> in the range of solutions.

5) Once you've drawn <u>all</u> the constraints on the graph, you'll be able to solve the problem. Don't forget the <u>non-negativity constraints</u> — they'll limit the graph to the <u>first quadrant</u>.

The **Unshaded Area** is the **Feasible Region**

Your finished graph should have an area, <u>bounded</u> by the lines of the <u>constraints</u>, that <u>hasn't</u> been <u>shaded</u>. This is the <u>feasible region</u> — the <u>coordinates</u> of any point inside the <u>unshaded area</u> will satisfy <u>all</u> the <u>constraints</u>.

EXAMPLE

On a graph, show the constraints $x + y \leq 5$, $3x - y \geq 2$, $y > 1$ and $x, y \geq 0$. Label the feasible region R.

Rearranging the inequalities into '$y = mx + c$' form and choosing the appropriate inequality sign gives: $y \leq 5 - x$, $y \leq 3x - 2$ and $y > 1$. The non-negativity constraints $x, y \geq 0$ are represented by the x- and y- axes.

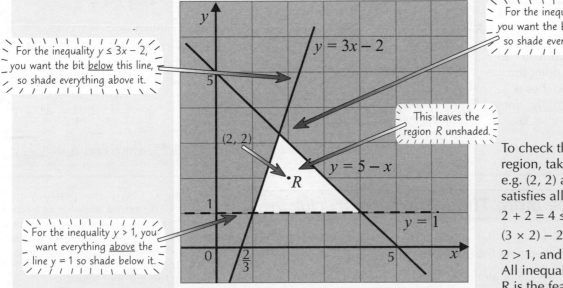

For the inequality $y \leq 3x - 2$, you want the bit <u>below</u> this line, so shade everything above it.

For the inequality $y \leq 5 - x$, you want the bit <u>below</u> this line — so shade everything above it.

This leaves the region R unshaded.

For the inequality $y > 1$, you want everything <u>above</u> the line $y = 1$ so shade below it.

To check that R is the feasible region, take a point inside R, e.g. (2, 2) and check that it satisfies all the inequalities:

$2 + 2 = 4 \leq 5$

$(3 \times 2) - 2 = 4 \geq 2$

$2 > 1$, and $2 \geq 0$.

All inequalities are satisfied so R is the feasible region.

It's feasible that I might become a film star...

...but not very likely — I'm a terrible actor. Anyway, although it's possible that you might get a linear programming problem with more than two variables, you won't have to draw graphs for these ones. You'll only have to graph 2-variable problems.

Optimal Solutions

Don't throw away your graphs just yet — you still need them for the next few pages. You're now getting on to the really useful bit — actually solving the linear programming problem.

Draw a line for the Objective Function

All the points in the feasible region (see previous page) satisfy all the constraints in the problem. You need to be able to work out which point (or points) also optimises the objective function. The objective function is usually of the form $Z = ax + by$, where Z either needs to be maximised (e.g. profit) or minimised (e.g. cost) to give the optimal solution.

The Objective Line Method

1. **Draw the straight line $Z = ax + by$, choosing a fixed value of Z (a and b will be given in the question). This is called the objective line.**

2. **Move the line to the right, keeping it parallel to the original line. As you do this, the value of Z increases (if you move the line to the left, the value of Z decreases).**

3. **If you're trying to maximise Z, the optimal solution will be the last point within the feasible region that the objective line touches as you slide it to the right.**

4. **If you're trying to minimise Z, the optimal solution will be the last point within the feasible region that the objective line touches as you slide it to the left.**

This is sometimes called the ruler method, as a good way to do it is to slide a ruler over the graph parallel to the objective line.

The objective lines have the Same Gradient

When you draw your first objective line, you can use any value for Z. Pick one that makes the line easier to draw — e.g. let Z be a multiple of both a and b so that the intercepts with the axes are easy to find.

EXAMPLE Using the example from the previous page, maximise the profit $P = 2x + 3y$.

First, choose a value for P, say $P = 6$. This means that the objective line goes through $(3, 0)$ and $(0, 2)$. Draw this line on the graph.

Slide the objective line to the right until it reaches the last point within R. At this point, P is maximised.

This is the first objective line, where $P = 6$.

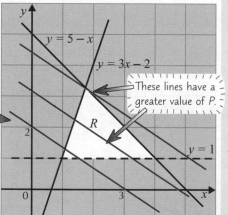

These lines have a greater value of P.

From the diagram, you can see that the last point the objective line touches is the intersection of the lines $y = 5 - x$ and $y = 3x - 2$.

To find the point of intersection, solve these simultaneous equations. This will give you the optimal solution.

Substituting $y = 5 - x$ into $y = 3x - 2$ gives: $5 - x = 3x - 2$

$$7 = 4x \Rightarrow x = \frac{7}{4}$$

Putting $x = \frac{7}{4}$ into $y = 5 - x$ gives $y = \frac{13}{4}$.

Now put these values into the objective function to find P: $P = 2x + 3y$

$$= 2\left(\frac{7}{4}\right) + 3\left(\frac{13}{4}\right)$$

$$= \frac{14}{4} + \frac{39}{4} = \frac{53}{4} = 13.25$$

So the maximum value of P is 13.25, which occurs at $\left(\frac{7}{4}, \frac{13}{4}\right)$.

There might be More Than One optimal solution

If the optimal solution is on a dotted line, the actual solution will just be a point very very close to it. You don't need to worry about this though.

1) If the objective line is parallel to one of the constraints, you might end up with a section of a line that gives the optimal solution.

2) If this happens, any point along the line is an optimal solution (as long as it's inside the feasible region).

3) This shows that there can be more than one optimal solution to a problem.

Maximise your chance of passing D1...

Make sure you're happy with solving simultaneous equations — if not, have a look back at page 24. You won't have to solve any tricky quadratic equations for this section, but you need to be able to solve linear simultaneous equations quick-smart.

Optimal Solutions

If you object to using the <u>objective line</u>, there is another method. This one uses a lot more <u>simultaneous equations</u>, but you don't have to worry about keeping the ruler <u>parallel</u> or <u>stopping global warming</u> or anything like that.

Optimal Solutions are found at Vertices

The <u>optimal solution</u> for the example on the previous page was found at a <u>vertex</u> of the <u>feasible region</u>. This isn't a coincidence — if you have a go at some more linear programming problems, you'll find that the optimal solutions <u>always</u> occur at a vertex (or an <u>edge</u>) of the feasible region. This gives you another way to solve the problem.

The Vertex Method

1. **Find the x- and y-values of the <u>vertices</u> of the <u>feasible region</u>. You do this by solving the <u>simultaneous equations</u> of the <u>lines</u> that <u>intersect</u> at each vertex.**

2. **Put these values into the <u>objective function</u> $Z = ax + by$ to find the value of Z.**

3. **Look at the Z values and work out which is the <u>optimal value</u>. Depending on your objective function, this might be either the <u>smallest</u> (if you're trying to <u>minimise</u> Z) or the <u>largest</u> (if you're trying to <u>maximise</u> Z).**

If two vertices A and B produce the same Z value, this means that all points along the edge AB are also optimal solutions.

Test Every vertex

Even if it looks <u>obvious</u> from the graph, you still have to <u>test</u> each vertex of the feasible region. Sometimes the <u>origin</u> will be one of the vertices — it's really easy to test, as the objective function will just be equal to <u>0</u> there. Don't forget vertices on the <u>x-</u> and <u>y-axes</u> too.

You can sometimes just read off the coordinates from your graph (as long as it's accurate).

EXAMPLE

Minimise $Z = 8x + 9y$, subject to the constraints $2x + y \geq 6$, $x - 2y \leq 2$, $x \leq 4$, $y \leq 4$ and $x, y \geq 0$.

Drawing these constraints on a graph produces the diagram below, where A, B, C and D are the vertices of the feasible region R:

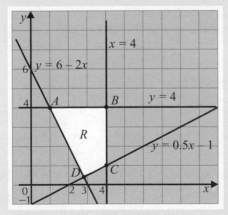

Point A is the intersection of the lines $y = 6 - 2x$ and $y = 4$, so A has coordinates $(1, 4)$.

Point B is the intersection of the lines $x = 4$ and $y = 4$, so B has coordinates $(4, 4)$.

Point C is the intersection of the lines $x = 4$ and $y = 0.5x - 1$, so C has coordinates $(4, 1)$.

Point D is the intersection of the lines $y = 6 - 2x$ and $y = 0.5x - 1$, which has coordinates $\left(\frac{14}{5}, \frac{2}{5}\right)$.

You need to use simultaneous equations to find the coordinates of D.

Putting these values into the objective function $Z = 8x + 9y$:

At A, $Z = (8 \times 1) + (9 \times 4) = 44$.

At B, $Z = (8 \times 4) + (9 \times 4) = 68$.

At C, $Z = (8 \times 4) + (9 \times 1) = 41$.

At D, $Z = (8 \times 2.8) + (9 \times 0.4) = 26$.

So the minimum value of Z is 26, which occurs when $x = \frac{14}{5}$ and $y = \frac{2}{5}$.

In this example, it was really <u>easy</u> to find the coordinates of A, B and C, as at least one of the values was <u>given</u> by the <u>equation of the line</u>. D was a bit harder, as it involved <u>simultaneous equations</u>, but it wasn't too bad.

I'm getting vertigo...

Some questions might give you two different objective functions and ask you to minimise cost and maximise profit for the same set of constraints. The vertex method is really useful here, as once you've worked out the coordinates of the vertices, you can easily put the values into both objective functions without having to do any more work (or draw on confusing lines).

Optimal Integer Solutions

The methods on the previous two pages are all very well and good, but I can't exactly make 2.5 teddy bears, even if it does <u>maximise my profit</u>. There must be a better way, one that doesn't involve <u>mutilating soft toys</u>...

Some problems need **Integer Solutions**

1) Sometimes it's fine to have <u>non-integer solutions</u> to linear programming problems — for example, if you were making different <u>fruit juices</u>, you could realistically have 3.5 litres of one type of juice and 4.5 litres of another.

2) However, if you were making <u>garden furniture</u>, you couldn't make 3.5 tables and 4.5 benches — so you need <u>integer solutions</u>.

3) You won't always be <u>told</u> whether a problem needs integer solutions — you might have to <u>work it out</u> for yourself. It's common sense really — just think about whether you can have <u>fractions</u> of the <u>decision variables</u>.

You can use the **Objective Line Method** or the **Vertex Method**

Both of the methods covered on pages 222-223 can be used to find an <u>optimal integer solution</u> — it just depends on how <u>clear</u> your <u>graph</u> is.

Some problems have optimal integer solutions that are far away from the vertices — but you don't need to worry about these for D1.

1) You use the <u>objective line method</u> in exactly the <u>same way</u> as before, but instead of looking for the last <u>vertex</u> the line touches, you need to look for the last <u>point</u> with <u>integer coordinates</u> in the <u>feasible region</u>. This might be hard to do if your graph isn't very <u>accurate</u>, or if the scale isn't <u>clear</u>.

2) The other way to find the optimal integer solution is to use the <u>vertex method</u> to find which vertex to use. Then, consider all the points with <u>integer coordinates</u> that are <u>close by</u>. Make sure you <u>check</u> whether these points still <u>satisfy</u> the <u>constraints</u> though — test this <u>before</u> you put the values into the objective function.

The **Optimal Integer Solution** must be **Inside** the **Feasible Region**

It's easy to forget that <u>not all</u> the solutions near the optimal vertex will be <u>inside</u> the <u>feasible region</u> — you can check either <u>by eye</u> on an <u>accurate graph</u>, or put the <u>coordinates</u> into each of the <u>constraints</u>.

> **EXAMPLE** The optimal solution to the problem on the previous page occurred at $\left(\frac{14}{5}, \frac{2}{5}\right)$ (= (2.8, 0.4)). Find the optimal integer solution.
>
> Looking at the integers nearby gives you the points (3, 0), (3, 1), (2, 0) and (2, 1) to test. However, the point (3, 0) doesn't satisfy the constraint $x - 2y \leq 2$, and (2, 0) and (2, 1) don't satisfy $2x + y \geq 6$, so the optimal integer solution is at (3, 1), where $Z = (8 \times 3) + (9 \times 1) = 33$.

EXAMPLE

A company makes baby clothes. It makes x sets of girls' clothes and y sets of boys' clothes, for a profit of £6 and £5 respectively, subject to the constraints $x + y \leq 9$, $3x - y \leq 9$, $y \leq 7$ and $x, y \geq 0$. Maximise the profit, $P = 6x + 5y$

This example uses the vertex method, but you could also use the objective line method.

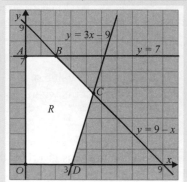

The feasible region is the area $OABCD$, with coordinates $O(0, 0)$, $A(0, 7)$, $B(2, 7)$, $C\left(\frac{9}{2}, \frac{9}{2}\right)$ and $D(3, 0)$.

The value of P at each vertex is O: £0, A: £35, B: £47, C: £49.50 and D: £18.

The maximum value of P is £49.50, which occurs at $\left(\frac{9}{2}, \frac{9}{2}\right)$ (= (4.5, 4.5)). However, making 4.5 sets of clothes is not possible, so an integer solution is needed.

The integer coordinates near C are (4, 5), (5, 5), (5, 4) and (4, 4). (5, 5) and (5, 4) don't satisfy the constraint $3x - y \leq 9$ so are outside the feasible region. At (4, 5), $P = £49$, and at (4,4), $P = £44$, so £49 is the maximum profit.

So the company needs to make 4 sets of girls' clothes and 5 sets of boys' clothes to make the maximum profit of £49.

My solution to a problem is to close my eyes and hope it'll go away...

Oo, I do like a nice integer now and then. Like a good cup of tea, integers can really brighten your morning. They're harder to dunk biscuits in, but they do provide sensible solutions to linear programming problems. Watch out for them in the exam.

D1 Section 5 — Practice Questions

This section isn't too bad really — once you've got your head round turning the <u>constraints</u> into <u>inequalities</u>, all you have to do is draw a <u>graph</u> and Bob's your uncle. I wish I had an Uncle Bob. Anyway, here are some warm-up questions for you to have a go at.

Warm-up Questions

1) Give a brief definition of:
   a) <u>decision variables</u>
   b) <u>objective function</u>
   c) <u>optimal solution</u>

2) What does a <u>dotted line</u> on a graph show?

3) What is an <u>objective line</u>?

4) Give one example of a problem that <u>doesn't</u> need <u>integer solutions</u>, and one example that <u>does</u>.

5) A company makes posters in two sizes, large and small. It takes 10 minutes to print each large poster, and 5 minutes to print each small poster. There is a total of 250 minutes per day allocated to printing. It takes 6 minutes to laminate a large poster and 4 minutes for a small poster, with a total of 200 minutes laminating time. The company want to sell at least as many large posters as small, and they want to sell at least 10 small posters. Large posters are sold for a profit of £6 and small posters are sold for a profit of £3.50.
   a) Write this out as a <u>linear programming problem</u>. Identify the <u>decision variables</u>, <u>constraints</u> and <u>objective function</u>.
   b) Show the <u>constraints</u> for this problem <u>graphically</u>. Label the <u>feasible region</u> R.
   c) <u>Maximise</u> the <u>profit</u>, using either the <u>objective line method</u> or the <u>vertex method</u>. Don't worry about integer solutions for now.
   d) Use your answer to part c) to find the <u>optimal integer solution</u>.

Exam questions try and <u>scare</u> you by throwing a lot of <u>information</u> at you all at once. Don't let them bully you, just take them <u>one bit at a time</u> and you'll soon get the better of them. Here are a few for you to do.

Exam Questions

1   Anna is selling red and white roses at a flower stall. She buys the flowers from a wholesaler, where red roses cost 75p each and white roses cost 60p each. Based on previous sales, she has come up with the following constraints:

   • She will sell both red roses and white roses.

   • She will sell more red roses than white roses.

   • She will sell a total of at least 100 flowers.

   • The wholesaler has 300 red roses and 200 white roses available.

Let x be the number of red roses she buys and y be the number of white roses she buys. Formulate this information as a linear programming problem. Write out the constraints as inequalities and identify a suitable objective function, stating how it should be optimised. You do not need to solve this problem.

(7 marks)

D1 Section 5 — Practice Questions

In the words of some little <u>Dickensian orphan</u>, 'please sir, I want some more'. Well, I'd be more than happy to oblige — here's another page full of <u>exciting exam questions</u>.

2   A company sells three packs of craft paper, bronze, silver and gold. Each pack is made up of three different types of paper, tissue paper, sugar paper and foil.

- The gold pack is made up of 6 sheets of foil, 15 sheets of sugar paper and 15 sheets of tissue paper.
- The silver pack is made up of 2 sheets of foil, 9 sheets of sugar paper and 4 sheets of tissue paper.
- The bronze pack is made up of 1 sheet of foil, 6 sheets of sugar paper and 1 sheet of tissue paper.
- Each day, there are 30 sheets of foil available, 120 sheets of sugar paper available and 60 sheets of tissue paper available.
- The company is trying to reduce the amount of foil used, so it uses at least three times as many sheets of sugar paper as of foil.

The company makes x gold packs, y silver packs and z bronze packs in a day.

(a)   Apart from the non-negativity constraints, write out the other four constraints as inequalities in terms of x, y and z. Simplify each inequality where possible.

(8 marks)

(b)   On Monday, the company decides to make the same number of silver packs as bronze packs.

   (i)   Show that your inequalities from part (a) become
$$2x + y \leq 10$$
$$x + y \leq 8$$
$$3x + y \leq 12$$
$$2y \geq x$$

(3 marks)

   (ii)   On graph paper, draw a graph showing the constraints from part (i) above, as well as the non-negativity constraints. Label the feasible region R.

(5 marks)

   (iii)   Use your graph to work out the maximum number of packs the company can make on Monday.

(2 marks)

   (iv)   Gold packs are sold for a profit of £3.50, silver packs are sold for a profit of £2 and bronze packs are sold for a profit of £1. Use your answers to parts (ii) and (iii) to maximise the profit they make, and state how many of each type of pack they need to sell.

(3 marks)

3   The graph below shows the constraints of a linear programming problem. The feasible region is labelled R.

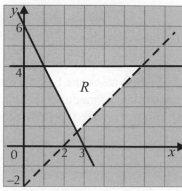

(a)   Including the non-negativity constraints, find the inequalities that produce R.

(5 marks)

(b)   Find the coordinates of each vertex of R.

(4 marks)

The aim is to minimise $C = 4x + y$.

(c)   Find the optimal solution and state where this value occurs.

(3 marks)

Matchings

Bipartite graphs are a really useful way of allocating people to tasks, taking into account what they want to do, what they can do, and what they won't do. For example, on the frequent occasions that I have to use my superpowers, I'd prefer to fly, am prepared to turn invisible but will _not_ dissolve into a puddle of water.

Bipartite Graphs have Two Sets of Nodes

The points in a graph are called <u>nodes</u> (or <u>vertices</u>) and the lines are called <u>arcs</u> (or <u>edges</u>) — this was covered in Section 2.

1) A <u>bipartite graph</u> is made up of <u>two sets</u> of <u>nodes</u> that are <u>linked</u> by <u>arcs</u>. The arcs go from <u>one set</u> of nodes to the <u>other</u> — nodes within the <u>same set</u> can't be joined to each other.

2) In lots of the examples you'll come across, one set of nodes will be the <u>people</u> and the other will be the <u>jobs</u> or <u>tasks</u> they have to do. You'll be told <u>who</u> can do which <u>job</u>.

3) There doesn't have to be the <u>same number</u> of nodes in each set.

Sometimes the information will be in a table instead of being written out.

EXAMPLE

Jenny, Katie, Lizzie, Martyn and Nikki are planning a picnic. Jenny can bring sandwiches and crisps, Katie can bring sandwiches and pork pies, Lizzie can bring drinks, Martyn can bring pork pies and biscuits and Nikki can bring biscuits, quiche and crisps. Draw a bipartite graph to show this information.

First, list all the people on one side of the graph, and all the food (and drink) on the other side. Then draw lines connecting each person to all the items they can bring.

The lines show what each person can bring.

In this example, there are more nodes on the right than on the left.

A Matching links One Person to One Task

Once you've drawn your bipartite graph showing who can do what, you need to work out a <u>solution</u> that assigns <u>one person</u> to <u>one job</u>. This is called a <u>matching</u>.

1) In a matching, you can only have <u>one arc</u> for <u>each node</u> — so each person only does <u>one job</u> (and each job only has <u>one person</u> doing it). Matchings are <u>one-to-one</u>.

2) You won't always be able to match <u>all</u> the nodes in one set with all the nodes in the other — it depends on who can do which task.

3) If there are the <u>same number</u> of people as jobs, and each <u>job</u> is assigned to a <u>person</u>, the matching is said to be <u>complete</u>. So if there are x <u>nodes</u> in each set, a complete matching has x arcs.

4) It's <u>not always possible</u> to have a complete matching — e.g. if there are <u>two jobs</u> that can only be done by the <u>same person</u>, one of the jobs <u>won't</u> be included in the matching.

5) If a complete matching can't be done, you might have to find a <u>maximum (or 'maximal')</u> <u>matching</u> — a matching that has the <u>greatest number</u> of arcs possible (so as many jobs as possible are being done). There can be <u>more than one</u> possible maximum matching.

You don't need to worry about the difference between 'maximal' and 'maximum' for D1

EXAMPLE

Andy, Ben, Carys and Daniel are going to a theme park. They can only afford to go on one ride each, and they each want to try out a different ride. Andy wants to go on the teacups or the big wheel, Ben wants to go on the rollercoaster or the teacups, Carys want go on the log flume or the rollercoaster and Daniel only wants to go on the big wheel. Draw a bipartite graph to show this information, then use it to find a complete matching.

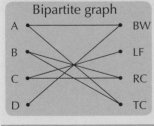

Bipartite graph

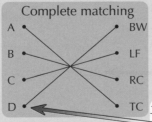

Complete matching

After a bad experience, Andy flatly refuses to go on the teacups again, so a complete matching is no longer possible. Find a maximum matching for their next visit to the theme park.

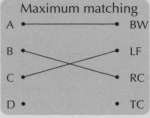

Maximum matching

Start with Daniel, who only wants to go on the big wheel, and work from there.

Add one teaspoon of bipartite of soda...

If you're thinking the graphs on this page look familiar, you're dead right — you came across bipartite graphs on page 195.

Alternating Paths

Sometimes your original matching can be improved. To do this, you need to be able to find an <u>alternating path</u>.

Start with an **Initial Matching** and find an **Alternating Path**

Draw <u>any</u> matching from a bipartite graph. This is the <u>initial matching</u> — you're trying to <u>improve</u> it.

The Alternating Path Method

1) An alternating path <u>starts</u> at a node on one side of the graph that <u>isn't included</u> in the <u>initial matching</u> and <u>finishes</u> at a node on the <u>other side</u> of the graph that also <u>isn't</u> in the initial matching.

2) To get from the start node to the finishing node, you have to <u>alternate</u> between arcs that are <u>not in</u> and <u>in</u> the initial matching. So the <u>first</u> arc you use (from the unmatched starting node) is <u>not in</u> the initial matching, the second arc is, the third one isn't and so on until you get to a finish node.

3) When you reach a finish node (one not in the initial matching), you can <u>stop</u> — you've made a 'breakthrough'.

4) Now use your alternating path to construct an <u>improved matching</u>. Take the path and <u>change</u> the <u>status</u> of the arcs, so any arcs <u>not in</u> the initial matching are <u>in</u> the new matching, and the arcs that were <u>in</u> the initial matching are <u>not in</u> the new one. Any arcs in the initial matching that aren't in the alternating path just <u>stay as they are</u>.

5) The <u>improved matching</u> should include <u>two nodes</u> that weren't in the initial matching, and have <u>one extra arc</u>.

The alternating path **Can't Change** the **Original Information**

It's really important that you <u>stick</u> to the <u>information</u> you were given in the first place — you can't make people do jobs in the <u>alternating path</u> that they weren't doing in the <u>bipartite graph</u>. That wouldn't be fair.

EXAMPLE

> Anne, Dick, George, Julian and Timmy are on an adventure holiday.
> They have a choice of five activities: abseiling, canoeing, diving, mountain biking and rock-climbing
>
> Anne wants to go mountain biking or rock-climbing, Dick wants to go canoeing or diving, George wants to go abseiling or mountain biking, Julian wants to go diving or rock-climbing and Timmy just wants to go abseiling.
>
> Find an alternating path that improves on this initial matching:
> Anne – mountain biking, Dick – canoeing, George – abseiling, Julian – rock climbing

1) Start by drawing a <u>bipartite graph</u> so you see all the preferences:

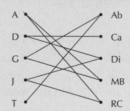

2) And draw the <u>initial matching</u>:

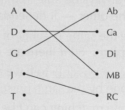

3) For the <u>alternating path</u>, start at T and find a path that connects it to Di, the other unmatched node (it won't be a direct path, as Timmy doesn't want to go diving).

One alternating path goes like this:

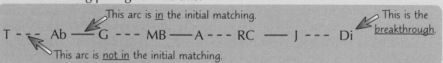

This arc is <u>in</u> the initial matching.

This is the breakthrough.

This arc is <u>not in</u> the initial matching.

> Don't forget that every arc in your alternating path has to be valid, i.e. in the original bipartite graph.

4) <u>Changing the status</u> of arcs gives:

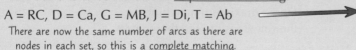

5) Now use this to construct the <u>improved matching</u>:
A = RC, D = Ca, G = MB, J = Di, T = Ab
There are now the same number of arcs as there are nodes in each set, so this is a <u>complete matching</u>.

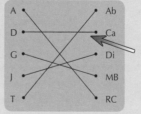

The arc D — Ca isn't in the alternating path so you can just leave it as it is.

You put your left hand in...

... your right hand not in. In, not in, in not in and shake it all about.

Maximum Matchings

Once you're happy with finding <u>alternating paths</u> and <u>improved matchings</u> (see previous page),
you can use that method to find a '<u>maximum matching</u>' — one that uses the greatest number of arcs possible.

Repeat the *Alternating Path* method to get a *Maximum Matching*

To find the <u>maximum matching</u>, you need to keep finding <u>alternating paths</u> and <u>improved matchings</u>
— this is the <u>maximum matching algorithm</u>.

The Maximum Matching Algorithm

1) **Start with <u>any</u> initial matching.**

2) **Try to find an <u>alternating path</u> (using the method on p.228). If you find one, use this to form
   an <u>improved matching</u>. If there isn't one, then this is a <u>maximum matching</u> — so <u>stop</u>.**

3) **If there are no <u>unmatched</u> nodes, <u>stop</u> — the matching is <u>complete</u>.**
   If there are still unmatched nodes, <u>repeat step 2)</u> using the
   <u>improved matching</u> as the new <u>initial matching</u>.

   There might be more than
   one maximum matching.

Use Trees to show different Paths

In some alternating paths, you have a <u>choice</u> between different nodes. You should draw a <u>tree diagram</u>
to show the possible paths, then pick the route that gets you to a <u>breakthrough</u> the <u>fastest</u>.

EXAMPLE

A music school offers tuition in the following instruments: clarinet, flute, piano, saxophone, trumpet and violin.
Six children want to start lessons, but the school only takes on one pupil per instrument.
Their preferences are:

Child	First choice	Second choice	Third choice
Chad	Saxophone	Trumpet	—
Elly	Trumpet	Piano	Clarinet
Karen	Flute	Piano	—
Mike	Trumpet	Saxophone	—
Peter	Clarinet	Violin	Piano
Stuart	Trumpet	Clarinet	—

Draw a bipartite
graph to show
this information.

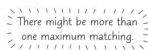

Initially, Chad, Elly, Karen and Peter are matched to their first choice. Draw this initial matching,
then use the maximum matching algorithm to improve the matching as much as possible.

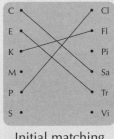

Initial matching

1) From the <u>initial matching</u>, find an
   <u>alternating path</u> from Mike to an
   unmatched instrument:

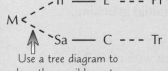

This route reaches
a breakthrough
first, so use this.

Use a tree diagram to
show the possible routes.

This produces the matching:
C = Sa, E = Pi, K = Fl, M = Tr
There are still <u>two unmatched nodes</u>,
we need another alternating path.

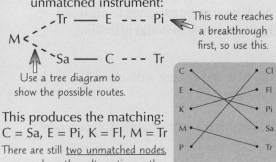

2) Using the <u>new matching</u>, find an
   alternating path from Stuart to an
   unmatched instrument:

$$S \underset{}{\overset{}{<}} \begin{matrix} \text{Tr} - \text{M} --- \text{Sa} \\ \text{Cl} - \text{P} --- \text{Vi} \end{matrix}$$

This route reaches a
breakthrough first.

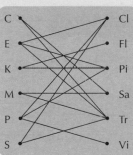

This produces the matching:
C = Sa, E = Pi, K = Fl,
M = Tr, P = Vi, S = Cl

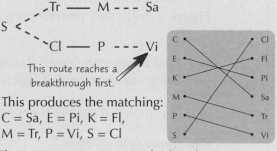

There are no more unmatched nodes,
so this is a <u>complete matching</u>.

If only marks grew on trees...

I bet you never thought AS-level Maths would be about joining the dots. Well, technically it's drawing arcs between nodes,
but I'd rather think of it as joining the dots — much more fun that way. Right, I'm off to do some potato printing now.

D1 Section 6 — Practice Questions

That's the <u>end</u> of <u>Section 6</u> — and it's also the <u>end</u> of <u>D1</u>. Just one page of <u>practice questions</u>, then you're ready to have a go at some <u>practice exams</u> to test your powers of decisiveness.

Warm-up Questions

1) Elizabeth, Jane, Kitty, Lydia and Mary are going to an art gallery. Elizabeth likes Renaissance art, portraits and sculptures, Jane likes portraits, sculptures and modern art, Kitty likes the cafe, Lydia likes modern art and the cafe and Mary likes Renaissance art.
   a) Draw this information on a <u>bipartite graph</u>.
   b) Use your bipartite graph to find a <u>complete matching</u>.

2) At a school dance, Alice, Bella, Charlotte, Daisy, Evie and Felicity have to be paired with Gerwyn, Hector, Iago, Jason, Kyle and Liam. The <u>bipartite graph</u> below shows who the girls want to dance with.

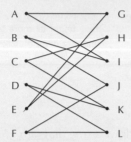

The <u>initial matching</u> pairs Alice with Gerwyn, Bella with Jason, Charlotte with Hector and Daisy with Liam.

   a) Draw the <u>initial matching</u>.
   b) Using the <u>maximum matching algorithm</u>, find <u>alternating paths</u> so that everyone has someone to dance with and they aren't left sitting by themselves and looking sad.

<u>Exam questions</u> on matchings are all pretty <u>similar</u>, so it's a good idea to get some practice in now.

Exam Question

1

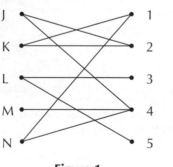

Figure 1

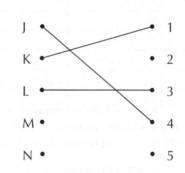

Figure 2

Five tutors, James, Kelly, Lee, Mia and Nick need to be assigned to classes 1 - 5. **Figure 1** shows their preferences and **Figure 2** shows an initial matching.

a) Find an alternating path starting from Mia and ending with class 2. Write out the improved matching that your path gives.

(3 marks)

b) A complete matching is not possible for this bipartite graph. Explain why.

(1 mark)

Mia agrees to teach class 3.

c) Starting with the matching found in part (a), use the maximum matching algorithm to find a complete matching. Write out the alternating path and the final matching.

(3 marks)

General Certificate of Education
Advanced Subsidiary (AS) and Advanced Level

Decision Mathematics D1 — Practice Exam One

Time Allowed: 1 hour 30 min

Calculators may be used for this exam (except those with
facilities for symbolic algebra, differentiation or integration).

Give any non-exact numerical answers to an appropriate degree of accuracy.

There are 75 marks available for this paper.

1   Frederick

Vicky

Dexter

Kamui

Janet

Nahla

Sophia

Anthony

Luke

Paul

Samira

a)   Sort these names into alphabetical order using a quick sort. Make all your pivots clear.

(5 marks)

b)   Use the binary search algorithm on the ordered list from part (a) to try to find the name Sebastian.

(4 marks)

2   Stephanie is packing to go to university. She has boxes that can hold a maximum of 6 kg.
The items she needs to pack into the boxes have the following weights (in kg):

1.8   3.5   2.6   4.1   1.2   0.8   2.0   2.4   3.1

a)   Use the first fit method to pack the items into the boxes.
State how many boxes are used and how much space is wasted.

(3 marks)

b)   Use the first fit decreasing method to pack the items again. You don't need to use an algorithm
to sort the items. State how many boxes are used and how much space is wasted.

(3 marks)

c)   Suggest a reason why the answer to part (a) might be a more practical way of packing.

(1 mark)

3   Kalim, Louisa, Melissa, Nathan, Olivia and Priya need to be
matched to six jobs numbered 1 - 6.

The table on the right shows the jobs that each person is willing to do.

Name	Jobs
Kalim	2, 5
Louisa	1, 4
Melissa	2, 6
Nathan	1, 3
Olivia	3, 5
Priya	2, 4

a)   Show this information on a bipartite graph.

(2 marks)

Initially, Kalim is matched to job 5, Louisa is matched to job 1, Melissa is matched to job 2,
Nathan is matched to job 3 and Priya is matched to job 4.

b)   Starting with the initial matching, use an alternating path to find a complete matching for
the information above.

(4 marks)

4   Natalie makes greetings cards to sell at craft fairs. She makes x birthday cards and y blank cards.

Each birthday card costs 40p to make and each blank card costs 50p to make.
She has a total of £8 she can spend on materials for the cards.

a)   Write out this constraint as an inequality in terms of x and y.

(1 mark)

b)   Another two constraints are represented by the inequalities $y \leq 8$ and $y \leq x$.

Write out in words what these constraints mean in terms of the number of birthday cards
and blank cards.

(2 marks)

c)   On graph paper, draw these three constraints, together with the
non-negativity constraints $x, y \geq 0$ and label the feasible region R.

(4 marks)

Natalie makes a profit of £0.75 on each birthday card and £1.50 on each blank card.

Assume she sells all the cards she makes. She wants to maximise the profit, P.

d)   Write out the objective function, P, in terms of x and y.

(1 mark)

e)   Use the graph from part (c) to optimise the objective function. State how many birthday cards
and how many blank cards Natalie should make, and write down the maximum profit.

(5 marks)

5   The network below represents the main roads between towns in an area.
The weights on the arcs represent distances in miles, and the vertices represent road junctions.

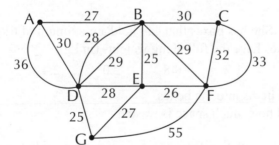

Total length of roads = 460 miles

a)   Duncan works in town C and lives in town G. On a copy of the diagram, use Dijkstra's
algorithm to find Duncan's shortest route home. Write down the route and give its length.

(7 marks)

b)   Duncan drives a truck which paints lines on roads. He is going to paint white lines on all
the roads. He uses the shortest route that travels down each road, starting from C and ending at G.
(i)   Which roads does Duncan need to drive along more than once?
(ii)   How many miles will Duncan drive?

(2 marks)

c)   The next day Duncan paints another set of lines on all the roads.
He has to start and finish at work (C).

Find the length of the shortest route that Duncan can use. Show all your working.

(5 marks)

6   The diagram shows the various cycle paths in a park. The bike hire shop is at G.
A floodlighting system is to be installed so that the cycle paths can be used after dark. There will be a light at each intersection. The number on each edge represents the distance, in metres, between two intersections.

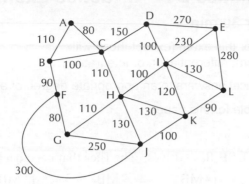

a)   Cabling must be laid along the paths so that each intersection is connected.
Starting from the bike hire shop (G), use Prim's algorithm to find a cabling layout that will use a minimum amount of cable. List the paths in the order they are added.

(5 marks)

b)   Draw your minimum spanning tree and state its length.

(4 marks)

c)   The warden used Kruskal's algorithm to find the same minimum spanning tree.
Find the tenth and the eleventh edges that the warden added to his spanning tree.

(3 marks)

7

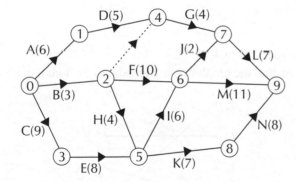

The network above shows the activities involved in a construction project.
Each activity is represented by an arc, and its duration is given in days.

a)   Calculate the early event times and the late event times for the activities.
Write these using the box notation on a copy of the network.

(5 marks)

b)   Calculate the total float on activity D.

(1 mark)

c)   State the length of the critical path in days.

(1 mark)

d)   The sum of all the activity times is 90 days. Calculate a lower bound for the number of workers needed to complete the project in the minimum time. You must show your working.

(2 marks)

e)   Given that each task requires one worker, schedule the activities so that the process is completed in the shortest time using the minimum number of workers.

(4 marks)

f)   Explain why a dotted line has been used to connect event 2 and event 4.

(1 mark)

General Certificate of Education
Advanced Subsidiary (AS) and Advanced Level

Decision Mathematics D1 — Practice Exam Two

Time Allowed: 1 hour 30 min

Calculators may be used for this exam (except those with
facilities for symbolic algebra, differentiation or integration).

Give any non-exact numerical answers to an appropriate degree of accuracy.

There are 75 marks available for this paper.

1   Computer files are being backed up on 16 MB USB flash drives. The files that need backing up are of sizes:

   9 MB   10 MB   6 MB   14 MB   7 MB   11 MB   3 MB

   a)   What is the lower bound for the number of drives needed?

   (2 marks)

   b)   Use the first fit bin packing algorithm to arrange the files onto the flash drives. State how many
        flash drives are used and how much space is wasted.

   (3 marks)

   c)   Use the full bin packing algorithm to find an optimal solution. Again, state how many
        flash drives are used and how much space is wasted.

   (3 marks)

2   Carlos, Diego, Edgar, Fatima and Gino are being assigned to tasks 1 - 5.
    The graph below left shows the tasks they are able to do, and the graph below right shows the initial matching.

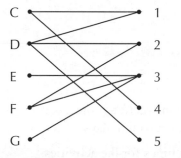 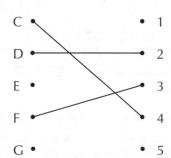

   a)   Find an alternating path from Edgar to task 1. Draw on a graph the improved matching
        given by the alternating path.

   (3 marks)

   b)   It is not possible to obtain a complete matching. Explain why.

   (1 mark)

   Edgar is trained to carry out task 5.

   c)   Using the matching found in part a), find a complete matching for this information.

   (2 marks)

3   An algorithm for finding the next multiple of 9 from any integer *A* is represented as the flow chart below:

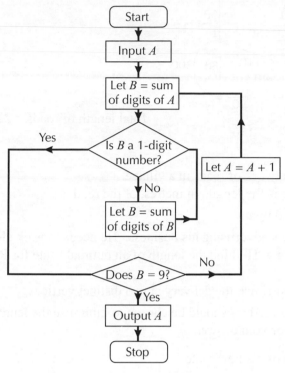

a)   Use the algorithm to show that the output for *A* = 977 is 981. State the values of *A* and *B* at each pass through the flow chart.

(4 marks)

b)   Write down the output when *A* = 978.

(1 mark)

c)   Explain why no more than 9 passes through the algorithm will ever be needed.

(1 mark)

4   a)   Draw the activity network described in this precedence table.
         You must use two dummies.

(4 marks)

Activity	Immediate preceding activities
A	—
B	—
C	A, B
D	A, B
E	C
F	C, D
G	E
H	F
I	G, H
J	F

b)   Explain why each dummy is necessary.

(2 marks)

5

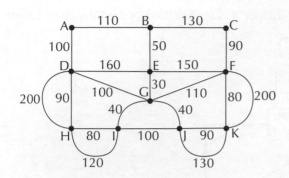

Total length of roads = 2200 m

The diagram shows the network of roads in a village.
The number on each edge is the length, in metres, of the road.

Rufus runs a roof repair company.

a) Rufus delivers leaflets advertising his business. He needs to walk along each road at least once, starting and finishing at H. Find the length of an optimal route for Rufus.

(6 marks)

b) If Rufus can start and finish the delivery at two distinct vertices,

(i) state which two vertices should be chosen to minimise the length of his new route. Give a reason for your answer.

(2 marks)

(ii) Find the length of his new route.

(1 mark)

c) Rufus lives at H and needs to repair a roof at C. Use Dijkstra's algorithm to find the minimum driving distance between H and C. State the corresponding route.

(6 marks)

6   The table shows the distances, in metres, between bird boxes in a reserve.

	A	B	C	D	E	F
A	–	140	167	205	150	173
B	140	–	145	148	159	170
C	167	145	–	210	195	180
D	205	148	210	–	155	178
E	150	159	195	155	–	185
F	173	170	180	178	185	–

a) Use Prim's algorithm, starting from A, to find a minimum spanning tree for this table of distances. You must list the arcs that form your tree in the order that they are selected.

(3 marks)

b) Draw your tree and calculate its total weight.

(2 marks)

c) One advantage of Prim's algorithm over Kruskal's is that you don't have to check for cycles. Explain why you don't have to check for cycles in Prim's algorithm.

(2 marks)

7

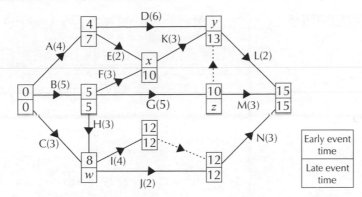

The network above shows the activities involved in a process. The activities are represented by the arcs. The number in brackets on each arc gives the time, in hours, to complete the activity. Some of the early and late times for events are shown.

a) Find the values of w, x, y and z.

(2 marks)

b) List the critical activities.

(1 mark)

c) Calculate the total float on activities D, E and K.

(3 marks)

d) Draw a cascade (Gantt) chart for the project.

(4 marks)

e) An inspector visits during hour 4 and during hour 10. Assuming the project is on schedule, which activities must be happening during these hours?

(3 marks)

8   A charity is selling tickets to its annual charity ball. It has two corporate packages available: business class and premier.

Each business class package includes 5 tickets and each premier package includes 10 tickets.

Each business class package includes 2 bottles of wine and each premier package includes 8 bottles of wine.

There are 300 tickets and 160 bottles of wine available.

At least 5 of each type of package are sold, and at least 20 packages are sold in total.

The charity sells x business class packages and y premier packages.

a) 5 constraints are required when expressing this information as a linear programming problem.

(i) Show why the following constraints are required: $x + 2y \le 60$, $x + 4y \le 80$

(2 marks)

(ii) State three other constraints that must also be used.

(2 marks)

b) Each business class package makes a profit of £150 and each premier package makes a profit of £200. The charity wants to calculate the minimum and maximum profit from the ball.

(i) On graph paper, draw a diagram to represent this linear programming problem. Clearly label the feasible region and draw on an objective line.

(6 marks)

(ii) Find the maximum profit from the ball. Write down how many of each type of package need to be sold to make this profit.

(2 marks)

(iii) Find the minimum profit from the ball. Write down how many of each type of package need to be sold to make this profit.

(2 marks)

Answers

C1 Section 1 — Algebra Fundamentals
Warm-up Questions

1) a) a & b are constants, x is a variable.

   b) a & b are constants, x is a variable.

   c) a, b & c are constants, y is a variable.

   d) a is a constant, x & y are variables.

2) Identity symbol is $\equiv$.

3) A, C and D are identities.

4) a) x^8 b) a^{15} c) x^6 d) a^8 e) x^4y^3z f) $\frac{b^2c^5}{a}$

5) a) 4   b) 2   c) 8   d) 1   e) $\frac{1}{7}$

6) a) $x = \pm\sqrt{5}$ b) $x = -2 \pm \sqrt{3}$

7) a) $2\sqrt{7}$ b) $\frac{\sqrt{5}}{6}$ c) $3\sqrt{2}$ d) $\frac{3}{4}$

8) a) $\frac{8}{\sqrt{2}} = \frac{8}{\sqrt{2}} \times \frac{\sqrt{2}}{\sqrt{2}} = \frac{8\sqrt{2}}{2} = 4\sqrt{2}$

   b) $\frac{\sqrt{2}}{2} = \frac{\sqrt{2}}{(\sqrt{2})^2} = \frac{1}{\sqrt{2}}$

9) $136 + 24\sqrt{21}$

10) $3 - \sqrt{7}$

11) a) $a^2 - b^2$ b) $a^2 + 2ab + b^2$

    c) $25y^2 + 210xy$ d) $3x^2 + 10xy + 3y^2 + 13x + 23y + 14$

12) a) $xy(2x + a + 2y)$ b) $a^2x(1 + b^2x)$

    c) $8(2y + xy + 7x)$ d) $(x - 2)(x - 3)$

13) a) $\frac{52x + 5y}{60}$ b) $\frac{5x - 2y}{x^2y^2}$ c) $\frac{x^3 + x^2 - y^2 + xy^2}{x(x^2 - y^2)}$

14) a) $\frac{3a}{2b}$ b) $\frac{2(p^2 + q^2)}{p^2 - q^2}$ c)

Exam Questions

1 a) $27^{\frac{1}{3}} = \sqrt[3]{27}$

      $= 3$ *[1 mark]*

   b) $27^{\frac{4}{3}} = (27^{\frac{1}{3}})^4$ *[1 mark]*

      $= 3^4 = 3 \times 3 \times 3 \times 3 = 9 \times 9$

      $= 81$ *[1 mark]*

2 a) $(5\sqrt{3})^2 = (5^2)(\sqrt{3})^2 = 25 \cdot 3$

      $= 75$ *[1 mark]*

   b) $(5 + \sqrt{6})(2 - \sqrt{6}) = 10 - 5\sqrt{6} + 2\sqrt{6} - 6$ *[1 mark]*

      $= 4 - 3\sqrt{6}$ *[1 mark]*

3 $10000\sqrt{10} = 10^4 \cdot 10^{\frac{1}{2}}$ *[1 mark]*

      $= 10^{4 + \frac{1}{2}}$ *[1 mark]*

      $= 10^{\frac{9}{2}}$

   so $k = \frac{9}{2}$ *[1 mark]*

4 Multiply top and bottom by $3 + \sqrt{5}$ to 'rationalise the denominator':

   $\frac{5 + \sqrt{5}}{3 - \sqrt{5}} = \frac{(5 + \sqrt{5})(3 + \sqrt{5})}{(3 - \sqrt{5})(3 + \sqrt{5})}$ *[1 mark]*

   $= \frac{15 + 5\sqrt{5} + 3\sqrt{5} + 5}{9 - 5}$ *[1 mark]*

   $= \frac{20 + 8\sqrt{5}}{4}$ *[1 mark]*

   $= 5 + 2\sqrt{5}$ *[1 mark]*

5 $2x^4 - 32x^2 = 2x^2(x^2 - 16)$

      $= 2x^2(x + 4)(x - 4)$

   [3 marks available in total — 1 mark for each correct factor]

6 $\frac{x + 5x^3}{\sqrt{x}} = x^{-\frac{1}{2}}(x + 5x^3)$ *[1 mark]*

      $= x^{\frac{1}{2}} + 5x^{\frac{5}{2}}$ *[1 mark]*

7 $\frac{(5 + 4\sqrt{x})^2}{2x} = \frac{25 + 40\sqrt{x} + 16x}{2x}$ *[1 mark]*

      $= \frac{1}{2}x^{-1}(25 + 40x^{\frac{1}{2}} + 16x)$ *[1 mark]*

      $= \frac{25}{2}x^{-1} + 20x^{-\frac{1}{2}} + 8$,

   so $P = 20$ and $Q = 8$ *[1 mark]*

C1 Section 2 — Quadratics & Polynomials
Warm-up Questions

1) a) $(x + 1)^2$ b) $(x - 10)(x - 3)$

   c) $(x + 2)(x - 2)$ d) $(3 - x)(x + 1)$

   e) $(2x + 1)(x - 4)$ f) $(5x - 3)(x + 2)$

2) a) $(x - 2)(x - 1) = 0$, so $x = 2$ or 1

   b) $(x + 4)(x - 3) = 0$, so $x = -4$ or 3

   c) $(2 - x)(x + 1) = 0$, so $x = 2$ or –1

   d) $(x + 4)(x - 4) = 0$, so $x = \pm 4$

   e) $(3x + 2)(x - 7) = 0$, so $x = -2/3$ or 7

   f) $(2x + 1)(2x - 1) = 0$, so $x = \pm 1/2$

   g) $(2x - 3)(x - 1) = 0$, so $x = 3/2$ or 1

3) a) $(x - 2)^2 - 7$; minimum value = –7 at $x = 2$, and this crosses the x-axis at $x = 2 \pm \sqrt{7}$

   b) $\frac{21}{4} - (x + \frac{3}{2})^2$; maximum value = 21/4 at $x = -3/2$, and this crosses the x-axis at $-\frac{3}{2} \pm \frac{\sqrt{21}}{2}$.

   c) $2(x - 1)^2 + 9$; minimum value = 9 at $x = 1$, and this doesn't cross the x-axis.

   d) $4(x - \frac{7}{2})^2 - 1$; minimum value = –1 at $x = 7/2$, crosses the x-axis at $x = \frac{7}{2} \pm \frac{1}{2}$ i.e. $x = 4$ or 3

Answers

4) a) $b^2 - 4ac = 16$, so 2 roots

b) $b^2 - 4ac = 0$, so 1 root

c) $b^2 - 4ac = -8$, so no roots

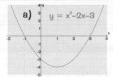 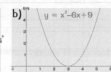 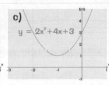

5) a) Using the quadratic formula with $a = 3$, $b = -7$ and $c = 3$:

$$x = \frac{-(-7) \pm \sqrt{(-7)^2 - (4 \times 3 \times 3)}}{(2 \times 3)}$$

$$= \frac{7 \pm \sqrt{49 - 36}}{6}$$

$$= \frac{7 \pm \sqrt{13}}{6}$$

so $x = \frac{7 + \sqrt{13}}{6}$, $x = \frac{7 - \sqrt{13}}{6}$.

b) Using the quadratic formula with $a = 2$, $b = -6$ and $c = -2$:

$$x = \frac{-(-6) \pm \sqrt{(-6)^2 - (4 \times 2 \times -2)}}{(2 \times 2)}$$

$$= \frac{6 \pm \sqrt{36 + 16}}{4}$$

$$= \frac{6 \pm \sqrt{52}}{4}$$

$$= \frac{6 \pm 2\sqrt{13}}{4}$$

$$= \frac{3 \pm \sqrt{13}}{2}$$

so $x = \frac{3 + \sqrt{13}}{2}$, $x = \frac{3 - \sqrt{13}}{2}$.

c) Using the quadratic formula with $a = 1$, $b = 4$ and $c = -6$:

$$x = \frac{-(4) \pm \sqrt{(4)^2 - (4 \times 1 \times -6)}}{(2 \times 1)}$$

$$= \frac{-4 \pm \sqrt{16 + 24}}{2}$$

$$= \frac{-4 \pm \sqrt{40}}{2}$$

$$= \frac{-4 \pm 2\sqrt{10}}{2}$$

$$= -2 \pm \sqrt{10}$$

so $x = -2 + \sqrt{10}$, $x = -2 - \sqrt{10}$.

6) $k^2 - (4 \times 1 \times 4) > 0$, so $k^2 > 16$ and so $k > 4$ or $k < -4$.

Exam Questions

1 For equal roots, $b^2 - 4ac = 0$ *[1 mark]*

$a = 1$, $b = 2k$ and $c = 4k$

so $(2k)^2 - (4 \times 1 \times 4k) = 0$ *[1 mark]*

$4k^2 - 16k = 0$

$4k(k - 4) = 0$ *[1 mark]*

so $k = 4$ (as k is non-zero). *[1 mark]*

2 a) For distinct real roots, $b^2 - 4ac > 0$ *[1 mark]*

$a = p$, $b = p + 3$ and $c = 4$

so $(p + 3)^2 - (4 \times p \times 4) > 0$ *[1 mark]*

$p^2 + 6p + 9 - 16p > 0$

$p^2 - 10p + 9 > 0$ *[1 mark]*

b) $p^2 - 10p + 9$ is a u-shaped quadratic, which crosses the x-axis when $p^2 - 10p + 9 = 0$, that is when $(p - 9)(p - 1) = 0$ *[1 mark]*, which occurs at $p = 9$ and $p = 1$ *[1 mark]*. As it's u-shaped, $p^2 - 10p + 9 > 0$ when p is outside these values *[1 mark]*, that is when $p < 1$ or $p > 9$ *[1 mark]*.

3 Expanding the brackets on the RHS gives the quadratic $mx^2 + 4mx + 4m + p$. Equating the coefficients of x^2 gives $m = 5$ *[1 mark]*. Equating the coefficients of x gives $n = 4m$, so $n = 20$ *[1 mark]*. Equating the constant terms gives $14 = 4m + p \Rightarrow p = -6$ *[1 mark]*.

4 a) $a = 12 \div 2 = 6$ *[1 mark]*, so the completed square is $(x - 6)^2 + b = x^2 - 12x + 36 + b$. Equating coefficients gives $15 = 36 + b$, so $b = 15 - 36 = -21$ *[1 mark]*. The final expression is $(x - 6)^2 - 21$.

b) (i) The minimum occurs when the expression in brackets is equal to 0, which means the minimum is the value of b, which from (a) above is -21 *[1 mark]*.

(ii) From above, the minimum occurs when the expression in brackets is equal to 0, i.e. when $x = 6$ *[1 mark]*.

Part (b) is dead easy once you've completed the square — you can just take your values straight from there.

5 a) Put $a = 1$, $b = -14$ and $c = 25$ into the quadratic formula:

$$x = \frac{-(-14) \pm \sqrt{(-14)^2 - (4 \times 1 \times 25)}}{(2 \times 1)}$$

$$= \frac{14 \pm \sqrt{196 - 100}}{2}$$

$$= \frac{14 \pm \sqrt{96}}{2}$$

$$= \frac{14 \pm 4\sqrt{6}}{2}$$

$$= 7 \pm 2\sqrt{6}$$

so $x = 7 + 2\sqrt{6}$, $x = 7 - 2\sqrt{6}$

[3 marks available — 1 mark for putting correct values of a, b and c into the quadratic formula, 1 mark each for final x-values.]

b)

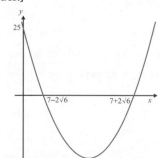

[3 marks available — 1 mark for drawing u-shaped curve, 1 mark for using answers from a) as x-axis intercepts and 1 mark for correct y-axis intercept (0, 25).]

c) As the graph is u-shaped, the inequality is < 0 when x is between the two intercepts, i.e. $7 - 2\sqrt{6} \leq x \leq 7 + 2\sqrt{6}$ *[1 mark]*.

Answers

6 a) (i) $m = 10 \div 2 = 5$ *[1 mark]*, so the expression becomes $-(5 - x)^2 + n = -25 + 10x - x^2 + n$. Equating coefficients gives $-27 = -25 + n$, so $n = -27 - (-25) = -2$ *[1 mark]*. The final expression is $-(5 - x)^2 - 2$.

(ii) $(5 - x)^2 \geq 0$ for all values of x, so $-(5 - x)^2 \leq 0$. Therefore $-(5 - x)^2 - 2 < 0$ for all x, i.e. the function is always negative. *[1 mark]*

b) (i) The y-coordinate is the maximum value, which is -2 *[1 mark]*, and this occurs when the expression in the brackets = 0.
The x-value that makes the expression in the brackets 0 is 5 *[1 mark]*, so the coordinates of the maximum point are $(5, -2)$.

(ii)

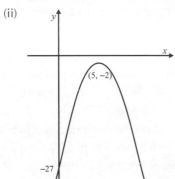

[2 marks available — 1 mark for drawing n-shaped curve that sits below the x-axis, 1 mark for correct y-axis intercept (0, –27)]

C1 Section 3 — Simultaneous Equations and Inequalities

Warm-up Questions

1) a) $x > -\frac{38}{5}$ b) $y \leq \frac{7}{8}$ c) $y \leq -\frac{3}{4}$

2) a) $x > \frac{5}{2}$ b) $x > -4$ c) $x \leq -3$

3) a) $-\frac{1}{3} \leq x \leq 2$ b) $x < 1 - \sqrt{3}$ or $x > 1 + \sqrt{3}$
   c) $x \leq -3$ or $x \geq -2$

4) a) $x \leq -3$ and $x \geq 1$ b) $x < -\frac{1}{2}$ and $x > 1$
   c) $-3 < x < 2$

5) a) $x = -3, y = -4$ b) $x = -\frac{1}{6}, y = -\frac{5}{12}$

6) a) The line and the curve meet at the points $(2, -6)$ and $(7, 4)$.
   b) The line is a tangent to the parabola at the point $(2, 26)$.
   c) The equations have no solution and so the line and the curve never meet.

7) a) $\left(\frac{1}{4}, -\frac{13}{4}\right)$ b) $(4, 5)$ c) $(-5, -2)$

Exam Questions

1 a) $3x + 2 \leq x + 6$
$\qquad 2x \leq 4$ *[1 mark]*
$\qquad x \leq 2$ *[1 mark]*

b) $\quad 20 - x - x^2 > 0$
$(4 - x)(5 + x) > 0$ *[1 mark]*

The graph crosses the x-axis at $x = 4$ and $x = -5$ *[1 mark]*. The coefficient of x^2 is negative so the graph is n-shaped *[1 mark]*.
So $20 - x - x^2 > 0$ when $-5 < x < 4$ *[1 mark]*.

c) From above, x will satisfy both inequalities when $-5 < x \leq 2$ *[1 mark]*.

For this bit, all you need to do is use your answers to parts a) and b) and work out which values of x fit in them both.

2 a) $3 \leq 2p + 5 \leq 15$
This inequality has 3 parts. Subtract 5 from each part to give: $-2 \leq 2p \leq 10$ *[1 mark]*.
Now divide each part by 2 to give: $-1 \leq p \leq 5$
[1 mark for $-1 \leq p$ and 1 mark for $p \leq 5$].

b) $q^2 - 9 > 0$
$(q + 3)(q - 3) > 0$ *[1 mark]*
The function is 0 at $q = -3$ and $q = 3$ *[1 mark]*.
The coefficient of x^2 is positive so the graph is u-shaped.
So $q^2 - 9 > 0$ when $q < -3$ or $q > 3$
[1 mark for each correct inequality].

Use D.O.T.S. (Difference of Two Squares) to factorise the quadratic — remember that $a^2 - b^2 = (a + b)(a - b)$.

3 a) $(3x + 2)(x - 5)$ *[1 mark]*

b) $(3x + 2)(x - 5) \leq 0$
The function is 0 at $x = -\frac{2}{3}$ and $x = 5$ *[1 mark]*.
The coefficient of x^2 is positive so the graph is u-shaped, meaning the function is less than or equal to 0 between these x-values.
I.e. $3x^2 - 13x - 10 \leq 0$ when $-\frac{2}{3} \leq x \leq 5$
[2 marks, 1 for $-\frac{2}{3} \leq x$ and 1 for $x \leq 5$].

4 First, take the linear equation and rearrange it to get x on its own: $x = 6 - y$ *[1 mark]*. Now substitute the equation for x into the quadratic equation:
$\qquad (6 - y)^2 + 2y^2 = 36$ *[1 mark]*
$36 - 12y + y^2 + 2y^2 = 36$
$\qquad\qquad 3y^2 - 12y = 0$
$\qquad\qquad y^2 - 4y = 0$
$\qquad\qquad y(y - 4) = 0$ *[1 mark]*
so $y = 0$ and $y = 4$. *[1 mark]*.
Now you've got the y-values, put them back into the equation for x ($x = 6 - y$) to find the x-values.
When $y = 0$, $x = 6 - y = 6 - 0 = 6$.
When $y = 4$, $x = 6 - y = 6 - 4 = 2$.
So solutions are $x = 6, y = 0$ *[1 mark]*
and $x = 2, y = 4$ *[1 mark]*.

Answers

5 a) At points of intersection, $-2x + 4 = -x^2 + 3$ *[1 mark]*
$$x^2 - 2x + 1 = 0$$
$$(x - 1)^2 = 0 \quad \textit{[1 mark]}$$
so $x = 1$ *[1 mark]*. When $x = 1$, $y = -2x + 4 = 2$, so there is one point of intersection at $(1, 2)$ *[1 mark]*.

b)
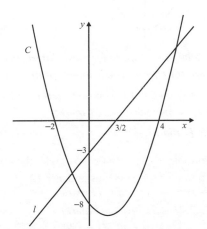

[5 marks available — 1 mark for drawing n-shaped curve, 1 mark for x-axis intercepts at ±√3, 1 mark for maximum point of curve and y-axis intercept at (0, 3). 1 mark for line crossing the y-axis at (0, 4) and the x-axis at (2, 0). 1 mark for line and curve touching in one place.]

6 a)
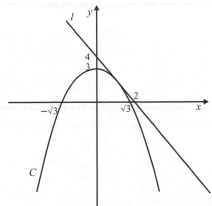

[5 marks available — 1 mark for drawing u-shaped curve, 1 mark for x-axis intercepts at –2 and 4 , 1 mark for y-axis intercept at (0, –8). 1 mark for line crossing the y-axis at (0, –3) and the x-axis at (3/2, 0). 1 mark for line and curve touching in two places.]

b) At points of intersection,
$$2x - 3 = (x + 2)(x - 4)$$
$$2x - 3 = x^2 - 2x - 8 \quad \textit{[1 mark]}$$
$$0 = x^2 - 4x - 5 \quad \textit{[1 mark]}$$

c) $x^2 - 4x - 5 = 0$
$(x - 5)(x + 1) = 0$ *[1 mark]*
so $x = 5$, $x = -1$ *[1 mark]*.
When $x = 5$, $y = (2 \times 5) - 3 = 7$ and when $x = -1$,
$y = (2 \times -1) - 3 = -5$, so the points of intersection are
$(5, 7)$ *[1 mark]* and $(-1, -5)$ *[1 mark]*.

C1 Section 4 —
Coordinate Geometry and Graphs
Warm-up Questions

1) a) (i) $y + 1 = 3(x - 2)$ (ii) $y = 3x - 7$ (iii) $3x - y - 7 = 0$

b) (i) $y + \frac{1}{3} = \frac{1}{5}x$ (ii) $y = \frac{1}{5}x - \frac{1}{3}$

(iii) $3x - 15y - 5 = 0$

2) a) $y = \frac{3}{2}x - 4$ b) $y = -\frac{1}{2}x + 4$

3) The equation of the required line is $y = \frac{3}{2}x + \frac{15}{2}$.

4) a)   b)
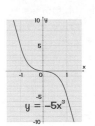
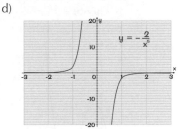

c)   d)
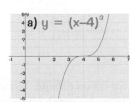
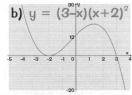

5)
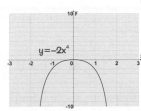
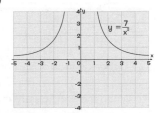

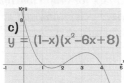
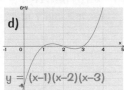

6) a)   b)
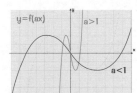
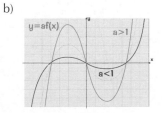

c)   d)
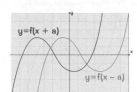
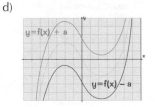

Answers

Exam Questions

1 a) Just rearrange into the form $y = mx + c$ and read off m:

$3y = 15 - 4x$

$y = -\frac{4}{3}x + 5$ *[1 mark]*

so the gradient of the line PQ is $-\frac{4}{3}$ *[1 mark]*.

b) Gradient of the line $= -1 \div -\frac{4}{3} = \frac{3}{4}$ *[1 mark]*

So $y = \frac{3}{4}x + c$.

Now use the x- and y- values of R to find C:

$1 = \frac{3}{4}(3) + c$

$1 = \frac{9}{4} + c$

$\Rightarrow c = -\frac{5}{4}$ *[1 mark]*

so the equation of the line is $y = \frac{3}{4}x - \frac{5}{4}$ *[1 mark]*

2 When the brackets are multiplied out, the first term is $2x^3$, so the graph is a positive cubic graph.
$y = 0$ when $x = 2$ or $x = -\frac{1}{2}$, so the graph touches the x-axis twice. When $x = 0$, $y = (1)(-2)^2 = 4$.

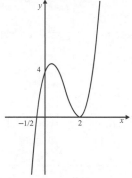

[4 marks available — 1 mark for correct shape, 1 mark for x-axis intercept at −1/2, 1 mark for graph touching the x-axis at 2 and 1 mark for correct y-axis intercept at 4.]

3 a)

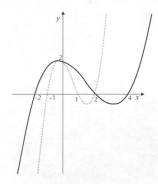

[3 marks available — 1 mark for horizontal stretch, 1 mark for x-axis intercepts at −2, 2 and 4, 1 mark for correct y-axis intercept at 2.]

b)

[2 marks available — 1 mark for horizontal translation to the right, 1 mark for x-axis intercepts at 3, 5 and 6]

4 a)

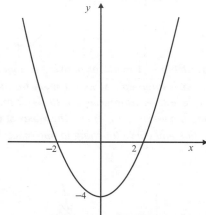

[2 marks available — 1 mark for x-axis intercepts at −2 and 2, 1 mark for correct y-axis intercept (0, −2)]

b) The curve is reflected in the x-axis *[1 mark]*, and stretched vertically by a scale factor of 2 *[1 mark]*.

c) $y = f(x) + 2$ *[1 mark]*

5 a) Using the formula $y - y_1 = m(x - x_1)$, with the coordinates of point S for the x- and y- values and $m = -2$,

$y - (-3) = -2(x - 7)$ *[1 mark]*

$y + 3 = -2x + 14$ *[1 mark]*

$y = -2x + 11$ *[1 mark]*

b) Putting $x = 5$ into $y = -2x + 11$ gives $y = 1$
[1 mark], so T does lie on the line.

6 a) Gradient of $LK = \frac{8-6}{5-2} = \frac{2}{3}$ *[1 mark]*

so gradient of $l_1 = -1 \div \frac{2}{3} = -\frac{3}{2}$ *[1 mark]*.

Now, putting this gradient and the x- and y- coordinates of L into the formula $y - y_1 = m(x - x_1)$ gives:

$y - 6 = -\frac{3}{2}(x - 2)$

$y = -\frac{3}{2}x + 3 + 6$

$y = -\frac{3}{2}x + 9$ *[1 mark]*

$\Rightarrow 3x + 2y - 18 = 0$ *[1 mark]*

b) Putting $x = 0$ into $y = -\frac{3}{2}x + 9$ gives $y = 9$ *[1 mark]*,
so M = (0, 9) *[1 mark]*.

c) Putting $y = 0$ into $3x + 2y - 18 = 0$ gives $x = 6$
[1 mark], so N = (6, 0) *[1 mark]*.

Answers

C1 Section 5 — Sequences and Series
Warm-up Questions

1) a) nth term $= 4n - 2$ b) nth term $= 0.5n - 0.3$

c) nth term $= -3n + 24$ d) nth term $= -6n + 82$

2) $a_{k+1} = a_k + 5$, $u_1 = 32$

3) Last term $(l) = 15$

4) Common difference $(d) = 0.75$

5) $S_8 = 168$

6) First work out n: $a = 5$, $l = 65$, $d = 3$

so, $65 = 5 + 3(n - 1)$

so, $n = 21$

Now use: $S_{21} = 21 \times \dfrac{(5 + 65)}{2}$

so, $S_{21} = 735$

They're not gonna hand it to you on a plate — you often have to work out a and d from the question. Luckily it's pretty simple. Phew.

7) a) Common difference $(d) = 4$

b) 15th term $= 63$ c) $S_{10} = 250$

8) First, work out d:

We know that the 7th term is 36 and the 10th is 30.

So the difference in 3 'moves' is –6 i.e. $d = -2$.

This gives the expression for the nth term as $-2n + 50$.

So, $a = 48$ (i.e. when $n = 1$)

and, $l = 40$ ($n = 5$). Now find the sum of the series:

$S_5 = 5 \times \dfrac{(48 + 40)}{2}$

so $S_5 = 220$

9) $S_{20} = 610$

10) $S_{10} = 205$

Exam Questions

1 a) $h_2 = h_{1+1} = 2 \times 5 + 2 = 12$ *[1 mark]*

$h_3 = h_{2+1} = 2 \times 12 + 2 = 26$ *[1 mark]*

$h_4 = 2h_3 + 2 = 54$ *[1 mark]*

b) $\sum_{r=3}^{6} h_r = h_3 + h_4 + h_5 + h_6$

$h_5 = 2h_4 + 2 = 110$ *[1 mark]*

$h_6 = 2(110) + 2 = 222$

so $\sum_{r=3}^{6} h_r = 26 + 54 + 110 + 222$ *[1 mark]*

$= 412$ *[1 mark]*

Nothing hard here, just pop the numbers in. Pop, pop, pop...

2 a) $a_2 = 3k + 11$ *[1 mark]*

$a_3 = 3a_2 + 11$

$= 3(3k + 11) + 11 = 9k + 33 + 11$

$= 9k + 44$ *[1 mark]*

$a_4 = 3a_3 + 11$

$= 3(9k + 44) + 11$

$= 27k + 143$ *[1 mark]*

b) $\sum_{r=1}^{4} a_r = k + (3k + 11) + (9k + 44) + (27k + 143)$

$= 40k + 198$ *[1 mark]*

$40k + 198 = 278$

$40k = 278 - 198 = 80$ *[1 mark]*

$k = 2$ *[1 mark]*

3 Use the nth term formula: $a_n = a_1 + (n - 1)d$:

$a + (7 - 1)d = a + 6d$ *[1 mark]*

You know that $a_7 = 580$ so

$a + 6d = 580$ *[1 mark]*

And you know that $S_{15} = 9525$, so using the series formula:

$S_{15} = \dfrac{15}{2}[2a + (15 - 1)d] = 9525$ *[1 mark]*

$= \dfrac{15}{2}(2a + 14d) = 9525$, i.e. $15a + 105d = 9525$

then you can divide everything by 15 to give:

$a + 7d = 635$ *[1 mark]*

then solve them simultaneously:

$(a + 7d) - (a + 6d) = d = 635 - 580$ *[1 mark]*

$d = 55$ *[1 mark]*

and finally use this value of d to find a:

$a + (6 \times 55) = 580$

$a = 580 - 330$

$= 250$ *[1 mark]*

A lot of steps needed for that one, but don't panic if the question seems complicated. If you're stuck, write down all the sequence and series formulas — then see what formulas you can fill in using the info in the question. A light bulb should go 'bing'. Hopefully...

4 a) $a_{31} = 22 + (31 - 1)(-1.1)$ *[1 mark]*

$= 22 + 30(-1.1)$

$= 22 - 33 = -11$ *[1 mark]*

b) $a_k = 0$

$a_1 + (k - 1)d = 0$

$22 + (k - 1) \times -1.1 = 0$ *[1 mark]*

$k - 1 = \dfrac{-22}{-1.1} = \dfrac{220}{11} = 20$

$k = 20 + 1 = 21$ *[1 mark]*

c) We want to find the first value of n for which $S_n < 0$.

Using the formula for sum of a series:

$S_n = \dfrac{n}{2}[2 \times 22 + (n - 1)(-1.1)] < 0$ *[1 mark]*

$S_n = \dfrac{n}{2}(44 - 1.1n + 1.1) < 0$

$\dfrac{n}{2}(45.1 - 1.1n) < 0$ *[1 mark]*

$\dfrac{n}{2}(45.1 - 1.1n) = 0$

$\Rightarrow \dfrac{n}{2} = 0$ or $45.1 - 1.1n = 0$

$\Rightarrow n = 0$ or $n = 41$

The coefficient of n^2 is negative so graph is n-shaped.

Need to find negative part, so $n < 0$ or $n > 41$.

Since n cannot be negative then $n > 41$.

Now we just want the first (i.e. lowest) value of n for which this is true, which is $n = 42$. *[1 mark]*

Answers

5 a) $a = 6$

$d = 8$ *[1 mark]*

$a_n = 6 + 8(n - 1)$ *[1 mark]*

$= 8n - 2$ *[1 mark]*

b) $S_{10} = \frac{10}{2}[2 \times 6 + (10 - 1)8]$ *[1 mark]*

$= 5 \times (12 + 72)$ *[1 mark]*

$= 420$ *[1 mark]*

c) First find an expression for S_k:

$S_k = \frac{k}{2}[2 \times 6 + 8(k - 1)]$

$= \frac{k}{2} \times (12 + 8k - 8)$

$= \frac{k}{2}(8k + 4)$ *[1 mark]*

$= \frac{8k^2 + 4k}{2} = 4k^2 + 2k$ *[1 mark]*

Then, you know that the total sum will be less than 2450, because he hadn't yet reached that limit by day k, so:

$4k^2 + 2k < 2450$

$\Rightarrow 2k^2 + k < 1225$

$\Rightarrow 2k^2 + k - 1225 < 0$

$\Rightarrow (2k - 49)(k + 25) < 0$ *[1 mark]*

Don't forget the difference between series and sequences. Part c is about a series — the cumulative sum. So don't use the wrong equations, or you'll get very muddled indeed.

d) Since $(2k - 49)(k + 25) < 0$,

$2k - 49 = 0$ or $k + 25 = 0$

$k = 24.5$ or $k = -25$ *[1 mark]*

Coefficient of k^2 is positive so graph is u-shaped.

Need negative part, so $-25 < k < 24.5$.

k will be the largest whole number that satisfies the inequality, i.e. $k = 24$. *[1 mark]*

Keep an eye on what you're being asked to find. Values in a sequence can be any number (oh, the possibilities), but the term positions are always whole numbers. So if you calculate a position and end up with a decimal number... something's not right.

C1 Section 6 — Differentiation
Warm-up Questions

1) $\frac{d}{dx}(x^n) = nx^{n-1}$

2) a) $\frac{dy}{dx} = 2x$ b) $\frac{dy}{dx} = 4x^3 + \frac{1}{2\sqrt{x}}$

c) $\frac{dy}{dx} = -\frac{14}{x^3} + \frac{3}{2\sqrt{x^3}} + 36x^2$

3) They're the same.

4) a) $\frac{dy}{dx} = 4x = 8$

b) $\frac{dy}{dx} = 8x - 1 = 15$

c) $\frac{dy}{dx} = 3x^2 - 14x = -16$

5) Differentiate $v = 17t^2 - 10t$ to give: $\frac{dv}{dt} = 34t - 10$

so, when $t = 4$, $\frac{dv}{dt} = 126$ ml/s.

6) The tangent and normal must go through (16, 6).

Differentiate to find $\frac{dy}{dx} = \frac{3}{2}\sqrt{x} - 3$, so gradient at (16, 6) is 3.

Therefore tangent can be written $y_T = 3x + c_T$;

putting $x = 16$ and $y = 6$ gives $6 = 3 \times 16 + c_T$, so $c_T = -42$,

and the equation of the tangent is $y_T = 3x - 42$.

The gradient of the normal must be $-\frac{1}{3}$, so the equation of

the normal is $y_N = -\frac{1}{3}x + c_N$

Substituting in the coordinates of the point (16, 6)

gives $6 = -\frac{16}{3} + c_N \Rightarrow c_N = \frac{34}{3}$; so the normal is

$y_N = -\frac{1}{3}x + \frac{34}{3} = \frac{1}{3}(34 - x)$.

7) For both curves, when $x = 4$, $y = 2$, so they meet at (4, 2). Differentiating the first curve gives

$\frac{dy}{dx} = x^2 - 4x - 4$, which at $x = 4$ is equal to –4.

Differentiating the other curve gives $\frac{dy}{dx} = \frac{1}{2\sqrt{x}}$, and so the gradient at (4, 2) is ¼. If you multiply these two gradients together you get –1, so the two curves are perpendicular at $x = 4$.

Those questions covered the basics of differentiation, so if you got them all correct, bravely venture into the murky realm of Exam Questions. If you struggled, have a cuppa to fuel your noggin — then read the section again until it all makes sense.

Exam Questions

1 a) Rewrite all the terms as powers of x:

$y = x^7 + \frac{2}{x^3} = x^7 + 2x^{-3}$ *[1 mark]*

and then differentiate each term:

$\frac{dy}{dx} = 7x^6 + (-3)2x^{-4}$

$= 7x^6 - \frac{6}{x^4}$ *[1 mark]*

b) This is a second-order derivative — just differentiate the answer for part a):

$\frac{d^2y}{dx^2} = \frac{d}{dx}(7x^6 - 6x^{-4})$ *[1 mark]*

$= (7 \times 6)x^5 - (6 \times -4)x^{-5}$

$= 42x^5 + \frac{24}{x^5}$ *[1 mark]*

2 a) $\frac{dy}{dx} = 6x^2 - 8x - 4$

[2 marks for all 3 terms correct or 1 mark for 2 terms.]

b) To find the gradient, put $x = 2$ into the answer to part (a):
$6(2^2) - 8(2) - 4 = 24 - 16 - 4 = 4$ *[1 mark]*.

c) The gradient of the normal is $-1 \div$ the gradient of the tangent $= -1 \div 4 = -¼$ *[1 mark]*. At $x = 2$, the y-value is $2(2^3) - 4(2^2) - 4(2) + 12 = 16 - 16 - 8 + 12 = 4$ *[1 mark]*.
Putting these values into the formula $(y - y_1) = m(x - x_1)$
gives $(y - 4) = -¼(x - 2) \Rightarrow y = -¼x + ½ + 4 \Rightarrow$
$y = -¼x + 4½$ *[1 mark]*.

You could also give your answer in the form $x + 4y = 18$ by multiplying through by 4 to get rid of the fractions.

Answers

3 Rewrite the expression in powers of x, so it becomes $x^{-\frac{1}{2}} + x^{-1}$ *[1 mark]*. Then differentiate to get $\frac{dy}{dx} = -\frac{1}{2}x^{-\frac{3}{2}} - x^{-2}$ *[1 mark for each correct term]*.

Putting $x = 4$ into the derivative gives:

$$-\frac{1}{2}4^{-\frac{3}{2}} - 4^{-2} = -\frac{1}{2}(\sqrt{4})^{-3} - \frac{1}{4^2}$$

$$= -\frac{1}{2} \cdot \frac{1}{2^3} - \frac{1}{16} = -\frac{1}{2} \cdot \frac{1}{8} - \frac{1}{16}$$

$$= -\frac{1}{16} - \frac{1}{16} = -\frac{1}{8}$$

[1 method mark, 1 answer mark]

4 a) Rewrite the expression in powers of x, so it becomes:

$$\frac{x^2 + 3x^{\frac{3}{2}}}{x^{\frac{1}{2}}} \quad \text{[1 mark]}$$

Then divide the top of the fraction by the bottom:

$$\frac{x^2}{x^{\frac{1}{2}}} + \frac{3x^{\frac{3}{2}}}{x^{\frac{1}{2}}}$$

$$= x^{\frac{3}{2}} + 3x$$

So $p = \frac{3}{2}$ *[1 mark]* and $q = 1$ *[1 mark]*.

b) Use your answer to part a) to rewrite the equation as:

$$y = 3x^3 + 5 + x^{\frac{3}{2}} + 3x \quad \text{[1 mark]}.$$

Then differentiate each term to give:

$$\frac{dy}{dx} = 9x^2 + \frac{3}{2}x^{\frac{1}{2}} + 3 \quad \text{[1 mark for each correct term]}.$$

5 a) $\frac{dy}{dx} = (3 \times mx^{(3-1)}) - (2 \times x^{(2-1)}) + 8(1 \times x^{(1-1)})$

$$= 3mx^2 - 2x + 8$$

[1 method mark, 1 answer mark]

b) Rearranging the equation of the line parallel to the normal gives the equation: $y = 3 - 4x$, so it has a gradient of -4 *[1 mark]*. The normal also has gradient -4 because it is parallel to this line *[1 mark]*. The gradient of the tangent is $-1 \div$ the gradient of the normal $= -1 \div -4 = \frac{1}{4}$ *[1 mark]*.

c) (i) So you know that when $x = 5$, the gradient $3mx^2 - 2x + 8 = \frac{1}{4}$. *[1 mark]*

Now find the value of m:

$$m(3 \times 5^2) - (2 \times 5) + 8 = \frac{1}{4} \quad \text{[1 mark]}$$

$$75m - 2 = \frac{1}{4}$$

$$m = \frac{9}{4} \times \frac{1}{75} = \frac{9}{300} = \frac{3}{100} = 0.03 \quad \text{[1 mark]}$$

(ii) When $x = 5$, then:

$$y = (\frac{3}{100} \times 5^3) - (5^2) + (8 \times 5) + 2 \quad \text{[1 mark]}$$

$$= \frac{375}{100} - 25 + 40 + 2$$

$$= \frac{375}{100} + 17 = \frac{2075}{100}$$

$$= 20.75 \quad \text{[1 mark]}$$

They've given you all the information you need, in a funny roundabout kinda way. Get comfortable figuring out gradients of normals and tangents and then applying them to curves — otherwise the exam will be very UNcomfortable. You have been warned...

6 a) Rearrange $2x - y = 6$ into an expression for y: $y = 2x - 6$. *[1 mark]*. Now find x^2y^2 in terms of x: $x^2y^2 = x^2(2x - 6)^2 = x^2(4x^2 - 24x + 36)$ $= 4x^4 - 24x^3 + 36x^2$ *[1 mark]*

b) (i) $\frac{d(4x^4 - 24x^3 + 36x^2)}{dx} = 16x^3 - 72x^2 + 72x$

[2 marks for all three terms, or 1 mark for two]

$$= 8(2x^3 - 9x^2 + 9x) \quad \text{[1 mark]}$$

So $k = 8$ *[1 mark]*

(ii) At $x = 1$, $8(2x^3 - 9x^2 + 9x) = 8(2 - 9 + 9)$ $= 16$ *[1 mark]*.

c) $\frac{d^2W}{dy^2} = 48x^2 - 144x + 72$ *[1 mark]*

So for $x = 1$: $\frac{d^2W}{dy^2} = 48 - 144 + 72 = -24$ *[1 mark]*

C1 Section 7 — Integration

Warm-up Questions

1) i) Increase the power of x by 1, ii) divide by the new power, iii) add a constant.

2) An integral without limits to integrate between. Because there's more than one right answer.

3) Differentiate your answer, and if you get back the function you integrated in the first place, your answer's right.

4 a) $2x^5 + C$ **b)** $\frac{3x^2}{2} + \frac{5x^3}{3} + C$ **c)** $\frac{3}{4}x^4 + \frac{2}{3}x^3 + C$

5) Integrating gives $y = 3x^2 - 7x + C$; then substitute $x = 1$ and $y = 0$ to find that $C = 4$. So the equation of the curve is $y = 3x^2 - 7x + 4$.

6 a) Integrate to get $y = \frac{3x^4}{4} + 2x + C$. Putting $x = 1$ and $y = 0$ gives $C = -\frac{11}{4}$, and so the required curve is $y = \frac{3x^4}{4} + 2x - \frac{11}{4}$.

b) If the curve has to go through $(1, 2)$ instead of $(1, 0)$, substitute the values $x = 1$ and $y = 2$ to find a different value for C; call this value C_1. Making these substitutions gives $C_1 = -\frac{3}{4}$, and the equation of the new curve is $y = \frac{3x^4}{4} + 2x - \frac{3}{4}$.

Well, that's covered the bog-standard integration skills — but you've got to know how to apply them in all sorts of fiddly ways, so get practising these exam questions...

Answers

Exam Questions

1 a) $f(x) = \dfrac{x^{\frac{1}{2}}}{\frac{1}{2}} + 4x - \dfrac{5x^4}{4} + C$

and then simplify each term further if possible...

$= 2\sqrt{x} + 4x - \dfrac{5x^4}{4} + C$

[3 marks available — 1 mark for each term. Lose 1 mark if C missing or terms not simplified, e.g. ÷½ not converted to ×2. Note — you don't need to put surds in for it to be simplified — indices are fine.]

b) First rewrite everything in terms of powers of x:

$f'(x) = 2x + 3x^{-2}$

Now you can integrate each term (don't forget to add C):

$f(x) = \dfrac{2x^2}{2} + \dfrac{3x^{-1}}{-1} + C$

Then simplify each term:

$f(x) = x^2 - \dfrac{3}{x} + C$

[2 marks available — 1 mark for each term. Lose 1 mark if C missing or terms not simplified.]

c) Following the same process as in part b):

$f'(x) = 6x^2 - \dfrac{1}{3}x^{-\frac{1}{2}}$

$f(x) = \dfrac{6x^3}{3} + \dfrac{1}{3}(x^{\frac{1}{2}} \div \dfrac{1}{2}) + C$

$f(x) = 2x^3 + \dfrac{2}{3}\sqrt{x} + C$

[2 marks available — 1 mark for each term. Lose 1 mark if C missing or terms not simplified.]

2 a) Multiply out the brackets and simplify the terms:

$(5 + 2\sqrt{x})^2 = (5 + 2\sqrt{x})(5 + 2\sqrt{x})$

$= 25 + 10\sqrt{x} + 10\sqrt{x} + 4x$

$= 25 + 20\sqrt{x} + 4x$

So $a = 25$, $b = 20$ and $c = 4$

[3 marks: one for each constant]

b) Integrate your answer from a), treating each term separately:

$\int (25 + 20\sqrt{x} + 4x)\,dx = 25x + \left(20x^{\frac{3}{2}} \div \dfrac{3}{2}\right) + \left(\dfrac{4x^2}{2}\right) + C$

$= 25x + \dfrac{40\sqrt{x^3}}{3} + 2x^2 + C$

[3 marks available — 1 for each term. Lose 1 mark if C missing or answers not simplified (surds not necessary)]

Don't forget to add C, don't forget to add C, don't forget to add C. Once, twice, thrice I beg of you, because it's very important.

3 To find $f(x)$ you integrate $f'(x)$, but it helps to write all terms in powers of x, so $5\sqrt{x} = 5x^{\frac{1}{2}}$ and $\dfrac{6}{x^2} = 6x^{-2}$ *[1 mark]*

Now integrate each term:

$\int (2x + 5x^{\frac{1}{2}} + 6x^{-2})\,dx = \dfrac{2x^2}{2} + \left(5x^{\frac{3}{2}} \div \dfrac{3}{2}\right) + \left(\dfrac{6x^{-1}}{-1}\right) + C$

$f(x) = x^2 + \dfrac{10\sqrt{x^3}}{3} - \dfrac{6}{x} + C$

[2 marks for correct terms, 1 mark for +C]

You've been given a point on the curve so you can calculate the value of C:

If $y = 7$ when $x = 3$, then

$3^2 - \dfrac{6}{3} + \dfrac{10\sqrt{3^3}}{3} + C = 7$ *[1 mark]*

$9 - 2 + 10\sqrt{3} + C = 7$

$7 + 10\sqrt{3} + C = 7$

$C = -10\sqrt{3}$

$f(x) = x^2 - \dfrac{6}{x} + \dfrac{10\sqrt{x^3}}{3} - 10\sqrt{3}$ *[1 mark]*

4 a) Rearrange the terms so each is written as a power of x, showing your working:

$\dfrac{1}{\sqrt{36x}} = \dfrac{1}{\sqrt{36}\sqrt{x}} = \dfrac{1}{6} \times \dfrac{1}{\sqrt{x}} = \dfrac{1}{6}x^{-\frac{1}{2}}$ *[1 mark]*

$2\left(\sqrt{\dfrac{1}{x^3}}\right) = 2\left(\dfrac{1}{x^3}\right)^{\frac{1}{2}} = 2(x^{-3})^{\frac{1}{2}}$

$= 2(x^{(-3 \times \frac{1}{2})}) = 2x^{-\frac{3}{2}}$ *[1 mark]*

This shows that $f'(x) = \dfrac{1}{6}x^{-\frac{1}{2}} - 2x^{-\frac{3}{2}}$ — so $A = \dfrac{1}{6}$ and $B = 2$ *[1 mark]*

b) Integrate $f'(x)$ to find $f(x)$:

$f(x) = \left(\dfrac{1}{6} \times x^{\frac{1}{2}} \div \dfrac{1}{2}\right) - \left(2x^{-\frac{1}{2}} \div -\dfrac{1}{2}\right) + C$ *[1 mark]*

$= \dfrac{1}{3}x^{\frac{1}{2}} + \left(-2 \div -\dfrac{1}{2}\right)\left(\dfrac{1}{\sqrt{x}}\right) + C$

$= \dfrac{\sqrt{x}}{3} + \dfrac{4}{\sqrt{x}} + C$ *[1 mark]*

Now use the coordinates $(1, 7)$ to find the value of C:

$7 = \dfrac{\sqrt{1}}{3} + \dfrac{4}{\sqrt{1}} + C$ *[1 mark]*

$7 - \dfrac{1}{3} - 4 = C$

$C = \dfrac{8}{3}$

So $y = \dfrac{\sqrt{x}}{3} + \dfrac{4}{\sqrt{x}} + \dfrac{8}{3}$ *[1 mark]*

Well, aren't we having lovely integrating fun. Keep toddling through, and it'll be time for tea and biscuits in no time.

5 a) The tangent at $(1, 2)$ has the same gradient as the curve at that point, so use $f'(x)$ to calculate the gradient:

$f'(1) = 1^3 - 2$ *[1 mark]*

$= -1$ *[1 mark]*

Put this into the straight-line equation $y - y_1 = m(x - x_1)$:

$y - 2 = -1(x - 1)$ *[1 mark]*

$y = -x + 1 + 2$

$y = -x + 3$ *[1 mark]*

No need to go off on a tangent here — just find the gradient and then find the equation. Boom. Done. And move swiftly on...

Answers

b) $f(x) = \int\left(x^3 - \frac{2}{x^2}\right)dx = \int(x^3 - 2x^{-2})dx$ *[1 mark]*

$= \frac{x^4}{4} - 2\frac{x^{-1}}{-1} + C = \frac{x^4}{4} + 2x^{-1} + C$ *[1 mark]*

$= \frac{x^4}{4} + \frac{2}{x} + C$

Now use the coordinates (1, 2) to find the value of C:

$2 = \frac{1^4}{4} + \frac{2}{1} + C$ *[1 mark]*

$2 - \frac{1}{4} - 2 = C$

$C = -\frac{1}{4}$ *[1 mark]*

So $f(x) = \frac{x^4}{4} + \frac{2}{x} - \frac{1}{4}$ *[1 mark]*

6 a) Multiply out the brackets in $f'(x)$
$(x - 1)(3x - 1) = 3x^2 - x - 3x + 1$
$= 3x^2 - 4x + 1$ *[1 mark]*
Now $f(x) = \int(3x^2 - 4x + 1)dx$
$= \frac{3x^3}{3} - \frac{4x^2}{2} + \frac{x}{1} + C$ *[1 mark]*
$= x^3 - 2x^2 + x + C$ *[1 mark]*
Input the x and y coordinates to find C:
$10 = 3^3 - 2(3^2) + 3 + C$ *[1 mark]*
$10 - 27 + 18 - 3 = C$
$C = -2$ *[1 mark]*
So $f(x) = x^3 - 2x^2 + x - 2$ *[1 mark]*

b) First calculate the gradient of $f(x)$ when $x = 3$:
$f'(3) = 3(3^2) - (4 \times 3) + 1$
$= 27 - 12 + 1$
$= 16$ *[1 mark]*
Use the fact that the tangent gradient multiplied by the normal gradient must equal –1 to find the gradient of the normal (n): $16 \times n = -1$ therefore $n = -\frac{1}{16}$ *[1 mark]*.
Put n and $P(3, 10)$ into the formula for the equation of a line and rearrange until it's in the form $y = \frac{a - x}{b}$:
$y - 10 = -\frac{1}{16}(x - 3)$ *[1 mark]*
$y = \frac{3}{16} - \frac{x}{16} + 10$
$y = \frac{163 - x}{16}$
So $a = 163$ and $b = 16$. *[1 mark]*
The End. Can I go home now?

C1 — Practice Exam One

1 a) Rearrange according to Laws of Indices (if you've forgotten, they've got their very own page in Section 1):
$36^{-\frac{1}{2}} = \frac{1}{36^{\frac{1}{2}}} = \frac{1}{\sqrt{36}}$ *[1 mark]*
$= \frac{1}{6}$ *[1 mark]*

b) First simplify the surd on the bottom of the fraction:
$\sqrt[m]{a^n} = a^{\frac{n}{m}}$ so $\sqrt{a^4} = a^2$ *[1 mark]*

Then rewrite the entire expression so that you're only multiplying things (remember that $\div a^n = \times a^{-n}$):
$\frac{a^6 \times a^3}{a^2} \div a^{\frac{1}{2}} = a^6 \times a^3 \times a^{-2} \times a^{-\frac{1}{2}}$ *[1 mark]*
Finally, add the powers together, because $a^m \times a^n = a^{m+n}$:
$= a^{6+3-2-\frac{1}{2}} = a^{\frac{13}{2}}$ *[1 mark]*

Lots of laws to remember there. Make sure you don't multiply powers when you should be adding, and vice versa.

2 a) Multiply out the brackets first:
$(5\sqrt{5} + 2\sqrt{3})^2 = (5\sqrt{5} + 2\sqrt{3})(5\sqrt{5} + 2\sqrt{3})$
$= (5\sqrt{5})^2 + 2(5\sqrt{5} \times 2\sqrt{3}) + (2\sqrt{3})^2$
So now you've got three terms to deal with, and they're all a little bit nasty. The first term is:
$(5\sqrt{5})^2 = 5\sqrt{5} \times 5\sqrt{5}$
$= 5 \times 5 \times \sqrt{5} \times \sqrt{5}$
$= 5 \times 5 \times 5$
$= 125$ *[1 mark]*
The second term is:
$2(5\sqrt{5} \times 2\sqrt{3}) = 2 \times 5 \times 2 \times \sqrt{5} \times \sqrt{3}$
$= 20\sqrt{15}$ *(don't forget, $\sqrt{5} \times \sqrt{3} = \sqrt{5 \times 3}$)* *[1 mark]*
And the third term is:
$2\sqrt{3} \times 2\sqrt{3} = 2 \times 2 \times \sqrt{3} \times \sqrt{3}$
$= 2 \times 2 \times 3 = 12$ *[1 mark]*
So all you have to do now is add the three terms together:
$125 + 20\sqrt{15} + 12 = 137 + 20\sqrt{15}$ *[1 mark]*
So $a = 137$, $b = 20$ and $c = 15$.

b) Rationalising the denominator means getting rid of surds on the bottom line of a fraction.
Multiply top and bottom lines by $\sqrt{5} - 1$:
$\frac{10 \times (\sqrt{5} - 1)}{(\sqrt{5} + 1) \times (\sqrt{5} - 1)}$ *[1 mark]*
$= \frac{10(\sqrt{5} - 1)}{5 - 1} = \frac{10(\sqrt{5} - 1)}{4} = \frac{5(\sqrt{5} - 1)}{2}$
[1 mark for simplifying top, 1 mark for simplifying bottom]

Mmm, look at all that lovely maths vocab. Rationalising surd denominators — look back at Section 1 if you got stuck with this.

3 a) $\frac{dy}{dx} = 2x$ *[1 mark]*

b) $\frac{dy}{dx} = 12x^3 - 2$ *[1 mark for each correct term]*

c) $y = x^3 - 2x^2 + 4x - 8$
$\frac{dy}{dx} = 3x^2 - 4x + 4$ *[1 mark for expanding brackets,*
2 marks for correct differentiation of all terms or 1 mark
for correct differentiation of 2 terms]

Answers

4 As one of these equations is a quadratic, you need the substitution method, so rearrange to get y on its own:

$y + x = 7$

$y = 7 - x$ *[1 mark]*

And now substitute that into the quadratic to get:

$7 - x = x^2 + 3x - 5$ *[1 mark]*

Rearrange again to get everything on one side of the equation, and then factorise it:

$0 = x^2 + 4x - 12$

$(x + 6)(x - 2) = 0$ *[1 mark]*

Which gives two values : $x = -6$ and $x = 2$ *[1 mark]*

Don't forget to find the corresponding values of y as well:

$y = 7 - x$

So when $x = -6$

$y = 7 - -6 = 7 + 6$

$y = 13$

And when $x = 2$

$y = 7 - 2$

$y = 5$ *[1 mark]*

So the equations meet at (–6, 13) and (2, 5).

Simultaneous equations — you should be able to do them standing on your head... although don't try that in the exam...

5 a) k is the y-value when $x = 4$, so substitute $x = 4$ into the equation of the line to find y:

$y + (2 \times 4) - 5 = 0$

$y = 5 - 8$

$y = -3$ so $k = -3$ *[1 mark]*

b) Remember — the gradients of two perpendicular lines multiply together to make –1. This means that the gradient of the new line will be: $\dfrac{-1}{\text{the gradient of AB}}$

The gradient of a straight line is the coefficient of x when the equation of the line is written in the form $y = mx + c$.

$y + 2x - 5 = 0$ becomes $y = -2x + 5$, so the gradient of AB = –2. *[1 mark]*

Now you know the gradient of the new line will be $\dfrac{-1}{-2}$, which equals $\dfrac{1}{2}$. *[1 mark]*

Finally, get the equation of the new line using $y - y_1 = m(x - x_1)$ and A (1, 3) — the point it passes through: $y - 3 = \dfrac{1}{2}(x - 1)$ *[1 mark]*

$y = \dfrac{1}{2}x - \dfrac{1}{2} + 3 = \dfrac{1}{2}x + \dfrac{5}{2}$

$y = \dfrac{x + 5}{2}$ *[1 mark]*

Remember the process, remember the formulas, and you'll be fine. But the only way to make sure you know it is practice. If you forgot anything, go back to the coordinate geometry section and go through it 'til it sinks in...

6 a) $4x + 7 > 7x + 4$

Subtract $4x + 4$ from both sides which gives

$3 > 3x$ *[1 mark]*

$1 > x$ or $x < 1$ *[1 mark]*

b) You've got to find the minimum value of $(x - 5)(x - 3)$ and when k is less than that, $(x - 5)(x - 3) > k$ will be true for all possible values of x.

As always, it helps to sketch a graph and think about what the function looks like:

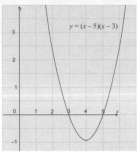

Remember, it's a symmetrical graph, so the minimum is halfway between $x = 3$ and $x = 5$ — i.e. when $x = 4$. *[1 mark]*

Put this x-value into the equation to find the lowest possible y-value: $(x - 5)(x - 3) = (4 - 5)(4 - 3)$ *[1 mark]*

$= -1 \times 1 = -1$ *[1 mark]*

So for the range of values $k < -1$ it is true that

$(x - 5)(x - 3) > k$ for all possible values of x.

You can also find the minimum value by completing the square — when you've got your quadratic in the form $(x + m)^2 + n$, the minimum value occurs at n.

7 $\displaystyle\int (4x^3 + 6x + 3)\, dx = \dfrac{4x^4}{4} + \dfrac{6x^2}{2} + 3x + C$

$= x^4 + 3x^2 + 3x + C$

[4 marks available in total — 1 mark for each correct term]

8 a) To find the coordinates of A, solve the two lines as simultaneous equations:

$l_1\colon x - y + 1 = 0$

$l_2\colon 2x + y - 8 = 0$

Add the two equations together to get rid of y:

$x + 2x + 1 - 8 - y + y = 0$ *[1 mark]*

$3x - 7 = 0$

$x = \dfrac{7}{3}$ *[1 mark]*

Now put $x = \dfrac{7}{3}$ back into l_1 to find y:

$\dfrac{7}{3} - y + 1 = 0$

$y = \dfrac{7}{3} + 1 = \dfrac{10}{3}$ So A is $\left(\dfrac{7}{3}, \dfrac{10}{3}\right)$ *[1 mark]*

Still with me? Deep breath... there's a whole lot more geometry a-coming your way...

Answers

b) There's a lot of information here, so draw a quick sketch to make things a bit clearer:

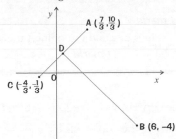

To find the equation of line BD, you need its gradient. But before you can find the gradient, you need to find the coordinates of point D — the midpoint of AC. To find the midpoint of two points, find the average of the x-values and the average of the y-values:

$$D = \left(\frac{x_a + x_c}{2}, \frac{y_a + y_c}{2}\right) = \left(\frac{\frac{7}{3} + \frac{-4}{3}}{2}, \frac{\frac{10}{3} + \frac{-1}{3}}{2}\right) \quad \textbf{[1 mark]}$$

$$D = \left(\frac{1}{2}, \frac{3}{2}\right) \quad \textbf{[1 mark]}$$

To find the gradient (m) of BD, use this rule: $m_{BD} = \frac{y_D - y_B}{x_D - x_B}$

$$\text{Gradient} = \frac{\frac{3}{2} - -4}{\frac{1}{2} - 6} = \frac{\frac{3}{2} + \frac{8}{2}}{\frac{1}{2} - \frac{12}{2}} = \frac{3 + 8}{1 - 12} = -1 \quad \textbf{[1 mark]}$$

Now you can find the equation of BD. Input the known values of x and y at B $(6, -4)$ and the gradient (-1) into $y = mx + c$, which gives:

$-4 = (-1 \times 6) + c \quad \textbf{[1 mark]}$

$-4 + 6 = c$

$c = 2$

So the equation for BD is $y = -x + 2$ \quad \textbf{[1 mark]}

But the question asks for it in the form $ax + by + c = 0$

So rearrange to get: $x + y - 2 = 0$ \quad \textbf{[1 mark]}

Line blah, point blah, midpoint blah... it's easy to feel bamboozled when reading a long list of geometry babble, which is why it helps to DRAW A GRAPH. It's up to you, but I know what I'd do...

c) Look at the sketch above. To prove triangle ABD is a right-angled triangle, you need to prove that lines AD and BD are perpendicular — in other words, prove the product of their gradients equals -1.

You already know the gradient of BD = -1.
Use the same rule to find the gradient of AD:

$$m_{AD} = \frac{y_D - y_A}{x_D - x_A} = \frac{\frac{3}{2} - \frac{10}{3}}{\frac{1}{2} - \frac{7}{3}} = \frac{\frac{9}{6} - \frac{20}{6}}{\frac{3}{6} - \frac{14}{6}} = \frac{9 - 20}{3 - 14} = 1 \quad \textbf{[1 mark]}$$

$m_{BD} \times m_{AD}$ \quad \textbf{[1 mark]}
$= -1 \times 1 = -1$ \quad \textbf{[1 mark]}

So triangle ABD is a right-angled triangle.

9 a) Start by finding a — it's the coefficient of x halved:

$x^2 - 6x + 5 = (x - 3)^2 + b$ \quad \textbf{[1 mark]}

Multiply out to get $x^2 - 6x + 9 + b = x^2 - 6x + 5$

Simplify to find b: $9 + b = 5$ so $b = 5 - 9 = -4$

So the answer is $x^2 - 6x + 5 = (x - 3)^2 - 4$ \quad \textbf{[1 mark]}

Completing the square should be like riding a bike — you never forget how to do it. So if you did forget, go back to the quadratics section and start pedalling...

b) To factorise $x^2 - 6x + 5$ you need two numbers that add up to -6 and multiply to give 5. Easy peasy...

$x^2 - 6x + 5 = (x - 1)(x - 5)$ \quad \textbf{[2 marks for correct answer, or 1 mark if there's a sign error]}

c) To sketch the graph, find where the curve cuts the two axes: When the line cuts the x-axis, $y = 0$. i.e. $x^2 - 6x + 5 = 0$

Use your answer to part b) to give...

$(x - 1)(x - 5) = 0$

so $x = 1$ or $x = 5$,

i.e. the graph cuts the x-axis at $(1, 0)$ and $(5, 0)$.

Now, when $x = 0$, $y = 0^2 - (6 \times 0) + 5$, so the graph cuts the y-axis at $(0, 5)$.

Don't just start scribbling any old curve, make sure you've plotted the important points and labelled them clearly — otherwise no lovely marks for you.

Now you're all ready to sketch the graph:

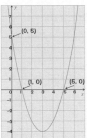

[3 marks available — 1 mark for correct shape, 1 mark for correct y-axis intercept, 1 mark for correct x-axis intercepts]

10 a) Integrate the derivative to find the equation of the curve:

$$\int (3x^2 + 6x - 4)dx = \frac{3x^3}{3} + \frac{6x^2}{2} - 4x + C$$

$y = x^3 + 3x^2 - 4x + C$

[2 marks for all four terms correct, 1 mark for 2 correct terms]

Use the point $(0, 0)$ to find C:

$0 = 0 + 0 - 0 + C \Rightarrow C = 0$ \textbf{[1 mark]}

So the equation of the curve is $y = x^3 + 3x^2 - 4x$ \textbf{[1 mark]}.

b) All the terms in $y = x^3 + 3x^2 - 4x$ contain 'x' so take out x as a factor: $y = x(x^2 + 3x - 4)$ \textbf{[1 mark]}. The remaining quadratic in the bracket will factorise:

$x^2 + 3x - 4 = (x + 4)(x - 1)$ \textbf{[1 mark]}.

So writing y as a product of all 3 factors gives:

$y = x(x + 4)(x - 1)$ \textbf{[1 mark]}.

11 If the quadratic $ax^2 + bx + c = 0$ has no real roots, this means the discriminant gives a negative value:

$b^2 - 4ac < 0$ \textbf{[1 mark]}. So:

$(-4)^2 - [4 \times 1 \times (k - 1)] < 0$ \quad \textbf{[1 mark]}

$\Rightarrow 16 - (4k - 4) < 0$

$\Rightarrow 20 - 4k < 0$ \quad \textbf{[1 mark]}

$20 < 4k$

$k > 5$ \quad \textbf{[1 mark]}

It's no good just remembering the discriminant formula — you also have to know what the result tells you. If you don't know the meaning of negative discriminants, head back to the quadratics section...

Answers

12 a) In an arithmetic sequence, the nth term is defined by the formula $a + (n - 1)d$. The 12th term is 79, so the equation is $79 = a + 11d$ *[1 mark]*, and the 16th term is 103, so the other equation is $103 = a + 15d$ *[1 mark]*. Solving these simultaneously (by taking the first equation away from the second) gives $24 = 4d$, so $d = 6$ *[1 mark]*. Putting this value of d into the first equation gives $79 = a + (11 \times 6)$, so $a = 13$ *[1 mark]*.

b) The formula to find the sum, S_n, is $S_n = \frac{n}{2}[2a + (n - 1)d]$ *[1 mark]*. Putting in the values of a and d from above gives the formula
$$S_n = \frac{n}{2}[(2 \times 13) + 6(n - 1)] \quad \text{[1 mark]}$$
$$= \frac{n}{2}[20 + 6n] \text{ or } 10n + 3n^2 \quad \text{[1 mark]}$$

c) $S_{15} = (10 \times 15) + 3(15^2)$
$$= 150 + 675$$
$$= 825 \quad \text{[1 mark]}$$

At the end of the exam, it's tradition to have a comforting cup of tea and vow never to differentiate the equation of a curve ever again. That was just a practice though, so don't celebrate just yet... Oh alright, you can have a cuppa I suppose...

C1 — Practice Exam Two

1 a) Just multiply out the brackets:
$$(\sqrt{3} + 1)(\sqrt{3} - 1) = 3 - \sqrt{3} + \sqrt{3} - 1 \quad \text{[1 mark]}$$
$$= 2 \quad \text{[1 mark]}$$

b) To rationalise the denominator, you want to get rid of the surd on the bottom line of the fraction. To do this, use the difference of two squares — e.g. if the denominator's $\sqrt{a} + b$, you multiply by $\sqrt{a} - b$:
$$\frac{\sqrt{3}}{\sqrt{3} + 1} \times \frac{\sqrt{3} - 1}{\sqrt{3} - 1} \quad \text{[1 mark]}$$
$$= \frac{3 - \sqrt{3}}{2} \quad \text{[1 mark]}$$
(the bottom bit is substituted in from part a) *[1 mark]*

2 a) $\frac{x^2 + 2x}{\sqrt{x}} = x^{-\frac{1}{2}}(x^2 + 2x) = x^{\frac{3}{2}} + 2x^{\frac{1}{2}}$ *[1 mark for each correct term]*, so $m = \frac{3}{2}$ and $n = \frac{1}{2}$.

b) From part (a) above, $y = x^{\frac{3}{2}} + 2x^{\frac{1}{2}} + 3x^3 - x$. Differentiating this gives
$$\frac{dy}{dx} = \frac{3}{2}x^{\frac{1}{2}} + \left(2 \cdot \frac{1}{2}\right)x^{-\frac{1}{2}} + (3 \cdot 3)x^2 - 1$$
$$= \frac{3}{2}x^{\frac{1}{2}} + x^{-\frac{1}{2}} + 9x^2 - 1$$
[4 marks available — 1 mark for each correct term]

3 a) Integrate the derivative $\left(\frac{dy}{dx}\right)$ to find the equation of the curve. But first multiply out the bracket, which gives:
$$4(1 - x) = 4 - 4x \quad \text{[1 mark]}$$

Now integrate, not forgetting the constant of integration:
$$\int (4 - 4x)dx = 4x - 4\frac{x^2}{2} + C$$
$$= 4x - 2x^2 + C \quad \text{[1 mark]}$$
Substitute the known values of x and y at point A to find C:
$$6 = (4 \times 2) - (2 \times 2^2) + C \quad \text{[1 mark]}$$
$$6 = 8 - 8 + C$$
$$6 = C$$
So the equation of the curve is: $y = 4x - 2x^2 + 6$ *[1 mark]*

b) First, find the points where the curve cuts the axes — Where it cuts the y-axis, $x = 0$.
$$y = (4 \times 0) - (2 \times 0^2) + 6 = 6,$$
so it crosses the y-axis at $(0, 6)$.
It cuts the x-axis at $y = 0$
$$4x - 2x^2 + 6 = 0$$
$$2x^2 - 4x - 6 = 0$$
$$2(x^2 - 2x - 3) = 0$$
$$2(x + 1)(x - 3) = 0$$
$$x = -1 \text{ or } x = 3$$
So the curve cuts the x-axis at $(-1, 0)$ and $(3, 0)$.

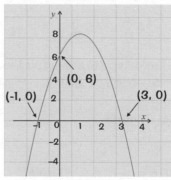

[4 marks available — 1 mark for factorising y, 1 mark for correct shape and correct y-axis intercept, 1 mark each for correct x-axis intercepts]

Pure Picasso... beautiful, curved, and all intercepts clearly labelled.

4 a) Complete the square by halving the coefficient of x to find the number in the brackets (m):
$$x^2 - 7x + 17 = \left(x - \frac{7}{2}\right)^2 + n$$
Now simplify this equation to find n:
$$n = x^2 - 7x + 17 - \left[\left(x - \frac{7}{2}\right)^2\right]$$
$$n = x^2 - 7x + 17 - x^2 + \frac{14x}{2} - \frac{49}{4}$$
$$n = 17 - \frac{49}{4} = \frac{19}{4} \quad \text{[1 mark]}$$
So you can express $x^2 - 7x + 17$ as:
$$\left(x - \frac{7}{2}\right)^2 + \frac{19}{4} \quad \text{[1 mark]}$$
The maximum value of $f(x)$ will be when the denominator is as small as possible — so you want the minimum value of $x^2 - 7x + 17$. Using the completed square above, you can see that the minimum value is $\frac{19}{4}$ because the squared part can equal but never be below 0 *[1 mark]*.
So max value of $f(x)$ is $\frac{1}{19/4} = 1 \times \frac{4}{19} = \frac{4}{19}$ *[1 mark]*.

You've got to put on your thinking cap for that one — but they give you a big hint by getting you to complete the square first.

b) If the function only has one root, then $b^2 - 4ac = 0$ *[1 mark]*. For this equation, $a = 3$, $b = k$ and $c = 12$. Use this formula to find k:

$k^2 - (4 \times 3 \times 12) = 0$ *[1 mark]*

$k^2 - 144 = 0$

$k^2 = 144$

$k = \pm 12$ *[1 mark]*

Of course, it's no good remembering the general formula if you don't know what a, b and c stand for. Imagine forgetting that it's $ax^2 + bx + c$ — you'd feel a right wally.

5 a) A sketch will definitely help here:

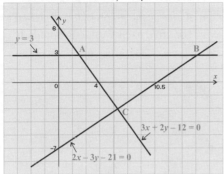

Finding the coordinates of points A, B and C is just a small matter of simultaneous equations. A and B are simple — you already know their y-value is 3, so:

Point A is the point where $y = 3$ and $3x + 2y - 12 = 0$ meet:

$3x + (2 \times 3) - 12 = 0$

$3x + 6 = 12$

$3x = 6$

$x = 2$

Point $A = (2, 3)$ *[1 mark]*

Point B is the point where $y = 3$ and $2x - 3y - 21 = 0$ meet:

$2x - (3 \times 3) - 21 = 0$

$2x = 30$

$x = 15$

Point $B = (15, 3)$ *[1 mark]*

Point C is slightly trickier — you've got to solve a pair of simultaneous equations:

Label the equations: $2x - 3y - 21 = 0$ (a)

$3x + 2y - 12 = 0$ (b)

It's easy to get muddled with all those 3s, 2s and 1s. Just a slip of concentration and then — whoopsy — you've solved $3x - 2y + 12 = 0$, and that's no use to anyone...

Multiply (a) by 2 and (b) by 3 to equalise the coefficients of y:

$4x - 6y - 42 = 0$

$9x + 6y - 36 = 0$

Now add together to get rid of y:

$4x + 9x - 6y + 6y - 42 - 36 = 0$ *[1 mark]*

$13x - 78 = 0$

$13x = 78$

$x = 6$ *[1 mark]*

Substitute this value of x into (a) or (b):

$(3 \times 6) + 2y - 12 = 0$

$18 - 12 + 2y = 0$

$2y = -6$

$y = -3$

So C has the coordinates $(6, -3)$. *[1 mark]*

b) To show that the triangle is right-angled, you need to prove that the gradients of two of the lines multiply together to make −1. Looking at the sketch, the right angle looks like it's at C, so use lines BC and AC:

BC: $2x - 3y - 21 = 0$

$3y = 2x - 21$

$y = \frac{2}{3}x - 7$

So the gradient of BC $= \frac{2}{3}$.

AC: $3x + 2y - 12 = 0$

$2y = 12 - 3x$

$y = 6 - \frac{3}{2}x$

So the gradient of AC $= -\frac{3}{2}$ *[1 mark]*.

$\frac{2}{3} \times -\frac{3}{2} = -1$ *[1 mark]*

The lines that meet at C are perpendicular, so the triangle must be right-angled.

That gradient rule is proving useful, isn't it — you can find equations of normals and tangents and prove right angles, and it might even be the password for the magical kingdom of Narnia...

c) Point D has the coordinates $(3, d)$ so it must lie on the line $x = 3$. If you draw this on your sketch, you can see D must lie above line AB or below line AC to lie outside the triangle.

For $(3, d)$ to be above the line $y = 3$, $d > 3$. *[1 mark]* Now work out the values of d that would give D as below the line AC. If D was <u>on</u> the line AC, then $(3, d)$ would satisfy the equation of AC:

$3x + 2y - 12 = 0$ when $x = 3$ and $y = d$

$(3 \times 3) + 2d - 12 = 0$ *[1 mark]*

$9 + 2d - 12 = 0$

$2d = 3$

$d = 1.5$

But D has to be below line AC, so d has to be less than 1.5. So you've shown that either $d > 3$ or $d < 1.5$ *[1 mark]*.

6 a) When $f(x) = ax^2 + bx + c$ has no real roots, you know that $b^2 - 4ac < 0$. Here, $a = -j$, $b = 3j$ and $c = 1$ *[1 mark]*.

Therefore $(3j)^2 - (4 \times -j \times 1) < 0$ *[1 mark]*

$9j^2 + 4j < 0$ *[1 mark]*

Answers

b) To find the values where $9j^2 + 4j < 0$, you need to start by solving $9j^2 + 4j = 0$: $j(9j + 4) = 0$, so $j = 0$ **[1 mark]** or $9j = -4 \Rightarrow j = -\frac{4}{9}$ **[1 mark]**. The graph looks like this:

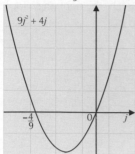

From the graph, you can see that $9j^2 + 4j < 0$ when $-\frac{4}{9} < j < 0$ **[1 mark]**.

If you made a mistake, do not pass GO, do not collect £200, and move your tiny top hat back to the pages on the quadratic formula...

7 a) $f'(x) = \dfrac{d(x^3 - 3x + 2)}{dx} = 3x^2 - 3$
    [2 marks — 1 mark for each correct term].

   b) Just multiply out the brackets to show that they give the correct equation:
   $(x - 1)^2(x + 2) = (x - 1)(x - 1)(x + 2)$ **[1 mark]**
   $= (x^2 - 2x + 1)(x + 2)$
   $= x^3 + 2x^2 - 2x^2 - 4x + x + 2$ **[1 mark]**
   $= x^3 - 3x + 2$

   c) (i) Use the factorisation from b) to find where the curve cuts the axes:
   When $y = 0$, $(x - 1)^2(x + 2) = 0 \Rightarrow x = 1$ or $x = -2$, so the curve cuts the x-axis at $(1, 0)$ and $(-2, 0)$.
   When $x = 0$, $y = 0^3 - (3 \times 0) + 2 = 2$, so the curve cuts the y-axis at $(0, 2)$.

   Put these with the basic shape of a cubic with a positive x^3 term and you can sketch the graph:

   [3 marks available — 1 mark for correct shape, 1 mark for correct x-axis intercepts, 1 mark for correct y-axis intercept]

(ii) The graph of $f(x - a)$ is $f(x)$ translated to the right. For $f(x - 3)$, each x-value needs to be increased by 3:

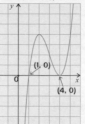

[2 marks available — 1 mark for moving to the right, 1 mark for correct new x-intercepts]

(iii) The graph of $af(x)$ is the graph of $f(x)$ stretched along the y-axis. For $2f(x)$, each y-value is doubled:

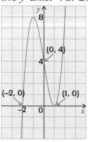

[2 marks available — 1 mark for stretching, 1 mark for correct new x- and y- intercepts]

It's no good sitting in the exam thinking 'hmm... what's the difference between af(x) and f(x + a)... err... umm...' — so learn it while you have the chance, and learn it good...

8   Solve $x^3 - 4x^2 - 7x + 10 = 0$ by factorising it. They've given you the factor $(x - 1)$, so first you need to find the quadratic that multiplies with that factor to give the original equation.
The first and last terms of the quadratic are simple to find — what multiplies with x to give x^3 and what multiplies with -1 to give $+10$?
$(x - 1)(x^2 ..... - 10)$

Now multiply that out and see if anything's missing:
$(x - 1)(x^2 - 10) = x^3 - x^2 - 10x + 10$
To get the correct coefficients for x^2 and x you need an additional: $-3x^2$ and $+3x$ **[1 mark]**.
$(-3x)(x - 1) = -3x^2 + 3x$

So you know that the quadratic needs $-3x$ in it, which gives: $(x - 1)(x^2 - 3x - 10)$ **[1 mark]**.
The quadratic further factorises into $(x - 5)(x + 2)$, so $x^3 - 4x^2 - 7x + 10 = (x - 1)(x - 5)(x + 2)$ **[1 mark]**
Now solve the equation $(x - 1)(x - 5)(x + 2) = 0$
So $x = 1$, 5 or -2 **[1 mark]**
No two ways about it — questions like those are a bit taxing. The only way is to practise and practise 'til you know exactly what you're looking for. And then practise some more...

Answers

9 a) To find the tangent you first need to find the gradient, so differentiate the equation of the curve:

$\dfrac{d(x^3 - 2x^2 + 4)}{dx} = 3x^2 - 4x$ ***[1 mark]***

When $x = 1$ then the gradient =

$(3 \times 1^2) - (4 \times 1) = 3 - 4$

$= -1$ ***[1 mark]***

When $x = 1$, $y = 1^3 - (2 \times 1^2) + 4 = 1 - 2 + 4 = 3$

Use these values and the equation $y - y_1 = m(x - x_1)$ to find the equation of the tangent at $x = 1$:

$y - 3 = -1(x - 1)$

$y = -x + 1 + 3$

$y = 4 - x$ ***[1 mark]***

b) First find the gradient of the tangent at $x = 2$:

$(3 \times 2^2) - (4 \times 2) = 12 - 8 = 4$

Using the gradient rule:

Gradient of normal $= \dfrac{-1}{\text{gradient of tangent}}$ ***[1 mark]***

So gradient of normal $= -\dfrac{1}{4}$ ***[1 mark]***

When $x = 2$, $y = 2^3 - (2 \times 2^2) + 4 = 4$

Use these values and the equation $y - y_1 = m(x - x_1)$ to find the equation of the normal at $x = 2$:

$y - 4 = -\dfrac{1}{4}(x - 2)$

$y = \dfrac{1}{2} - \dfrac{x}{4} + 4$

$y = \dfrac{9}{2} - \dfrac{x}{4}$ ***[1 mark]***

All this talk about tangents is making me want a tangerine... 'scuse me, just off to the grocer's...

c) First find where both lines cross the x-axis:

Tangent: $y = 4 - x$ so when $y = 0$:

$4 - x = 0 \Rightarrow x = 4$

The tangent cuts the x-axis at $(4, 0)$.

Normal: $y = \dfrac{9}{2} - \dfrac{x}{4}$ so when $y = 0$:

$\dfrac{9}{2} - \dfrac{x}{4} = 0 \Rightarrow \dfrac{9}{2} = \dfrac{x}{4} \Rightarrow x = 4 \times \dfrac{9}{2} = 18$

The normal cuts the x-axis at $(18, 0)$.

So the distance between the two intercepts

$= 18 - 4$

$= 14$ ***[2 marks for correct answer, or 1 mark for correct method]***

10 a) Work out the terms one by one:

$x_1 = 6$

$x_2 = 3x_1 - 4 = (3 \times 6) - 4 = 14$

$x_3 = 3x_2 - 4 = (3 \times 14) - 4 = 38$ ***[1 mark]***

$x_4 = 3x_3 - 4 = (3 \times 38) - 4 = 110$

$x_4 = 110$ ***[1 mark]***

b) (i) Use the information given and the general expression for the nth term in a sequence: $u_n = a + (n - 1)d$ to formulate expressions for the 3rd and 7th term:

$u_3 = a + 2d = 9$

$u_7 = a + 6d = 33$ ***[1 mark]***

Now solve them as simultaneous equations:

$u_7 - u_3 = a + 6d - a - 2d = 33 - 9$

$4d = 24$

$d = 6$ ***[1 mark]***

$a + 2d = 9$

$a + (2 \times 6) = 9$

$a + 12 = 9$

$a = 9 - 12 = -3$ ***[1 mark]***

So the first term is -3 and the common difference is 6.

(ii) $S_n = \dfrac{n}{2}[2a + (n - 1)d]$

So $S_{12} = \dfrac{12}{2}[(2 \times -3) + 6(12 - 1)]$ ***[1 mark]***

$= 6[-6 + 66]$ ***[1 mark]***

$= 6 \times 60$

So $S_{12} = 360$ ***[1 mark]***

(iii) $(6n + 1)$ is the rule for finding the nth term, so jot down the first few terms:

$u_1 = (6 \times 1) + 1 = 7$

$u_2 = (6 \times 2) + 1 = 13$

$u_3 = (6 \times 3) + 1 = 19$

So you can see that $d = 6$ and $a = 7$ ***[1 mark]***

It's the same as the sequence before, except the first term is 7 instead of -3. So...

$\displaystyle\sum_{1}^{12}(6n + 1) = S_{12} = \dfrac{12}{2}[(2 \times 7) + (11 \times 6)]$

$= 6(14 + 66) = 6 \times 80 = 480$ ***[1 mark]***

(Or you could note that each term is 10 more than the equivalent term in part (ii), so just add 10×12 to your answer for part (ii).)

Oh my giddy goat, we've finished. You'd better go and have a lie down, otherwise all the numbers will fall out of your head and we'd have to start again...

Answers

C2 Section 1 — Algebra and Functions
Warm-up Questions

1) a) $f(x) = (x + 2)(3x^2 - 10x + 15) - 36$

   b) $f(x) = (x + 2)(x^2 - 3) + 10$

   c) $f(x) = (x + 2)(2x^2 - 4x + 14) - 31$

2) a) (i) You just need to find f(–1).
         This is $-6 - 1 + 3 - 12 = -16$.

      (ii) Now find f(1). This is $6 - 1 - 3 - 12 = -10$.

      (iii) Now find f(2). This is $48 - 4 - 6 - 12 = 26$.

   b) (i) $f(-1) = -1$

      (ii) $f(1) = 9$

      (iii) $f(2) = 38$

   c) (i) $f(-1) = -2$

      (ii) $f(1) = 0$

      (iii) $f(2) = 37$

3) a) You need to find f(–2). This is
      $(-2)^4 - 3(-2)^3 + 7(-2)^2 - 12(-2) + 14$
      $= 16 + 24 + 28 + 24 + 14 = 106$.

   b) You need to find $f(-4/2) = f(-2)$. You found this in part a,
      so remainder = 106. You might also have noticed that
      $2x + 4$ is a multiple of $x + 2$ (from part a), so the remainder
      must be the same.

   c) You need to find f(3). This is
      $(3)^4 - 3(3)^3 + 7(3)^2 - 12(3) + 14$
      $= 81 - 81 + 63 - 36 + 14 = 41$.

   d) You need to find $f(6/2) = f(3)$. You found this in part c),
      so remainder = 41. You might also have noticed that $2x - 6$
      is a multiple of $x - 3$, so the remainder must be the same.

4) a) You need to find f(1) — if f(1) = 0, then $(x - 1)$ is a factor:
      $f(1) = 1 - 4 + 3 + 2 - 2 = 0$, so $(x - 1)$ is a factor.

   b) You need to find f(–1) — if f(–1) = 0, then $(x + 1)$ is a factor:
      $f(-1) = -1 - 4 - 3 + 2 - 2 = -8$, so $(x + 1)$ is not a factor.

   c) You need to find f(2) — if f(2) = 0, then $(x - 2)$ is a factor:
      $f(2) = 32 - (4 \times 16) + (3 \times 8) + (2 \times 4) - 2$
      $= 32 - 64 + 24 + 8 - 2 = -2$, so $(x - 2)$ is not a factor.

   d) The remainder when you divide by $(2x - 2)$ is the same as
      the remainder when you divide by $x - 1$.
      $(x - 1)$ is a factor (i.e. remainder = 0), so $(2x - 2)$ is also a
      factor.

5) If $f(x) = 2x^4 + 3x^3 + 5x^2 + cx + d$, then to make sure f(x) is
    exactly divisible by $(x - 2)(x + 3)$, you have to make sure
    $f(2) = f(-3) = 0$.
    $f(2) = 32 + 24 + 20 + 2c + d = 0$, i.e. $\underline{2c + d = -76}$.
    $f(-3) = 162 - 81 + 45 - 3c + d = 0$, i.e. $\underline{3c - d = 126}$.
    Add the two underlined equations to get: $5c = 50$,
    and so $c = 10$. Then $d = -96$.

Exam Questions

1 a) (i) Remainder = $f(1) = 2(1)^3 - 5(1)^2 - 4(1) + 3$ *[1 mark]*
        $= -4$ *[1 mark]*.

     (ii) Remainder = $f\left(-\frac{1}{2}\right) = 2\left(-\frac{1}{8}\right) - 5\left(\frac{1}{4}\right) - 4\left(-\frac{1}{2}\right) + 3$
        [1 mark] $= \frac{7}{2}$ *[1 mark]*.

   b) If $f(-1) = 0$ then $(x + 1)$ is a factor.
      $f(-1) = 2(-1)^3 - 5(-1)^2 - 4(-1) + 3$ *[1 mark]*
      $= -2 - 5 + 4 + 3 = 0$, so $(x + 1)$ is a factor of f(x).
      [1 mark]

   c) $(x + 1)$ is a factor, so divide $2x^3 - 5x^2 - 4x + 3$ by $x + 1$:
      $2x^3 - 5x^2 - 4x + 3 - \underline{2x^2}(x + 1) = 2x^3 - 5x^2 - 4x + 3 - 2x^3 - 2x^2$
      $= -7x^2 - 4x + 3$.
      $-7x^2 - 4x + 3 - (\underline{-7x})(x + 1) = -7x^2 - 4x + 3 + 7x^2 + 7x$
      $= 3x + 3$. Finally $3x + 3 - \underline{3}(x + 1) = 0$.
      so $2x^3 - 5x^2 - 4x + 3 = (2x^2 - 7x + 3)(x + 1)$.
      This is the method from p.54.

      Factorising the quadratic expression gives:
      $f(x) = (2x - 1)(x - 3)(x + 1)$.

      [4 marks available — 1 mark for dividing by x + 1 to find
      quadratic factor, 1 mark for correct quadratic factor,
      1 mark for attempt to factorise quadratic, 1 mark for
      correct factorisation of quadratic.]

2 a) $f(p) = (4p^2 + 3p + 1)(p - p) + 5$
      $= (4p^2 + 3p + 1) \times 0 + 5$
      $= 5$ *[1 mark]*.

   b) $f(-1) = -1$.
      $f(-1) = (4(-1)^2 + 3(-1) + 1)((-1) - p) + 5$
           $= (4 - 3 + 1)(-1 - p) + 5$
           $= 2(-1 - p) + 5 = 3 - 2p$ *[1 mark]*
      So: $3 - 2p = -1$, $p = 2$ *[1 mark]*.

   c) $f(x) = (4x^2 + 3x + 1)(x - 2) + 5$
      $f(1) = (4 + 3 + 1)(1 - 2) + 5 = -3$ *[1 mark]*.

C2 Section 2 — Circles and Trigonometry
Warm-up Questions

1) (a) 3, (0, 0)

   (b) 2, (2, –4)

   (c) 5, (–3, 4)

2) $\cos 30° = \frac{\sqrt{3}}{2}$, $\sin 30° = \frac{1}{2}$, $\tan 30° = \frac{1}{\sqrt{3}}$

    $\cos 45° = \frac{1}{\sqrt{2}}$, $\sin 45° = \frac{1}{\sqrt{2}}$, $\tan 45° = 1$

    $\cos 60° = \frac{1}{2}$, $\sin 60° = \frac{\sqrt{3}}{2}$, $\tan 60° = \sqrt{3}$

3) Sine Rule: $\frac{x}{\sin X} = \frac{y}{\sin Y} = \frac{z}{\sin Z}$

    Cosine Rule: $x^2 = y^2 + z^2 - 2yz\cos X$

    Area: $\frac{1}{2}xy\sin Z$

Answers

4) $\tan x = \frac{\sin x}{\cos x}$; $\cos^2 x = 1 - \sin^2 x$

5) (a) B = 125°, a = 3.66 m, c = 3.10 m, area is 4.64 m²

(b) r = 20.05 km, P = 1.49°, Q = 168.51°

6) Freda's angles are 22.3°, 49.5°, 108.2°

7) One triangle: c = 4.98, C = 72.07°, B = 72.93°
    Other possible triangle: c = 3.22, C = 37.93°, B = 107.07°

8)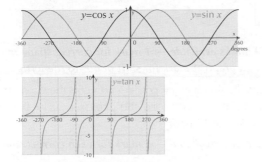

9) (a)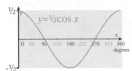

(b)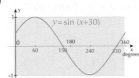

(c)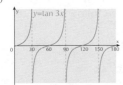

10) (a) (i) θ = 240°, 300°.

(ii) θ = 135°, 315°.

(iii) θ = 135°, 225°.

(b) (i) θ = 33.0°, 57.0°, 123.0°, 147.0°, –33.0°, –57.0°,
         –123.0°, –147.0°

(ii) θ = –17.5°, 127.5°

(iii) θ = 179.8°

11) x = 70.5°, 120°, 240°, 289.5°.

12) x = –30°

13) $(\sin y + \cos y)^2 + (\cos y - \sin y)^2$
    $\equiv (\sin^2 y + 2\sin y\cos y + \cos^2 y) +$
    $\quad (\cos^2 y - 2\cos y\sin y + \sin^2 y)$
    $\equiv 2(\sin^2 y + \cos^2 y) \equiv 2$

14) $\frac{\sin^4 x + \sin^2 x\cos^2 x}{\cos^2 x - 1} \equiv -1$

LHS:
$\frac{\sin^2 x(\sin^2 x + \cos^2 x)}{(1 - \sin^2 x) - 1}$

$\equiv \frac{\sin^2 x}{-\sin^2 x} \equiv -1 \equiv$ RHS

Exam Questions

1 a) Rearrange equation and complete the square:
       $x^2 - 2x + y^2 - 10y + 21 = 0$ *[1 mark]*
       $(x - 1)^2 - 1 + (y - 5)^2 - 25 + 21 = 0$ *[1 mark]*
       $(x - 1)^2 + (y - 5)^2 = 5$ *[1 mark]*
       Compare with $(x - a)^2 + (y - b)^2 = r^2$:
       centre = (1, 5) *[1 mark]*, radius = $\sqrt{5}$ = 2.24 (to 3 s.f.)
       [1 mark].

b) The point (3, 6) and centre (1, 5) both lie on the diameter.
    Gradient of the diameter = $\frac{6 - 5}{3 - 1}$ = 0.5.
    Q (q, 4) also lies on the diameter, so $\frac{4 - 6}{q - 3}$ = 0.5.
    $-2 = 0.5q - 1.5$
    So $q = (-2 + 1.5) \div 0.5 = -1$.
    [3 marks available — 1 mark for finding the gradient of the diameter, 1 mark for linking this with the point Q, and 1 mark for correct calculation of q.]

c) Tangent at Q is perpendicular to the diameter at Q, so
    gradient $m = -\frac{1}{0.5} = -2$
    $y - y_1 = m(x - x_1)$, and (–1, 4) is a point on the line, so:
    $y - 4 = -2(x + 1)$
    $y - 4 = -2x - 2$
    $2x + y - 2 = 0$ is the equation of the tangent.

    [5 marks available — 1 mark for gradient = –1 ÷ gradient of diameter, 1 mark for correct value for gradient, 1 mark for substituting Q in straight-line equation, 2 marks for correct substitution of values in the correct form, or 1 mark if not in the form ax + by + c = 0.]

2 a) 3 cos x = 2 sin x, and $\tan x = \frac{\sin x}{\cos x}$,
       You need to substitute tan in somewhere, so look at how you can rearrange to get sin/cos in the equation...
       Divide through by cos x to give:
       $3\frac{\cos x}{\cos x} = 2\frac{\sin x}{\cos x}$
       $\Rightarrow 3 = 2 \tan x$
       $\Rightarrow \tan x = \frac{3}{2}$ (or = 1.5).
       [2 marks available — 1 mark for correct substitution of tan x, 1 mark for correct final answer.]

b) Using $\tan x = 1.5$
       $x = 56.3°$ *[1 mark]*
    and a 2nd solution can be found from
       $x = 180° + 56.3° = 236.3°$ *[1 mark]*.

    Don't forget the other solution! Either use the CAST diagram or sketch a graph to help.

3 a) Centre of C = (–4, 4), so a = –4 and b = 4 *[1 mark]*.
       Radius r is the length of MJ.
       Using Pythagoras, $r^2 = (0 - -4)^2 + (7 - 4)^2 = 25$
       [1 mark]
       So substituting into the equation gives:
       $(x + 4)^2 + (y - 4)^2 = 25$ *[1 mark]*

b) Angle MJH between the radius and tangent is a right angle, so use trig ratios to calculate angle JMH (or θ).
    E.g. Length MJ = $r = \sqrt{25}$ = 5.

Answers

Using Pythagoras,
Length $JH = \sqrt{(6-0)^2 + (-1-7)^2} = 10$.
$\tan\theta = \frac{10}{5} = 2$
$\theta = \tan^{-1}2 = 1.1071$ rad

[4 marks available — 1 mark for identifying that MJH is a right angle and that trig ratios can be used, 1 mark for calculation of two lengths of the triangle, 1 mark for correct substitution into a trig ratio, and 1 mark for correct answer in radians.]

c) Arc length $S = r\theta$ *[1 mark]*
Angle of sector = $\theta = 1.1071$ rad and $r = 5$ (both from (b))
so $S = 5 \times 1.1071 = 5.54$ to 3 s.f. *[1 mark]*.

4 a) Using the cosine rule with $\triangle AMC$:
$a^2 = b^2 + c^2 - 2bc\cos A$
$2.30^2 = 2.20^2 + 2.20^2 - (2 \times 2.20 \times 2.20 \times \cos\theta)$
$\cos\theta = \frac{5.29 - 4.84 - 4.84}{-9.68} = 0.4535...$
$\theta = \cos^{-1} 0.4535... = 1.10$ rad to 3 s.f.

[2 marks available — 1 mark for correct substitution into cosine rule formula, and 1 mark for correct answer in radians.]

...or, you could divide it into 2 right-angled triangles and then use: $\theta = 2\sin^{-1}\frac{1.15}{2.20} = 1.10$ rad. Whatever works.

b) Arc length $S = r\theta = 2.20 \times 1.10$ *[1 mark]*
= 2.42 m *[1 mark]*.
Perimeter of slab = 2.42 + 1.5 + 1.5 = 5.42 m
[1 mark].

c) Area of slab =
Area $\triangle ABC$ + Area $\triangle AMC$ – Area sector AMC.
Area of the triangles can be found using $\frac{1}{2}ab\sin C$.
Area of sector can be found using $\frac{1}{2}r^2\theta$.
So area of slab = $\left(\frac{1}{2} \times 1.50 \times 1.50 \times \sin 1.75\right) + \left(\frac{1}{2} \times 2.20 \times 2.20 \times \sin 1.10\right) - \left(\frac{1}{2} \times 2.20^2 \times 1.10\right)$
= 1.107 + 2.157 – 2.662 = 0.602 m² to 3 s.f.

[5 marks available — 1 mark for correct calculation of each of the three shapes, 1 mark for combining the three shapes in the correct way, 1 mark for correct final answer.]

5 a) Area of cross-section = $\frac{1}{2}r^2\theta$
= $\frac{1}{2} \times 20^2 \times \frac{\pi}{4} = 50\pi$ cm².
Volume = area of cross-section × height, so
$V = 50\pi \times 10 = 500\pi$ cm³.

[3 marks available — 1 mark for correct use of area formula, 1 mark for 50π, and 1 mark for correct final answer.]

b) Surface area is made up of 2 × cross-sectional area + 2 × side rectangles + 1 curved end rectangle.
Cross-sectional area = 50π (from part (a))
Area of each side rectangle = 10 × 20 = 200
Area of end rectangle = 10 × arc length
= $10 \times (20 \times \frac{\pi}{4})$
= 50π.
$S = (2 \times 50\pi) + (2 \times 200) + 50\pi = (150\pi + 400)$ cm².

[5 marks available — 1 mark for each correct shape area, 1 mark for correct combination, and 1 mark for correct final answer.]

6 a) (i)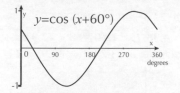

[2 marks available — 1 mark for correct shape of cos x graph, 1 mark for shift of 60° to the left.]

(ii) E.g. the graph of $y = \cos(x + 60°)$ cuts the x-axis at 30° and 210°, so for $0 \le x \le 360°$, $\cos(x + 60°) = 0$ when $x = 30°$ and 210°.

[2 marks available — 1 mark for each correct solution.]

b)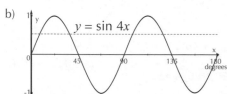

[2 marks available — 1 mark for correct $y = \sin x$ graph, 1 mark for 2 repetitions of the sine wave between 0 and 180°.]

c) E.g. $\sin 4x = 0.5$
$4x = 30°$
$x = 7.5°$ is one solution. *[1 mark]*
The graph in (b) shows there are 4 solutions between 0 and 180°, which, by the symmetry of the graph, lie 7.5° from where the graph cuts the x-axis, as follows:
$x = 45° - 7.5° = 37.5°$. *[1 mark]*
$x = 90° + 7.5° = 97.5°$. *[1 mark]*
$x = 135° - 7.5° = 127.5°$. *[1 mark]*

7 a) $\tan\left(x + \frac{\pi}{6}\right) = \sqrt{3}$
$x + \frac{\pi}{6} = \frac{\pi}{3}$ *[1 mark]*
Tan is positive in the 3rd quadrant of our good friend the CAST diagram, so we can work out the other solution as follows...
2nd solution between 0 and 2π can be found from:
$x + \frac{\pi}{6} = \pi + \frac{\pi}{3} = \frac{4}{3}\pi$ *[1 mark]*. So subtracting $\frac{\pi}{6}$ from each solution gives: $x = \frac{\pi}{6}$ *[1 mark]*, $x = \frac{7\pi}{6}$ *[1 mark]*

b) For this one, it's helpful to change the range — so look for solutions in the range $-\frac{\pi}{4} \le x - \frac{\pi}{4} \le 2\pi - \frac{\pi}{4}$.
$2\cos\left(x - \frac{\pi}{4}\right) = \sqrt{3}$
so $\cos\left(x - \frac{\pi}{4}\right) = \frac{\sqrt{3}}{2}$.
Solving this gives $x - \frac{\pi}{4} = \frac{\pi}{6}$, which is in the range — so it's a solution *[1 mark]*. From the symmetry of the cos graph there's another solution at $2\pi - \frac{\pi}{6} = \frac{11\pi}{6}$. But this is outside the range for $x - \frac{\pi}{4}$, so you can ignore it *[1 mark]*.
Using symmetry again, there's also a solution at $-\frac{\pi}{6}$ — and this one is in your range. So solutions for $x - \frac{\pi}{4}$ are $-\frac{\pi}{6}$ and $\frac{\pi}{6}$ *[1 mark]* $\Rightarrow x = \frac{\pi}{12}$ and $\frac{5\pi}{12}$. *[1 mark]*

You might find it useful to sketch the graph for this one — or you could use the CAST diagram if you prefer.

c) $\sin 2x = -\frac{1}{2}$, so look for solutions in the range $0 \leq 2x \leq 4\pi$ *[1 mark].*

It's easier to see what's going on by drawing a graph for this one:

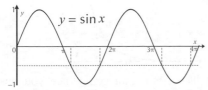

The graph shows there are 4 solutions between 0 and 4π. Putting $\sin 2x = -\frac{1}{2}$ into your calculator gives you the solution $2x = -\frac{\pi}{6}$, but this is outside the range *[1 mark]*. From the graph, you can see that the solutions within the range occur at $\pi + \frac{\pi}{6}$, $2\pi - \frac{\pi}{6}$, $3\pi + \frac{\pi}{6}$ and $4\pi - \frac{\pi}{6}$ *[1 mark]*, so $2x = \frac{7\pi}{6}$, $\frac{11\pi}{6}$, $\frac{19\pi}{6}$ and $\frac{23\pi}{6}$ *[1 mark]*.
Dividing by 2 gives:
$x = \frac{7\pi}{12}$, $\frac{11\pi}{12}$, $\frac{19\pi}{12}$ and $\frac{23\pi}{12}$ *[2 marks for all 4 correct, 1 mark for 2 correct]*

8 a) $2(1 - \cos x) = 3 \sin^2 x$, and $\sin^2 x = 1 - \cos^2 x$.
$\Rightarrow 2(1 - \cos x) = 3(1 - \cos^2 x)$ *[1 mark]*.

You need to get the whole equation into either sin or cos to get something useful at the end, so get used to spotting places to use the trig identities.

$\Rightarrow 2 - 2\cos x = 3 - 3\cos^2 x$
$\Rightarrow 3\cos^2 x - 2\cos x - 1 = 0$ *[1 mark]*.

b) From (a), the equation can be written as:
$3\cos^2 x - 2\cos x - 1 = 0$

Now this looks suspiciously like a quadratic equation, which can be factorised...

$(3\cos x + 1)(\cos x - 1) = 0$
$\Rightarrow \cos x = -\frac{1}{3}$ *[1 mark]* or $\cos x = 1$ *[1 mark]*
For $\cos x = -\frac{1}{3}$
$x = 109.5°$ (to 1 d.p.) *[1 mark]*,
and a 2nd solution can be found from
$x = (360° - 109.5°) = 250.5°$ *[1 mark]*.
For $\cos x = 1$
$x = 0°$ *[1 mark]* and $360°$ *[1 mark]*.

9 a) Using the cosine rule:
$a^2 = b^2 + c^2 - 2bc\cos A$
If XY is a, then angle $A = 180° - 100° = 80°$.
$XY^2 = 150^2 + 250^2 - (2 \times 150 \times 250 \times \cos 80°)$
$XY^2 = 71976.3867$
$XY = \sqrt{71976.3867} = 268.28$ m (to 2 d.p.)
$= 268$ m to the nearest m.

[2 marks available — 1 mark for correct substitution into cosine rule formula, and 1 mark for correct answer.]

b) Using the sine rule:
$\frac{a}{\sin A} = \frac{b}{\sin B}$, so $\frac{250}{\sin \theta} = \frac{268.2842 \text{ (from (a))}}{\sin 80°}$.

Rearranging gives:
$\frac{\sin \theta}{\sin 80°} = \frac{250}{268.2842} = 0.93$ to 2 d.p.
[3 marks available — 1 mark for correct substitution into sine rule formula, 1 mark for rearrangement into the correct form, and 1 mark for correct final answer.]

10   $2 - \sin x = 2 \cos^2 x$, and $\cos^2 x = 1 - \sin^2 x$
$\Rightarrow 2 - \sin x = 2(1 - \sin^2 x)$
$\Rightarrow 2 - \sin x = 2 - 2\sin^2 x$
$\Rightarrow 2\sin^2 x - \sin x = 0$

Now simply factorise and all will become clear...

$\sin x(2 \sin x - 1) = 0 \Rightarrow \sin x = 0$ or $\sin x = \frac{1}{2}$.
For $\sin x = 0$, $x = 0$, π and 2π.
For $\sin x = \frac{1}{2}$, $x = \frac{\pi}{6}$ and $\pi - \frac{\pi}{6} = \frac{5\pi}{6}$.
[6 marks available — 1 mark for correct substitution using trig identity, 1 mark for factorising quadratic in sin x, 1 mark for finding correct values of sin x, 1 mark for all three solutions when sin x = 0, 1 mark for each of the other 2 correct solutions.]

11 a) $(1 + 2\cos x)(3\tan^2 x - 1) = 0$
$\Rightarrow 1 + 2\cos x = 0 \Rightarrow \cos x = -\frac{1}{2}$.
OR:
$3\tan^2 x - 1 = 0 \Rightarrow \tan^2 x = \frac{1}{3} \Rightarrow \tan x = \frac{1}{\sqrt{3}}$ or $-\frac{1}{\sqrt{3}}$.
For $\cos x = -\frac{1}{2}$
$x = \frac{2\pi}{3}$ and $-\frac{2\pi}{3}$.
Drawing the cos x graph helps you find the second one here, and don't forget the limits are $-\pi \leq x \leq \pi$.

For $\tan x = \frac{1}{\sqrt{3}}$
$x = \frac{\pi}{6}$ and $-\pi + \frac{\pi}{6} = -\frac{5\pi}{6}$.
For $\tan x = -\frac{1}{\sqrt{3}}$
$x = -\frac{\pi}{6}$ and $-\frac{\pi}{6} + \pi = \frac{5\pi}{6}$.
Again, look at the graph of tan x if you're unsure.
[6 marks available — 1 mark for each correct solution.]

b) $3\cos^2 x = \sin^2 x \Rightarrow 3 = \frac{\sin^2 x}{\cos^2 x}$
$\Rightarrow 3 = \tan^2 x \Rightarrow \tan x = \pm\sqrt{3}$
For $\tan x = \sqrt{3}$, $x = \frac{\pi}{3}$ and $\frac{\pi}{3} - \pi = -\frac{2\pi}{3}$.
For $\tan x = -\sqrt{3}$, $x = -\frac{\pi}{3}$ and $-\frac{\pi}{3} + \pi = \frac{2\pi}{3}$.
[4 marks available — 1 mark for each correct solution.]

12 a) The line through P is a diameter, and as such is perpendicular to the chord AB at the midpoint M.

Gradient of AB = Gradient of $AM = \frac{(7 - 10)}{(11 - 9)} = -\frac{3}{2}$.
$\Rightarrow$ Gradient of PM $= \frac{2}{3}$.

Gradient of PM $= \frac{(7 - 3)}{(11 - p)} = \frac{2}{3}$

$\Rightarrow 3(7 - 3) = 2(11 - p)$
$\Rightarrow 12 = 22 - 2p \Rightarrow p = 5$.

Answers

[5 marks available — 1 mark for identifying that PM and AM are perpendicular, 1 mark for correct gradient of AM, 1 mark for correct gradient of PM, 1 mark for substitution of the y-value of P into the gradient or equation of the line, and 1 mark for correct final answer.]

b) The centre of C is $P(5, 3)$, so $a = 5$ *[1 mark]* and $b = 3$ *[1 mark]* in the equation $(x - a)^2 + (y - b)^2 = r^2$. The radius is the length of AP, which can be found using Pythagoras as follows:
$r^2 = AP^2 = (9 - 5)^2 + (10 - 3)^2 = 65$ *[1 mark]*
So the equation of C is:
$(x - 5)^2 + (y - 3)^2 = 65$ *[1 mark]*.

C2 Section 3 — Logs and Exponentials
Warm-up Questions

1) a) $3^3 = 27$ so $\log_3 27 = 3$

   b) To get fractions you need negative powers
   $3^{-3} = 1/27$
   $\log_3 (1/27) = -3$

   c) Logs are subtracted so divide
   $\log_3 18 - \log_3 2 = \log_3 (18 \div 2)$
   $= \log_3 9$
   $= 2 \ (3^2 = 9)$

2) a) Logs are added so you multiply —
   remember $2 \log 5 = \log 5^2$.
   $\log 3 + 2 \log 5 = \log (3 \times 5^2)$
   $= \log 75$

   b) Logs are subtracted so you divide and the power half means square root
   $\frac{1}{2} \log 36 - \log 3 = \log (36^{\frac{1}{2}} \div 3)$
   $= \log (6 \div 3)$
   $= \log 2$

   c) Logs are subtracted so you divide and the power quarter means fourth root
   $\log 2 - \frac{1}{4} \log 16 = \log (2 \div 16^{\frac{1}{4}})$
   $= \log (2 \div 2)$
   $= \log 1 = 0$

3) This only looks tricky because of the algebra, just remember the laws: $\log_b (x^2 - 1) - \log_b (x - 1) = \log_b \{(x^2 - 1)/(x - 1)\}$
   Then use the difference of two squares:
   $(x^2 - 1) = (x - 1)(x + 1)$ and cancel to get
   $\log_b (x^2 - 1) - \log_b (x - 1) = \log_b (x + 1)$

4) a) Filling in the answers is just a case of using the calculator

x	−3	−2	−1	0	1	2	3
y	0.0156	0.0625	0.25	1	4	16	64

   Check that it agrees with what we know about the graphs. It goes through the common point (0, 1), and it follows the standard shape.

b) Then you just need to draw the graph, and use a scale that's just right.

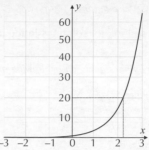

c) The question tells you to use the graph to get your answer, so you'll need to include the construction lines, but check the answer with a calculator.
   $x = \log 20 / \log 4 = 2.16$, but you can't justify this accuracy if your graph's not up to it, so 2.2 is a good estimate.

5) a) $x = \log_{10} 240 / \log_{10} 10 = \log_{10} 240 = 2.380$

   b) $x = 10^{5.3} = 199526.2... = 200000$ (to 3 s.f.)

   c) $2x + 1 = \log_{10} 1500 = 3.176$, so $2x = 2.176$, so $x = 1.088$

   d) $(x - 1) \log 4 = \log 200$, so $x - 1 = \log 200 / \log 4 = 3.822$, so $x = 4.822$

6) First solve for $1.5^P > 1\ 000\ 000$
   $P \times \log_{10} 1.5 > \log_{10} 1\ 000\ 000$,
   so $P > (\log_{10} 1\ 000\ 000) / (\log_{10} 1.5)$, $P > 34.07$.
   We need the next biggest integer, so this will be $P = 35$.

Exam Questions

1 a) (i) $\log_a 20 - 2 \log_a 2$
   $= \log_a 20 - \log_a (2^2)$ *[1 mark]*
   $= \log_a (20 \div 2^2)$ *[1 mark]*
   $= \log_a 5$ *[1 mark]*

   (ii) $\frac{1}{2} \log_a 16 + \frac{1}{3} \log_a 27$
   $= \log_a (16^{\frac{1}{2}}) + \log_a (27^{\frac{1}{3}})$ *[1 mark]*
   $= \log_a (16^{\frac{1}{2}} \times 27^{\frac{1}{3}})$ *[1 mark]*
   $= \log_a (4 \times 3) = \log_a 12$ *[1 mark]*

b) (i) $\log_2 64 = 6$ *[1 mark]* (since $2^6 = 64$)

   (ii) $2 \log_3 9 = \log_3 (9^2) = \log_3 81$ *[1 mark]*
   $\log_3 81 = 4$ *[1 mark]* (since $3^4 = 81$)

c) (i) $\log_6 25 = \frac{\log 25}{\log 6} = 1.7965$ to 4 d.p. *[1 mark]*

   (ii) $\log_3 10 + \log_3 2 = \log_3 (10 \times 2) = \log_3 20$ *[1 mark]*
   $\log_3 20 = \frac{\log 20}{\log 3} = 2.7268$ to 4 d.p. *[1 mark]*

2 a) $2^x = 9$, so taking logs of both sides gives
   $\log 2^x = \log 9$

   This is usually the first step in getting x on its own — then you can use your trusty log laws...

   $\Rightarrow x \log 2 = \log 9$
   $\Rightarrow x = \frac{\log 9}{\log 2} = 3.17$ to 2 d.p.
   [3 marks available — 1 mark for taking logs of both sides, 1 mark for x log 2 = log 9, and 1 mark for correct final answer.]

b) $2^{2x} = (2^x)^2$ (from the power laws) *[1 mark]*,
so let $y = 2^x$ and $y^2 = 2^{2x}$. This gives a quadratic in y:
$y^2 - 13y + 36 = 0$

Now the big question is — will it factorise? You betcha...

$(y - 9)(y - 4) = 0$, so $y = 9$ or $y = 4$, that is,
$\Rightarrow 2^x = 9$ *[1 mark]* or $2^x = 4$ *[1 mark]*

From (a), for $2^x = 9$, $x = 3.17$ to 2 d.p. *[1 mark]*
and for $2^x = 4$, $x = 2$ (since $2^2 = 4$) *[1 mark]*.

3    $\log_7 (y + 3) + \log_7 (2y + 1) = 1$
$\Rightarrow \log_7 ((y + 3)(2y + 1)) = 1$

To remove the $\log_7$, do 7 to the power of each side:
$(y + 3)(2y + 1) = 7^1 = 7$

Multiply out, rearrange, and re-factorise:
$2y^2 + 7y + 3 = 7$
$\Rightarrow 2y^2 + 7y - 4 = 0$
$\Rightarrow (2y - 1)(y + 4) = 0$
$\Rightarrow y = \frac{1}{2}$ or $y = -4$,
but since $y > 0$, $y = \frac{1}{2}$ is the only solution.

*[5 marks available — 1 mark for combining the
two logs, 1 mark for 7 to the power of each
side, 1 mark for the correct factorisation of the
quadratic, 1 mark for correct solutions and 1 mark
for stating that only y = ½ is a valid solution.]*

4 a) $\log_3 x = -\frac{1}{2}$, so do 3 to the power of each side to
remove the log:
$x = 3^{-\frac{1}{2}}$ *[1 mark]*
$\Rightarrow x = \frac{1}{3^{\frac{1}{2}}}$ *[1 mark]* $\Rightarrow x = \frac{1}{\sqrt{3}}$ *[1 mark]*.

b) $2 \log_3 x = -4$
$\Rightarrow \log_3 x = -2$, and 3 to the power of each side gives:
$x = 3^{-2}$ *[1 mark]*
$\Rightarrow x = \frac{1}{9}$ *[1 mark]*

5 a) $6^{(3x + 2)} = 9$, so taking logs of both sides gives:
$(3x + 2) \log 6 = \log 9$ *[1 mark]*
$\Rightarrow 3x + 2 = \frac{\log 9}{\log 6} = 1.2262...$ *[1 mark]*
$\Rightarrow x = (1.2262... - 2) \div 3 = -0.258$ to 3 s.f. *[1 mark]*

b) $3^{(y^2 - 4)} = 7^{(y + 2)}$, so taking logs of both sides gives:
$(y^2 - 4) \log 3 = (y + 2) \log 7$ *[1 mark]*
$\Rightarrow \frac{(y^2 - 4)}{(y + 2)} = \frac{\log 7}{\log 3} = 1.7712...$ *[1 mark]*

*The top of the fraction is a 'difference of two squares' so it will
simplify as follows...*

$\frac{(y - 2)(y + 2)}{(y + 2)} = 1.7712...$ *[1 mark]*
$\Rightarrow y - 2 = 1.7712...$ *[1 mark]* $\Rightarrow y = 3.77$ to 3 s.f. *[1 mark]*

6 a) $\log_4 p - \log_4 q = \frac{1}{2}$, so using the log laws:
$\log_4 \left(\frac{p}{q}\right) = \frac{1}{2}$
Doing 4 to the power of both sides gives:
$\frac{p}{q} = 4^{\frac{1}{2}} = \sqrt{4} = 2$

$\Rightarrow p = 2q$

*[3 marks available — 1 mark for combining the
two logs, 1 mark for 4 to the power of each
side, 1 mark for the correct final working.]*

b) Since $p = 2q$ (from (a)), the equation can be written:
$\log_2 (2q) + \log_2 q = 7$ *[1 mark]*
This simplifies to:
$\log_2 (2q^2) = 7$ *[1 mark]*
Doing 2 to the power of both sides gives:
$2q^2 = 2^7 = 128$ *[1 mark]*
$\Rightarrow q^2 = 64$, $\Rightarrow q = 8$ (since p and q are positive) *[1 mark]*
$p = 2q \Rightarrow p = 16$ *[1 mark]*

C2 Section 4 — Sequences and Series
Warm-up Questions

Oh goody, let's get started...

1 a) $a = 2, r = -3$

*You find r by putting the information you're given into the formula
for u_2: $u_2 = u_1 \times r$, so $-6 = 2 \times r \Rightarrow r = -3$.*

10th term, $u_{10} = ar^9$
$= 2 \times (-3)^9 \quad = -39366$

b) $S_{10} = \frac{2(1 - (-3)^{10})}{1 - (-3)} = \frac{1 - (-3)^{10}}{2} = -29524$

2 a) $a = 2, r = 4$, so $S_{12} = \frac{2(4^{12} - 1)}{4 - 1} = 11,184,810$

b) $a = 30, r = ½$, so $S_{12} = \frac{30\left(1 - \left[\frac{1}{2}\right]^{12}\right)}{1 - \frac{1}{2}} = 59.985$ (to 3 d.p.)

3 a) $r = 2$, so series is divergent.
b) $r = 1/3$, so series is convergent.
c) $r = 1/3$, so series is convergent.
d) $r = 1/4$, so series is convergent.

4 a) $r = $ 2nd term $\div$ 1st term
$r = 12 \div 24 = ½$

b) 7th term $= ar^6$
$= 24 \times (½)^6$
$= 0.375$ (or $\frac{3}{8}$)

c) $S_{10} = \frac{24\left(1 - \left[\frac{1}{2}\right]^{10}\right)}{1 - \frac{1}{2}} = 47.953$ (to 3 d.p.)

d) $S_\infty = \frac{a}{1 - r} = \frac{24}{1 - \frac{1}{2}} = 48$

5    $a = 2, r = 3$
You need $ar^{n-1} = 1458$, i.e. $2 \times 3^{n-1} = 1458$, i.e. $3^{n-1} = 729$.
Then, use logs to find that:
$\log 3^{n-1} = \log 729$
$(n - 1) \log 3 = \log 729$
$n - 1 = \frac{\log 729}{\log 3}$
$n - 1 = 6$, i.e. $n = 7$, the 7th term $= 1458$.

Answers

6   1   5   10   10   5   1

7     $1 + 12x + 66x^2 + 220x^3$

8     $29120x^4$

9     $(2 + 3x)^5 = 2^5\left(1 + \frac{3}{2}x\right)^5$

$$= 2^5\left[1 + \frac{5}{1}\left(\frac{3}{2}x\right) + \frac{5 \times 4}{1 \times 2}\left(\frac{3}{2}x\right)^2 + \ldots\right]$$

x^2 term is $2^5 \times \frac{5 \times 4}{1 \times 2}\left(\frac{3}{2}x\right)^2$

so coefficient is $2^5 \times \frac{5 \times 4}{1 \times 2} \times \frac{3^2}{2^2} = 720$

Exam Questions

1     $(4 + 3x)^{10}$

$$= 4^{10}\left[1 + \frac{10}{1}\left(\frac{3}{4}x\right) + \frac{10 \times 9}{1 \times 2}\left(\frac{3}{4}x\right)^2 + \frac{10 \times 9 \times 8}{1 \times 2 \times 3}\left(\frac{3}{4}x\right)^3\right.$$
$$\left. + \frac{10 \times 9 \times 8 \times 7}{1 \times 2 \times 3 \times 4}\left(\frac{3}{4}x\right)^4 + \ldots\right]$$

So the x coefficient $= 4^{10} \times \frac{10}{1} \times \frac{3}{4} = 7864320$ *[1 mark]*

x^2 coefficient $= 4^{10} \times \frac{90}{2} \times \frac{9}{16} = 26542080$ *[1 mark]*

x^3 coefficient $= 4^{10} \times \frac{720}{6} \times \frac{27}{64} = 53084160$ *[1 mark]*

x^4 coefficient $= 4^{10} \times \frac{5040}{24} \times \frac{81}{256} = 69672960$ *[1 mark]*

No problems there... as long as you've got binomial expansion straight in your head. If it's still a tangle of factors, powers and garden gnomes, go back and sort it out — you'll be glad you did.

2   a)   The ratio = 1.3, which is > 1, so the sequence is divergent. *[1 mark]*

   b) $u_3 = 12 \times 1.3^2 = 20.28$ *[1 mark]*

       $u_{10} = 12 \times 1.3^9 = 127.25$ *[1 mark]*

3   a) $S_\infty = \frac{a}{1 - r} = \frac{20}{1 - \frac{3}{4}} = \frac{20}{\frac{1}{4}} = 80$

      [2 marks available — 1 mark for formula, 1 mark for correct answer]

   b) $u_{15} = ar^{14} = 20 \times \left(\frac{3}{4}\right)^{14} = 0.356$ (to 3 sig. fig.)

      [2 marks available — 1 mark for formula,
      1 mark for correct answer]

   c)   Use the formula for the sum of a geometric series to write an expression for S_n:

$$S_n = \frac{a(1 - r^n)}{1 - r} = \frac{20\left(1 - \frac{3^n}{4}\right)}{1 - \frac{3}{4}} \quad \text{[1 mark]}$$

so $\frac{20\left(1 - \frac{3^n}{4}\right)}{1 - \frac{3}{4}} > 79.76$

Now rearrange and use logs to get n on its own:

$$\frac{20\left(1 - \frac{3^n}{4}\right)}{1 - \frac{3}{4}} > 79.76 \Rightarrow 20\left(1 - \frac{3^n}{4}\right) > 19.94$$

$\Rightarrow 1 - \frac{3^n}{4} > 0.997 \Rightarrow 0.003 > 0.75^n$ *[1 mark]*

$\Rightarrow \log 0.003 > n\log 0.75$ *[1 mark]*

$\Rightarrow \frac{\log 0.003}{\log 0.75} < n$ *[1 mark]*

Bit tricky that last bit. If x < 1, then log x has a negative value — and when dividing by a negative value on either side of an inequality sign you need to change the direction of the inequality.

$\frac{\log 0.003}{\log 0.75} = 20.1929....$

so $n > 20.1929....$

But n must be an integer, as it is a term not a value, therefore $n = 21$ *[1 mark]*

Don't get too calculator-happy and forget that n will be an integer — you'd have lost that mark if you'd just put 20.1929...

4   a) $u_n = ar^{n-1}$ where $a = 1$ and $r = 1.5$,

      so $u_5 = 1 \times (1.5)^4$ *[1 mark]*

      $= 5.06$ km to the nearest 10m   *[1 mark]*

      Make sure you get the ratios right here. If the values increase by 0.5 each time then the ratio is 1.5 — a ratio of 0.5 would decrease the value.

   b) $a = 2$ and $r = 1.2$ *[1 mark]*

      $u_9 = 2 \times (1.2)^8 = 8.60$ km

      $u_{10} = 2 \times (1.2)^9 = 10.32$ km   *[1 mark]*

      $u_9 < 10$ km and $u_{10} > 10$ km

      so day 10 is the first day Chris runs more than 10km. *[1 mark]*

   c)   In 10 days, Alex ran a total of 30 km   *[1 mark]*

      Use the formula for the sum of first n terms: $S_n = \frac{a(1 - r^n)}{1 - r}$.

      Chris ran a total of: $\frac{2(1 - 1.2^{10})}{1 - 1.2} = 51.917$ km   *[1 mark]*

      Heather ran a total of: $\frac{1(1 - 1.5^{10})}{1 - 1.5} = 113.330$ km
      [1 mark]

      So they raised £(30 + 51.917 + 113.330)

               = £195.25   *[1 mark]*

5   a) $S_\infty = \frac{a}{1 - r}$ and $u_2 = ar$ *[1 mark]*

      So $36 = \frac{a}{1 - r}$ i.e. $36 - 36r = a$ *[1 mark]*

      and $5 = ar$. *[1 mark]*

      Substituting for a gives: $5 = (36 - 36r)r = 36r - 36r^2$

      i.e. $36r^2 - 36r + 5 = 0$ *[1 mark]*

   b)   Factorising gives: $(6r - 1)(6r - 5) = 0$

      So $r = \frac{1}{6}$ or $r = \frac{5}{6}$. *[1 mark for each correct value]*

      If $r = \frac{1}{6}$ and $ar = 5$ then $\frac{a}{6} = 5$ i.e. $a = 30$

      If $r = \frac{5}{6}$ and $ar = 5$ then $\frac{5a}{6} = 5$ i.e. $a = 6$

      [1 mark for each correct value]

C2 Section 5 — Differentiation
Warm-up Questions

1)   a)   A point (on a graph) where $\frac{dy}{dx}$ (the gradient) = 0.

   b) $y = x^3 - 6x^2 - 63x + 21 \Rightarrow \frac{dy}{dx} = 3x^2 - 12x - 63$, so set this equal to zero (and divide by 3) to get that the stationary points are where $x^2 - 4x - 21 = 0$, i.e. $(x - 7)(x + 3) = 0$, and so the stationary points are $(7, -371)$ and $(-3, 129)$.

Answers

2) Differentiate again to find $\frac{d^2y}{dx^2}$.
If this is positive, stationary point is a minimum;
if it's negative, stationary point is a maximum.

3) $\frac{dy}{dx} = 3x^2 - \frac{3}{x^2}$; this is zero at (1, 4) and (–1, –4).

$\frac{d^2y}{dx^2} = 6x + \frac{6}{x^3}$; at $x = 1$ this is positive
so (1, 4) is a minimum; at $x = -1$ this is negative,
so (–1, –4) is a maximum.

4) a) $\frac{dy}{dx} = 12x - 6$, so function is increasing when $x > 0.5$, and
decreasing when $x < 0.5$.

b) $\frac{dy}{dx} = -\frac{2}{x^3}$, so function is increasing when $x < 0$, and
decreasing when $x > 0$.

5) Differentiate to find the stationary points:
$\frac{dy}{dx} = 3x^2 - 4 = 0$, so $3x^2 = 4$ and $x = \pm 1.15$
When $x = \pm 1.15$, $y = \mp 3.08$, so the graph has 2 stationary

point with coordinates (1.15, –3.08) and (–1.15, 3.08).
Differentiate again to decide if
they're minimum or maximum: $\frac{d^2y}{dx^2} = 6x$.

When $x = 1.15$, $\frac{d^2y}{dx^2} = 6.9$, so it's a minimum.

When $x = -1.15$, $\frac{d^2y}{dx^2} = -6.9$, so it's a maximum.
It crosses the y-axis when $x = 0$:
$y = 0^3 - (4 \times 0) = 0$

It crosses the x-axis when $y = 0$:
$x^3 - 4x = 0 \Rightarrow x(x^2 - 4) = 0$
$\Rightarrow x(x + 2)(x - 2) = 0$

So, $y = 0$ when $x = -2$, 0 and 2.
The graph looks like this ⟹

(-1.15, 3.08)

$y = x^3 - 4x$

(1.15, -3.08)

6) Maximum is when $\frac{dh}{dm} = 0$.

So, $\frac{m}{5} - \frac{3m^2}{800} = m\left(\frac{1}{5} - \frac{3m}{800}\right) = 0$
Since $m \neq 0$, $\frac{1}{5} - \frac{3m}{800} = 0 \Rightarrow \frac{3m}{800} = \frac{1}{5}$

$\Rightarrow m = \frac{800}{15} = 53.3\,\text{g}$ to 3 s.f.

So, $h_{max} = \frac{53.3^2}{10} - \frac{53.3^3}{800} = 94.8\,\text{m}$ to 3 s.f.

Exam Questions

1 a) $y = 6 + \frac{4x^3 - 15x^2 + 12x}{6} = 6 + \frac{2}{3}x^3 - \frac{5}{2}x^2 + 2x$

$\frac{dy}{dx} = 2x^2 - 5x + 2$

[1 mark for each correct term]

b) Stationary points occur when $2x^2 - 5x + 2 = 0$. *[1 mark]*
Factorising the equation gives: $(2x - 1)(x - 2) = 0$

So stationary points occur when $x = 2$ *[1 mark]* and

$x = \frac{1}{2}$ *[1 mark]*.
When $x = 2$:

$y = 6 + \frac{4(2^3) - 15(2^2) + (12 \times 2)}{6} = 5\frac{1}{3}$ *[1 mark]*

When $x = \frac{1}{2}$:

$y = 6 + \frac{4\left(\frac{1}{2}\right)^3 - 15\left(\frac{1}{2}\right)^2 + 12\left(\frac{1}{2}\right)}{6} = 6\frac{11}{24}$ *[1 mark]*
So coordinates of the stationary points on the
curve are $\left(2, 5\frac{1}{3}\right)$ and $\left(\frac{1}{2}, 6\frac{11}{24}\right)$.

c) Differentiate again to find $\frac{d^2y}{dx^2} = 4x - 5$. *[1 mark]*
When $x = 2$ this gives $\Rightarrow 4(2) - 5 = 3$, which is positive,
therefore the curve is minimum at $\left(2, 5\frac{1}{3}\right)$. *[1 mark]*
When $x = \frac{1}{2}$ this gives $\Rightarrow 4\left(\frac{1}{2}\right) - 5 = -3$, which is negative,
so the maximum is at $\left(\frac{1}{2}, 6\frac{11}{24}\right)$. *[1 mark]*

2 a) Find the value of x that gives the minimum
value of y — the stationary point of curve y.

Differentiate, and then solve $\frac{dy}{dx} = 0$:
$\frac{dy}{dx} = \frac{1}{\sqrt{x}} - \frac{27}{x^2}$ *[1 mark for each term]*

$\frac{1}{\sqrt{x}} - \frac{27}{x^2} = 0$ *[1 mark]* $\Rightarrow \frac{1}{\sqrt{x}} = \frac{27}{x^2}$
$x^2 \div x^{1/2} = 27 \Rightarrow x^{3/2} = 27$ *[1 mark]*
$x = \sqrt[3]{27^2} = 9$. *[1 mark]*

So, 9 miles per hour gives the minimum coal consumption.
$x = 0$ also satisfies dy/dx, but this is outside the range of x.

b) $\frac{d^2y}{dx^2} = \frac{54}{x^3} - \frac{1}{2\sqrt{x^3}}$ *[1 mark]*
At the stationary point $x = 9$,

so $\frac{54}{9^3} - \frac{1}{2\sqrt{9^3}} = 0.05555...$ which is positive,

therefore the stationary point is a minimum. *[1 mark]*

c) $y = 2\sqrt{9} + \frac{27}{9} = 9$ *[1 mark]*

3 a) First step, multiply out function to get $y = 3x^3 - 8x^2 + 3x + 2$
$\frac{dy}{dx} = 9x^2 - 16x + 3 = 0$ at the stationary point. *[1 mark]*
Solve using the quadratic formula:
$x = \frac{16 \pm \sqrt{(-16)^2 - (4 \times 9 \times 3)}}{2 \times 9} = \frac{16 \pm 2\sqrt{37}}{18}$
$x = 1.56$ and 0.213 *[1 mark]*. Substituting these values for
x into the original equation for y gives: $y = -1.40$ and 2.31
So the stationary points have coordinates:
(1.56, –1.40) *[1 mark]* and (0.213, 2.31) *[1 mark]*

b) $\frac{d^2y}{dx^2} = 18x - 16$ *[1 mark]*
At $x = 1.56$, $\frac{d^2y}{dx^2} = 12.1$ is > 0,

so it's a minimum *[1 mark]*
At $x = 0.213$, $\frac{d^2y}{dx^2} = -12.2$ is < 0,

so it's a maximum *[1 mark]*

c) y is a positive cubic function, with stationary points as
found in parts a) and b). The curve crosses the y-axis
when $x = 0$, so $y = 2$ *[1 mark]*. The initial cubic equation
can be factorised to find where it intersects the x-axis:
$y = (x - 1)(3x^2 - 5x - 2)$
$y = (x - 1)(3x + 1)(x - 2)$

so $y = 0$ when $x = 1, -\frac{1}{3}$ and 2 *[1 mark]*.

Answers

The sketch looks like this:

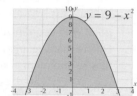

[1 mark]

$y = 3x^3 - 8x^2 + 3x + 2$

4 a) Surface area $= [2 \times (d \times x)] + [2 \times (d \times \frac{x}{2})] + [x \times \frac{x}{2}]$

$= 2dx + \frac{2dx}{2} + \frac{x^2}{2}$ *[1 mark]*

surface area $= 72$ so $3dx + \frac{x^2}{2} = 72$

$\Rightarrow x^2 + 6dx = 144$ *[1 mark]*

$d = \frac{144 - x^2}{6x}$ *[1 mark]*

Volume = width × height × depth $= \frac{x}{2} \times x \times d$

$= \frac{x^2}{2} \times \frac{144 - x^2}{6x} = \frac{144x^2 - x^4}{12x} = 12x - \frac{x^3}{12}$ *[1 mark]*

b) Differentiate V and then solve for when $\frac{dV}{dx} = 0$:

$\frac{dV}{dx} = 12 - \frac{x^2}{4}$ *[1 mark for each correct term]*

$12 - \frac{x^2}{4} = 0$ *[1 mark]* $\frac{x^2}{4} = 12$

$x^2 = 48$

$x = \sqrt{48} = 4\sqrt{3}$ m *[1 mark]*

c) $\frac{d^2V}{dx^2} = -\frac{x}{2}$ *[1 mark]*

so when $x = 4\sqrt{3}$, $\frac{d^2V}{dx^2} = -2\sqrt{3}$ *[1 mark]*

$\frac{d^2V}{dx^2}$ is negative, so it's a maximum point. *[1 mark]*

$x = 4\sqrt{3}$ at V_{max}, so $V_{max} = (12 \times 4\sqrt{3}) - \frac{(4\sqrt{3})^3}{12}$

$V_{max} = 55.4$ m³ *[1 mark]*

5 a) $f'(x) = 2x^3 - 3 = 0$ at the stationary point. *[1 mark]*

$2x^3 = 3 \Rightarrow x = 1.145$ *[1 mark]*, which gives:

$y = f(1.145) = \frac{1}{2}(1.145)^4 - 3(1.145) = -2.576$ *[1 mark]*

So coordinates of the stationary point are (1.145, –2.576).

b) $f''(x) = 6x^2$ *[1 mark]* so at the stationary point:

$f''(1.145) = 7.87$, which is positive, so it is a minimum. *[1 mark]*

c) (i) As the stationary point is a minimum, $f'(x) > 0$ to the right of the stationary point. So the function is increasing when $x > 1.145$ *[1 mark]*

(ii) As the stationary point is a minimum, $f'(x) < 0$ to the left of the stationary point. So the function is decreasing when $x < 1.145$ *[1 mark]*

d) Intersects the x-axis when:

$y = \frac{1}{2}x^4 - 3x = 0$

$x(\frac{1}{2}x^3 - 3) = 0$

$\Rightarrow x = 0$

or $x = \sqrt[3]{6} = 1.82$ *[1 mark]*

So the graph looks like this ✍

[1 mark]

f(x) = ½x⁴ – 3x

(1.145, -2.576)

C2 Section 6 — Integration
Warm-up Questions

1) Check whether there are limits to integrate between. If there are, then it's a definite integral; if not, it's an indefinite integral.

2) The area between a curve and the x-axis, starting at the lower limit and going up to the upper limit.

3) a) $\int_0^1 (4x^3 + 3x^2 + 2x + 1)dx$

$= [x^4 + x^3 + x^2 + x]_0^1$

$= 4 - 0 = 4$

b) $\int_1^2 (\frac{8}{x^5} + \frac{3}{\sqrt{x}})dx = [-\frac{2}{x^4} + 6\sqrt{x}]_1^2$

$= (-\frac{2}{16} + 6\sqrt{2}) - (-2 + 6) = -\frac{33}{8} + 6\sqrt{2}$

c) $\int_1^6 \frac{3}{x^2} dx = [\frac{-3}{x}]_1^6 = -\frac{1}{2} - (-3) = \frac{5}{2}$

4) a) $\int_{-3}^3 (9 - x^2)dx = [9x - \frac{x^3}{3}]_{-3}^3$

$= 18 - (-18) = 36$

$y = 9 - x^2$

b) $\int_1^\infty \frac{3}{x^2}dx = [-\frac{3}{x}]_1^\infty$

$= 0 - (-3) = 3$

$y = \frac{3}{x^2}$

Integral extends this way forever.

5) $\int_1^8 y\, dx = \int_1^8 x^{-\frac{1}{3}}dx = [\frac{3}{2}x^{\frac{2}{3}}]_1^8$

$= (\frac{3}{2} \times 8^{\frac{2}{3}}) - (\frac{3}{2} \times 1^{\frac{2}{3}}) = (\frac{3}{2} \times 4) - (\frac{3}{2} \times 1) = \frac{9}{2}$

6) a) $h = \frac{(3 - 0)}{3} = 1$

$x_0 = 0: y_0 = \sqrt{9} = 3$

$x_1 = 1: y_1 = \sqrt{8} = 2.8284$

$x_2 = 2: y_2 = \sqrt{5} = 2.2361$

$x_3 = 3: y_3 = \sqrt{0} = 0$

$\int_a^b y\, dx \approx \frac{1}{2}[(3 + 0) + 2(2.8284 + 2.2361)]$

$= 6.5645 \approx 6.56$

b) $h = \frac{(1.2 - 0.2)}{5} = 0.2$

$x_0 = 0.2:$ $y_0 = 0.2^{0.04} = 0.93765$

$x_1 = 0.4:$ $y_1 = 0.4^{0.16} = 0.86363$

$x_2 = 0.6:$ $y_2 = 0.6^{0.36} = 0.83202$

$x_3 = 0.8:$ $y_3 = 0.8^{0.64} = 0.86692$

$x_4 = 1:$ $y_4 = 1^1 = 1$

$x_5 = 1.2:$ $y_5 = 1.2^{1.44} = 1.30023$

$\int_a^b y\, dx \approx \frac{0.2}{2}[(0.93765 + 1.30023) + 2(0.86363 + 0.83202 + 0.86692 + 1)]$

$= 0.1 \times 9.36302 \approx 0.936$

7) a) $A = \int_0^2 (x^3 - 5x^2 + 6x)dx$

$= [\frac{x^4}{4} - \frac{5}{3}x^3 + 3x^2]_0^2 = \frac{8}{3}$

b) $A = \int_1^4 2\sqrt{x}\, dx = [\frac{4}{3}x^{\frac{3}{2}}]_1^4 = \frac{28}{3}$

Answers

c) $A = \int_0^2 2x^2\,dx + \int_2^6 (12-2x)\,dx$

$= \left[\frac{2}{3}x^3\right]_0^2 + \left[12x - x^2\right]_2^6$

$= \frac{16}{3} + 16 = \frac{64}{3}$

d) $A = \int_1^4 (x+3)\,dx - \int_1^4 (x^2 - 4x + 7)\,dx$

$= \left[\frac{x^2}{2} + 3x\right]_1^4 - \left[\frac{x^3}{3} - 2x^2 + 7x\right]_1^4$

$= \frac{33}{2} - 12 = \frac{9}{2}$

Exam Questions

1   $\int_2^7 (2x - 6x^2 + \sqrt{x})\,dx = \left[x^2 - 2x^3 + \frac{2\sqrt{x^3}}{3}\right]_2^7$

   [1 mark for each correct term]

   $= \left(7^2 - (2 \times 7^3) + \frac{2\sqrt{7^3}}{3}\right) - \left(2^2 - (2 \times 2^3) + \frac{2\sqrt{2^3}}{3}\right)$

   [1 mark]

   $= -624.6531605 - (-10.11438192)$

   $= -614.5387786 = -614.5388$ to 4 d.p. *[1 mark]*

2 a) (i) $h = \frac{8-2}{3} = 2$ *[1 mark]*

   $x_0 = 2 \ y_0 = \sqrt{(3 \times 2^3)} + \frac{2}{\sqrt{2}} = 6.31319$

   $x_1 = 4 \ y_1 = \sqrt{(3 \times 4^3)} + \frac{2}{\sqrt{4}} = 14.85641$

   $x_2 = 6 \ y_2 = \sqrt{(3 \times 6^3)} + \frac{2}{\sqrt{6}} = 26.27234$

   $x_3 = 8 \ y_3 = \sqrt{(3 \times 8^3)} + \frac{2}{\sqrt{8}} = 39.89894$ *[1 mark]*

   $\int_2^8 y\,dx \approx \frac{2}{2}[6.31319 + 2(14.85641 + 26.27234) + 39.89894]$ *[1 mark]*

   $= 128.46963 \approx 128.47$ to 2 d.p. *[1 mark]*

   (ii) $h = \frac{5-1}{4} = 1$ *[1 mark]*

   $x_0 = 1 \ y_0 = \frac{1^3 - 2}{4} = -0.25$

   $x_1 = 2 \ y_1 = \frac{2^3 - 2}{4} = 1.5$

   $x_2 = 3 \ y_2 = \frac{3^3 - 2}{4} = 6.25$

   $x_3 = 4 \ y_3 = \frac{4^3 - 2}{4} = 15.5$

   $x_4 = 5 \ y_4 = \frac{5^3 - 2}{4} = 30.75$ *[1 mark]*

   $\int_1^5 y\,dx \approx \frac{1}{2}[-0.25 + 2(1.5 + 6.25 + 15.5) + 30.75]$

   [1 mark] ≈ 38.5 *[1 mark]*

  b) Increase the number of intervals calculated. *[1 mark]*

3   The limits are the x-values when $y = 0$, so first solve
   $(x-3)^2(x+1) = 0$: *[1 mark]*
   $(x-3)(x-3)(x+1) = 0$
   $x = 3$ *[1 mark]* and $x = -1$ *[1 mark]*
   Hence, to find the area, calculate:

   $\int_{-1}^3 (x-3)^2(x+1)\,dx = \int_{-1}^3 (x^3 - 5x^2 + 3x + 9)\,dx$ *[1 mark]*

   $= \left[\frac{x^4}{4} - \frac{5}{3}x^3 + \frac{3}{2}x^2 + 9x\right]_{-1}^3$ *[1 mark]*

   $= \left(\frac{3^4}{4} - \frac{5}{3}3^3 + \frac{3}{2}3^2 + (9 \times 3)\right) -$

   $\left(\frac{(-1)^4}{4} - \left(\frac{5}{3} \times (-1)^3\right) + \left(\frac{3}{2} \times (-1)^2\right) + (9 \times (-1))\right)$ *[1 mark]*

   $= 15\frac{3}{4} - -5\frac{7}{12}$ *[1 mark]*

   $= 21\frac{1}{3}$ *[1 mark]*

4   $n = 5, \ h = \frac{4 - 1.5}{5} = 0.5$ *[1 mark]*
   $x_1 = 2.0, \ x_2 = 2.5, \ x_3 = 3.0$ *[1 mark]*

   $y_0 = 2.8182$ *[1 mark]*, $y_3 = 6.1716$ *[1 mark]*,
   $y_4 = 7.1364$ *[1 mark]*

   $\int_{1.5}^4 y\,dx \approx \frac{0.5}{2}[2.8182 + 2(4 + 5.1216 + 6.1716 + 7.1364) + 8]$
   [1 mark] $= 13.91935 = 13.9$ to 3 s.f. *[1 mark]*

5 a) $m = \frac{y_2 - y_1}{x_2 - x_1} = \frac{0 - -5}{-1 - 4} = -1$ *[1 mark]*
   $y - y_1 = m(x - x_1)$
   $y - -5 = -1(x - 4)$ so $y + 5 = 4 - x$
   $y = -x - 1$ *[1 mark]*

  b) Multiply out the brackets and then integrate:
   $(x+1)(x-5) = x^2 - 4x - 5$ *[1 mark]*
   $\int_{-1}^4 (x^2 - 4x - 5)\,dx = \left[\frac{x^3}{3} - 2x^2 - 5x\right]_{-1}^4$ *[1 mark]*

   $= \left(\frac{4^3}{3} - 2(4^2) - (5 \times 4)\right)$

   $- \left(\frac{(-1)^3}{3} - 2(-1)^2 - (5 \times -1)\right)$ *[1 mark]*

   $= -30\frac{2}{3} - 2\frac{2}{3}$ *[1 mark]* $= -33\frac{1}{3}$ *[1 mark]*

  c) Subtract the area under the curve from the area under the
   line to leave the area in between. The area under the line is
   a triangle (where $b = 5$ and $h = -5$), so use the formula for
   the area of a triangle to calculate it *[1 mark]*:

   $A = \frac{1}{2}bh = \frac{1}{2} \times 5 \times -5 = -12.5$ *[1 mark]*

   $A = -12\frac{1}{2} - -33\frac{1}{3}$ *[1 mark]*

   $= 20\frac{5}{6}$ *[1 mark]*

   *You could also have integrated the line y = −x − 1 to find the area
   under the line — you'd have got the same answer.*

C2 Practice Exam 1

1 a) The graph of $y = \tan 2t$ is going to be the same shape as
   $\tan x$, but squashed horizontally by a factor of 2. So the
   graph will be periodic with a period of 90° instead of 180°.
   It asks for solutions in the range $0° \leq t < 360°$ — so that
   would also be a suitable range for your sketch:

   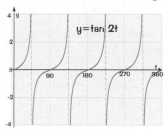

   So, for any value of k there are 4 solutions.

   *[3 marks available — 1 mark for shape, 1 mark for 90°
   period, 1 mark for 4 solutions in range.]*

Answers

b) Divide by cos $2t$ to get $\tan 2t = \sqrt{2}$ (as $\tan = \frac{\sin}{\cos}$):

$\sin 2t = \sqrt{2}\cos 2t$

$\Rightarrow \tan 2t = \sqrt{2}$ **[1 mark]**

$2t = \tan^{-1}\sqrt{2} = 54.7356...$

$t = 27.37°$ (to 2 d.p.) **[1 mark]**

That's only one solution, but you know from part (a) that there's got to be four. You also know from part (a) that the graph repeats every 90°, so just add 90° on three times to get the other answers:

$t = 117.37°, 207.37°$ and $297.37°$ **[1 mark]**

2 a) In the laws of logs: $\log_a a = 1$. So $\log_3 3 = 1$. **[1 mark]**

b) Use the laws of logs to rewrite the expression:

$\log_a 4 + 3\log_a 2 = \log_a(4 \times 2^3) = \log_a 32$. **[1 mark]**

Therefore $\log_a \chi = \log_a 32$ so $\chi = 32$. **[1 mark]**

3 a) c represents the coefficient of x^3, so find an expression for the coefficient of x^3 using the binomial expansion formula:

$(j + kx)^6 = j^6\left(1 + \frac{k}{j}x\right)^6$

Coefficient of $x^3 = j^6 \times \frac{6 \times 5 \times 4}{1 \times 2 \times 3} \times \left(\frac{k}{j}\right)^3$ **[1 mark]**

so $j^6 \times \frac{6 \times 5 \times 4}{1 \times 2 \times 3} \times \left(\frac{k}{j}\right)^3 = 20,000$

$j^6 \times 20 \times \left(\frac{1}{j^3}\right) \times k^3 = 20,000$

$j^6 \times j^{-3} \times k^3 = 1000$ **[1 mark]**

$= j^3 \times k^3 = 1000$

$(jk)^3 = 1000$ so $jk = \sqrt[3]{1000} = 10$ **[1 mark]**

b) Write an expression for the coefficient of x and then solve simultaneously with the equation $jk = 10$:

coefficient of $x = j^6 \times \frac{6}{1} \times \frac{k}{j} = 37500$

$= j^6 \times j^{-1} \times k \times 6 = 37500 \Rightarrow kj^5 = 6250$ **[1 mark]**

From a) $jk = 10$ so $k = \frac{10}{j}$

$kj^5 = \frac{10}{j} \times j^5 = 6250$ so $10 \times j^{-1} \times j^5 = 6250$ **[1 mark]**

$j^4 = 625$

$j = 5$ **[1 mark]**

Now input $j = 5$ into $jk = 10$ to find: $k = 2$ **[1 mark]**

Funny how part a) helps you work out part b).
It's almost as if the examiners are trying to help you... weird.

c) Coefficient of x^2, $b = 5^6 \times \frac{6 \times 5}{1 \times 2} \times \left(\frac{2}{5}\right)^2 = 37500$

[2 marks available — 1 mark for formula, 1 mark for correct answer]

4 a) Find $\frac{dy}{dx}$ then input the known values of x and the gradient:

$\frac{dy}{dx} = 6x^2 + z$ **[1 mark]**

when $x = 20$ $6(20^2) + z = 0$ **[1 mark]**

$2400 + z = 0$

$z = -2400$ **[1 mark]**

$y = w$ when $x = 20$, so

$w = 2(20^3) + (-2400 \times 20) - 5$ **[1 mark]**

$w = 16\,000 - 48\,000 - 5 = -32\,005$ **[1 mark]**

b) Find $\frac{d^2y}{dx^2}$ then input $x = 20$ to find if it's a negative or positive value:

$\frac{d^2y}{dx^2} = 12x$ **[1 mark]**

so when $x = 20$, $\frac{d^2y}{dx^2} = 240$ **[1 mark]**

This is positive, so it is a minimum. **[1 mark]**

5 a) The curve and the line intersect where:

$2x - 4 = (x - 2)(x - 4)$

$\Rightarrow 2x - 4 = x^2 - 6x + 8$

Rearrange to $= 0$ and then factorise:

$x^2 - 8x + 12 = 0$

$(x - 6)(x - 2) = 0$

$\Rightarrow x = 6$ and $x = 2$

To find the y-coordinates put the x-values in $y = 2x - 4$:

when $x = 2$, $y = (2 \times 2) - 4 = 0$

when $x = 6$, $y = (6 \times 2) - 4 = 8$

so the two points of intersection are $(2, 0)$ and $(6, 8)$.

Find where the line and the curve cut the axes by solving the equations for $y = 0$ (x-axis intersection) and $x = 0$ (y-axis intersection):

$2x - 4 = 0$ when $x = 2$. When $x = 0$, $y = -4$.

$(x - 2)(x - 4) = 0$ when $x = 2$ and $x = 4$. When $x = 0$, $y = -2 \times -4 = 8$.

The quadratic is U-shaped because the coefficient of x^2 is positive.

Now you're ready to sketch the graph.

Phew, at last. I thought we'd never get there...

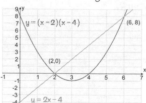

[3 marks available — 1 mark for intersection points, 1 mark for axes intercepts, 1 mark for shape.]

b) $\int_2^4 (x - 2)(x - 4)\,dx = \int_2^4 (x^2 - 6x + 8)\,dx$ **[1 mark]**

$= \left[\frac{x^3}{3} - 3x^2 + 8x\right]_2^4$ **[1 mark]**

$= \left(\frac{64}{3} - 48 + 32\right) - \left(\frac{8}{3} - 12 + 16\right)$

$= -\frac{4}{3}$ **[1 mark]**

The area's negative because it's below the x-axis.

c) Look at your sketch for part a):

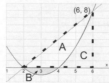

Area enclosed by line and curve $= A + B$, where $B = \frac{4}{3}$ (from part b)).

Answers

Triangle AC $= \frac{1}{2} \times$ base $\times$ height

$= \frac{4 \times 8}{2} = 16$ *[1 mark]*

C is the area beneath $y = (x-2)(x-4)$ between $x = 4$ and $x = 6$:

$\int_4^6 (x^2 - 6x + 8)dx = \left[\frac{x^3}{3} - 3x^2 + 8x\right]_4^6$ *[1 mark]*

$= \left(\frac{216}{3} - 108 + 48\right) - \left(\frac{64}{3} - 48 + 32\right)$

$= \frac{20}{3}$ *[1 mark]*

So A $= 16 - \frac{20}{3} = \frac{28}{3}$,

Total area: A + B $= \frac{28}{3} + \frac{4}{3} = \frac{32}{3}$. *[1 mark]*

6 a) $360° = 2\pi$ radians.

So $120° = 120° \times \frac{2\pi}{360°}$ radians $= \frac{120°}{360°} \times 2\pi$ radians

$= \frac{2\pi}{3}$ radians *[1 mark]*

b) Arc length S $= r\theta$, so

$40 = \frac{2\pi}{3} \times r$ *[1 mark]*

$r = 40 \div \frac{2\pi}{3} = 19.0985...$

$r = 19.1$ cm (3 s.f.) *[1 mark]*

c) $A = \frac{1}{2}r^2\theta$ *[1 mark]*

$= \frac{1}{2} \times (19.0985...)^2 \times \frac{2\pi}{3} = 381.9718...$

$= 382$ cm^2 (to the nearest cm^2) *[1 mark]*

7 a) The centre of the circle must be the midpoint of AB, since AB is a diameter. Midpoint of AB is:

$\left(\frac{2+0}{2}, \frac{1+-5}{2}\right)$ *[1 mark]* $= (1, -2)$ *[1 mark]*

The radius is half of the diameter, so half of length AB.
Use Pythagoras' theorem to find the distance:

AB $= \sqrt{(2-0)^2 + (1-(-5))^2}$

AB $= \sqrt{40} = 2\sqrt{10}$ *[1 mark]*

Radius $= \frac{2\sqrt{10}}{2} = \sqrt{10}$ *[1 mark]*

b) The general equation for a circle with centre (a, b) and radius r is: $(x-a)^2 + (y-b)^2 = r^2$.
So for a circle centre $(1, -2)$ and radius $\sqrt{10}$ that gives:

$(x-1)^2 + (y+2)^2 = 10$. *[1 mark]*

Multiply out to get the form given in the question:

$(x-1)(x-1) + (y+2)(y+2) = 10$ *[1 mark]*

$x^2 - 2x + 1 + y^2 + 4y + 4 = 10$

$x^2 + y^2 - 2x + 4y - 5 = 0$ *[1 mark]*

c) Find the equations of the tangent at A and the normal at C and then solve them simultaneously to find where the lines cross.
The tangent at A is at right angles to the diameter at A. The diameter at A has the gradient:

$\frac{1 - -5}{2 - 0} = 3$

so the tangent has the gradient $-\frac{1}{3}$ *[1 mark]*

Put the gradient $-\frac{1}{3}$ and point A $(2, 1)$ into the formula for the equation of a straight line and rearrange:

$y - y_1 = m(x - x_1)$

$y - 1 = -\frac{1}{3}(x - 2)$

$\Rightarrow y = \frac{5}{3} - \frac{x}{3}$ *[1 mark]*

A normal passes through the centre, so the gradient of the normal through C is the gradient of the line from the centre to C

$= \frac{-1 - -2}{4 - 1} = \frac{1}{3}$ *[1 mark]*

Using the straight line formula at C $(4, -1)$:

$y - -1 = \frac{1}{3}(x - 4)$

$y = \frac{x}{3} - \frac{7}{3}$ *[1 mark]*

Solve the two equations simultaneously to find the point of intersection:

$\frac{5}{3} - \frac{x}{3} = \frac{x}{3} - \frac{7}{3}$ *[1 mark]*

$\frac{5}{3} + \frac{7}{3} = \frac{2x}{3}$

$12 = 2x$

$6 = x$ *[1 mark]*

When $x = 6$, $y = \frac{6}{3} - \frac{7}{3} = -\frac{1}{3}$ *[1 mark]*

so D has coordinates $\left(6, -\frac{1}{3}\right)$

8 a) $u_{n+1} = ru_n \Rightarrow u_2 = ru_1$

$\Rightarrow u_1 = \frac{u_2}{r}$

$u_1 = -2 \div -\frac{1}{2} = 4$

Using the nth term formula $u_n = ar^{(n-1)}$, where $a = u_1 = 4$:

$u_{13} = 4 \times \left(-\frac{1}{2}\right)^{12} = \frac{1}{1024}$

[3 marks available — 1 mark for a = 4,
1 mark for inputting u into nth term formula,
1 mark for correct value of u_{13}.]

b) The sum to infinity of a converging geometric series is given by: $S_\infty = \frac{a}{1 - r}$ *[1 mark]*

So just plug in the numbers to get:

$S_\infty = \frac{4}{1 - (-\frac{1}{2})} = 4 \div \frac{3}{2} = 4 \times \frac{2}{3}$ *[1 mark]*

$= 2\frac{2}{3}$ *[1 mark]*

To infinity and $2\frac{2}{3}$! Hmm... doesn't sound quite so adventurous when you put it like that...

9 a) Multiply out the brackets and rearrange to get zero on one side:

$(x-1)(x^2 + x + 1) = 2x^2 - 17$

$\Rightarrow x^3 - 1 = 2x^2 - 17$ *[1 mark]*

$\Rightarrow x^3 - 2x^2 + 16 = 0$ *[1 mark]*

Answers

b) To show whether $(x + 2)$ is a factor of f(x) you need the factor theorem, which says that $(x - a)$ is a factor of a polynomial f(x) if and only if f(a) = 0. So if $(x + 2)$ is a factor of f(x), f(–2) = 0. *[1 mark]*
f(x) = $x^3 - 2x^2 + 16$
f(-2) = $(-2)^3 - 2 \times (-2)^2 + 16$
= $-8 - 8 + 16$
= 0   *[1 mark]*
f(-2) = 0, therefore $(x + 2)$ *is a factor of* f(x)   *[1 mark]*

c) From part b) you know that $(x + 2)$ is a factor of f(x).
Dividing f(x) by $(x + 2)$ gives:
$x^3 - 2x^2 + 16 - \underline{x^2}(x + 2) = x^3 - 2x^2 + 16 - x^3 - 2x^2$
$= -4x^2 + 16$
$-4x^2 + 16 - (\underline{-4x})(x + 2) = -4x^2 + 16 + 4x^2 + 8x = 8x + 16$
$8x + 16 - \underline{8}(x + 2) = 0$.
So $x^3 - 2x^2 + 16 = (x + 2)(x^2 - 4x + 8)$
[3 marks available — 1 mark for each correct term in the quadratic.]

d) From b) you know that $x = -2$ is a root. From c),
f(x) = $(x + 2)(x^2 - 4x + 8)$. So for f(x) to equal zero,
either $(x + 2) = 0$ (so $x = -2$) or $(x^2 - 4x + 8) = 0$ *[1 mark]*.
Completing the square of $(x^2 - 4x + 8)$ gives
$x^2 - 4x + 8 = (x - 2)^2 + something$
$= (x - 2)^2 + 4$
The equation $(x - 2)^2 + 4 = 0$ has no real roots. So f(x) = 0 has no solutions other than $x = -2$. *[1 mark]*

You could also have shown that $x^2 - 4x + 8$ has no real roots by finding the discriminant — the discriminant is $(-4)^2 - (4 \times 1 \times 8)$ = 16 − 32 = −16, which is < 0 so it has no real roots.

e) To find $(x^3 - 2x^2 + 3x - 3) \div (x - 1)$ keep subtracting lumps of $(x - 1)$ to get rid of all powers of x.
First get rid of the x^3 term by subtracting x^2 lots of $(x - 1)$:
$(x^3 - 2x^2 + 3x - 3) - \underline{x^2}(x - 1)$
$= x^3 - 2x^2 + 3x - 3 - x^3 + x^2$
$= -x^2 + 3x - 3$
Now do the same with the bit you've got left to get rid of the x^2 term:
$-x^2 + 3x - 3 + \underline{x}(x - 1)$
$= -x^2 + 3x - 3 + x^2 - x$
$= 2x - 3$
Finally, get rid of the x term in the bit that's left:
$2x - 3 - \underline{2}(x - 1) = 2x - 3 - 2x + 2$
$= -1$
Which all means that $(x^3 - 2x^2 + 3x - 3) \div (x - 1)$
$= x^2 - x + 2$ with a remainder –1.
[4 marks available — 1 mark for each correct term and 1 mark for remainder.]

C2 Practice Exam 2

1 a) $\cos(x - 60°)$ is the graph of $\cos x$ translated horizontally by 60°. Because it's minus 60° it's shifted to the right:

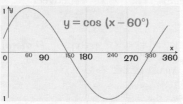

[2 marks available — 1 mark for shape, 1 mark for shifted right by 60°.]

b) Remember that $\sin^2 x + \cos^2 x = 1$ so with a little rearranging you can replace the $\sin^2$ with a $1 - \cos^2$:
$\sin^2(x - 60°) + \cos^2(x - 60°) = 1$
$\sin^2(x - 60°) = 1 - \cos^2(x - 60°)$ *[1 mark]*
Replace $\sin^2(x - 60°)$ in the original equation:
$2[1 - \cos^2(x - 60°)] = 1 + \cos(x - 60°)$ *[1 mark]*
Now multiply out the bracket and rearrange:
$2[1 - \cos^2(x - 60°)] = 1 + \cos(x - 60°)$
$\Rightarrow 2 - 2\cos^2(x - 60°) = 1 + \cos(x - 60°)$ *[1 mark]*
$\Rightarrow 2\cos^2(x - 60°) + \cos(x - 60°) - 1 = 0$ *[1 mark]*

c) Rewrite the quadratic so it looks a bit friendlier. Substitute y for $\cos(x - 60°)$ and the quadratic becomes:
$2y^2 + y - 1 = 0$.
This factorises to give: $(2y - 1)(y + 1) = 0$
so $y = \frac{1}{2}$ or $y = -1$ *[1 mark]*
Remember that $y = \cos(x - 60°)$, so
$\cos(x - 60°) = \frac{1}{2}$ or $\cos(x - 60°) = -1$
And by taking the inverse cosine of these you get:
$x - 60 = \cos^{-1}\left(\frac{1}{2}\right) = 60°$
$\Rightarrow x = 120°$ *[1 mark]*
$x - 60° = \cos^{-1}(-1) = 180°$
$\Rightarrow x = 240°$ *[1 mark]*
The graph from part a) shows there are four possible solutions, i.e. the curve intersects the line $y = \frac{1}{2}$ three times. The other two solutions are at the extreme left and the extreme right, $x = 0°$ and $x = 360°$.   *[1 mark]*
So the four solutions are:
$x = 0°$, $x = 120°$, $x = 240°$ and $x = 360°$.

2 a) Using the binomial expansion formula:
$(1 + x)^n =$
$1 + \frac{n}{1}x + \frac{n(n-1)}{1 \times 2}x^2 + \frac{n(n-1)(n-2)}{1 \times 2 \times 3}x^3 + ... + x^n$
Expand the expression $(1 + ax)^{10}$ into this form:
$1 + \frac{10}{1}(ax) + \frac{10 \times 9}{1 \times 2}(ax)^2 + \frac{10 \times 9 \times 8}{1 \times 2 \times 3}(ax)^3 + ...$ *[1 mark]*

Then simplify each coefficient:
$(1 + ax)^{10} = 1 + 10ax + 45a^2x^2 + 120a^3x^3 + ...$ *[1 mark]*

b) *Pay attention to the number at the front of the bracket. If it's not a 1, you have to rearrange everything so it is a 1. Fiddly but important.*

First take a factor of 2 to get it in the form $(1 + ax)^n$:
$(2 + 3x)^5 = \left[2\left(1 + \frac{3}{2}x\right)\right]^5 = 2^5\left(1 + \frac{3}{2}x\right)^5 = 32\left(1 + \frac{3}{2}x\right)^5$

Now expand:
$32\left[1 + \frac{5}{1}\left(\frac{3}{2}x\right) + \frac{5 \times 4}{1 \times 2}\left(\frac{3}{2}x\right)^2 + ...\right]$ **[1 mark]**

You only need the x^2 term, so simplify that one:
$32 \times \frac{20}{2} \times \left(\frac{3}{2}\right)^2 \times x^2 = 720x^2$
So the coefficient of x^2 is 720 **[1 mark]**

c) Look back to the x^2 term from part a) — $45a^2x^2$. This is equal to $720x^2$ so just rearrange the formula to find a:
$45a^2 = 720$
$a^2 = 16$
$a = \pm 4$ **[1 mark]**
And remember that part a) tells you that $a > 0$, so $a = 4$. **[1 mark]**

3 a) Look up the trapezium formula in the nice formula booklet they give you:
$\int_a^b y \, dx \approx \frac{h}{2}[y_0 + 2(y_1 + y_2 + ...y_{n-1}) + y_n]$
and remember that n is the number of intervals (in this case 4), and h is the width of each strip:
$h = \frac{b-a}{n} = \frac{2-0}{4} = 0.5$ **[1 mark]**
Work out each y value:
$x_0 = 0 \qquad y_0 = 2^{0^2} = 2^0 = 1$
$x_1 = 0.5 \qquad y_1 = 2^{0.5^2} = 2^{0.25} = 1.189 \,(3\,\text{d.p.})$
$x_2 = 1 \qquad y_2 = 2^{1^2} = 2^1 = 2$
$x_3 = 1.5 \qquad y_3 = 2^{1.5^2} = 2^{2.25} = 4.757 \,(3\,\text{d.p})$
$x_4 = 2 \qquad y_4 = 2^{2^2} = 2^4 = 16$ **[1 mark]**
And put all these values into the formula:
$\int_0^2 2^{x^2} dx \approx \frac{0.5}{2}[1 + 2(1.189 + 2 + 4.757) + 16]$ **[1 mark]**
$= \frac{1}{4}(17 + 15.892)$
$= 8.22 \,(3\,\text{s.f.})$ **[1 mark]**

b) Look at the diagram of the curve — it's U-shaped. A trapezium on each strip goes higher than the curve and has a greater area than that under the curve. **[1 mark]**
So the trapezium rule gives an overestimate for the area. **[1 mark]**

4 a) First put the two known terms into the formula for the nth term of a geometric series: $u_n = ar^{n-1}$:
$u_3 = ar^2 = \frac{5}{2} \qquad u_6 = ar^5 = \frac{5}{16}$
Divide the expression for u_6 by the expression for u_3 to get an expression just containing r and solve it:
$\frac{ar^5}{ar^2} = \frac{5}{16} \div \frac{5}{2} \Rightarrow r^3 = \frac{5}{16} \times \frac{2}{5} = \frac{10}{80}$ **[1 mark]**
$r^3 = \frac{1}{8} \qquad r = \sqrt[3]{\frac{1}{8}} \qquad r = \frac{1}{2}$ **[1 mark]**

Put this value back into the expression for u_3 to find a:
$a\left(\frac{1}{2}\right)^2 = \frac{5}{2} \Rightarrow \frac{a}{4} = \frac{5}{2} \Rightarrow a = 10$ **[1 mark]**

The nth term is $u_n = ar^{n-1}$, where $r = \frac{1}{2}$ and $a = 10$
$u_n = 10 \times \left(\frac{1}{2}\right)^{n-1} = 10 \times \frac{1}{2^{n-1}} = \frac{10}{2^{n-1}}$ **[1 mark]**

b) This asks you to find the sum of the first ten terms, so input the known values of n, a and r (from part a)) into the formula for the sum of the first n terms:
$S_n = \frac{a(1 - r^n)}{1 - r}, \quad S_{10} = \frac{10\left(1 - \left(\frac{1}{2}\right)^{10}\right)}{1 - \left(\frac{1}{2}\right)}$ **[1 mark]**
$= 10 \times 2 \times \left(1 - \frac{1}{2^{10}}\right)$
$= 20\left(1 - \frac{1}{1024}\right) = 20 \times \frac{1023}{1024} = 5 \times \frac{1023}{256}$ **[1 mark]**
$= \frac{5115}{256}$ **[1 mark]**

c) Just pop the known values of a and r into the formula for the sum to infinity:
$S_\infty = \frac{a}{1 - r} = \frac{10}{1 - \left(\frac{1}{2}\right)}$ **[1 mark]**
$= 10 \div \frac{1}{2} = 10 \times 2$
$= 20$ **[1 mark]**

5 a) Use the cosine rule: $a^2 = b^2 + c^2 - 2bc\cos A$, where $b = 10$, $c = 7$ and angle A = 60°:
$a^2 = 10^2 + 7^2 - 2 \times 10 \times 7 \times \cos 60°$ **[1 mark]**
$\Rightarrow a^2 = 149 - 140\cos 60°$
$\Rightarrow a^2 = 149 - 140(0.5)$
$\rightarrow a^2 - 79$
$\Rightarrow a = \sqrt{79} = 8.89$ cm to 3 s.f. **[1 mark]**

b) Now you can use the sine rule to find the angles:
$\frac{a}{\sin A} = \frac{b}{\sin B} = \frac{c}{\sin C}$
θ is the angle opposite the 10 cm side. So if we call the 10 cm side 'side b', then θ = angle B. Putting the known values into the sine rule gives:
$\frac{\sqrt{79}}{\sin 60°} = \frac{10}{\sin \theta}$ **[1 mark]**
$\sin\theta = \frac{10 \times \sin 60°}{\sqrt{79}} = 0.9744$
$\Rightarrow \theta = \sin^{-1}0.9744 = 77.0°$ to 3 s.f. **[1 mark]**

Now, you know 2 of the angles in the triangle, and as the angles in a triangle add up to 180°,
$\phi = 180 - 60 - 77 = 43°$ **[1 mark]**.

To summarise: $\theta = 77°$ and $\phi = 43°$.

6 a) (i) In the diagram, x is the adjacent side of a right-angled triangle with an angle θ and hypotenuse r — so use the cos formula:
$\cos\theta = \frac{\text{adjacent}}{\text{hypotenuse}} = \frac{x}{r}$
and so $x = r\cos\theta$ **[1 mark]**

Answers

(ii) As the stage is symmetrical, you know that distance y is the same on both triangles. Distance y is the opposite side of the right-angled triangle — so use the sine formula:

$$\sin\theta = \frac{\text{opposite}}{\text{hypotenuse}} = \frac{y}{r}$$

and so $y = r\sin\theta$ *[1 mark]*

b) Looking at the diagram, most of the perimeter is simple — $q + q + 2r$. But the top and curved lengths need a bit of thinking.

First the top length — you can see this is $2x$ so, using the expression for x found in a), you can write this as $2r\cos\theta$.

For the curved lengths — the shaded areas are sectors of circles, and the formula for the length of one arc is given by: $r\theta$

Now add them all up to get the total perimeter:
$q + q + 2r + 2r\cos\theta + r\theta + r\theta = 2[q + r(1 + \theta + \cos\theta)]$.

And do the same sort of thing for the area — break it down into a rectangle, a triangle and two sectors:

Area of rectangle = width × height = $2qr$

Area of triangle = $\frac{1}{2}$ × width × height = $\frac{1}{2}(2r\cos\theta)(r\sin\theta)$
$= r^2\cos\theta \sin\theta$

Area of 1 shaded sector = $\frac{1}{2}r^2\theta$

So the total area = $2qr + r^2\cos\theta\sin\theta + r^2\theta$
$= 2qr + r^2(\cos\theta\sin\theta + \theta)$.

[4 marks available — 1 mark for all individual lengths correct, 1 mark for all individual areas correct, 1 mark for each correct expression.]

Crumbs, that looked very intimidating. The best thing to do with questions like that is stay calm and break them into small chunks. And knowing all the formulas doesn't hurt...

c) Substitute the given values of P and θ into the equation for the perimeter: $P = 2[q + r(1 + \theta + \cos\theta)]$

$\Rightarrow 40 = 2\left[q + r\left(1 + \frac{\pi}{3} + \cos\frac{\pi}{3}\right)\right]$

$\Rightarrow 20 = q + r\left(\frac{3}{2} + \frac{\pi}{3}\right)$ *[1 mark]*

And then into the equation for the area:
$A = 2qr + r^2(\cos\theta\sin\theta + \theta)$

$\Rightarrow A = 2qr + r^2\left(\cos\frac{\pi}{3}\sin\frac{\pi}{3} + \frac{\pi}{3}\right)$

$= 2qr + r^2\left(\frac{\sqrt{3}}{4} + \frac{\pi}{3}\right)$ *[1 mark]*

To rearrange this formula for area into the form shown in the question, you need to get rid of q. Rearrange the perimeter formula to get an expression for q in terms of r, then substitute that into the area equation:

$q = 20 - r\left(\frac{3}{2} + \frac{\pi}{3}\right)$

$A = 2qr + r^2\left(\frac{\sqrt{3}}{4} + \frac{\pi}{3}\right)$

$= 2r\left[20 - r\left(\frac{3}{2} + \frac{\pi}{3}\right)\right] + r^2\left(\frac{\sqrt{3}}{4} + \frac{\pi}{3}\right)$

$= 40r - 2r^2\left(\frac{3}{2} + \frac{\pi}{3}\right) + r^2\left(\frac{\sqrt{3}}{4} + \frac{\pi}{3}\right)$

$= 40r - r^2\left[2\left(\frac{3}{2} + \frac{\pi}{3}\right) - \left(\frac{\sqrt{3}}{4} + \frac{\pi}{3}\right)\right]$ *[1 mark]*

And since $\left[2\left(\frac{3}{2} + \frac{\pi}{3}\right) - \left(\frac{\sqrt{3}}{4} + \frac{\pi}{3}\right)\right] = 3.614$ to 3 d.p.

this means: $A = 40r - 3.614r^2$ *[1 mark]*

7 a) (i) Remainder = $f(-3) = (-3)^3 - 6(-3)^2 - (-3) + 30$ *[1 mark]*
$= -48$ *[1 mark]*.

(ii) Remainder = $f\left(\frac{1}{4}\right) = \left(\frac{1}{64}\right) - 6\left(\frac{1}{16}\right) - \left(\frac{1}{4}\right) + 30$ *[1 mark]*
$= 29.4$ (3 s.f.) *[1 mark]*.

b) If $(x - 3)$ is a factor then $f(3) = 0$.
$f(3) = (3)^3 - 6(3)^2 - (3) + 30$ *[1 mark]*
$= 27 - 54 - 3 + 30 = 0$,
so $(x - 3)$ is a factor of $f(x)$. *[1 mark]*

c) $(x - 3)$ is a factor, so divide $x^3 - 6x^2 - x + 30$ by $x - 3$:
$x^3 - 6x^2 - x + 30 - \underline{x^2}(x - 3) = x^3 - 6x^2 - x + 30 - x^3 + 3x^2$
$= -3x^2 - x + 30$.
$-3x^2 - x + 30 - (\underline{-3x})(x - 3) = -3x^2 - x + 30 + 3x^2 - 9x$
$= -10x + 30$. Finally $-10x + 30 - (\underline{-10})(x - 3) = 0$.
so $x^3 - 6x^2 - x + 30 = (x^2 - 3x - 10)(x - 3)$.

This is the method from p.54.

Factorising the quadratic expression gives:
$f(x) = (x - 5)(x + 2)(x - 3)$.

[4 marks available — 1 mark for dividing by x – 3 to find quadratic factor, 1 mark for correct quadratic factor, 1 mark for attempt to factorise quadratic, 1 mark for correct factorisation of quadratic.]

8 a) A is on the y-axis, so the x-coordinate is 0. Just put $x = 0$ into the equation and solve:

$0^2 - (6 \times 0) + y^2 - 4y = 0$ *[1 mark]*
$y^2 - 4y = 0$
$y(y - 4) = 0$
$y = 0$ or $y = 4$
$y = 0$ is the origin, so A is at $(0, 4)$ *[1 mark]*

b) You're basically completing the square for x and y separately, so halve the coefficient of x to find a, and halve the coefficient of y to find b. For each one you'll end up with a number you don't want, which you need to take away each time:

Completing the square for $x^2 - 6x$ gives $(x - 3)^2$ but $(x - 3)^2 = x^2 - 6x + 9$ so you need to take 9 away:
$(x - 3)^2 - 9$ *[1 mark]*
Now the same for $y^2 - 4y$: $(y - 2)^2 = y^2 - 4y + 4$ so take away 4 which gives: $(y - 2)^2 - 4$ *[1 mark]*

Put these new expressions back into the original equation:
$(x - 3)^2 - 9 + (y - 2)^2 - 4 = 0$

$\Rightarrow (x - 3)^2 + (y - 2)^2 = 13$ *[1 mark]*

Answers

c) In the general equation for a circle $(x - a)^2 + (y - b)^2 = r^2$, the centre is (a, b) and the radius is r. So for the equation in part b) — $a = 3$, $b = 2$, $r = \sqrt{13}$.
Hence, the centre is (3, 2) *[1 mark]*
and the radius is $\sqrt{13}$. *[1 mark]*

d) The tangent is perpendicular to the radius.
The radius between A (0, 4) and centre (3, 2)
has a gradient of: $\frac{y_2 - y_1}{x_2 - x_1} = \frac{2 - 4}{3 - 0} = -\frac{2}{3}$ *[1 mark]*
Using the gradient rule, the tangent at A has a gradient of:
$\frac{-1}{-\frac{2}{3}} = \frac{3}{2}$ *[1 mark]*
Finally use $y - y_1 = m(x - x_1)$ to find the equation of the tangent to the circle at point A:
$y - 4 = \frac{3}{2}(x - 0)$ *[1 mark]*
$y = \frac{3}{2}x + 4$ *[1 mark]*

The tangent is perpendicular to the radius? That sounds like a useful life fact — like how to boil an egg or tie your laces. If you didn't know it, go back to the circles section and learn it...

9 a) $\frac{1}{x^2}$ tends towards infinity as x gets closer to 0, and $\frac{1}{x^2}$ gets smaller as x gets bigger. So the graph looks like this:

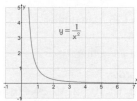

[1 mark]

b) $\int_1^\infty \frac{1}{x^2}dx = \int_1^\infty x^{-2}dx$
$= \left[\frac{x^{-1}}{-1}\right]_1^\infty = \left[-\frac{1}{x}\right]_1^\infty$ *[1 mark]*
$= (-0) - \left(-\frac{1}{1}\right) = 1$ *[1 mark]*

c) The tangent's gradient will be the same as the curve's gradient, so find $\frac{dy}{dx}$ at $x = 1$:
$y = \frac{1}{x^2} = x^{-2}$
$\frac{dy}{dx} = -2x^{-3} = -\frac{2}{x^3}$
when $x = 1$:
$= -\frac{2}{1^3} = -2$ *[1 mark]*
When $x = 1$, $y = \frac{1}{1^2} = 1$, so the tangent goes through the point (1, 1) and has a gradient of –2. Using the equation of a straight line: $y - y_1 = m(x - x_1)$, to find f(x):
$y - 1 = -2(x - 1)$
$y = -2x + 2 + 1$
$\Rightarrow f(x) = 3 - 2x$ *[1 mark]*

d) Find a value less than 1 for k.
Just do the integration normally, but write k instead of a number:
$\int_0^k f(x)dx = \int_0^k (3 - 2x)dx$
$= [3x - x^2]_0^k$
$= (3k - k^2) - 0$
$= 3k - k^2$
This integral has to be equal to 1, so you need to solve:
$3k - k^2 = 1$
$\Rightarrow k^2 - 3k + 1 = 0$ *[1 mark]*
This is a quadratic but it doesn't look like it's going to factorise, so using the quadratic formula:
$k = \frac{3 \pm \sqrt{(-3)^2 - 4 \times 1 \times 1}}{2 \times 1}$
$= \frac{3 \pm \sqrt{5}}{2}$ *[1 mark]*
The question asks for a value of less than 1 for k, but using surds. So use a calculator to find which is < 1 but leave in its surd form:
$\frac{3 + \sqrt{5}}{2} = 2.618$ $\frac{3 - \sqrt{5}}{2} = 0.382$
therefore $k = \frac{3 - \sqrt{5}}{2}$ *[1 mark]*

Gee to the whizzle, I thought we'd never finish C2, but hurrah! Run for your lives, before the maths sucks you back in again...

Answers

S1 Section 1 — Data
Warm-up Questions

1)   12.8, 13.2, 13.5, 14.3, 14.3, 14.6, 14.8, 15.2, 15.9, 16.1, 16.1, 16.2, 16.3, 17.0, 17.2   (all in cm)

2)

Length of call	Lower class boundary (lcb)	Upper class boundary (ucb)	Class width	Frequency	Frequency density = Height of column
0 - 2	0	2.5	2.5	10	4
3 - 5	2.5	5.5	3	6	2
6 - 8	5.5	8.5	3	3	1
9 - 15	8.5	15.5	7	1	0.143

Lots of fiddly details here — a table helps you get them right.

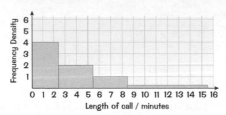

3) $\Sigma f = 16$, $\Sigma fx = 22$, so mean = $22 \div 16 = 1.375$
Median position = $17 \div 2 = 8.5$, so median = 1
Mode = 0.

4)

Speed	mid-class value x	Number of cars f	fx
30 - 34	32	12 (12)	384
35 - 39	37	37 (49)	1369
40 - 44	42	9	378
45 - 50	47.5	2	95
Totals		60	2226

Estimated mean = $2226 \div 60 = \underline{37.1 \text{ mph}}$
Median position is $61 \div 2 = 30.5$.
This is in class 35 - 39.
$30.5 - 12 = 18.5$, so median is 18.5th value in class.
Class width = 5, so median is:

$$34.5 + \left(18.5 \times \frac{5}{37}\right) = \underline{37 \text{ mph}}$$

Modal class is <u>35 - 39 mph</u>.

Easy eh? It doesn't hurt to double-check your mid-class values though.

5) a)

$$\text{Mean} = \frac{11 + 12 + 14 + 17 + 21 + 23 + 27}{7} = \frac{125}{7} = 17.9 \text{ to 3 sig. fig.}$$

b)

$$s.d. = \sqrt{\frac{11^2 + 12^2 + 14^2 + 17^2 + 21^2 + 23^2 + 27^2}{7} - \left(\frac{125}{7}\right)^2} = \sqrt{30.98} = 5.57 \text{ to 3 sig. fig.}$$

Just numbers and a formula. Simple.

6)

Score	Mid-class value, x	x^2	f	fx	fx^2
100 - 106	103	10609	6	618	63654
107 - 113	110	12100	11	1210	133100
114 - 120	117	13689	22	2574	301158
121 - 127	124	15376	9	1116	138384
128 - 134	131	17161	2	262	34322
Totals			50 ($= \Sigma f$)	5780 ($= \Sigma fx$)	670618 ($= \Sigma fx^2$)

$$\text{Mean} = \frac{5780}{50} = 115.6$$

$$s^2 = \frac{670618}{50} - 115.6^2 = 49$$

7) Let $y = x - 20$.
Then
$\bar{y} = \bar{x} - 20$ or $\bar{x} = \bar{y} + 20$
$\sum y = 125$ and $\sum y^2 = 221$
So $\bar{y} = \frac{125}{100} = 1.25$ and $\bar{x} = 1.25 + 20 = \underline{21.25}$
$s_y^2 = \frac{221}{100} - 1.25^2 = 0.6475$ and so $s_y = 0.805$ to 3 sig. fig.
Therefore $\underline{s_x = 0.805}$ to 3 sig. fig.

If you got in a muddle, look back at stuff about coding on the standard deviation pages.

8)

Time	Mid-class x	$y = x - 35.5$	f	fy	fy^2
30 - 33	31.5	-4	3	-12	48
34 - 37	35.5	0	6	0	0
38 - 41	39.5	4	7	28	112
42 - 45	43.5	8	4	32	256
Totals			20 ($= \Sigma f$)	48 ($= \Sigma fy$)	416 ($= \Sigma fy^2$)

$\bar{y} = \frac{48}{20} = 2.4$
So $\bar{x} = \bar{y} + 35.5 = 2.4 + 35.5 = \underline{37.9 \text{ minutes}}$
$s_y^2 = \frac{416}{20} - 2.4^2 = 15.04$, and so $s_y = 3.88$ minutes, to 3 sig. fig.
But $s_x = s_y$, and so $\underline{s_x = 3.88 \text{ minutes}}$, to 3 sig. fig.

9)   IQR = $88 - 62 = 26$, so $3 \times$ IQR = 78.
So upper fence = $88 + 78 = 166$.
This means that:
a) 161 is not an outlier.   **b)** 176 is an outlier.
Lower fence = $62 - 78 = -16$.
This means that: **c)** 0 is not an outlier.

10)   Put the 20 items of data in order:
1, 4, 5, 5, 5, 5, 5, 6, 6, 7, 7, 8, 10, 10, 12, 15, 20, 20, 30, 50
Then the median position is 10.5, and since the 10th and the 11th items are both 7, the median = <u>7</u>.
Lower quartile = <u>5</u>.
Upper quartile = $(12 + 15) \div 2 = \underline{13.5}$.

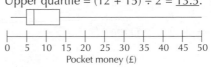

This data is positively skewed. Most 15-year-olds earned a small amount of pocket money. A few got very large amounts.

11)   Quartile coefficient of skewness =
$$\frac{(150 - 132) - (132 - 86)}{150 - 86} = \frac{-28}{64} = -0.438, \text{ to 3 s.f.}$$

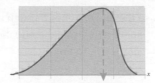 — possible shape.

This is negatively skewed.

Aah, I do love a good sketch. Isn't it pretty?

Answers

Exam Questions

1 a)

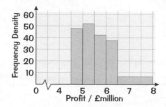

Profit	Class width	Frequency	Frequency density = Height of column
4.5 - 5.0	0.5	24	48
5.0 - 5.5	0.5	26	52
5.5 - 6.0	0.5	21	42
6.0 - 6.5	0.5	19	38
6.5 - 8.0	1.5	10	6.67

[1 mark for correct axes, plus 2 marks if all bars drawn correctly, or 1 mark for at least 3 bars correct.]

b) The distribution is positively skewed — only a few businesses make a high profit. The modal profit is between £5 million and £5.5 million.
[1 mark per sensible comment, up to a maximum of 2.]

Not too tricky — a nice gentle start.

2 a) Let $y = x - 30$.
$\bar{y} = \frac{228}{19} = 12$ and so $\bar{x} = \bar{y} + 30 = 42$ *[1 mark]*
$s_y^2 = \frac{3040}{19} - 12^2 = 16$ *[1 mark]*, and so $s_y = 4$
But $s_x = s_y$ and so $\underline{s_x = 4}$ *[1 mark]*

b) $\bar{x} = \frac{\sum x}{19} = 42$
And so $\sum x = 42 \times 19 = \underline{798}$ *[1 mark]*
$s_x^2 = \frac{\sum x^2}{19} - \bar{x}^2 = \frac{\sum x^2}{19} - 42^2 = 16$ *[1 mark]*
And so $\sum x^2 = (16 + 42^2) \times 19 = \underline{33820}$ *[1 mark]*
A bit harder...

c) New $\sum x = 798 + 32 = 830$ *[1 mark]*
So new $\bar{x} = \frac{830}{20} = \underline{41.5}$ *[1 mark]*
New $\sum x^2 = 33820 + 32^2 = 34844$ *[1 mark]*
So new $s_x^2 = \frac{34844}{20} - 41.5^2 = 19.95$
and new $\underline{s_x = 4.47}$ to 3 sig. fig. *[1 mark]*

3 a) (i) Times = 2, 3, 4, 4, 5, 5, 5, 7, 10, 12
Median position = 5.5, so median = 5 minutes
[1 mark]

(ii) Lower quartile = 4 minutes *[1 mark]*
Upper quartile = 7 minutes *[1 mark]*

b)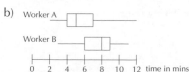

[6 marks available overall — 1 mark for each median in the right place, 1 mark for each pair of quartiles shown correctly, and 1 mark for each pair of lines showing the extremes correctly drawn.]

c) Various statements could be made,
e.g. the times for Worker B are longer than those for Worker A, on average.
The IQR for both workers is the same — generally they both work with the same consistency.
The range for Worker A is larger than that for Worker B. Worker A had a few items he/she could iron very quickly and a few which took a long time.
[1 mark for any sensible answer]

d) Worker A would be better to employ. The median time is less than for Worker B, and the upper quartile is less than the median of Worker B. Worker A would generally iron more items in a given time than worker B.
[1 mark for any sensible answer]

Don't be put off by these questions — you just have to show you understand what the data is telling you

4 a) $\bar{a} = \frac{60.3}{20} = 3.015\,\text{g}$ *[1 mark]*

b) $s_A^2 = \frac{219}{20} - 3.015^2$ *[1 mark]* $= 1.860\,\text{g}^2$ *[1 mark]*
So $s_A = 1.36\,\text{g}$ to 3 sig. fig. *[1 mark]*

c) Brand A chocolate drops are heavier on average than brand B. Brand B chocolate drops are much closer to their mean weight than brand A.
[1 mark for each of 2 sensible statements]

"Mmm, chocolate drops" does not count as a sensible statement...

d) Mean of A and B $= \frac{\sum a + \sum b}{50} = \frac{60.3 + (30 \times 2.95)}{50}$
$= 2.976\,\text{g}$ *[1 mark]*
$\frac{\sum b^2}{30} - 2.95^2 = 1$, and so $\sum b^2 = 291.075$ *[1 mark]*
Variance of A and $B = \frac{\sum a^2 + \sum b^2}{50} - 2.976^2$
$= \frac{219 + 291.075}{50} - 2.976^2$
$= 1.3449$ *[1 mark]*

So $s.d. = \sqrt{1.3449} = 1.16\,\text{g}$ to 3 sig. fig *[1 mark]*

Work through each step carefully so you don't make silly mistakes and lose any lovely marks.

5 a) There are 30 males, so median is in $31 \div 2 = 15.5$th position. Take the mean of the 15th and 16th readings to get median $= (62 + 65) \div 2 = 63.5$ *[1 mark]*

b) The female median is 64.5 (halfway between the 8th and 9th readings). The female median is higher than the male median. The females scored better than the males on average.
Female range $= 79 - 55 = 24$.
Male range $= 79 - 43 = 36$
The female range is less than the male range. Their scores are more consistent than the males'.

[Up to 2 marks available for any sensible comments]

Answers

6 a) Total number of people = 38
Median position = $(38 + 1) \div 2 = 19.5$ *[1 mark]*
19th value = 15; 20th value = 16,
so median = <u>15.5 hits</u> *[1 mark]*.
Mode = <u>15 hits</u> *[1 mark]*

b) Lower quartile = 10th value = 14
Upper quartile = 29th value = 17
[1 mark for both]

So interquartile range = 17 – 14 = 3,
and upper fence = $17 + (1.5 \times 3) = 21.5$.
This means that 25 is outlier *[1 mark]*.

c)

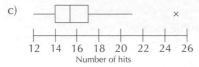

12 14 16 18 20 22 24 26
Number of hits

[1 mark]

As the whisker on the right-hand side is much longer than the left-hand whisker, the distribution seems to be positively skewed. *[1 mark]*

If you've forgotten what goes where on a box plot, just look at page 110.

d) If 25 was removed then the right-hand 'tail' of the box plot would be much shorter, and the distribution would be more symmetrical *[1 mark]*.

7 Find the total area underneath the histogram using the grid squares *[1 mark]*:
$2 + 1.5 + 2 + 2 + 1.5 + 4 + 5 + 3 + 4 = 25$ *[1 mark]*
So each grid square represents 2 lions *[1 mark]*.
The number of squares for lengths above 220 cm is 7
[1 mark], which represents $7 \times 2 = 14$ lions *[1 mark]*.

S1 Section 2 — Probability

Warm-up Questions

1) a) The sample space would be as below:

		Dice				
		1	2	3	4	5	6
Coin	H	2	4	6	8	10	12
	T	5	6	7	8	9	10

b) There are 12 outcomes in total, and 9 of these are more than 5, so P(score >5) = 9/12 = 3/4

c) There are 6 outcomes which have a tail showing, and 3 of these are even, so P(even score given that you throw a tail) = 3/6 = 1/2

Hmm, a bit fiddly but not too bad.

2) a) 20% of the people eat chips, and 10% of these is 2% — so 2% eat both chips and sausages.

Now you can draw the Venn diagram:

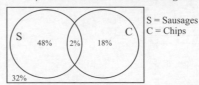

By reading the numbers in the appropriate sets from the diagram you can see...

b) 18% eat chips but not sausages.

c) 18% + 48% = 66% eat chips or sausages, but not both.

These questions do make you work up an appetite... mmm, sausages...

3) Draw a sample space diagram

```
        6 ┤ 7  8  9  10 11 12
        5 ┤ 6  7  8  9  10 11
2ⁿᵈ Dice 4 ┤ 5  6  7  8  9  10
        3 ┤ 4  5  6  7  8  9
        2 ┤ 3  4  5  6  7  8
        1 ┤ 2  3  4  5  6  7
          └─┼──┼──┼──┼──┼──┼── 1ˢᵗ Dice
            1  2  3  4  5  6
```

There are 36 outcomes altogether.

a) 15 outcomes are prime (since 2, 3, 5, 7 and 11 are prime), so P(prime) = 15/36 = 5/12

b) 7 outcomes are square numbers (4 and 9), so P(square) = 7/36

c) Being prime and a square number are exclusive events, so P(prime or square) = 15/36 + 7/36 = 22/36 = 11/18

You have to think outside the probability box for this one, but it's basic maths really.

4) a)

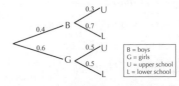

b) Choosing an upper school pupil means either 'boy and upper' or 'girl and upper'.
P(boy and upper) = $0.4 \times 0.3 = 0.12$.
P(girl and upper) = $0.6 \times 0.5 = 0.30$.
So P(Upper) = 0.12 + 0.30 = 0.42.

5) Draw a tree diagram:

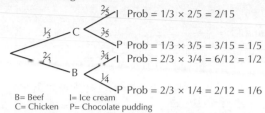

B= Beef   I= Ice cream
C= Chicken   P= Chocolate pudding

a) P(chicken or ice cream but not both) = P(C∩P) + P(B∩I)
= 1/5 + 1/2 = 7/10

b) P(ice cream) = P(C∩I) + P(B∩I) = 2/15 + 1/2 = 19/30

c) P(chicken|ice cream) = P(C∩I) ÷ P(I)

= (2/15) ÷ (19/30) = 4/19

You aren't asked to draw a tree diagram, but it makes it a lot easier if you do.

Exam Questions

1 a) The Venn diagram would look something like this:

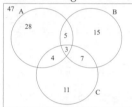

[1 mark for the central figure correct, 2 marks for '5', '7' and '4' correct (get 1 mark for 2 correct), 1 mark for '28', '15' and '11' correct, plus 1 mark for a box with '47' outside the circles.]

b) (i) Add up the numbers in all the circles to get 73 people out of 120 buy at least 1 type of soap *[1 mark]*. So the probability = 73/120 *[1 mark]*

(ii) Add up the numbers in the intersections to get 5 + 3 + 4 + 7 = 19, meaning that 19 people buy at least two soaps *[1 mark]*, so the probability a person buys at least two types = 19/120 *[1 mark]*.

(iii) 28 + 11 + 15 = 54 people buy only 1 soap *[1 mark]*, and of these 15 buy soap B *[1 mark]*.

So probability of a person who only buys one type of soap buying type B is 15/54 = 5/18 *[1 mark]*

2 a)

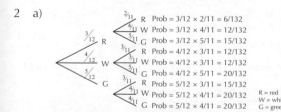

R = red
W = white
G = green

[3 marks available — 1 mark for each set of 3 branches on the right-hand side correct]

b) The second counter is green means one of three outcomes 'red then green' or 'white then green' or 'green then green'. So P(2nd is green) = 15/132 + 20/132 + 20/132 *[1 mark]* = 55/132 = 5/12 *[1 mark]*

c) For both to be red there's only one outcome: 'red then red' *[1 mark]*. P(both red) = 6/132 = 1/22 *[1 mark]*

d) 'Both same colour' is the complementary event of 'not both same colour'. So P(not same colour) = 1 − P(both same colour) *[1 mark]*. Both same colour is either R and R or W and W or G and G.
P(not same colour) = 1 − [6/132 + 12/132 + 20/132] *[1 mark]* = 1 − 38/132 = 94/132 = 47/66 *[1 mark]*
(Alternatively, 1 mark for showing P(RW or RG or WR or WG or GR or GW), 1 mark for adding the 6 correct probabilities and 1 mark for the correct answer.)

Ooh, that was a long one. Shouldn't be too tricky though, as long as your tree diagram was nice and clear.

3 a) (i) J and K are independent, so
P(J ∩ K) = P(J) × P(K) = 0.7 × 0.1 = 0.07 *[1 mark]*

(ii) P(J ∪ K) = P(J) + P(K) − P(J ∩ K) *[1 mark]*
= 0.7 + 0.1 − 0.07 = 0.73 *[1 mark]*

b) Drawing a quick Venn Diagram often helps:

P(L|K′) = P(L ∩ K′) ÷ P(K′)
Now L ∩ K′ = L — think about it — all of L is contained in K′, so L ∩ K′ (the 'bits in both L and K′) are just the bits in L. Therefore P(L ∩ K′) = P(L) = 1 − P(K ∪ J) = 1 − 0.73 = 0.27 *[1 mark]*
P(K′) = 1 − P(K) = 1 − 0.1 = 0.9 *[1 mark]*
And so P(L|K′) = 0.27 ÷ 0.9 = 0.3 *[1 mark]*

That was a bit complicated — you just need to put your thinking cap on and DON'T PANIC.

4 Draw a tree diagram:

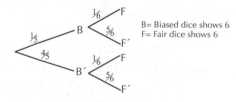

B= Biased dice shows 6
F= Fair dice shows 6

a) P(B′) = 0.8 *[1 mark]*

b) Either at least one of the dice shows a 6 or neither of them do, so these are complementary events. Call F the event 'the fair dice shows a 6'.
Then P(F ∪ B) = 1 − P(F′ ∩ B′) *[1 mark]*

= 1 − (4/5 × 5/6) = 1 − 2/3 = 1/3 *[1 mark]*

Answers

c) P(exactly one 6 | at least one 6)
= P(exactly one 6 ∩ at least one 6) ÷ P(at least one 6).
The next step might be a bit easier to get your head round if you draw a Venn diagram:

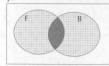

'exactly one 6' ∩ 'at least one 6' = 'exactly one 6'
(Look at the diagram — 'exactly one 6' is the cross-hatched area, and 'at least one 6' is the cross-hatched area <u>plus</u> the grey bit. So the bit in common to both is just the cross-hatched area.)
Now, that means P(exactly one 6 ∩ at least one 6) = P(B∩F') + P(B'∩F) — this is the cross-hatched area in the Venn diagram,
i.e. P(exactly one 6 ∩ at least one 6) = (1/5 × 5/6) + (4/5 × 1/6) = 9/30 = 3/10 (using the fact that B and F are independent) *[1 mark]*
P(at least one 6) = 1/3 (from b)).

And all of this means P(exactly one 6 | at least one 6)
= 3/10 ÷ 1/3 *[1 mark]* = 9/10 *[1 mark]*

Blauuurgh — the noise of a mind boggling. Part C is difficult to get your head round, but it's just a matter of remembering the right formula (the one on page 117), breaking it down into separate parts and working through it step by step. Yay.

S1 Section 3 — Discrete Random Variables
Warm-up Questions

1) a) All the probabilities have to add up to 1.
So $0.5 + k + k + 3k = 0.5 + 5k = 1$, i.e. $5k = 0.5$, i.e. $k = 0.1$.

b) $P(Y < 2) = P(Y = 0) + P(Y = 1) = 0.5 + 0.1 = 0.6$.

2) $P(W \leq 0.2) = P(W = 0.2) = 0.2$
$P(W \leq 0.3) = P(W = 0.2) + P(W = 0.3) = 0.4$
$P(W \leq 0.4) = P(W = 0.2) + P(W = 0.3)$
$\qquad\qquad\qquad + P(W = 0.4) = 0.7$
$P(W \leq 0.5) = P(W = 0.2) + P(W = 0.3) + P(W = 0.4)$
$\qquad\qquad\qquad\qquad + P(W = 0.5) = 1$

So the cumulative distribution function of W is

w	0.2	0.3	0.4	0.5
P(W ≤ w)	0.2	0.4	0.7	1

3) $P(R = 0) = P(R \leq 0) = F(0) = 0.1$
$P(R = 1) = P(R \leq 1) - P(R \leq 0) = 0.5 - 0.1 = 0.4$
$P(R = 2) = P(R \leq 2) - P(R \leq 1) = 1 - 0.5 = 0.5$

So the PF of R is:

r	0	1	2
P(R = r)	0.1	0.4	0.5

$P(0 \leq R \leq 1) = 0.5$

Nothing too difficult there, but lots of little tiddly bits to think about. Don't lose concentration — it's easy to make silly mistakes if you do.

4) There are 5 possible outcomes, and the probability of each of them is k, so $k = 1 \div 5 = 0.2$.
Mean of X $= \dfrac{0 + 4}{2} = 2$.
Variance of X $= \dfrac{(4 - 0 + 1)^2 - 1}{12} = \dfrac{24}{12} = 2$

5) a) As always, the probabilities have to add up to 1, so
$k = 1 - \left(\dfrac{1}{6} + \dfrac{1}{2} + \dfrac{5}{24}\right) = 1 - \dfrac{21}{24} = \dfrac{3}{24} = \dfrac{1}{8}$

b) $E(X) = \left(1 \times \dfrac{1}{6}\right) + \left(2 \times \dfrac{1}{2}\right) + \left(3 \times \dfrac{1}{8}\right) + \left(4 \times \dfrac{5}{24}\right)$
$= \dfrac{4 + 24 + 9 + 20}{24} = \dfrac{57}{24} = \dfrac{19}{8}$

$E(X^2) = \left(1^2 \times \dfrac{1}{6}\right) + \left(2^2 \times \dfrac{1}{2}\right) + \left(3^2 \times \dfrac{1}{8}\right) + \left(4^2 \times \dfrac{5}{24}\right)$
$= \dfrac{4 + 48 + 27 + 80}{24} = \dfrac{159}{24} = \dfrac{53}{8}$

$Var(X) = E(X^2) - [E(X)]^2 = \dfrac{53}{8} - \left(\dfrac{19}{8}\right)^2$
$= \dfrac{424 - 361}{64} = \dfrac{63}{64}$

c) $E(2X - 1) = 2E(X) - 1 = 2 \times \dfrac{19}{8} - 1 = \dfrac{30}{8} = \dfrac{15}{4}$

$Var(2X - 1) = 2^2 Var(X) = 4 \times \dfrac{63}{64} = \dfrac{63}{16}$

6) a) $E(X) = (1 \times 0.1) + (2 \times 0.2) + (3 \times 0.25) + (4 \times 0.2)$
$\qquad\qquad + (5 \times 0.1) + (6 \times 0.15) = 3.45$

b) $Var(X) = E(X^2) - (E(X))^2$
$E(X^2) = (1 \times 0.1) + (4 \times 0.2) + (9 \times 0.25) + (16 \times 0.2)$
$\qquad\qquad + (25 \times 0.1) + (36 \times 0.15) = 14.25$
So $Var(X) = 14.25 - 3.45^2 = 2.3475$

Exam Questions

1 a) The probability of getting 3 heads is: $\dfrac{1}{2} \times \dfrac{1}{2} \times \dfrac{1}{2} = \dfrac{1}{8}$
[1 mark]

The probability of getting 2 heads is: $3 \times \dfrac{1}{2} \times \dfrac{1}{2} \times \dfrac{1}{2} = \dfrac{3}{8}$
(multiply by 3 because any of the three coins could be the tail — the order in which the heads and the tail occur isn't important).
[1 mark]

Similarly the probability of getting 1 head is:
$3 \times \dfrac{1}{2} \times \dfrac{1}{2} \times \dfrac{1}{2} = \dfrac{3}{8}$

And the probability of getting no heads is $\dfrac{1}{2} \times \dfrac{1}{2} \times \dfrac{1}{2} = \dfrac{1}{8}$

So the probability of 1 or no heads $= \dfrac{3}{8} + \dfrac{1}{8} = \dfrac{1}{2}$ *[1 mark]*

Hence the probability distribution of X is:

x	20p	10p	nothing
P(X = x)	⅛	⅜	½

[1 mark]

b) You need the probability that X >10p *[1 mark]*

This is just $P(X = 20p) = \dfrac{1}{8}$ *[1 mark]*

Easy peasy. The difficult question is — why would anyone play such a rubbish game?

2 a) All the probabilities must add up to 1, so
$2k + 3k + k + k = 1$, i.e. $7k = 1$, and so $k = \frac{1}{7}$. *[1 mark]*

b) $P(X \leq 0) = P(X = 0) = \frac{2}{7}$ *[1 mark]*

$P(X \leq 1) = P(X = 0) + P(X = 1) = \frac{5}{7}$ *[1 mark]*

$P(X \leq 2) = P(X = 0) + P(X = 1) + P(X = 2) = \frac{6}{7}$
[1 mark]

$P(X \leq 3) = P(X = 0) + P(X = 1) + P(X = 2) + P(X = 3) = 1$
[1 mark]

So the cumulative distribution function is as in the following
table:

x	0	1	2	3
$P(X \leqslant x)$	$\frac{2}{7}$	$\frac{5}{7}$	$\frac{6}{7}$	1

c) $P(X > 2) = 1 - P(X \leq 2) = 1 - \frac{6}{7} = \frac{1}{7}$ *[1 mark]*

(Or $P(X > 2) = P(X = 3) = \frac{1}{7}$, using part a).)

*All probabilities add up to one! OK, that's the last time I'm going
to say it. Scout's honour.*

3 a)

x	0	1	2	3	4	5	6	7	8	9
$P(X = x)$	0.1	0.1	0.1	0.1	0.1	0.1	0.1	0.1	0.1	0.1

[1 mark]

b) mean $= \frac{0 + 9}{2} = 4.5$ *[1 mark]*

Variance $= \frac{(9 - 0 + 1)^2 - 1}{12}$ *[1 mark]* $= \frac{99}{12}$

$= 8.25$ *[1 mark]*

c) $P(X < 4.5) = P(X = 0) + P(X = 1)$
$+ P(X = 2) + P(X = 3) + P(X = 4)$ *[1 mark]*
$= 0.5$ *[1 mark]*

4 a) $P(X = 1) = a$, $P(X = 2) = 2a$, $P(X = 3) = 3a$.
Therefore the total probability is $3a + 2a + a = 6a$.
This must equal 1, so $a = \frac{1}{6}$. *[1 mark]*

b) $E(X) = \left(1 \times \frac{1}{6}\right) + \left(2 \times \frac{2}{6}\right) + \left(3 \times \frac{3}{6}\right) = \frac{1 + 4 + 9}{6}$ *[1 mark]*
$= \frac{7}{3}$ *[1 mark]*

c) $E(X^2) = Var(X) + [E(X)]^2 = \frac{5}{9} + \left(\frac{7}{3}\right)^2 = \frac{5 + 49}{9}$ *[1 mark]*
$= \frac{54}{9} = 6$ *[1 mark]*

d) $E(3X + 4) = 3E(X) + 4 = 3 \times \frac{7}{3} + 4 = 11$ *[1 mark]*
$Var(3X + 4) = 3^2 Var(X) = 9 \times \frac{5}{9}$ *[1 mark]*
$= 5$ *[1 mark]*

5 a) $E(X) = (0 \times 0.4) + (1 \times 0.3) + (2 \times 0.2) + (3 \times 0.1)$
$= 0 + 0.3 + 0.4 + 0.3$ *[1 mark]*
$= 1$ *[1 mark]*

b) $E(6X + 8) = 6E(X) + 8 = 6 + 8$ *[1 mark]*
$= 14$ *[1 mark]*

c) The formula for variance is $Var(X) = E(X^2) - [E(X)^2]$
So first work out $E(X^2)$:
$E(X^2) = (0^2 \times 0.4) + (1^2 \times 0.3) + (2^2 \times 0.2) + (3^2 \times 0.1)$
$= 0.3 + 0.8 + 0.9$ *[1 mark]*
$= 2$ *[1 mark]*

Then complete the formula also using your answer to part a):
$E(X^2) - [E(X)^2] = 2 - (1^2)$ *[1 mark]*
$= 1$ *[1 mark]*

d) $Var(aX + b) = a^2 Var(X)$
$Var(5 - 3X) = (-3)^2 Var(X) = 9Var(X) = 9 \times 1$
[1 mark]
$= 9$ *[1 mark]*

S1 Section 4 —
Correlation and Regression
Warm-up Questions

1) a)

b) First you need to find these values:
$\sum x = 1880$, $\sum y = 247$, $\sum x^2 = 410400$,
$\sum y^2 = 6899$ and $\sum xy = 40600$.
Then put these values into the PMCC formula:

$$\frac{40600 - \frac{[1880][247]}{10}}{\sqrt{\left(410400 - \frac{[1880]^2}{10}\right)\left(6899 - \frac{[247]^2}{10}\right)}}$$

$$= \frac{40600 - 46436}{\sqrt{(410400 - 353440)(6899 - 6100.9)}}$$

$$= \frac{-5836}{\sqrt{56960 \times 798.1}} = \frac{-5836}{6742.3865} = -0.866 \text{ (to 3 sig.fig.)}$$

c) The PMCC tells you that there is a strong negative
correlation between drink volume and alcohol
concentration — cocktails with smaller volumes tend to
have higher concentrations of alcohol.

*Don't panic about that nasty ol' PMCC equation. You need to
know how to USE it, but they give you the formula in the exam, so
you don't need to REMEMBER it. Hurrah.*

2) a) **Independent**: the annual number of sunny days
Dependent: the annual number of volleyball-related
injuries

b) **Independent**: the annual number of rainy days
Dependent: the annual number of Monopoly-related
injuries

c) **Independent**: a person's disposable income
Dependent: a person's spending on luxuries

d) **Independent**: the number of cups of tea drunk per day
Dependent: the number of trips to the loo per day

e) **Independent**: the number of festival tickets sold
Dependent: the number of pairs of Wellington boots
bought

Answers

3) a) (i) $S_{rr} = 26816.78 - \dfrac{517.4^2}{10} = 46.504$

 (ii) $S_{rw} = 57045.5 - \dfrac{517.4 \times 1099}{10} = 183.24$

 b) $b = \dfrac{S_{rw}}{S_{rr}} = \dfrac{183.24}{46.504} = 3.94$

 c) $a = \overline{w} - b\overline{r}$, where $\overline{w} = \dfrac{\sum w}{10} = 109.9$

 and $\overline{r} = \dfrac{\sum r}{10} = 51.74$

 So $a = 109.9 - 3.94 \times 51.74 = -94.0$

 d) The equation of the regression line is: $w = 3.94r - 94.0$

 e) When $r = 60$, the regression line gives an estimate for w of:
 $w = 3.94 \times 60 - 94.0 = 142.4$ g

 f) This estimate might not be very reliable because it uses an
 r-value from outside the range of the original data.
 It is extrapolation.

 *You'll be given the equations for finding a regression line — but you
 still need to know how to use them, otherwise the formula booklet
 will just be a blur of incomprehensible squiggles. Oh, and you need
 to practise USING them of course...*

4) Substitute expressions for w and r into the regression line's
 equation from Question 3: $8Q = 3.94(P + 5) - 94.0$
 Now rearrange to get an equation of the form $Q = a + bP$:
 $8Q = 3.94P + (3.94 \times 5 - 94.0)$
 i.e. $Q = 0.493P - 9.29$.

Exam Questions

1) a)

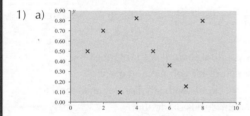

 **[2 marks for all points plotted correctly, or 1 mark
 if at least 3 points are plotted correctly.]**

 *Aren't scatter diagrams pretty... Just make sure you're
 not so distracted by their artistic elegance that you
 forget to be accurate and lose easy marks.*

 b) You need to work out these sums:

 $\sum x = 36, \sum y = 3.94,$
 $\sum x^2 = 204, \sum y^2 = 2.4676, \sum xy = 17.66$

 Then:

 $S_{xx} = \sum x^2 - \dfrac{(\sum x)^2}{n} = 204 - \dfrac{36^2}{8} = 42$

 $S_{yy} = \sum y^2 - \dfrac{(\sum y)^2}{n} = 2.4676 - \dfrac{3.94^2}{8} = 0.52715$

 $S_{xy} = \sum xy - \dfrac{(\sum x)(\sum y)}{n} = 17.66 - \dfrac{36 \times 3.94}{8} = -0.07$

 [3 marks available — 1 for each correct term]

This means:

$r = \dfrac{S_{xy}}{\sqrt{S_{xx}S_{yy}}} = \dfrac{-0.07}{\sqrt{42 \times 0.52715}} = -0.015$ (to 3 d.p.).

[1 mark]

c) This very small value for the correlation coefficient tells
 you that there appears to be only a very weak linear
 relationship between the two variables (or perhaps no
 linear relationship at all) **[1 mark]**.

2) a)

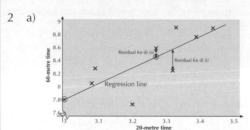

**[2 marks for all points plotted correctly, or 1 mark
if at least 3 points are plotted correctly.]**

b) It's best to make a table like this one, first:

									Totals
20-metre time, x	3.39	3.2	3.09	3.32	3.33	3.27	3.44	3.08	26.12
60-metre time, y	8.78	7.73	8.28	8.25	8.91	8.59	8.9	8.05	67.49
x^2	11.4921	10.24	9.5481	11.0224	11.0889	10.6929	11.8336	9.4864	85.4044
xy	29.7642	24.736	25.5852	27.39	29.6703	28.0893	30.616	24.794	220.645

**[2 marks for at least three correct sums,
or 1 mark if one total found correctly.]**

Then: $S_{xy} = 220.645 - \dfrac{26.12 \times 67.49}{8} = 0.29015$

[1 mark]

$S_{xx} = 85.4044 - \dfrac{26.12^2}{8} = 0.1226$

[1 mark]

Then the gradient b is given by:

$b = \dfrac{S_{xy}}{S_{xx}} = \dfrac{0.29015}{0.1226} = 2.3666$

[1 mark]

And the intercept a is given by:

$a = \overline{y} - b\overline{x} = \dfrac{\sum y}{n} - b\dfrac{\sum x}{n}$

$= \dfrac{67.49}{8} - 2.3666 \times \dfrac{26.12}{8} = 0.709$

[1 mark]

So the regression line has equation: $y = 2.367x + 0.709$

[1 mark]

To plot the line, find two points that the line passes through.
A regression line always passes through $(\overline{x}, \overline{y})$, which here
is (3.27, 8.44). Then put $x = 3$ (say) to find that the line also
passes through (3, 7.81).
Now plot these points (in circles) on your scatter diagram,
and draw the regression line through them
[1 mark for plotting the line correctly].

*Hmm, lots of fiddly things to calculate there. Remember, you get
marks for method as well as correct answers, so take it step by step
and show all your workings.*

Answers

c) (i) $y = 2.367 \times 3.15 + 0.709 = 8.17$ (to 3 sig. fig.), (8.16 if $b = 2.3666$ used) *[1 mark]*

This should be reliable, since we are using interpolation within the range of x for which we have data *[1 mark]*.

(ii) $y = 2.367 \times 3.88 + 0.709 = 9.89$ (to 3 sig. fig.) *[1 mark]*

This could be unreliable, since we are extrapolating beyond the range of the data *[1 mark]*.

d) (i) residual $= 8.25 - (2.367 \times 3.32 + 0.709)$
$= -0.317$ (3 sig. fig.), (-0.316 if $b = 2.3666$ used)

[1 mark for calculation, 1 mark for plotting residual correctly]

(ii) residual $= 8.59 - (2.367 \times 3.27 + 0.709)$
$= 0.141$ (3 sig. fig.), (0.142 if $b = 2.3666$ used)

[1 mark for calculation, 1 mark for plotting residual correctly]

3 a) Put the values into the correct PMCC formula:

$$\text{PMCC} = \frac{S_{xy}}{\sqrt{S_{xx}S_{yy}}} = \frac{12666}{\sqrt{310880 \times 788.95}}$$
$$= \frac{12666}{15661.059} = 0.809$$

[1 mark for correctly substituting the values into the PMCC formula, and 1 mark for the correct final answer.]

b) There is a strong positive correlation between the miles cycled in the morning and calories consumed for lunch. Generally, the further they have cycled, the more they eat *[1 mark]*.

c) 0.809
[1 mark]

Remember — the PMCC won't be affected if you multiply all the variables, so changing the data from miles to km doesn't change it.

4 a) At $x = 12.5$, $y = 211.599 + (9.602 \times 12.5) = 331.624$

At $x = 14.7$, $y = 211.599 + (9.602 \times 14.7) = 352.748$

[1 mark for each value of y correctly calculated]

b) Using the equation: 'Residual = Observed y-value – Estimated y-value':

At $x = 12.5$: Residual $= 332.5 - 331.624 = 0.876$ *[1 mark]*
At $x = 14.7$: Residual $= 352.1 - 352.748 = -0.648$ *[1 mark]*

And that's the end of that — Section 4 done and dusted. Bring on Section 5...

S1 Section 5 — The Normal Distribution

Warm-up Questions

1) Use the Z-tables:

a) $P(Z < 0.84) = 0.7995$

b) $P(Z < 2.95) = 0.9984$

c) $P(Z > 0.68) = 1 - P(Z \leq 0.68) = 1 - 0.7517 = 0.2483$

d) $P(Z \geq 1.55) = 1 - P(Z < 1.55)$
$= 1 - P(Z \leq 1.55) = 1 - 0.9394 = 0.0606$

e) $P(Z < -2.10) = P(Z > 2.10) = 1 - P(Z \leq 2.10)$
$= 1 - 0.9821 = 0.0179$

f) $P(Z \leq -0.01) = P(Z \geq 0.01)$
$= 1 - P(Z < 0.01) = 1 - 0.5040 = 0.4960$

g) $P(Z > 0.10) = 1 - P(Z \leq 0.10) = 1 - 0.5398 = 0.4602$

h) $P(Z \leq 0.64) = 0.7389$

i) $P(Z > 0.23) = 1 - P(Z \leq 0.23) = 1 - 0.5910 = 0.4090$

j) $P(0.10 < Z \leq 0.50) = P(Z \leq 0.50) - P(Z \leq 0.10)$
$= 0.6915 - 0.5398 = 0.1517$

k) $P(-0.62 \leq Z < 1.10) = P(Z < 1.10) - P(Z < -0.62)$
$= P(Z < 1.10) - P(Z > 0.62) = P(Z < 1.10) - (1 - P(Z \leq 0.62))$
$= 0.8643 - (1 - 0.7324) = 0.5967$

l) $P(-0.99 < Z \leq -0.74) = P(Z \leq -0.74) - P(Z \leq -0.99)$
$= P(Z \geq 0.74) - P(Z \geq 0.99)$
$= (1 - P(Z < 0.74)) - (1 - P(Z < 0.99))$
$= (1 - 0.7704) - (1 - 0.8389) = 0.0685$

I know... these all get a bit fiddly. As always — take it nice and slow, and double-check each step as you do it.

2) a) If $P(Z < z) = 0.9131$, then from the Z-table, $z = 1.36$.

b) If $P(Z < z) = 0.5871$, then from the Z-table, $z = 0.22$.

c) If $P(Z > z) = 0.0359$, then $P(Z \leq z) = 0.9641$.
From the Z-table, $z = 1.80$.

d) If $P(Z > z) = 0.01$, then from the percentage-points table, $z = 2.3263$.

e) If $P(Z \leq z) = 0.4013$, then z must be negative (and so won't be in the Z-table).
But this means $P(Z < -z) = 1 - 0.4013 = 0.5987$.
Using the Z-table, $-z = 0.25$, so $z = -0.25$.

It's getting a bit trickier here, with all the z and $-z$ business. If you need to draw a graph here to make it a bit clearer what's going on, then draw one.

f) If $P(Z \geq z) = 0.995$, then $P(Z \leq -z) = 1 - 0.995 = 0.005$.
From the percentage points table, $-z = 2.5758$.
So $z = -2.5758$.

When you've answered a question like this, always ask yourself whether your answer looks 'about right'. Here, you need a number that Z is very very likely to be greater than... so your answer is going to be negative, and it's going to be pretty big.
So $z = -2.5758$ looks about right.

3) a) $P(X < 55) = P\left(Z < \frac{55 - 50}{\sqrt{16}}\right) = P(Z < 1.25) = 0.8944$

b) $P(X < 42) = P\left(Z < \frac{42 - 50}{\sqrt{16}}\right) = P(Z < -2)$
$= P(Z > 2) = 1 - P(Z \leq 2) = 1 - 0.9772 = 0.0228$

c) $P(X > 56) = P\left(Z > \frac{56 - 50}{\sqrt{16}}\right) = P(Z > 1.5)$
$= 1 - P(Z \leq 1.5) = 1 - 0.9332 = 0.0668$

d) $P(47 < X < 57) = P(X < 57) - P(X \leq 47)$
$= P(Z < 1.75) - P(Z \leq -0.75)$
$= 0.9599 - P(Z \geq 0.75)$
$= 0.9599 - (1 - P(Z < 0.75))$
$= 0.9599 - (1 - 0.7734) = 0.7333$

Answers

4) a) $P(X < 0) = P\left(Z < \frac{0-5}{7}\right) = P(Z < -0.71) = P(Z > 0.71)$
   $= 1 - P(Z \leq 0.71) = 1 - 0.7611 = 0.2389$

   b) $P(X < 1) = P\left(Z < \frac{1-5}{7}\right) = P(Z < -0.57) = P(Z > 0.57)$
   $= 1 - P(Z \leq 0.57) = 1 - 0.7157 = 0.2843$

   c) $P(X > 7) = P\left(Z > \frac{7-5}{7}\right) = P(Z > 0.29)$
   $= 1 - P(Z \leq 0.29) = 1 - 0.6141 = 0.3859$

   d) $P(2 < X < 4) = P(X < 4) - P(X \leq 2)$
   $= P(Z < -0.14) - P(Z \leq -0.43)$
   $= P(Z > 0.14) - P(Z \geq 0.43)$
   $= (1 - P(Z \leq 0.14)) - (1 - P(Z < 0.43))$
   $= (1 - 0.5557) - (1 - 0.6664)) = 0.1107$

5) $P(X < 8) = 0.8925$ means $P\left(Z < \frac{8-\mu}{\sqrt{10}}\right) = 0.8925$.

   From tables, $\frac{8-\mu}{\sqrt{10}} = 1.24$.

   So $\mu = 8 - 1.24 \times \sqrt{10} = 4.08$ (to 3 sig. fig.).

6) $P(X > 221) = 0.3085$ means $P\left(Z > \frac{221-\mu}{8}\right) = 0.3085$,

   or $P\left(Z \leq \frac{221-\mu}{8}\right) = 1 - 0.3085 = 0.6915$.

   From tables, $\frac{221-\mu}{8} = 0.5$.

   So $\mu = 221 - 8 \times 0.5 = 217$

7) $P(X < 13) = 0.6$ means $P\left(Z < \frac{13-11}{\sigma}\right) = 0.6$,

   or $P\left(Z \geq \frac{13-11}{\sigma}\right) = 1 - 0.6 = 0.4$.

   From the percentage points table, $\frac{13-11}{\sigma} = 0.2533$.

   So $\sigma = \frac{13-11}{0.2533} = 7.90$.

8) $P(X \leq 110) = 0.9678$ means $P\left(Z < \frac{110-108}{\sigma}\right) = 0.9678$,

   or $P\left(Z < \frac{2}{\sigma}\right) = 0.9678$.

   From tables, $\frac{2}{\sigma} = 1.85$.

   So $\sigma = \frac{2}{1.85} = 1.08$.

9) $P(X < 15.2) = 0.9783$ means $P\left(Z < \frac{15.2-\mu}{\sigma}\right) = 0.9783$.

   From tables, $\frac{15.2-\mu}{\sigma} = 2.02$, or $2.02\sigma + \mu = 15.2$.

   $P(X > 14.8) = 0.1056$ means $P\left(Z > \frac{14.8-\mu}{\sigma}\right) = 0.1056$,

   or $P\left(Z \leq \frac{14.8-\mu}{\sigma}\right) = 1 - 0.1056 = 0.8944$.

   From tables, $\frac{14.8-\mu}{\sigma} = 1.25$, or $1.25\sigma + \mu = 14.8$.

   Solving these simultaneous equations gives $\sigma = 0.52$

   and $\mu = 14.15$.

10) If X represents the mass of an item, then:

   a) $P(X < 55) = P\left(Z < \frac{55-55}{4.4}\right) = P(Z < 0) = 0.5$.

   b) $P(X < 50) = P\left(Z < \frac{50-55}{4.4}\right) = P(Z < -1.14)$
   $= P(Z > 1.14) = 1 - P(Z \leq 1.14)$
   $= 1 - 0.8729 = 0.1271$

   c) $P(X > 60) = P\left(Z > \frac{60-55}{4.4}\right) = P(Z > 1.14)$
   $= 1 - P(Z \leq 1.14) = 1 - 0.8729 = 0.1271$

11) If X represents the mass of an egg, then:

   a) $P(X < 1) = P\left(Z < \frac{1-1.4}{0.3}\right) = P(Z < -1.33)$
   $= P(Z > 1.33) = 1 - P(Z \leq 1.33)$
   $= 1 - 0.9082 = 0.0918$

   b) $P(X > 1.5) = P\left(Z > \frac{1.5-1.4}{0.3}\right) = P(Z > 0.33)$
   $= 1 - P(Z \leq 0.33)$
   $= 1 - 0.6293 = 0.3707$

   c) $P(1.3 < X < 1.6) = P(X < 1.6) - P(X < 1.3)$
   $= P\left(Z < \frac{1.6-1.4}{0.3}\right) - P\left(Z < \frac{1.3-1.4}{0.3}\right)$
   $= P(Z < 0.67) - P(Z < -0.33)$
   $= 0.7486 - (1 - P(Z < 0.33))$
   $= 0.7486 - (1 - 0.6293) = 0.3779$

12) a) Here X ~ N(80, 15). You need to use your percentage points table for these.

   If $P(X < a) = 0.99$, then $P\left(Z \geq \frac{a-80}{\sqrt{15}}\right) = 0.01$

   So $\frac{a-80}{\sqrt{15}} = 2.3263$ (using the table).

   Rearrange this to get $a = 80 + 2.3263 \times \sqrt{15}$
   $= 89.01$

   b) $|X - 80| < b$ means that X is 'within b' of 80,
   i.e. $80 - b < X < 80 + b$.
   Since 80 is the mean of X, and since a normal distribution is symmetrical,
   $P(80 - b < X < 80 + b) = 0.8$ means that
   $P(80 < X < 80 + b) = 0.4$
   i.e. $P\left(\frac{80-80}{\sqrt{15}} < Z < \frac{80+b-80}{\sqrt{15}}\right) = 0.4$
   i.e. $P\left(0 < Z < \frac{b}{\sqrt{15}}\right) = 0.4$

   This means that $P\left(Z < \frac{b}{\sqrt{15}}\right) - P(Z \leq 0) = 0.4$
   i.e. $P\left(Z < \frac{b}{\sqrt{15}}\right) - 0.5 = 0.4$, or $P\left(Z < \frac{b}{\sqrt{15}}\right) = 0.9$
   Use your percentage points table to find that
   $\frac{b}{\sqrt{15}} = 1.2816$, or $b = 1.2816 \times \sqrt{15} = 4.964$

   An elephant never forgets... that a normal distribution is symmetrical. It's often very useful for figuring out these sorts of questions, and elephants love calculating probability distributions more than they love peanuts.

Answers

Exam Questions

1 **a)** Let X represent the exam marks. Then X ~ N(50, 30^2).

$$P(X \geq 41) = P\left(Z \geq \frac{41-50}{30}\right)$$

$$= P\left(Z \geq -\frac{9}{30}\right) = P(Z \geq -0.3) \text{ [1 mark]}$$

$$= P(Z \leq 0.3) = 0.6179 \text{ [1 mark]}$$

So 0.6179 × 1000 = 618 is the expected number who passed the exam *[1 mark]*.

b) If *a* is the mark needed for a distinction, then:

$$P(X \geq a) = 0.1 \text{ means } P\left(Z \geq \frac{a-50}{30}\right) = 0.1 \text{ [1 mark]}.$$

From tables, $\frac{a-50}{30} = 1.2816$ *[1 mark]*,

or *a* = 88.4 *[1 mark]*.

2 Assume that the lives of the batteries are distributed as: $N(\mu, \sigma^2)$. Then P(X < 20) = 0.4 and P(X < 30) = 0.8.
[1 mark]

Transform these 2 equations to get:

$$P\left[Z < \frac{20-\mu}{\sigma}\right] = 0.4 \text{ and } P\left[Z < \frac{30-\mu}{\sigma}\right] = 0.8 \text{ [1 mark]}$$

Now you need to use your percentage points table to get:

$$\frac{20-\mu}{\sigma} = -0.2533 \text{ [1 mark] and } \frac{30-\mu}{\sigma} = 0.8416 \text{ [1 mark]}$$

Now rewrite these as:

$$20 - \mu = -0.2533\sigma \text{ and } 30 - \mu = 0.8416\sigma. \text{ [1 mark]}$$

Subtract these two equations to get:

$$10 = (0.8416 + 0.2533)\sigma$$

i.e. $\sigma = \frac{10}{0.8416 + 0.2533} = 9.1333$ *[1 mark]*

Now use this value of σ in one of the equations above:

$$\mu = 20 + 0.2533 \times 9.1333 = 22.31 \text{ [1 mark]}$$

So X~N(22.31, 9.13^2) i.e. X~N(22.31, 83.4)

This might feel a bit repetitive but if you're comfortable doing probability distribution questions in lots of ever-so-very-slightly different ways then the exam'll be a doddle. Nearly finished....

3 **a)** $P(X > 145) = P\left(Z > \frac{145-120}{25}\right) = P(Z > 1)$ *[1 mark]*

$$= 1 - P(Z \leq 1) \text{ [1 mark]}$$

$$= 1 - 0.8413 = 0.1587 \text{ [1 mark]}.$$

b) P(120 < X < *j*) = 0.4641 means

$$P(X < j) - P(X \leq 120) = 0.4641 \text{ [1 mark]}.$$

$$P(X < j) - P(X \leq 120)$$

$$= P\left(Z < \frac{j-120}{25}\right) - P\left(Z \leq \frac{120-120}{25}\right) \text{ [1 mark]}$$

$$= P\left(Z < \frac{j-120}{25}\right) - 0.5 = 0.4641,$$

or $P\left(Z < \frac{j-120}{25}\right) = 0.9641$ *[1 mark]*

From tables, $\frac{j-120}{25} = 1.80$,

or *j* = 1.8 × 25 + 120 = 165 *[1 mark]*

4 **a)** For a normal distribution, mean = median, so median = 12 inches *[1 mark]*.

b) Let *X* represent the base diameters. Then P(X > 13) = 0.05 *[1 mark]*.

$$P(X > 13) = P\left(Z > \frac{13-12}{\sigma}\right) = 0.05 \text{ [1 mark]}.$$

From the percentage points table, $\frac{1}{\sigma} = 1.6449$ *[1 mark]*.

This means $\sigma = \frac{1}{1.6449} = 0.61$ (to 2 sig. fig.) *[1 mark]*.

c) $P(X < 10.8) = P\left(Z < \frac{10.8-12}{0.61}\right) = P(Z < -1.97)$ *[1 mark]*

$$= 1 - P(Z \leq 1.97) = 1 - 0.9756 = 0.0244 \text{ [1 mark]}$$

So you would expect 0.0244 × 100 ≈ 2 pizza bases to be discarded *[1 mark]*.

d) P(at least 1 base too small) = 1 – P(no bases too small).
P(base not too small) = 1 – 0.0244 = 0.9756.
P(no bases too small) = 0.9756^3 = 0.9286 *[1 mark]*
P(at least 1 base too small) = 1 – 0.9286 *[1 mark]*
= 0.0714 *[1 mark]*.

5 **a)** Let *X* represent the volume of compost in a bag. Then $X \sim$ N(50, 0.4^2).

$$P(X < 49) = P\left(Z < \frac{49-50}{0.4}\right) = P(Z < -2.50) \text{ [1 mark]}$$

$$= 1 - P(Z \leq 2.50) \text{ [1 mark]}$$

$$= 1 - 0.9938 = 0.0062 \text{ [1 mark]}$$

b) $P(X > 50.5) = P\left(Z > \frac{50.5-50}{0.4}\right) = P(Z > 1.25)$ *[1 mark]*

$$= 1 - P(Z \leq 1.25) \text{ [1 mark]}$$

$$= 1 - 0.8944 = 0.1056 \text{ [1 mark]}$$

So in 1000 bags, 0.1056 × 1000 *[1 mark]*
≈ 106 bags *[1 mark]* (approximately) would be expected to contain more than 50.5 litres of compost.

c) $P(Y < 74) = 0.10$ means $P\left(Z < \frac{74-75}{\sigma}\right) = 0.1$.

To use the percentage points table, rewrite this as

$$P\left(Z > \frac{75-74}{\sigma}\right) = 0.1 \text{ [1 mark]},$$

which gives $\frac{1}{\sigma} = 1.2816$ *[1 mark]*.

So, $\sigma = \frac{1}{1.2816} = 0.78$ litres *[1 mark]*

S1 — Practice Exam One

1) **a)** 4.1 metres (the upper quartile) *[1 mark]*.

b) They represent outliers *[1 mark]* — data values that are a long way away from the rest of the readings *[1 mark]*.

c)

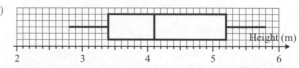

[1 mark for the median shown correctly, 1 mark for the upper quartile, 1 mark for the lower quartile, and 1 mark for both the minimum and the maximum shown correctly.]

Answers

d) The giraffes in the zoo are generally taller, with a higher median *[1 mark]*, and higher upper and lower quartiles *[1 mark]*. The two populations seem similarly varied, since they have similar ranges (ignoring the outliers), although the IQR for the giraffes in the zoo is greater than for the giraffes in the game reserve *[1 mark]*. The skew of the distributions is very different, with the zoo's population having a positive skew and the game reserve's population having a negative skew *[1 mark]*.

2) a) The events are mutually exclusive, so they can't both happen. Hence $P(A \cap B) = 0$ *[1 mark]*.

b) For mutually exclusive events, $P(A \cup B) = P(A) + P(B)$ *[1 mark]*. So $P(A \cup B) = 0.3 + 0.4 = 0.7$ *[1 mark]*.

c) The probability that neither event happens is equal to $1 - P(A \cup B)$ *[1 mark]* $= 1 - 0.7 = 0.3$ *[1 mark]*.

d) $P(A \mid B)$ is the probability of A, given that B has already happened. But since A and B are mutually exclusive, they can't both happen — so $P(A \mid B)$ must equal zero. You can use the formula for conditional probability to get the same answer:
$$P(A \mid B) = \frac{P(A \cap B)}{P(B)}.$$
[1 mark for the correct answer, and 1 mark for a reasonable explanation.]

3) a) $p = 0.2$, since the probabilities have to add up to 1 *[1 mark]*.

Easy.

b) Discrete uniform distribution *[1 mark]*.

Easy.

c) $F(25) = P(X \le 25)$, but this is just the same as $P(X \le 20)$ *[1 mark]*, because the distribution is discrete. This is equal to $P(X = 0) + P(X = 10) + P(X = 20) = 0.6$ *[1 mark]*.

Getting harder.

d) $E(X) = \sum xP(X = x)$
$= 0.2 \times (0 + 10 + 20 + 50 + 100)$ *[1 mark]*
$= 36$ *[1 mark]*
$Var(X) = \sum x^2 P(X = x) - \{E(X)\}^2$
$= 0.2 \times (0 + 100 + 400 + 2500 + 10\,000) - 36^2$ *[1 mark]*
$= 1304$ *[1 mark]*

A bit harder.

e) $E(3X - 4) = 3E(X) - 4$ *[1 mark]*
$= 3 \times 36 - 4 = 104$ *[1 mark]*
$Var(3X - 4) = 3^2 \times Var(X)$ *[1 mark]*
$= 9 \times 1304 = 11\,736$ *[1 mark]*

Trickier — especially the variance bit.

f) X is a random variable that shows what's paid out. The expected value of X is 36p, so a charge of 40p will average a profit of 4p per game *[1 mark]*. This is unlikely to be sufficient to cover the owner's costs *[1 mark for any sensible comment]*.

Ooh — easy again.

g) Probably not. It would be usual for small prizes to have high probabilities and big prizes to have low probabilities. *[1 mark for saying that this model is unlikely, and 1 mark for any sensible explanation as to why.]*

Sigh of relief — wasn't as bad as I thought.

4) This is one of those where you have to 'standardise' the normal variable (i.e. subtract the mean and divide by the standard deviation) and use tables for the standard normal variable Z.

a) $P(X < 7.5) = P\left(\frac{X - 8}{\sqrt{1.2}} < \frac{7.5 - 8}{\sqrt{1.2}}\right)$
$= P(Z < -0.46)$ *[1 mark]*
$= 1 - P(Z \le 0.46)$ *[1 mark]*
$= 1 - 0.6772 = 0.3228$ *[1 mark]*

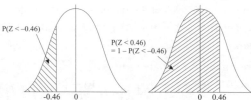

A quick sketch always helps.

b) You need to find the probability that the duration is less than 7 minutes or more than 9 minutes. This is $P(X > 9) + P(X < 7)$. By symmetry, these are equal, so:
$$P(X < 7) + P(X > 9) = 2 \times P\left(Z < \frac{7 - 8}{\sqrt{1.2}}\right) \text{ [1 mark]}$$
$= 2 \times P(Z < -0.91)$ *[1 mark]*
$= 2 \times (1 - P(Z \le 0.91))$ *[1 mark]*
$= 2 \times (1 - 0.8186) = 0.3628$ *[1 mark]*

These do definitely get a bit tricky. Just have a good think about what you need to find out before you launch into a load of working out. Getting loads of practice at this kind of question is dead helpful — it gets you used to working with those tables.

c) You need to find d where: $P(X > d) = 0.01$.
$$P(X > d) = 0.01, \text{ so } P\left(Z > \frac{d - 8}{\sqrt{1.2}}\right) = 0.01 \text{ [1 mark]}.$$
Using the percentage points table: $\frac{d - 8}{\sqrt{1.2}} = 2.3263$ *[1 mark]*.
So $d = \sqrt{1.2} \times 2.3263 + 8$ *[1 mark]*,
i.e. $d = 10.5$ minutes (to 3 sig. fig.) *[1 mark]*

These 'standardise the normal variable' questions get everywhere. They look hard, but you soon get used to them. And once you get your head round the basic idea, you'll be able to do pretty much anything they ask you.

5) a) If C is the event 'has had a crash' and G is the event 'wears glasses', then the tree diagram is as follows:

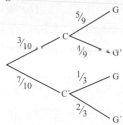

[1 mark for each correct pair of branches, but lose 1 mark for not cancelling down fractions.]

b) The easiest way is to work out the probabilities of the two branches ending in G by multiplying along each branch, and then adding the results (i.e. $P(G) = P(G \cap C) + P(G \cap C')$).

$P(G \cap C) = \frac{3}{10} \times \frac{5}{9} = \frac{15}{90} = \frac{1}{6}$ *[1 mark]*,

and $P(G \cap C') = \frac{7}{10} \times \frac{1}{3} = \frac{7}{30}$ *[1 mark]*.

Adding these together you get:

$P(G) = \frac{1}{6} + \frac{7}{30} = \frac{5+7}{30} = \frac{12}{30} = \frac{2}{5}$ *[1 mark]*.

c) You need to find: $P(C\,|\,G) = \frac{P(C \cap G)}{P(G)}$ *[1 mark]*.

You've just worked out $P(C \cap G)$ and $P(G)$, so

$P(C\,|\,G) = \frac{1}{6} \div \frac{2}{5}$ *[1 mark]*

$= \frac{1}{6} \times \frac{5}{2} = \frac{5}{12}$ *[1 mark]*.

Tree diagrams are things you've definitely seen before, and they don't get any harder. The question may sound more complicated, but it's not really.

6) a) The mean (μ) is given by $\mu = \frac{\sum x}{10} = \frac{500}{10} = 50$ *[1 mark]*.

The variance (σ^2) is 'the mean of the squares minus the square of the mean'. And the standard deviation (σ) is just the square root of the variance.

So the variance is given by

$\sigma^2 = \frac{\sum x^2}{10} - 50^2$

$= \frac{25622}{10} - 2500 = 62.2$ *[1 mark]*

So $\sigma = \sqrt{62.2} = 7.89$ *[1 mark]*.

b) (i) The mean will be unchanged *[1 mark]*, because the new value is equal to the original mean *[1 mark]*.

(ii) The standard deviation will decrease *[1 mark]*. This is because the standard deviation measures the deviation of values from the mean. So by adding a new value that's equal to the mean, you're not adding to the total deviation from the mean, but as you have an extra reading, you now have to divide by 11 (not 10) when you work out the variance *[1 mark]*.

Understanding what the standard deviation actually is can help you get your head round questions like this.

7) a) $S_{RR} = \sum R^2 - \frac{(\sum R)^2}{n}$

$= 847 - \frac{79^2}{8} = 66.875 = 66.9$ (to 3 sig. fig.) *[1 mark]*

$S_{MM} = \sum M^2 - \frac{(\sum M)^2}{n}$

$= 45884 - \frac{600^2}{8} = 884$ *[1 mark]*

$S_{RM} = \sum RM - \frac{\sum R \sum M}{n}$

$= 6143 - \frac{79 \times 600}{8} = 218$ *[1 mark]*

b) $r = \frac{S_{RM}}{\sqrt{S_{RR} \times S_{MM}}} = \frac{218}{\sqrt{66.875 \times 884}}$ *[1 mark]*

$= 0.897$ (to 3 sig. fig.) *[1 mark]*

c) The amount of revision (R) *[1 mark]*, since the exam mark depends on the amount of revision done, not the other way around *[1 mark]*.

The explanatory variable is also known as the independent variable (and the response variable as the dependent variable). The dependent variable depends on the value of the independent variable.

d) The correlation coefficient is very close to 1, so if the points were plotted on a scatter graph, they would all lie close to a straight line *[1 mark]*.

e) $b = \frac{S_{RM}}{S_{RR}} = \frac{218}{66.875} = 3.26$ (to 3 sig. fig.) *[1 mark]*

$a = \overline{M} - b\overline{R} = \frac{600}{8} - 3.26 \times \frac{79}{8}$ *[1 mark]*

$= 42.8$ (to 3 sig. fig.) *[1 mark]*

So $M = 42.8 + 3.26R$ *[1 mark]*.

f) Using the regression line to estimate the mark:
$M = 42.8 + 3.26 \times 4 = 55.84 \approx 56$ marks *[1 mark]*.

g) 4 is outside the range for which there is data — extrapolation is needed so the estimate is unreliable / This estimate may be unreliable, since this student may not fit the pattern generated by the others
[1 mark for any sensible comment].

S1 — *Practice Exam Two*

1) a) Let H be hard centre, N be nutty and S be soft centre.

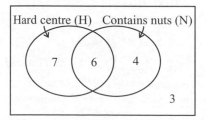

Other answers are possible, e.g. showing S and N instead of H and N.

[1 mark for two appropriate sets (e.g. H and N, or S and N, etc.), plus 2 marks for all numbers correctly marked on the diagram, or 1 mark if only 1 or 2 numbers are correctly marked.]

Answers

b) (i) $P(S) = \dfrac{\text{Number of soft centres}}{\text{Total number of chocolates}}$

$= \dfrac{4+3}{20} = \dfrac{7}{20}$ *[1 mark]*

(ii) $P(H \mid N)$

$= \dfrac{\text{Number of chocolates in both H and N}}{\text{Number of chocolates in N}}$ *[1 mark]*

$= \dfrac{6}{10} = \dfrac{3}{5}$ *[1 mark]*

In part (ii), you're "given that the chocolate contains a nut", so you only need to look at the circle containing nutty chocolates — you can ignore the rest of the diagram.

c) $P(\text{pick hard centre with 1st pick}) = \dfrac{13}{20} \times \dfrac{7}{19} \times \dfrac{6}{18}$

$= \dfrac{546}{6840} = \dfrac{91}{1140}$ *[1 mark]*

This is P(HSS), but you also need to add P(SHS) and P(SSH). In fact, P(HSS) = P(SHS) = P(SSH), so you need to multiply the above answer by 3 *[1 mark]*.

So $P(\text{pick hard centre}) = 3 \times \dfrac{91}{1140} = \dfrac{91}{380}$ *[1 mark]*.

Don't forget that the hard centre could be the first one picked out of the box, or the second or the third — and you need to take the fact that there are 'different arrangements' into account when you work out your probability.

2) a) The total probability must be 1, so go through all the possible values of x and add the probabilities *[1 mark]*:

$\dfrac{k}{6} + \dfrac{2k}{6} + \dfrac{3k}{6} + \dfrac{3k}{6} + \dfrac{2k}{6} + \dfrac{k}{6} = \dfrac{12k}{6} = 2k$ *[1 mark]*.

This must equal 1, so k must be $\dfrac{1}{2}$ *[1 mark]*.

b) $F(3) = P(X \le 3)$, i.e. F(3) is the probability that X takes a value less than or equal to 3.

So $F(3) = P(X = 1) + P(X = 2) + P(X = 3)$ *[1 mark]*

$= \dfrac{1}{12} + \dfrac{2}{12} + \dfrac{3}{12}$

$= \dfrac{6}{12} = \dfrac{1}{2}$ *[1 mark]*

c) $E(X) = \sum x P(X = x)$

$= \left(1 \times \dfrac{1}{12}\right) + \left(2 \times \dfrac{2}{12}\right) + \left(3 \times \dfrac{3}{12}\right)$

$+ \left(4 \times \dfrac{3}{12}\right) + \left(5 \times \dfrac{2}{12}\right) + \left(6 \times \dfrac{1}{12}\right)$ *[1 mark]*

$= \dfrac{1 + 4 + 9 + 12 + 10 + 6}{12} = \dfrac{42}{12} = 3.5$ *[1 mark]*

d) (i) $E(2 - 3X) = 2 - 3E(X)$ *[1 mark]*

$= 2 - (3 \times 3.5) = -8.5$ *[1 mark]*

$\text{Var}(2 - 3X) = 3^2 \text{Var}(X)$ *[1 mark]*

$= 9 \times \dfrac{23}{12} = \dfrac{69}{4} = 17.25$ *[1 mark]*

The expected value formula is kind of what you'd expect. But be careful with the variance formula — it's a bit weird.

(ii) $\text{Var}(X) = E(X^2) - \{E(X)\}^2$

(i.e. the variance is the 'mean of the squares minus the square of the mean').

Rearranging you get: $E(X^2) = \text{Var}(X) + \{E(X)\}^2$ *[1 mark]*

So $E(X^2) = \dfrac{23}{12} + (3.5)^2 = \dfrac{23}{12} + \left(\dfrac{7}{2}\right)^2$ *[1 mark]*

$= \dfrac{23 + 147}{12} = \dfrac{170}{12} = \dfrac{85}{6}$ *[1 mark]*.

3) a) $\text{Mean} = \dfrac{\sum x}{n} = \dfrac{66.5}{12}$ *[1 mark]*

$= 5.54 \text{ (or £5540) (to 3 sig. fig.)}$ *[1 mark]*.

$\text{Variance} = \dfrac{\sum x^2}{n} - \left(\dfrac{\sum x}{n}\right)^2 = \dfrac{390.97}{12} - \left(\dfrac{66.5}{12}\right)^2$ *[1 mark]*

$= 1.87 \text{ (to 3 sig. fig.)}$ *[1 mark]*.

b) The ordered list of the 12 data points is:

3.8, 4.1, 4.2, 4.6, 4.9, 5.5, 5.8, 5.9, 6.0, 6.2, 6.4, 9.1.

The position of the median is $\dfrac{1}{2}(n + 1) = \dfrac{1}{2}(12 + 1) = 6.5$, so take the average of the 6th and 7th values.

So the median Q_2 is: $\dfrac{1}{2}(5.5 + 5.8) = 5.65$ *[1 mark]*.

Since $12 \div 4 = 3$, the lower quartile is the average of the 3rd and 4th values.

So the lower quartile Q_1 is: $\dfrac{1}{2}(4.2 + 4.6) = 4.4$ *[1 mark]*.

Since $12 \div 4 \times 3 = 9$, the upper quartile is the average of the 9th and 10th values.

So the upper quartile Q_3 is: $\dfrac{1}{2}(6.0 + 6.2) = 6.1$ *[1 mark]*.

c) (i) $\dfrac{Q_3 - 2Q_2 + Q_1}{Q_3 - Q_1} = \dfrac{6.1 - 2 \times 5.65 + 4.4}{6.1 - 4.4}$

$= -0.471 \text{(to 3 sig.fig.)}$ *[1 mark]*

(ii) The data is slightly negatively skewed *[1 mark]*, which means that the values between the median and the upper quartile are closer together than the values between the median and the lower quartile *[1 mark]*.

Make sure you get the direction of the skew correct... a negative skew means that there's a tail on the 'low side' of the distribution (dragging the mean downwards), which means there are a lot of values <u>higher</u> than the mean.

d) The lower fence is given by:

$Q_1 - 1.5 \times (Q_3 - Q_1) = 4.4 - 1.5 \times (6.1 - 4.4) = 1.85$ *[1 mark]*. So there are no outliers below the lower fence *[1 mark]*. The upper fence is given by:

$Q_3 + 1.5 \times (Q_3 - Q_1) = 6.1 + 1.5 \times (6.1 - 4.4) = 8.65$ *[1 mark]*. So there is one outlier — the value of 9.1 *[1 mark]*.

4) a) A normal distribution is symmetrical, so the median (i.e. the value that 50% of values fall below) is the same as the mean *[1 mark]*. So the median = 93 °C *[1 mark]*.

It's an easy question really, but not 100% obvious when you first look at it.

b) $P(X \ge 95) = 0.2$.

Transform this to a statement about the standard normal distribution Z by subtracting the mean (93) and dividing by the standard deviation (σ) *[1 mark]*:

$P\left(Z > \dfrac{95 - 93}{\sigma}\right) = 0.2$, i.e. $P\left(Z > \dfrac{2}{\sigma}\right) = 0.2$ *[1 mark]*.

Using the percentage points table: $\dfrac{2}{\sigma} = 0.8416$ *[1 mark]*.

So $\sigma = \dfrac{2}{0.8416} = 2.3764... = 2.38 \text{ (to 3 sig.fig.)}$ *[1 mark]*.

c) You need to find P(X < 88). This equals:

$P\left(Z < \dfrac{88 - 93}{2.3764}\right) = P(Z < -2.10) = 1 - P(Z < 2.10)$

[1 mark]. From tables, $P(Z < 2.10) = 0.9821$.

So $P(X < 88) = 1 - 0.9821 = 0.0179$ *[1 mark]*.

Answers

d) You need to find a value d with
$P(93 - d \leq T \leq 93 + d) = 0.99$. Because of the symmetry,
you can say that you need to find d with
$P(T \geq 93 + d) = 0.005$ *[1 mark]*, or
$P\left(Z \geq \frac{93 + d - 93}{2.3764}\right) = P\left(Z \geq \frac{d}{2.3764}\right) = 0.005$ *[1 mark]*.
Using the percentage points table, this gives:
$\frac{d}{2.3764} = 2.5758$ *[1 mark]*, or $d = 6.12$ *[1 mark]*.

*With a normal distribution, if you transform the variable to Z, then
chances are you're on the right track.*

5) a)

*[2 marks for all points correctly plotted, or 1 mark
if at least 4 points are correctly plotted.]*

b) $S_{TT} = \sum T^2 - \frac{\left(\sum T\right)^2}{n} = 828 - \frac{64^2}{8} = 316$ *[1 mark]*
$S_{Ty} = \sum Ty - \frac{\sum T \sum y}{n} = 219.05 - \frac{64 \times 26.2}{8} = 9.45$
[1 mark].

*These formulas will be given to you on the formula sheet, but they're
written in terms of x and y. Make sure you can still use them when
you're given data using variables other than x and y.*

c) If $y = a + bT$, then
$b = \frac{S_{Ty}}{S_{TT}} = \frac{9.45}{316} = 0.029905... = 0.0299$ (to 3 sig.fig.).
[1 mark].
And $a = \overline{y} - b\overline{T}$
$= \frac{26.2}{8} - 0.0299 \times \frac{64}{8}$ *[1 mark]*
$= 3.0358 = 3.04$ (to 3 sig. fig.) *[1 mark]*.

So the equation of the regression line is:
$y = 3.04 + 0.0299T$ *[1 mark]*.

*These formulas for the regression line coefficients will also be on your
formula sheet, but again make sure you know how to use them with
variables that are called something other than x and y.*

d) a represents the length of the cable when it is not under
tension (i.e. when $T = 0$) *[1 mark]*.
b represents the extra extension of the cable when the
tension is increased by 1 kN *[1 mark]*.

e) $y = 3.04 + 0.0299T$, so when $T = 30$,
$y = 3.04 + 0.0299 \times 30 = 3.937$ *[1 mark]*
$= 3.94$ metres (to 3 sig. fig.) *[1 mark]*
(or 3.93 m if $a = 3.0358$ used)

f) This estimate may be unreliable as it involves extrapolating
beyond the range of the experimental data *[1 mark]*.

6) a) (i) The probabilities must sum to 1, so:
$0.1 + 0.2 + p + q + 0.2 = 1$, or $p + q = 0.5$ *[1 mark]*.
$E(X) = 6.3$, so $(0.1 \times 2) + (0.2 \times 4) +$
$\qquad (p \times 6) + (q \times 8) + (0.2 \times 10) = 6.3$,
or $6p + 8q = 3.3$ *[1 mark]*.

(ii) Rearrange the first equation to give an expression for p,
then substitute this into the second equation.
$p = 0.5 - q$, so
$6 \times (0.5 - q) + 8q = 3.3$ *[1 mark]*,
which gives $2q = 0.3$. This means $q = 0.15$ *[1 mark]*.
Then $p = 0.5 - q = 0.5 - 0.15$, i.e. $p = 0.35$ *[1 mark]*.

*Even though simultaneous equations aren't part of S1, you're still
expected to know how to solve them.*

b) (i) $E(X^2) = (0.1 \times 2^2) + (0.2 \times 4^2) + (0.35 \times 6^2)$
$\qquad + (0.15 \times 8^2) + (0.2 \times 10^2)$ *[1 mark]*
$\quad = 45.8$ *[1 mark]*

(ii) $Var(X) = E(X^2) - E(X)^2 = 45.8 - 6.3^2$ *[1 mark]*
$\qquad = 6.11$ *[1 mark]*.

c) $F(2) = P(X \leq 2) = P(X = 2) = 0.1$.
$F(4) = P(X \leq 4) = P(X \leq 2) + P(X = 4) = 0.1 + 0.2 = 0.3$.
$F(6) = P(X \leq 6) = P(X \leq 4) + P(X = 6) = 0.3 + 0.35 = 0.65$.
$F(8) = P(X \leq 8) = P(X \leq 6) + P(X = 8) = 0.65 + 0.15 = 0.8$.
$F(10) = P(X \leq 10) = P(X \leq 8) + P(X = 10) = 0.8 + 0.2 = 1$.

x	2	4	6	8	10
$F(x)$	0.1	0.3	0.65	0.8	1

*[1 mark for at least one correct value, 2 marks if at
least two values are correct, 3 marks for at least three
correct, and 4 marks if all values are correct.]*

*An easy one to finish... as long as you remember what a cumulative
distribution function is.*

Answers

M1 Section 1 — Kinematics
Warm-up Questions

1) $u = 3$; $v = 9$; $a = a$; $s = s$; $t = 2$. Use $s = \frac{1}{2}(u + v)t$
$s = \frac{1}{2}(3 + 9) \times 2$ $s = \frac{1}{2}(12) \times 2 = 12$ m

2)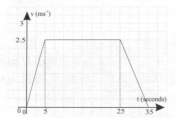

distance = $(5 \times 2.5) \div 2 + (20 \times 2.5) + (10 \times 2.5) \div 2$
= 68.75 m

3) velocity = area under (t, a) graph

a) $t = 3$, velocity = $(3 \times 5) \div 2 = 7.5$ ms^{-1}

b) $t = 5$, velocity = $7.5 + (2 \times 5) = 17.5$ ms^{-1}

c) $t = 6$, velocity = $17.5 + (1 \times 5) \div 2 = 20$ ms^{-1}

Exam Questions

1 a) Using $u = v - at$
$u = 17 - (9.8 \times 1.2)$
So, $u = 5.24$ ms^{-1}

[3 marks available in total]:
- *1 mark for using appropriate equation*
- *1 mark for correct workings*
- *1 mark for correct value of u*

These kind of questions are trivial if you've memorised all those constant acceleration equations. If you haven't, you know what to do now (turn to p. 147).

b) Using $s = ut + \frac{1}{2}at^2$
$s = (17 \times 2.1) + \frac{1}{2}(9.8 \times 2.1^2)$
So, $s = 57.3$
$h = \frac{s}{14} = 4.09$ m ≈ 4.1 m

[4 marks available in total]:
- *1 mark for using appropriate equation*
- *1 mark for correct value of s*
- *1 mark for correct workings*
- *1 mark for correct value of h*

2 a)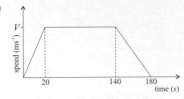

[3 marks available in total]:
- *1 mark for the correct shape*
- *1 mark for the correct times*
- *1 mark for correctly marking V on the vertical axis*

b) Area under graph (area of trapezium) = distance
$\frac{1}{2}(120 + 180)V = 2100$
$V = \frac{2100}{150} = 14$ ms^{-1}

[3 marks available in total]:
- *1 mark for using Area under graph = distance*
- *1 mark for correct workings*
- *1 mark for correct value of V*

The area could also be worked out in other ways, say, using two triangles and a rectangle. Best to use whatever's easiest for you.

c) Distance = area under graph
= $\frac{1}{2} \times 40 \times 14 = 280$ m

[2 marks available in total]:
- *1 mark for using Area under graph = distance*
- *1 mark for correct value*

d) Using $a = \frac{v - u}{t}$.
Acceleration period:
$t = 20$, $v = 14$, $u = 0$
$a = (14 - 0) \div 20 = 0.7$ ms^{-2}
Deceleration period:
$t = 40$, $v = 0$, $u = 14$
$a = (0 - 14) \div 40 = -0.35$ ms^{-2}

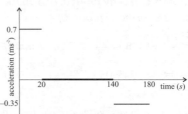

[3 marks available in total]:
- *1 mark for correct shape*
- *1 mark for correct value of acceleration (0.7 ms^{-2})*
- *1 mark for correct value of deceleration (0.35 ms^{-2})*

Yep, sometimes you're not given the values you need to label a graph and you have to work them out yourself. All good practice.

3 a) Using $v^2 = u^2 + 2as$
$20^2 = u^2 + (2 \times 9.8 \times 8)$
$u = \pm\sqrt{400 - 156.8}$
The initial velocity is negative, as the rocket is projected upwards.
So $u = -15.59... = -15.6$ ms^{-1} (3 s.f.)

[3 marks available in total]:
- *1 mark for using appropriate equation*
- *1 mark for correct workings*
- *1 mark for correct value of u*

b) Using $v = u + at$
$20 = -15.6 + 9.8t$
So, $9.8t = 35.6$
hence $t = 3.6$ s

[3 marks available in total]:
- *1 mark for using appropriate equation*
- *1 mark for correct workings*
- *1 mark for correct value of t*

You could also have done this using $s = ut + \frac{1}{2}at^2$. It's okay to use other equations as long as you get the right answer in the end.
Constant use of those constant acceleration equations...

Answers

M1 Section 2 — Vectors and Statics
Warm-up Questions

1) Displacement = $(15 \times 0.25) - (10 \times 0.75) = -3.75$ km

Time taken = 1 hour

Average velocity = -3.75 kmh^{-1} (i.e. 3.75 kmh^{-1} south)

2) $(3\mathbf{i} + 7\mathbf{j}) + 2 \times (-2\mathbf{i} + 2\mathbf{j}) - 3 \times (\mathbf{i} - 3\mathbf{j})$

$= (3 - 4 - 3)\mathbf{i} + (7 + 4 + 9)\mathbf{j} = -4\mathbf{i} + 20\mathbf{j}$

3)

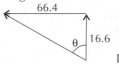

Resolving East: $60\sin45° + 70\cos70° = 66.4$ miles

Resolving North: $-40 - 60\cos45° + 70\sin70°$

$\qquad\qquad\qquad = -16.6$ miles

Magnitude of $\mathbf{r}$ = $\sqrt{66.4^2 + 16.6^2} = 68.4$ miles

Direction = $\theta = \tan^{-1}\left(\dfrac{66.4}{16.6}\right) = 76.0°$

Bearing is $360° - 76.0° = 284°$

Top-lop-tip: When answering questions concerning vectors or forces always draw a diagram. I promise it'll make it simpler.

4) a) Small point mass, no air resistance, no wind, released from rest.

b) Small point mass, no air resistance, no wind, released from rest.

c) Same assumptions as in a) and b), although it might not be safe to ignore wind if outside as table tennis balls are very light.

You need to get familiar with modelling and all the terminology used in M1 — it's going to be really tricky to figure out M1 questions if you're not.

5) 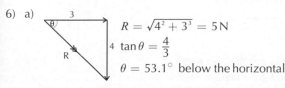 Assumptions: Point mass, one point of contact with ground, constant driving force D from engine, constant friction, F, includes road resistance and air resistance, acceleration = 0 as it's moving at 25mph (in a straight line).

6) a)

$R = \sqrt{4^2 + 3^3} = 5$ N

$\tan\theta = \dfrac{4}{3}$

$\theta = 53.1°$ below the horizontal

b)

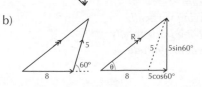

$R = \sqrt{(8 + 5\cos60°)^2 + (5\sin60°)^2} = 11.4$ N

$\tan\theta = \dfrac{5\sin60°}{8 + 5\cos60°} = 0.412$

So $\theta = 22.4°$ above the horizontal

c) Total force up $= 6 - 4\sin10° - 10\sin20° = 1.885$ N

Total force left $= 10\cos20° - 4\cos10° = 5.458$ N

$R = \sqrt{1.885^2 + 5.458^2} = 5.77$ N

$\theta = \tan^{-1}\dfrac{1.885}{5.458}$

$\qquad = 19.1°$ above the horizontal

7)

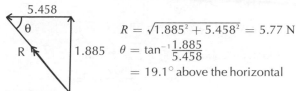

a) $\tan30° = \dfrac{20}{T_B}$

$T_B = \dfrac{20}{\tan30°}$

$\quad = 34.6$ N

b) $\sin30° = \dfrac{20}{mg}$

$mg = \dfrac{20}{\sin30°}$

$m = 4.08$ kg

8)

Huge hint: The angle of the plane to the horizontal (in this case 30°) will always be the angle in here.

Sine rule: $\dfrac{mg}{\sin100°} = \dfrac{70}{\sin30°}$

So $mg = \dfrac{70\sin100°}{\sin30°}$

$m = 14.1$ kg

$\dfrac{R}{\sin50°} = \dfrac{70}{\sin30°}$

So $R = \dfrac{70\sin50°}{\sin30°} = 107$ N

Yet another example of how triangles might just save your life mark for the M1 module. Although you could also have solved it by resolving forces parallel and perpendicular to the slope if that floats your boat.

9)

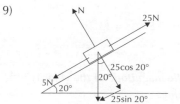

Force perpendicular to the slope: $N = 25\cos20° = 23.5$ N

Force parallel to the slope: $25 - 25\sin20° - 5 = 11.4$ N.
So the resultant force is 11.4 N up the slope.

Answers

10) a)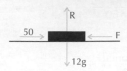

Resolve vertically: $R = 12g$
Use formula: $F \leq \mu R$
$$F \leq \frac{1}{2}(12g)$$
$$F \leq 58.8 \text{ N}$$

50 N isn't big enough to overcome friction — so it has no overall motion.

b) Force would have to be > 58.8 N

11) Moments about B: $60g \times 3 = T_2 \times 8$
So $T_2 = \frac{180g}{8} = 220.5$ N

Vertically balanced forces, so $T_1 + T_2 = 60g$
$$T_1 = 367.5 \text{ N}$$

Ah, I remember my first moments question, it was a very memorable moment...

Exam Questions

1   Resolve horizontally: $0 + 5\cos30° = 4.33$ N
Resolve vertically: $4 - 5\sin30° = 1.5$ N

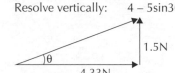

1.5N
4.33N
$\theta = \tan^{-1}\left(\frac{1.5}{4.33}\right) = 19.1°$

i.e. $\theta = 19.1°$ above the horizontal
Magnitude $= \sqrt{1.5^2 + 4.33^2} = 4.58$ N

[4 marks available in total]:
- *1 mark for resolving horizontally*
- *1 mark for resolving vertically*
- *1 mark for calculating the direction*
- *1 mark for calculating the magnitude*

Triangles — how do I love thee? Let me count the ways...

2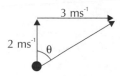

Magnitude $= \sqrt{2^2 + 3^2} = \sqrt{13} = 3.61$ ms^{-1}
$\theta = \tan^{-1}\left(\frac{3}{2}\right) = 56.3°$
So angle to river bank is $90° - 56.3° = 33.7°$

[4 marks available in total]:
- *1 mark for a diagram*
- *1 mark for calculating the magnitude of the velocity*
- *1 mark for correct workings*
- *1 mark for calculating the angle from the bank*

...one big way, really. They're just super useful when resolving things.

3   a)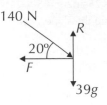

[2 marks available in total]:
- *1 mark for drawing 4 correct arrows*
- *1 mark for correctly labelling the arrows*

b) Resolve vertically:
$R = 39g + 140\sin20° = 430$ N
Resolve horizontally:
$F = 140\cos20° = 132$ N

[4 marks available in total]:
- *1 mark for resolving vertically*
- *1 mark for resolving horizontally*
- *1 mark for correct reaction magnitude*
- *1 mark for correct friction magnitude*

No need to panic if friction is involved — it's just another thing to consider when resolving (and an extra arrow to draw).

4   a) $R = A + B = (2\mathbf{i} - 11\mathbf{j}) + (7\mathbf{i} + 5\mathbf{j}) = (9\mathbf{i} - 6\mathbf{j})$

[2 marks available in total]:
- *1 mark for correct workings*
- *1 mark for correct resultant*

b) $|R| = \sqrt{9^2 + (-6)^2} = 10.8$ N

[2 marks available in total]:
- *1 mark for correct workings*
- *1 mark for correct magnitude*

5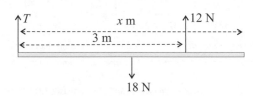

Taking moments about end string:
$12 \times 3 = 18 \times \frac{x}{2}$ so, $36 = 9x$
$x = 4$ m

[3 marks available in total]:
- *1 mark for diagram*
- *1 mark for correct workings*
- *1 mark for showing that x = 4 m*

6   a)

Resolve vertically:
$4\sin30° = 2$
Resolve horizontally:
$7 + 4\cos30° = 10.5$

$R = \sqrt{2^2 + 10.5^2}$
$= 10.7$ N

[4 marks available in total]:
- *1 mark for diagram*
- *1 mark for resolving vertically*
- *1 mark for resolving horizontally*
- *1 mark for correct magnitude*

Answers

b) $\tan \alpha = \dfrac{2}{10.5} = 0.19$

$\alpha = 10.8°$

[2 marks available in total]:
- *1 mark for correct workings*
- *1 mark for correct value of α*

7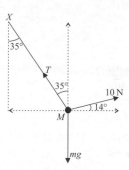

Resolving vertically: $mg = 10\sin14° + T\cos35°$
Need to find T.
Resolving horizontally: $T\sin35° = 10\cos14°$
So, $T = \dfrac{10\cos14°}{\sin35°} = 16.9$ N
So $mg = 10\sin14° + 16.9\cos35° = 16.3$ N
Therefore, the mass of $M = \dfrac{16.3}{g} = 1.66$ kg

[4 marks available in total]:
- *1 mark for resolving vertically*
- *1 mark for resolving horizontally*
- *1 mark for correct value of T*
- *1 mark for correct mass of M*

I think this question is sort of fun, but then I also think rods are sort of pretty... Anyway, a good clear diagram here will simplify matters no end.

8 a)

$\sin\theta = \dfrac{12}{15}$, so, $\theta = 53.1°$

[2 marks available in total]:
- *1 mark for correct workings*
- *1 mark for the correct value of θ*

b) $15^2 = W^2 + 12^2$
$W = \sqrt{15^2 - 12^2} = 9$ N

[2 marks available in total]:
- *1 mark for correct workings*
- *1 mark for correct value of W*

c) Remove W and the particle moves in the opposite direction to W, i.e. upwards. This resultant of the two remaining forces is 9 N upwards, because the particle was in equilibrium beforehand.

[2 marks available in total]:
- *1 mark for correct magnitude*
- *1 mark for correct direction*

The word 'state' in an exam question means that you shouldn't need to do any extra calculation to answer it.

9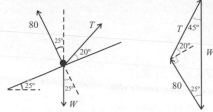

a) $\dfrac{T}{\sin25°} = \dfrac{80}{\sin45°}$

So, $T = \dfrac{80\sin25°}{\sin45°} = 47.8$ N

[3 marks available in total]:
- *1 mark for diagram*
- *1 mark for correct workings*
- *1 mark for correct value of T*

b) $\dfrac{W}{\sin110°} = \dfrac{80}{\sin45°}$

So, $W = \dfrac{80\sin110°}{\sin45°} = 106$ N

[2 marks available in total]:
- *1 mark for correct workings*
- *1 mark for correct value of W*

Do you remember the sine rule? If not, go and revise it (try p. 61) — a bit of trigonometry never hurt anyone, and it'll serve you well in M1. You can solve pretty much anything with triangles...

10 A = original position vector + tv = $(\mathbf{i} + 2\mathbf{j}) + 8(3\mathbf{i} + \mathbf{j})$
  = $(\mathbf{i} + 2\mathbf{j}) + (24\mathbf{i} + 8\mathbf{j}) = (25\mathbf{i} + 10\mathbf{j})$ m
  $B = A + tv = (25\mathbf{i} + 10\mathbf{j}) + 5(-4\mathbf{i} + 2\mathbf{j})$
  = $(25\mathbf{i} + 10\mathbf{j}) + (-20\mathbf{i} + 10\mathbf{j}) = (5\mathbf{i} + 20\mathbf{j})$ m

[4 marks available in total]:
- *1 mark for correct workings for A*
- *1 mark for correct value of A*
- *1 mark for correct workings for B*
- *1 mark for correct value of B*

Argh. My head hurts with the sheer quantity of $\mathbf{i}$s and $\mathbf{j}$s in those workings. Better to take it a step at a time (it's good for your health).

11 a)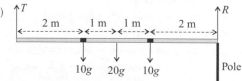

[2 marks available in total]:
- *1 mark for diagram*
- *1 mark for correct labelling*

b) Take moments about the pole:
$6T = (2 \times 10g) + (3 \times 20g) + (4 \times 10g)$
$6T = 20g + 60g + 40g = 120g$
So, $T = 20g$

[3 marks available in total]:
- *1 mark for taking moments about the pole*
- *1 mark for correct workings*
- *1 mark for correct value of T*

Why work around the pole? Because we have no idea what the magnitude of R is and working there allows us to ignore it.

Answers

c) Resolve vertically:

$T + R = 10g + 20g + 10g$

$20g + R = 40g$

So, $R = 20g$

[2 marks available in total]:
- *1 mark for resolving vertically*
- *1 mark for correct value of R*

I suppose we could have found R first and then T, but that's not the order the questions are in, so it's best just to run with it...

12 a) Resolving horizontally:

$A\cos50° = 58$ N

so, $A = \dfrac{58}{\cos 50°} = 90.2$ N

[3 marks available in total]:
- *1 mark for resolving horizontally*
- *1 mark for correct workings*
- *1 mark for correct value of A*

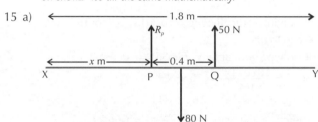

b) Resolving vertically:

$mg = A\sin50° = 90.2\sin50°$

so, $mg = 69.1$ and $m = 7.05$ kg

[3 marks available in total]:
- *1 mark for resolving vertically*
- *1 mark for correct workings*
- *1 mark for correct value of m*

13 a) Speed = $\sqrt{7^2 + (-3)^2} = \sqrt{58}$

$= 7.62$ ms^{-1}

[2 marks available in total]:
- *1 mark for correct workings*
- *1 mark for correct value*

A bit of a classic here, and easy if you remember that for a vector (xi + yj) then the magnitude = $\sqrt{x^2 + y^2}$.

b) Angle from horizontal = $\tan^{-1}\left(\dfrac{-3}{7}\right)$

$= -23.2°$ (i.e. 23° below the horizontal)

The bearing is measured from north

so, bearing = $90 + 23.2 = 113°$

[3 marks available in total]:
- *1 mark for correct workings*
- *1 mark for calculating angle from horizontal*
- *1 mark for correct bearing*

Bearings are measured from the north, because that's where Polar bears live, and where the Be[a]ring Strait is — maybe...

c) Position at $t = 4$:

$(\mathbf{i} + 5\mathbf{j}) + 4(7\mathbf{i} - 3\mathbf{j}) = (29\mathbf{i} - 7\mathbf{j})$ m

Displacement to $15\mathbf{i}$:

$15\mathbf{i} - (29\mathbf{i} - 7\mathbf{j}) = (-14\mathbf{i} + 7\mathbf{j})$ m

Velocity = $\dfrac{s}{t} = (-14\mathbf{i} + 7\mathbf{j})/3.5 = (-4\mathbf{i} + 2\mathbf{j})$

So, $a = -4$ and $b = 2$

[4 marks available in total]:
- *1 mark for calculating position at t = 4*
- *1 mark for calculating displacement to 15i*
- *1 mark for correct workings*
- *1 mark for correct values of a and b*

14 Resolve horizontally: $S\cos40° = F$

Resolve vertically: $R = 2g + S\sin40°$

It's limiting friction so $F = \mu R$

So, $S\cos40° = \dfrac{3}{10}(2g + S\sin40°)$

$S\cos40° = 0.6g + 0.3S\sin40°$

$S\cos40° - 0.3S\sin40° = 0.6g$

$S(\cos40° - 0.3\sin40°) = 0.6g$

$S = 10.3$ N

[4 marks available in total]:
- *1 mark for resolving horizontally*
- *1 mark for resolving vertically*
- *1 mark for correct workings*
- *1 mark for correct value of S*

A ring on a rod — it might be a car on a road, or a sled on snow... it's all the same mathematically.

15 a)

[2 marks available in total]:
- *1 mark for diagram (ensure that 0.6 < x < 0.9)*
- *1 mark for labelling*

b) Moments about P:

$80(0.9 - x) = 0.4(50) \Rightarrow 72 - 80x = 20$

So, $80x = 52$ and thus $x = 0.65$ m

[3 marks available in total]:
- *1 mark for taking moments about P*
- *1 mark for correct workings*
- *1 mark for correct value of x*

c) E.g. The log is a rod (i.e. a long particle)

The rocks are smooth (i.e. no friction)

The bird is a particle (i.e. dimensionless)

[3 marks available in total]:
- *1 mark for log as rod*
- *1 mark for rocks as smooth*
- *1 mark for bird as a particle*

d) Distance $QY = 1.8 - (0.4 + 0.65) = 0.75$ m

At the point of tipping about Q, so $R_p = 0$

Taking moments about point Q:

$0.75mg = 80(0.9 - 0.75) = 12$

So, $mg = 16$ N and therefore $m = 1.63$ kg

[4 marks available in total]:
- *1 mark for resolving about Q*
- *1 mark for realising $R_p = 0$*
- *1 mark for correct workings*
- *1 mark for correct value of m*

It's about to tip over about Q, so there's no pressure on P to hold the log up any more, hence $R_p = 0$.

Answers

e) Resolving vertically:

$R_P + R_Q = 80 + 16 = 96$ N

As $R_P = 0$ (about to tip about Q), $R_Q = 96$ N

[2 marks available in total]:
- **1 mark for resolving vertically**
- **1 mark for correct value of R_Q**

...and, you're done. Ace. I hope that all went swimmingly, but a bit more revision never hurt anyone, so if you're unsure about anything head back to the relevant section and have another look.

M1 Section 3 — Dynamics
Warm-up Questions

1) Resolve horizontally: $F_{net} = ma$

$2 = 1.5a$ so $a = 1\frac{1}{3}$ ms^{-2}

$v = u + at$ $v = 0 + (1\frac{1}{3} \times 3) = 4$ ms^{-1}

And we're off. I hope that bit of resolving was simple enough (if not you might want to go revise).

2) $F_{net} = (24\mathbf{i} + 18\mathbf{j}) + (6\mathbf{i} + 22\mathbf{j}) = 30\mathbf{i} + 40\mathbf{j}$

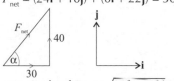

magnitude of $F_{net} = \sqrt{30^2 + 40^2} = 50$ N

$F_{net} = ma$, so $50 = 8a$, which gives $a = 6.25$ ms^{-2}

$\tan\alpha = \frac{40}{30}$, i.e. $\alpha = 53.1°$

$s = ut + \frac{1}{2}at^2$, $s = 0 \times 3 + \frac{1}{2} \times 6.25 \times 3^2 = 28.1$ m

3) Resolve horizontally:

$F_{net} = ma$ $P - 1 = 2 \times 0.3$

So, $P = 1.6$ N

Resolve vertically: $R = 2g$

Limiting friction: $F = \mu R$, so $1 = \mu \times 2g$,

which gives $\mu = 0.05$ (to 2 d.p.)

4) Resolving in ↖ direction:

$F_{net} = ma$

$R - 1.2g\cos25° = 1.2 \times 0$

$R = 1.2g\cos25°$

$R = 10.66$ N

Resolving in ↙ direction:

$F_{net} = ma$

$1.2g\sin25° - F = 1.2 \times 0.3$

So, $F = 1.2g\sin25° - 1.2 \times 0.3 = 4.61$ N

Limiting friction, so:

$F = \mu R$

$4.61 = \mu \times 10.66$

$\mu = 0.43$ (to 2 d.p.)

Assumptions:

i) brick slides down line of greatest slope

ii) acceleration is constant

iii) no air resistance

iv) point mass / particle

5) Resolving in ↖ direction:

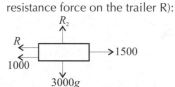

$F_{net} = ma$

$R - 600\cos30° = \left(\frac{600}{g}\right) \times 0$

$R = 600\cos30°$

Sliding, so $F = \mu R$

$F = 0.5 \times 600\cos30° = 259.8$

Resolving in ↙ direction:

$600\sin30° - F = \left(\frac{600}{g}\right)a$

$600\sin30° - 259.8 = 61.22 \times a$

$a = 0.657$ ms^{-2}

$u = 0$ ⎫ $v^2 = u^2 + 2as$

$s = 20$ ⎪ $v^2 = 0^2 + 2 \times 0.657 \times 20$

$a = 0.657$ ⎬ $v^2 = 26.28$

$v = ?$ ⎭ $v = 5.13$ ms^{-1}

6) Taking tractor and trailer together (and calling the resistance force on the trailer R):

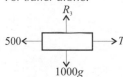

Resolving horizontally: $F_{net} = ma$

$1500 - R - 1000 = 3000 \times 0$

$R = 500$ N

For trailer alone:

$500 \leftarrow$ [ ] $\rightarrow T$

R_3 ↑ $1000g$ ↓

Resolving horizontally:

$F_{net} = ma$

$T - 500 = 1000 \times 0$

$T = 500$ N

T could be found instead by looking at the horizontal forces acting on the tractor alone.

7) Resolving downwards for A:

$F_{net} = ma$

$4g - T = 4 \times 1.2$

$T = 4g - 4.8$ ①

Resolving upwards for B:

$F_{net} = ma$

$T - W = \frac{W}{g} \times 1.2$ ②

Sub ① into ② :

$(4g - 4.8) - W = \frac{W}{g}(1.2)$

$4g - 4.8 = W(1 + \frac{1.2}{g})$

So $W = 30.6$ N

Answers

Here the particles are connected over a pulley, rather than in a straight line, but the key is still resolving — downwards and upwards instead of horizontally and vertically (or parallel and perpendicular to a plane, as in the next question). No need for any panic then. Phew.

8) For B, resolving vertically:

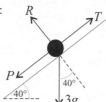

$F_{net} = ma$

$4g - T = 4a$

$T = 4g - 4a$ ①

For A, resolving in ↗ direction:

$F_{net} = ma$

$T - 3g\sin40° = 3a$ ②

Sub ① into ②:

$4g - 4a - 3g\sin40° = 3a$

$4g - 3g\sin40° = 7a$

$a = 2.90 \text{ ms}^{-2}$ (to 3 s.f.)

Sub into ①:

$T = 4g - (4 \times 2.9) = 27.6 \text{ N}$ (to 3 s.f.)

If equilibrium, then for B:

$T = 4g$

Then for A:

Resolving in ↗ direction:

$F_{net} = ma$

$T - 3g\sin40 - P = 0$

$4g - 3g\sin40 = P$

$P = 20.3 \text{ N}$ (to 3 s.f.)

9) a) $(5 \times 3) + (4 \times 1) = (5 \times 2) + (4 \times v)$

$19 = 10 + 4v$

$v = 2¼ \text{ ms}^{-1}$ to the right

b) $(5 \times 3) + (4 \times 1) = 9v$

$19 = 9v$

$v = 2\frac{1}{9} \text{ ms}^{-1}$ to the right

c) $(5 \times 3) + (4 \times -2) = (5 \times -v) + (4 \times 3)$

$7 = -5v + 12$

$5v = 5$

$v = 1 \text{ ms}^{-1}$ to the left

d) $(m \times 6) + (8 \times 2) = (m \times 2) + (8 \times 4)$

$6m + 16 = 2m + 32$

$4m = 16$

$m = 4 \text{ kg}$

Collision questions have me bouncing off the ceiling... Be careful with your directions (positive and negative) and it'll all be okay.

10) Impulse acts against motion, so $I = -2 \text{ Ns}$

$I = mv - mu$

$-2 = 0.3v - (0.3 \times 5)$

$v = -1\frac{2}{3} \text{ ms}^{-1}$

Impulse has 2 different equations, I = mv – mu and I = Ft. Remember both, but it's the first one that comes up most in exams.

11) You need to find the particle's velocities just before and just after impact. Falling (down = +ve):

$\left. \begin{array}{l} u = 0 \\ s = 2 \\ a = 9.8 \\ v = ? \end{array} \right\}$ $\begin{array}{l} v^2 = u^2 + 2as \\ v = \sqrt{2 \times 9.8 \times 2} \\ v = 6.261 \text{ ms}^{-1} \end{array}$

Rebound (this time, let up = +ve):

$\left. \begin{array}{l} v = 0 \\ u = ? \\ a = -9.8 \\ s = 1\frac{1}{3} \end{array} \right\}$ $\begin{array}{l} v^2 = u^2 + 2as \\ 0 - u^2 = 2 \times -9.8 \times 1\frac{1}{3} \\ u = 5.112 \text{ ms}^{-1} \end{array}$

Taking up = +ve:

Impulse $= mv - mu$

$= (0.45 \times 5.112) - (0.45 \times -6.261)$

$= 5.12 \text{ Ns}$

Exam Questions

1 a) Constant velocity, so, $a = 0$

Resolve horizontally:

$F_{net} = ma$

$T_2\cos40° - T_1\cos40° = 300 \times 0$

$T_2\cos40° = T_1\cos40°$

$T_2 = T_1$

Resolve vertically:

$F_{net} = ma$

$T_1\sin40° + T_2\sin40° - 300g = 300 \times 0$

Let $T_1 = T_2 = T$: $2T\sin40° = 300g$

$T = 2290 \text{ N}$ (to 3 s.f.)

[4 marks available in total]:
- *1 mark for resolving horizontally*
- *1 mark for resolving vertically*
- *1 mark for substituting T_1 or T_2*
- *1 mark for the correct value of T*

More triangles, that's what I like to see...

b) Resolve horizontally: $F_{net} = ma$

$T_2\cos40° - T_1\cos40° = 300 \times 0.4$

$T_2 - T_1 = 156.65 \text{ N}$ ①

Resolve vertically:

$F_{net} = ma$

$T_1\sin40° + T_2\sin40° - 300g = 300 \times 0$

$T_1\sin40° + T_2\sin40° = 300g$

So $T_1 + T_2 = 4573.83 \text{ N}$ ②

from ①: $T_2 = T_1 + 156.65$

into ②: $T_1 + (T_1 + 156.65) = 4573.83$

so $2T_1 = 4417.18$

$T_1 = 2210 \text{ N}$ (to 3 s.f.)

and, $T_2 = 2370 \text{ N}$ (to 3 s.f.)

[6 marks available in total]:
- *1 mark for resolving horizontally*
- *1 mark for finding ①*
- *1 mark for resolving vertically*
- *1 mark for finding ②*
- *1 mark for correct value of T_1*
- *1 mark for correct value of T_2*

Answers

c) E.g. cables are inextensible, particle is considered as a point mass, there's no air resistance.

[2 marks available in total]:
• **1 mark each for any 2 relevant assumptions.**

If you got these right, I will make the assumption that you've done some revision...

2 a) Resolving in $\nearrow$ direction:

$F_{net} = ma$
$8\cos15° + F - 7g\sin15° = 7 \times 0$
$F = 7g\sin15° - 8\cos15°$
$F = 10.03$ N

Resolving in $\nwarrow$ direction:
$F_{net} = ma$
$R - 8\sin15° - 7g\cos15° = 7 \times 0$
$R = 8\sin15° + 7g\cos15° = 68.33$ N

Limiting friction:
$F = \mu R$, i.e. $10.03 = \mu \times 68.33$,
which gives $\mu = 0.15$ (2 d.p.)

[5 marks available in total]:
• **1 mark for resolving in $\nearrow$ direction**
• **1 mark for correct value of F_{net} in $\nearrow$ direction**
• **1 mark for resolving in $\nwarrow$ direction**
• **1 mark correct value of R**
• **1 mark for correct value of μ**

b) 8 N removed:
Resolving in $\swarrow$ direction:
$7g\sin15° - F = 7a$ ①
Resolving in $\nwarrow$ direction:
$R - 7g\cos15° = 7 \times 0$
$R = 7g\cos15° = 66.26$ N
$F = \mu R$
$F = 0.147 \times 66.26 = 9.74$ N
① : $7g\sin15° - 9.74 = 7a$
$a = \dfrac{8.01}{7} = 1.14$ ms^{-2} (to 2 d.p.)
$s = 3; u = 0; a = 1.14; t = ?$ $\quad s = ut + \frac{1}{2}at^2$
$3 = 0 + \frac{1}{2} \times 1.14 \times t^2$ $\quad t = \sqrt{\dfrac{6}{1.14}} = 2.3$ s (to 2 s.f.)

[7 marks available in total]:
• **1 mark for resolving in $\swarrow$ direction**
• **1 mark for resolving in $\nwarrow$ direction**
• **1 mark for correct value of R**
• **1 mark for correct value of F**
• **1 mark for correct value of a**
• **1 mark for appropriate calculation method for t**
• **1 mark for correct value of t**

3 a) Considering the car and the caravan together:

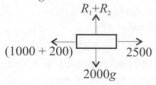

Resolving horizontally:
$F_{net} = ma$
$2500 - 1200 = 2000a$
$a = 0.65$ ms^{-2}

[3 marks available in total]:
• **1 mark for resolving horizontally**
• **1 mark for correct workings**
• **1 mark for correct value of a**

b) Either: *Caravan*
Resolving horizontally:

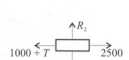

$F_{net} = ma$
$T - 200 = 500 \times 0.65$
$T = 525$ N

[2 marks available in total]:
• **1 mark for resolving horizontally**
• **1 mark for correct value of T**

Or: *Car*
Resolving horizontally:
$F_{net} = ma$
$2500 - (1000 + T) = 1500 \times 0.65$
$2500 - 1000 - T = 975$
$1500 - 975 = T$
$T = 525$ N

[2 marks available in total]:
• **1 mark for resolving horizontally**
• **1 mark for correct value of T**

Two different methods, one correct answer. At the end of the day, it doesn't matter which you use (but show your diagrams and workings), although it's certainly a bonus if you manage to pick the simpler way and save a bit of time in the exam.

4 a) For Q:
$F_{net} = ma$
Resolving vertically:
$mg - T = 0$, so $T = mg$
For P:
$F_{net} = ma$
Resolving in $\nwarrow$ direction:
$R - 1g\cos20° = 1 \times 0$, so $R = g\cos20°$
Limiting friction:
$F = \mu R = 0.1 \times g\cos20°$
Resolving in $\swarrow$ direction:
$1g\sin20° - F - T = 1 \times 0$
$1g\sin20° - 0.1g\cos20° - mg = 0$
$\sin20° - 0.1\cos20° = m$
$m = 0.25$ kg (to 2 s.f.)

[5 marks available in total]:
• **1 mark for resolving vertically**
• **1 mark for resolving in $\nwarrow$ direction**
• **1 mark for correct value of F**
• **1 mark for resolving in $\swarrow$ direction**
• **1 mark for correct value of m**

Answers

b) If Q = 1kg:

$F_{net} = ma$

For Q:

Resolving vertically:

$1g - T = 1a$, so $T = g - a$ ①

For P:

Resolving in ↖ direction:

$R = g\cos20°$

$F = \mu R = 0.1g\cos20°$

Resolving in ↗ direction:

$T - 1g\sin20° - F = 1a$

$T - 1g\sin20° - 0.1g\cos20° = 1a$ ②

Sub ① into ②: $(g - a) - g\sin20° - 0.1g\cos20° = a$

$g - g\sin20° - 0.1g\cos20° = 2a$

$5.527 = 2a$, so $a = 2.76$ ms^{-2} (to 3 s.f.)

i.e. the masses move with an acceleration of 2.76 ms^{-2}.

[5 marks available in total]:
- *1 mark for resolving vertically*
- *1 mark for resolving in ↖ direction*
- *1 mark for resolving in ↗ direction*
- *1 mark for substituting ① into ②*
- *1 mark for correct value of a*

5 Before   After

$(0.8 \times 4) + (1.2 \times 2) = (0.8 \times 2.5) + 1.2v$

$3.2 + 2.4 = 2.0 + 1.2v$

$v = 3$ ms^{-1}

Before   After

$(1.2 \times 3) + (m \times -4) = (1.2 + m) \times 0$

$3.6 = 4m$

$m = 0.9$ kg

[4 marks available in total]:
- *1 mark for using conservation of momentum*
- *1 mark for correct value of v*
- *1 mark for correct workings*
- *1 mark for correct value of m*

Diagrams are handy for collision questions too, partly because they make the question clearer for you, but they also make it easier for the examiner to see how you're going about answering the question.

6 a) Before   After

$(4000 \times 2.5) + (1000 \times 0) = 5000v$

$v = 2$ ms^{-1}

[2 marks available in total]:
- *1 mark for using conservation of momentum*
- *1 mark for correct value of v*

b) Impulse = $mv - mu$

$= (4000 \times 2) - (4000 \times 2.5)$

$= -2000$ Ns

[2 marks available in total]:
- *1 mark for correct workings*
- *1 mark for correct value of impulse*

c) E.g. track horizontal; no resistance (e.g. friction) to motion; wagons can be modelled as particles.

[2 marks available in total]:
- *1 mark each for any of the above assumptions or any other relevant assumption*

If you've forgotten what 'any other valid assumption' might be, I recommend you have a look at the table on p. 155 again...

7 a) $(8\mathbf{i} - 3\mathbf{j}) = (x\mathbf{i} + y\mathbf{j}) + (5\mathbf{i} + \mathbf{j})$

So, $x\mathbf{i} + y\mathbf{j} = (8\mathbf{i} - 3\mathbf{j}) - (5\mathbf{i} + \mathbf{j})$, so $x = 3$ and $y = -4$

[2 marks available in total]:
- *1 mark for correct value of x*
- *1 mark for correct value of y*

b) Magnitude of resultant force = $\sqrt{8^2 + (-3)^2}$

$= \sqrt{73} = 8.5$ N

Using $F = ma$:

$8.5 = 2.5a$, so $a = 3.4$ ms^{-2} (to 2 s.f.)

[3 marks available in total]:
- *1 mark using F = ma*
- *1 mark for correct workings*
- *1 mark for correct value of a*

8 a) Resolving forces acting on A:

$7g - T = 7a$

Resolving forces acting on B:

$T - 3g = 3a$, so $T = 3a + 3g$

Substituting T:

$7g - 3a - 3g = 7a$, so $4g = 10a$

hence $a = 3.92$ ms^{-2}

Using $t = \frac{(v - u)}{a}$:

$t = (5.9 - 0) \div 3.92$

So, $t = 1.51$ s (to 3 s.f.)

[4 marks available in total]:
- *1 mark for resolving forces*
- *1 mark for correct value of a*
- *1 mark for correct workings*
- *1 mark for correct value of t*

b) Using $s = \frac{v^2 - u^2}{2a}$

$s = (5.9^2 - 0^2) \div (2 \times 3.92)$

$s = 4.44$ m (to 3 s.f.)

[2 marks available in total]:
- *1 mark for correct workings*
- *1 mark for correct value of s*

c) When A hits the ground, speed of A = speed of B = 5.9 ms^{-1}

B will then continue to rise, momentarily stop and then fall freely under gravity. String will be taut again when displacement of B = 0.

So, $a = -9.8$, $s = 0$, $u = 5.9$

Using $s = ut + \frac{1}{2}at^2$:

$0 = 5.9t + \frac{1}{2}(-9.8)t^2 = 5.9t - 4.9t^2$

Answers

Solve for t:

$4.9t^2 = 5.9t$, so $t(4.9t - 5.9) = 0$

and so $t = 0$ or $t = 5.9 \div 4.9 = 1.20$.

So the string becomes taut again at $t = 1.20$ s (to 3 s.f.)

[4 marks available in total]:
- *1 mark for using $s = ut + \frac{1}{2}at^2$*
- *1 mark for correct workings*
- *1 mark for solving for t*
- *1 mark for correct value of t*

M1 Practice Exam 1

1 a) With uniform motion questions, always start by writing down the data you have and the data you need.

$u = 15$, $v = 40$, $a = ?$, $t = 4$, $s = ?$

You need to find a, so '$v = u + at$' is the equation you need. Rearrange to make a the subject and substitute in:

$a = \dfrac{v - u}{t} = \dfrac{40 - 15}{4} = 6.25$ ms^{-2}

[2 marks available in total]:
- *1 mark for using '$v = u + at$' or equivalent*
- *1 mark for correct value of a*

b) $u = 40$, $v = 26$, $a = -2.8$, $t = ?$, $s = ?$

You need t, so it's '$v = u + at$'. Rearrange and substitute:

$t = \dfrac{v - u}{a} = \dfrac{26 - 40}{-2.8} = 5$ s

[2 marks available in total]:
- *1 mark for using '$v = u + at$' or equivalent*
- *1 mark for correct value of t*

c) You've now got all the other quantities except the two distances, so you can use any of the formulas with s in. I'm going for '$s = \frac{1}{2}(u + v)t$' because it's nice and easy:

A to B: $s = \frac{1}{2}(u + v)t = \frac{1}{2}(15 + 40) \times 4 = 110$ m

B to C: $s = \frac{1}{2}(u + v)t = \frac{1}{2}(40 + 26) \times 5 = 165$ m

AC = 275 m

[3 marks available in total]:
- *1 mark for using '$s = \frac{1}{2}(u + v)t$' or equivalent*
- *1 mark for either intermediate distance correct*
- *1 mark for the correct value of AC*

You REALLY need to know those equations.

2 a) Start by adding as much information as possible to the diagram:

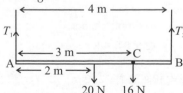

The rod is uniform, so you assume that all the weight acts at the centre. Whoops, given away part (c). Anyway, to solve "suspended rods" questions, you usually have to balance forces (resolve vertically) and balance moments.

Start by resolving vertically (balancing upward and downward forces):

$T_1 + T_2 = 36$ — equation (1)

Then take moments about A, which gives:

moments clockwise = moments anticlockwise

$(20 \times 2) + (16 \times 3) = T_2 \times 4$

So, $T_2 = 88 \div 4 = 22$

Substituting back into equation (1) gives:

$T_1 + 22 = 36$,

So, $T_1 = 14$ N and $T_2 = 22$ N

[3 marks available in total]:
- *1 mark for resolving vertically/balancing forces*
- *1 mark for taking moments about A or B*
- *1 mark for correct values for tensions*

b) Start by drawing a new diagram:

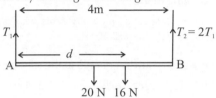

Then substitute $T_2 = 2T_1$ into equation (1)

$T_1 + 2T_1 = 36$

So, $T_1 = 12$ and $T_2 = 2T_1 = 24$

Taking moments about A again gives:

$20 \times 2 + 16 \times d = 24 \times 4$

So, $16d = 96 - 40 = 56$

and $d = 56 \div 16 = 3.5$

Distance of bead from A = 3.5 m

[3 marks available in total]:
- *1 mark for substituting '$T_2 = 2T_1$' into equation (1)*
- *1 mark for taking moments about A*
- *1 mark for correct value of A*

c) The rod is uniform, so you can assume all the weight acts at the centre.

[1 mark for correct assumption]

Just for a MOMENT I could've sworn there was some TENSION in the air... ho ho ho... oh dear, only 2 questions in and my jokes are this bad already... eesh...

3 a) The forces are given in terms of components, so to find the resultant force you just add them as vectors:

$\mathbf{F} = \mathbf{F}_1 + \mathbf{F}_2 = (3\mathbf{i} + 2\mathbf{j})$ N $+ (2\mathbf{i} - \mathbf{j})$ N $= (5\mathbf{i} + \mathbf{j})$ N

To find the angle of the resultant force, it's just trig as usual...

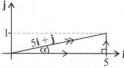

angle $= \alpha = \tan^{-1}\left(\dfrac{1}{5}\right) = 11.3°$

[3 marks available in total]:
- *1 mark for adding vectors to get resultant force*
- *1 mark for using $\tan^{-1}$ or equivalent correctly*
- *1 mark for correct answer*

Answers

b) There are a couple of ways of doing this, but both involve Pythagoras and '$F = ma$'. First way:

Find the magnitude of the resultant force:

$|\mathbf{F}| = \sqrt{(5^2 + 1^2)} = \sqrt{26} = 5.10$ N

Then use Newton's second law — $F = ma$:

Rearrange to give: $a = \dfrac{F}{m} = \dfrac{5.10}{0.5} = 10.2$ ms^{-2}

(Second way uses $\mathbf{F} = m\mathbf{a}$ to find vector form of acc^{n}, then Pythagoras to find magnitude. Either way is fine.)

Magnitude of acceleration = 10.2 ms^{-2}

[4 marks available in total]:
- *1 mark for using '$F = ma$'*
- *1 for correct calculation of <u>either</u> magnitude of force <u>or</u> vector form of acceleration (whichever method is used)*
- *1 mark for correct workings*
- *1 for correct value of a*

Here we go again — "split vector into components, use Pythagoras to find magnitude, ya de ya de ya". Yawn...

4 a) For collision questions, always decide at the start (and label) which direction you're going to use as positive.

Diagram:

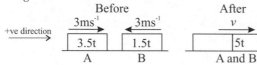

Use Conservation of Momentum to solve:

'total momentum of A and B before collision = momentum of combined trucks after collision'

$(3.5 \times 3) + (1.5 \times -3) = 5v$

So, $v = 6 \div 5 = 1.2$ ms^{-1} in the positive direction.

After the collision, the combined trucks move at 1.2 ms^{-1} in the same direction as A was moving before the collision.

[4 marks available in total]:
- *1 mark for using conservation of momentum*
- *1 mark for correct workings*
- *1 mark for correct speed*
- *1 mark for correct direction*

b) Impulse on B = change in momentum of B

Taking momentum in the positive direction as in part (a):

change in momentum
= final momentum of B – initial momentum of B
= $(1.5 \times 1.2) - (1.5 \times -3)$
= $1.8 - (-4.5) = 6.3$ Ns

Impulse on B = 6.3 Ns

[3 marks available in total]:
- *1 mark for 'impulse = change in momentum'*
- *1 mark for correct workings*
- *1 mark for correct answer*

c) Bit tricky — you need to work out the acceleration first, then you can use '$F = ma$' to find the force.

You're told the braking force is constant, so acceleration is constant, which means you need the speed/time equations.

You've got: $u = 1.2$, $v = 0$, $a = ?$, $s = 18$

So use $v^2 = u^2 + 2as$:

rearrange to get: $a = \dfrac{v^2 - u^2}{2s} = \dfrac{0 - 1.2^2}{36} = -0.04$ ms^{-2}

Now stick that into:

$F = ma = 5000 \times 0.04 = 200$ N

Braking force = 200 N

[4 marks available in total]:
- *1 mark for using '$v^2 = u^2 + 2as$' or equivalent*
- *1 mark for correct calculation of acceleration*
- *1 mark for using '$F = ma$'*
- *1 mark for correct value of F*

Not sure why, but I find these questions very satisfying when I've solved them. Maybe I'm just a freak.

5 a) $u = 0$, $v = ?$, $a = 1.5$, $t = 8$, $s = ?$

So you're going to need '$v = u + at$'

$v = 0 + (1.5 \times 8) = 12$ ms^{-1}

Greatest speed = 12 ms^{-1}

[2 marks available in total]:
- *1 mark for using '$v = u + at$' or equivalent*
- *1 mark for correct value of speed*

b) Use the information from part (a) to mark when the cyclist stops accelerating and travels at a constant speed:

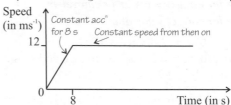

[2 marks available in total]:
- *1 mark for correct shape*
- *1 mark for correct numbers*

c)

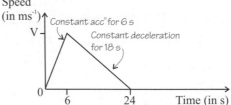

[2 marks available in total]:
- *1 mark for correct shape*
- *1 mark for correct numbers*

d) First work out how long the cyclist was travelling at a constant speed:

total time until overtaking = 24 s,

so cyclist travelled at a constant speed for $24 - 8 = 16$ s.

Total distance travelled = area under speed-time graph.

The area is split into two simple shapes:

area of triangle = $\frac{1}{2} \times 12 \times 8 = 48$

area of rectangle = $12 \times 16 = 192$

Distance travelled = $48 + 192 = 240$ m

[3 marks available in total]:
- *1 mark for using 'distance = area under graph' or equivalent*
- *1 mark for correct working*
- *1 mark for the correct value for distance travelled*

e) The cycle and car travel the same distance, so you can use the result from part (d). Again, the distance travelled is the area under the graph, so you can work back from there. This time the area under the graph is a simple triangle, so it's a bit easier.

distance = area under graph = $240 = \frac{1}{2} \times 24 \times V$

So, $V = 20$ ms^{-1}

[3 marks available in total]:
- *1 mark for 'distance = area under graph' or equivalent*
- *1 mark for recognising both cycle and car travelled the same distance*
- *1 mark for correct value of V*

This is as easy as Mechanics gets, so there's no excuse for getting questions like this wrong. If you found this hard, get it learnt before the exam. These questions never vary much — if you can do one, you can do them all.

6 a) Vector questions are almost always easier with a diagram, even if it's just a scribbled sketch.

Draw a sketch so it's clear in your head what's going on:

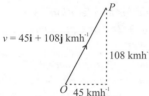

$v = 45\mathbf{i} + 108\mathbf{j}$ kmh^{-1}

Speed of $P = |V| = \sqrt{45^2 + 108^2} = 117$ kmh^{-1}

[2 marks available in total]:
- *1 mark for correct workings*
- *1 mark for correct speed*

b) Q is travelling at $(-70\mathbf{i} + 90\mathbf{j})$ kmh^{-1}, which means that in 1 hour it'll be at $(-70\mathbf{i} + 90\mathbf{j})$ km.
So, at 12:40 (after half an hour), it'll have travelled half of that, i.e. $(-35\mathbf{i} + 45\mathbf{j})$ km.
Position of Q at 12:40 $= -35\mathbf{i} + 45\mathbf{j}$

[2 marks available in total]:
- *1 mark for correct method*
- *1 mark for correct position*

c) After 40 minutes (two thirds of an hour) P will have travelled $\frac{2}{3}$ of $(45\mathbf{i} + 108\mathbf{j}) = (30\mathbf{i} + 72\mathbf{j})$ km from O.
Draw a quick sketch:

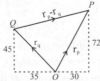

Position vector of P relative to $Q = r_p - r_q$
$= (30 - (-35))\mathbf{i} + (72 - 45)\mathbf{j} = 65\mathbf{i} + 27\mathbf{j}$

[3 marks available in total]:
- *1 mark for position of P relative to O*
- *1 mark for using 'r$_p$ – r$_q$'*
- *1 mark for correct position of P*

d)

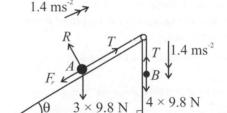

$\theta = \tan^{-1}\frac{65}{27} = 67.4°$, so bearing = 067°

[3 marks available in total]
- *1 mark for writing an equation in terms of tan*
- *1 mark for correct use of tan^{-1}*
- *1 mark for correct answer*

e) distance $= \sqrt{65^2 + 27^2} = \sqrt{4954} = 70.385$ km

time taken $= \dfrac{\text{distance}}{\text{speed}} = \dfrac{70.385}{160} = 0.4399$ hours

$= 26$ min, to the nearest min.
So, estimated time of arrival = 13:06

[4 marks available in total]:
- *1 mark for using Pythagoras to work out distance*
- *1 mark for 'time = distance ÷ speed'*
- *1 mark for adding onto original time*
- *1 mark for correct answer*

No, I didn't like that one either.

7 a) Hold onto your hats — this is a long, long question...

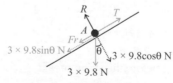

where $\theta = \tan^{-1}\frac{3}{4}$
[1 mark for adding all the correct forces]

b) Particle A remains on the angled plane, so the normal reaction must be equal to the component of the weight force in the opposite direction:

$\tan\theta = \frac{3}{4}$ so $\cos\theta = \frac{4}{5}$
So $R = 3 \times 9.8 \times \frac{4}{5} = 23.52$ N
So, normal reaction = 23.52 N

[2 marks available in total]:
- *1 mark for resolving perpendicular*
- *1 mark for correct value of R*

Answers

c) You don't know anything about the friction force on *A* yet, so start by looking at *B*, since that's only affected by weight and tension:

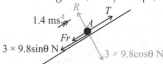

Use '*F* = *ma*' for *B*:
$(4 \times 9.8) - T = 4 \times 1.4$
So, *T* = 33.6 N

[2 marks available in total]:
- **1 mark for using '*F* = *ma*'**
- **1 mark for correct value of *T***

d) We now know the tension, so we can use '*F* = *ma*' on particle *A* to find the friction (in case you get confused — the '*F*' in '*F* = *ma*' means 'resultant force', not 'friction'). Start with a diagram (always helps):

For *A*: Resultant force = *ma*
$T - F_r - (3 \times 9.8 \sin\theta) = 3 \times 1.4$
Note that $\tan\theta = \frac{3}{4}$ so $\sin\theta = \frac{3}{5}$
Substitute in value of *T* from part (c):
$33.6 - F_r - 17.64 = 4.2$
So, $F_r = 11.76$ N

[3 marks available in total]:
- **1 mark for using '*F* = *ma*'**
- **1 mark for correct expression for resultant force**
- **1 mark for correct answer**

e) The system's moving, so friction is 'limiting', i.e. $F_r = \mu R$.
Rearranging and substituting values in gives:
$\mu = \frac{F_r}{R} = \frac{11.76}{23.52} = 0.5$

[2 marks available in total]:
- **1 mark for using '$F_r = \mu R$'**
- **1 mark for correct value of μ**

f) This is fairly straightforward — it's just uniform acceleration, so it's the usual equations:
Motion of *B*: *u* = 0, *v* = ?, *a* = 1.4, *s* = ?, *t* = 2
It's *s* you're after, so use '$s = ut + \frac{1}{2}at^2$':
$s = 0 + \left(\frac{1}{2} \times 1.4 \times 4\right) = 2.8$ m
Particle *B* moves 2.8 m before the string breaks.

[2 marks available in total]:
- **1 mark for using '$s = ut + \frac{1}{2}at^2$' or equivalent**
- **1 mark for correct value of *s***

g) This is easier than it looks (phew). Remember, while *A* and *B* are attached, they move together at the same speed. So you can use the data from part (f) to find the speed of (both) *A* and *B* when the string breaks:
$v = u + at = 0 + 1.4 \times 2 = 2.8$ ms^{-1}
Then draw yourself another diagram to show particle *A* immediately after the string breaks:

You need to work out the acceleration, so you need '*F* = *ma*':
$-11.76 - \left(3 \times 9.8 \times \frac{3}{5}\right) = 3a$
So, *a* = −9.8 ms^{-2}
Don't stop there — now you've got to find how far *A* travels before it comes to rest. It's the usual equations:
For *A*: *u* = 2.8, *v* = 0, *a* = −9 8, *s* = ?, *t* = ?
You're after *s*, so use '$v^2 = u^2 + 2as$':
Rearrange to give:
$s = \frac{v^2 - u^2}{2a} = \frac{0 - 2.8^2}{2 \times -9.8} = 0.4$ m
So particle A moves 0.4 m from the instant the string breaks until it comes to rest.

[5 marks available in total]:
- **1 mark for correct value for speed**
- **1 mark for using '*F* = *ma*'**
- **1 mark for correct value for acceleration**
- **1 mark for using '$v^2 = u^2 + 2as$' or equivalent**
- **1 mark for correct answer**

That was one beast of a question. But you made it. And honestly — this is the best kind of practice you can get. There's balancing forces, resolving forces, Newton's 2nd law and friction. All in one big question. It might feel like you've gone through the mill a bit, but if you got that question all right, I reckon you've got a pretty good understanding of mechanics.

M1 Practice Exam 2

1 a) Resolve along *x*-axis:
$A\sin40° - 20\cos(105-90)° = 0$
so, $A = \frac{20\cos15°}{\sin40°} = 30.1$ N

[2 marks available in total]:
- **1 mark for correct workings**
- **1 mark for correct value of A**

Something simple to get you started. I bet you'd never have guessed it would involve resolving forces.

b) Resolve along *y*-axis:
$R = 30.1\cos40° + 35 - 20\sin15° = 52.9$ N up the *y*-axis

[3 marks available in total]:
- **1 mark for correct workings**
- **1 mark for correct value of R**
- **1 mark for correct direction**

2 a) Using $a = \frac{F}{m}$:
$a = \frac{34\,000\,000}{1\,400\,000} = 24.3$ ms^{-2}

[2 marks available in total]:
- **1 mark for using '*F* = *ma*'**
- **1 mark for correct value of a**

See all those zeros cancel out... 34 ÷ 1.4 (or 340 ÷ 14) would give you the same answer. It's pretty satisfying to cross them all out — my maths teacher used to love doing that.

b) Using $F = ma$:
$34\,000\,000 - R = 1\,400\,000 \times 12$
so, $R = 34\,000\,000 - 16\,800\,000 = 17\,200\,000$ N

[2 marks available in total]:
* *1 mark using 'F = ma'*
* *1 mark correct value of R*

c) Using $s = ut + \frac{1}{2}at^2$:
$50\,000 = (0 \times t) + \frac{1}{2}(12 \times t^2)$
so, $t^2 = 50\,000 \div 6 = 8333.33$, and $t = 91.3$ s

[3 marks available in total]:
* *1 mark for using 's = ut + $\frac{1}{2}$at²'*
* *1 mark for correct workings*
* *1 mark for correct value of t*

d) Speed at 50 km, using $v^2 = u^2 + 2as$:
$v^2 = 0^2 + 2(12)50\,000$
$\Rightarrow v^2 = 1\,200\,000$

Additional height reached, using $s = \frac{v^2 - u^2}{2a}$:
$s = \frac{0 - 1200\,000}{2 \times (-9.6)} = 62\,500$ km
Add to initial height:
$62\,500 + 50\,000 = 112\,500$ m $= 112.5$ km

[4 marks available in total]:
* *1 mark for using 'v² = u² + 2as' or equivalent*
* *1 mark for calculating speed*
* *1 mark for calculating additional altitude*
* *1 mark for correct value of maximum altitude*

You could have used other motion equations here, but it's best to use one which includes the values that you're given in the question, not values that you've found yourself.

3 a)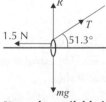

[2 marks available in total]:
* *1 mark for correct force arrows*
* *1 mark for correctly labelling the forces*

b) Resolving horizontally:
$F = T\cos51.3°$ so $T = \frac{1.5}{\cos 51.3} = 2.40$ N
Resolving vertically:
$mg = R + T\sin51.3°$
$R = \frac{F}{\mu} = 1.5 \div 0.6 = 2.5$ N
so $mg = 2.5 + 2.40\sin51.3°$
and $m = 0.446$ kg ≈ 0.45 kg (to 2 s.f.)

[6 marks available in total]:
* *1 mark for resolving horizontally*
* *1 mark for correct value of T*
* *1 mark for resolving vertically*
* *1 mark for using F = μR*
* *1 mark for correct workings*
* *1 mark for correct value of m*

Unless the question asks for a specific number of significant figures, put the whole answer down and then reduce it to an appropriate number (e.g. 2 or 3 s.f.).

4 a)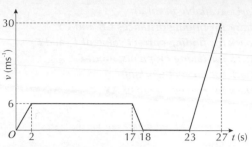

[2 marks available in total]:
* *1 mark for correct shape*
* *1 mark for correct numbers*

b) a = gradient of graph
$t = 0 - 2$ s: $a = 6 \div 2 = 3$ ms⁻²
$t = 17 - 18$ s: $a = -6 \div 1 = -6$ ms⁻²
$t = 23 - 27$ s: $a = 30 \div 4 = 7.5$ ms⁻²
Greatest acceleration is 7.5 ms⁻².

[3 marks available in total]:
* *1 mark for finding gradients or equivalent*
* *1 mark for correct workings*
* *1 mark for correct statement of greatest acceleration*

c) Distance = area under the graph
$t = 0 - 2$ s: $s = \frac{1}{2}(6)2 = 6$ m   (triangle: area = $\frac{1}{2}bh$)
$t = 2 - 17$ s: $s = 6 \times 15 = 90$ m   (rectangle)
$t = 17 - 18$ s: $s = \frac{1}{2}(6)1 = 3$ m   (triangle)
$t = 18 - 23$ s: $s = 0 \times 5 = 0$ m   (rectangle)
$t = 23 - 27$ s: $s = \frac{1}{2}(30)4 = 60$ m   (triangle)
Total distance $= 6 + 90 + 3 + 0 + 60 = 159$ m

[3 marks available in total]:
* *1 mark for using distance = area under graph*
* *1 mark for correct workings*
* *1 mark for correct value of total distance*

I don't really like roller coasters, but they do have some great maths involved. I'd give up writing books to go and design them, unless I had to test them. I feel sick thinking about that.

5 a)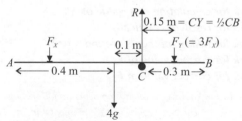

$CY = \frac{1}{2}CB = \frac{1}{2}(0.3) = 0.15$ m
$F_X + F_Y = 80g$, $F_Y = 3 \times F_X$,
so, $4F_X = 80g$, $F_X = 20g$ and $F_Y = 60g$
Taking moments about C:
$F_Y \times CY = (4g \times 0.1) + (F_X \times CX)$
so $(60g \times 0.15) = (4g \times 0.1) + (20g \times CX)$
and $CX = \frac{9g - 0.4g}{20g} = 0.43$ m
$AX = 0.5 - CX = 0.07$ m

Answers

[5 marks available in total]:
- **1 mark for finding distances CB and CY**
- **1 mark for finding correct value of F_X and F_Y**
- **1 mark for taking moments about C**
- **1 mark for correct workings**
- **1 mark for correct value of AX**

It's easy to think that you haven't been told the force at X and Y, but you have really. Keep an eye out for those stealth variables.

b) $F_X + F_Y = 80g$
Resolving vertically:
$R = F_X + F_Y + 4g = 84g = 823$ N

[2 marks available in total]:
- **1 mark for resolving vertically**
- **1 mark for correct value of R**

c)

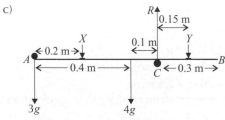

$AX = 0.07 + 0.13 = 0.2$ m
$F_X + F_Y = 80g$
Resolving vertically:
$R = 3g + F_X + 4g + F_Y = 87g$
Taking moments about point X:
$(0.2 \times 3g) + (0.3 \times 87g) = (0.2 \times 4g) + (0.45 \times F_Y)$
so $F_Y = 564$ N
and $F_X = 80g - 564 = 784 - 564 = 220$ N
OR:
Taking moments about point Y:
$(0.65 \times 3g) + (0.45 \times F_X) + (0.25 \times 4g) = (0.15 \times 87g)$
so $F_X = 220$ N
and $F_Y = 80g - 220 = 784 - 220 = 564$ N

[5 marks available in total]:
- **1 mark for calculating new value of AX**
- **1 mark for resolving vertically**
- **1 mark for taking moments around X or Y**
- **1 mark for correct value of F_X**
- **1 mark for correct value of F_Y**

There are two unknowns, so taking moments about one lets you find the other. A choice of points to take moments about. Nice.

6 a) Using conservation of momentum:
$(3 \times 7) + (0.4 \times 0) = 3v + (0.4 \times 9)$
so, $v = \dfrac{21 - 3.6}{3} = 5.8$ ms⁻¹

[3 marks available in total]:
- **1 mark for using conservation of momentum**
- **1 mark for correct workings**
- **1 mark for correct value of v**

b) Impulse = $mv - mu$
so, impulse exerted on $P = (0.4 \times 9) - (0.4 \times 0) = 3.6$ Ns

[3 marks available in total]:
- **1 mark for using impulse = change in momentum**
- **1 mark for correct workings**
- **1 mark for correct impulse**

c) Impulse = $mv - mu$
so, $2 = (0.1 \times 25) - 0.1u$
hence, $u = \dfrac{2.5 - 2}{0.1} = 5$ ms⁻¹

[3 marks available in total]:
- **1 mark for using impulse = change in momentum**
- **1 mark for correct workings**
- **1 mark for correct value of u**

I love moments — what a great name for a bit of maths. I was once in a musical about moments, it was set in Mexico. True story.

7 a)
$\tan\theta = \dfrac{6}{9}$
so, $\theta = 33.7°$ and so bearing = 034°

[2 marks available in total]:
- **1 mark for correct workings**
- **1 mark for correct bearing**

b) Speed of pirate ship = $\sqrt{6^2 + 9^2} = 10.8$ kmh⁻¹
Constant velocity, so, $s = vt$, so $t = 81 \div 10.8 = 7.5$ h
so time of arrival = 1350 + 7.5 = 2120 hours

[3 marks available in total]:
- **1 mark for correct speed**
- **1 mark for correct time taken**
- **1 mark for correct time of arrival**

c) 1950 − 1350 = 6 h
Position of pirate ship at 1950 = 6(6**i** + 9**j**) = (36**i** + 54**j**) km
Distance travelled by naval ship = (36**i** + 54**j**) − (−18**i** + 6**j**)
= (36 + 18)**i** + (54 − 6)**j** = (54**i** + 48**j**) km
Velocity of naval ship = (54**i** + 48**j**) ÷ 6 = (9**i** + 8**j**) kmh⁻¹

[4 marks available in total]:
- **1 mark for correct time taken**
- **1 mark for correct position of pirate ship**
- **1 mark for correct distance travelled by naval ship**
- **1 mark for correct velocity of naval ship**

*Argh, mateys, we've been caught by those naval scallywags...
If only we'd studied M1 harder we might've escaped capture.
Learn from those scurvy pirates — revise.
Disclaimer: CGP does not condone piracy.*

8 a)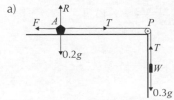

If mass of A = 0.2 kg, mass of W = 1.5 × 0.2 = 0.3 kg
For W, $F = ma$:
$0.3g - T = 0.3(4)$
$T = 2.94 - 1.2 = 1.74$
For A, $F = ma$:
$T - F = 4(0.2) = 0.8$

$F = \mu R$,

resolving vertically: $R = 0.2g = 1.96$

so $F = 1.96\mu$,

so $T - 1.96\mu = 0.8$,

$1.74 - 1.96\mu = 0.8$

so $\mu = 0.48$

[5 marks available in total]:
- *1 mark for using correct masses of A and W*
- *1 mark for using 'F = ma'*
- *1 mark for correct value of T*
- *1 mark for correct value of R*
- *1 mark for correct value of μ*

b) Speed of A at h = speed of W at h (where it impacts ground)

Calculate speed of A at h using $v^2 = u^2 + 2as$:

$v^2 = 0^2 + 2(4 \times h)$

$v^2 = 8h$ **equation 1**

Distance travelled by A beyond $h = \frac{3}{4}h$

Calculate frictional force slowing A after h using $F = ma$,

$F = \mu R$ and $R = mg = 0.2g = 1.96$

so $F = 0.48 \times 1.96 = 0.94$

so $a = -0.94 \div 0.2 = -4.7$

Speed of A at h using $u^2 = v^2 - 2as$:

$u^2 = 3^2 - 2(-4.7 \times \frac{3}{4}h) = 9 + 7.05h$ **equation 2**

Substituting **equation 1** into **equation 2** (where $v^2 = u^2$):

$8h = 9 + 7.05h$

so $h = 9.47$ m

Time taken to reach h using $s = ut + \frac{1}{2}at^2$:

$9.47 = (0 \times t) + \frac{1}{2}(4 \times t^2) = 2t^2$

so $t^2 = 4.74$ and $t = 2.18$ s

[7 marks available in total]:
- *1 mark for calculating speed of*
  A at h using initial acceleration
- *1 mark for calculating speed of*
  A at h using frictional force
- *1 mark for correct value of deceleration*
- *1 mark for equating speeds of A at h*
- *1 mark for correct value of h*
- *1 mark for using 's = ut + \frac{1}{2}at²' or equivalent*
- *1 mark for correct value of t*

Tricky. The first thing is to realise that you can calculate the time taken for W to fall indirectly using the connected particle, A. The second thing to realise is that A carries on moving after W hits the ground and is slowed by friction. Then, find that you can calculate the speed of A when it's moved a distance of h in two different ways. Finally, you can find h and use that to find t. Blimey...

c) E.g. string is inextensible — so A and W have the same acceleration when W is falling.

[1 mark for correct explanation]

A simple question to finish off. Well, you deserved a break after part b). That's M1 in the bag, so practise, practise and practise some more until the exam. If the next thing you have to do is the exam, then good luck — Mechanics loves you.

Answers

D1 Section 1 — Algorithms
Warm-up Questions

1) a) Input: raw ingredients (e.g. vegetables, water etc.)
Output: vegetable soup.

   b) Input: starting point (Leicester Square)
Output: final destination (the Albert Hall)

   c) Input: components (e.g. shelves, screws etc.)
Output: finished TV cabinet (*in theory — I'm not much good at flat-pack*)

2)

x	y
17	56
~~8~~	~~112~~
~~4~~	~~224~~
~~2~~	~~448~~
1	896
Total	952

So 17 × 56 = 952.

3) Diamond-shaped boxes are used for decisions — they'll ask a question.

4) $a = 16$

n	b	Output	n = a?
1	16	1	No
2	8	2	No
3	5⅓		No
4	4	4	No
5	3⅕		No
6	2⅔		No
7	2²⁄₇		No
8	2	8	No
9	1⁷⁄₉		No
10	1⅗		No
11	1⁵⁄₁₁		No
12	1⅓		No
13	1³⁄₁₃		No
14	1⅐		No
15	1¹⁄₁₅		No
16	1	16	Yes

So the factors of 16 are 1, 2, 4, 8 and 16.

5)
<u>72, 57</u>, 64, 54, 68, 71	swap
57, <u>72, 64</u>, 54, 68, 71	swap
57, 64, <u>72, 54</u>, 68, 71	swap
57, 64, 54, <u>72, 68</u>, 71	swap
57, 64, 54, 68, <u>72, 71</u>	swap
57, 64, 54, 68, 71, 72	end of first pass.

At the end of the second pass, the list is:
57, 54, 64, 68, 71, 72.
At the end of the third pass, the list is:
54, 57, 64, 68, 71, 72.
There are no swaps on the fourth pass, so the list is in order.

6) The maximum number of comparisons is 9 + 8 + 7 + 6 + 5 + 4 + 3 + 2 + 1 = 45 (or ½ × 9 × 10 = 45).

7) There are 6 numbers in the list, so the pivot is the ½(6 + 2) = 4th item in the list, so the pivot is 108.

8) Pivot is the ½(7 + 1) = 4th item = 0.5, so the list becomes:
0.4, 0.1, <u>0.5</u>, 0.8, 1.2, 0.7, 1.0.
The pivot for the first smaller list is the ½(2 + 2) = 2nd item = 0.1, and the pivot for the second list is ½(4 + 2) = 3rd item = 0.7. The list becomes:
<u>0.1</u>, 0.4, <u>0.5</u>, <u>0.7</u>, 0.8, 1.2, 1.0.
0.4 is in a list on its own, so it's in the right place. In the other list, the pivot is the ½(3 + 1) = 2nd item = 1.2, so the list becomes:
<u>0.1</u>, <u>0.4</u>, <u>0.5</u>, <u>0.7</u>, 0.8, 1.0, <u>1.2</u>.
The final list has 2 items, so the pivot is the ½(2 + 2) = 2nd item = 1.0, so the final (ordered) list is:
<u>0.1</u>, <u>0.4</u>, <u>0.5</u>, <u>0.7</u>, 0.8, <u>1.0</u>, <u>1.2</u>.

9) The middle item in the list is the ½(5 + 1) = 3rd item = Manhattan. Brooklyn ≠ Manhattan, and Brooklyn is before Manhattan, so throw away the second half of the list, including Manhattan. There are 2 items left in the list, so the middle item is the ½(2 + 2) = 2nd item = Brooklyn. This is the item you're looking for, so the search is complete. Brooklyn is the 2nd item in the list.

10) a) 5 + 11 + 8 + 9 + 12 + 7 = 52. 52 ÷ 15 = 3.467, so the lower bound is 4 boxes.

   b) Box 1: 5, 8   space left: ~~10 kg~~ 2 kg
Box 2: 11   space left: 4 kg
Box 3: 9   space left: 6 kg
Box 4: 12   space left: 3 kg
Box 5: 7   space left: 8 kg.
So the items are packed in 5 boxes, with 2 + 4 + 6 + 3 + 8 = 23 kg wasted space. The lower bound is 4 boxes, which means we can't say yet if this solution is optimal or not.

   c) First, reorder the numbers in descending order:
12, 11, 9, 8, 7, 5. Then,
Box 1: 12   space left: 3 kg
Box 2: 11   space left: 4 kg
Box 3: 9, 5   space left: ~~6 kg~~ 1 kg
Box 4: 8, 7   space left: ~~7 kg~~ 0 kg
So the items are packed in 4 boxes, with 3 + 4 + 1 + 0 = 8 kg wasted space. The lower bound for this problem is 4, so this solution is optimal (and now we can say for definite that the solution in b) was not optimal).

   d) By eye, 8 + 7 = 15, so this fills one box.
Box 1: 8, 7   space left: 0 kg
Box 2: 5, 9   space left: ~~10 kg~~ 1 kg
Box 3: 11   space left: 4 kg
Box 4: 12   space left: 3 kg
So the items are packed in 4 boxes, with 0 + 1 + 4 + 3 = 8 kg wasted space. The lower bound for this problem is 4, so this solution is also optimal.

This is actually the same as the answer to part c), but in a slightly different order — sometimes different methods produce the same solution.

Answers

Exam Questions

1 a) There are 10 items in the list, so the pivot is the $\frac{1}{2}(10 + 2)$ = 6th item = 89 *[1 mark]*. The list becomes:
77, 83, 78, 80, <u>89</u>, 96, 105, 112, 98, 94 *[1 mark]*
The first small list has 4 items, so the pivot is the $\frac{1}{2}(4 + 2)$ = 3rd item = 78, and the second list has 5 items, so the pivot is the $\frac{1}{2}(5 + 1)$ = 3rd item = 112. The list becomes:
77, <u>78</u>, 83, 80, <u>89</u>, 96, 105, 98, 94, <u>112</u> *[1 mark]*
77 is in a list on its own, so it's in the right place.
The second list now has 2 items, so the pivot is the 2nd item = 80, and the third list has 4 items, so the pivot is the 3rd item = 98. The list becomes:
<u>77</u>, <u>78</u>, <u>80</u>, 83, <u>89</u>, 96, 94, <u>98</u>, 105, <u>112</u> *[1 mark]*
83 and 105 are in lists on their own, so they are in the correct place. The remaining list has 2 items in it, so the pivot is the 2nd item = 94. The list becomes:
<u>77</u>, <u>78</u>, <u>80</u>, <u>83</u>, <u>89</u>, <u>94</u>, 96, <u>98</u>, <u>105</u>, <u>112</u> *[1 mark]*

b) (i) After the first pass, the final (6th) number in the list will be in the correct position. *[1 mark]*

(ii) There are 6 items, so the maximum number of passes is 6 *[1 mark]*. The maximum number of swaps is 5 + 4 + 3 + 2 + 1 = 15 *[1 mark]*.

2 a)

N	C	D	Output	N = A?
1	8	12	1	No
2	4	6	2	No
3	$2\frac{2}{3}$	4		No
4	2	3	4	No
5	$1\frac{3}{5}$	$2\frac{2}{5}$		No
6	$1\frac{1}{3}$	2		No
7	$1\frac{1}{7}$	$1\frac{5}{7}$		No
8	1	$1\frac{1}{2}$		Yes

The results are 1, 2 and 4.

[3 marks available in total:

• 1 mark for correct values of C;

• 1 mark for correct values of D;

• 1 mark for correct outputs (there should be 3 outputs)]

b) (i) This algorithm produces the common factors of the inputs. *[1 mark]*

(ii) The output would be 1 *[1 mark]*, as 19 and 25 have no common factors. *[1 mark]*

3 a) Calculate the lower bound by adding up all the lengths, dividing by the length of a plank and rounding up.
So 1.2 + 2.3 + 0.6 + 0.8 + 1.5 + 1.0 + 0.9 + 2.5 = 10.8 *[1 mark]*. 10.8 ÷ 3 = 3.6, so the lower bound is 4. *[1 mark]*

b) Plank 1: 1.2, 0.6, 0.8   space left: ~~1.8~~ ~~1.2~~ 0.4
Plank 2: 2.3   space left: 0.7
Plank 3: 1.5, 1.0   space left: ~~1.5~~ 0.5
Plank 4: 0.9   space left: 2.1
Plank 5: 2.5   space left: 0.5   *[1 mark]*
So 5 planks are used *[1 mark]* and there is 0.4 + 0.7 + 0.5 + 2.1 + 0.5 = 4.2 m wasted wood. *[1 mark]*

c) (i) By eye, 1.2 + 0.8 + 1.0 = 3 and 0.6 + 1.5 + 0.9 = 3 *[1 mark]*, so there are 2 full planks. The rest are placed using the first fit algorithm.
Plank 1: 1.2, 0.8, 1.0   space left: 0
Plank 2: 0.6, 1.5, 0.9   space left: 0
Plank 3: 2.3   space left: 0.7
Plank 4: 2.5   space left: 0.5 *[1 mark]*
So 4 planks are used and there is 0.7 + 0.5 = 1.2 m wasted wood *[1 mark]*.

(ii) This solution is optimal as it uses 4 planks and the lower bound is 4 *[1 mark]*.

4 a) There are 7 names in the list, so the pivot is the $\frac{1}{2}(7 + 1)$ = 4th name = James *[1 mark]*. The list becomes:
Adam, Dan, Helen, <u>James</u>, Mark, Stella, Robert *[1 mark]*
There are 3 names in the first list, so the pivot is the 2nd name = Dan. There are 3 names in the second list, so the pivot is the 2nd name = Stella. The list becomes:
Adam, <u>Dan</u>, Helen, <u>James</u>, Mark, Robert, <u>Stella</u> *[1 mark]*
Adam and Helen are in a list on their own, so they are in the correct place. The remaining list has 2 names in it, so the pivot is the 2nd name = Robert. The list is:
<u>Adam</u>, <u>Dan</u>, <u>Helen</u>, <u>James</u>, Mark, <u>Robert</u>, <u>Stella</u> *[1 mark]*
There are no more items to choose as a pivot so the list is in order.

b) (i) The list is now: 1. Adam, 2. Dan, 3. Helen, 4. James, 5. Mark, 6. Robert, 7. Stella
From the method above, the middle name is the 4th name = James. Adam ≠ James and Adam comes before James so throw away the second half of the list, including James *[1 mark]*. The remaining list is:
1. Adam, 2. Dan, 3. Helen. The middle name is the 2nd name = Dan. Adam ≠ Dan and Adam comes before Dan so throw away the second half of the list, including Dan *[1 mark]*. This leaves the name Adam, which is the name you're searching for. Adam is the first name in the list *[1 mark]*.

(ii) As above, the middle name is the 4th name = James. Laura ≠ James and Laura comes after James so throw away the first half of the list, including James *[1 mark]*. The remaining list is: 5. Mark, 6. Robert, 7. Stella. The middle name is the $\frac{1}{2}(5 + 7)$ = 6th name = Robert. Laura ≠ Robert and Laura comes before Robert so throw away the second half of the list, including Robert *[1 mark]*. This leaves the name Mark, and Mark ≠ Laura, so Laura is not in the list *[1 mark]*.

Answers

D1 Section 2 — Algorithms on Graphs
Warm-up Questions

1) a) A graph which has a number associated with each edge.

b) A graph in which one or more of the edges have a direction associated with them.

c) A connected graph with no cycles.

d) A subgraph which contains all the vertices of the original graph and is also a tree.

2) E.g.

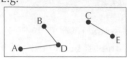

3) 5

4) E.g. ADBC

5) E.g. BCDB

6) E.g.

7) A = 1, B = 2, C = 3, D = 3, E = 1
Sum of degrees = double number of edges

8)

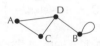

9) a) ABDEF. Shortest route A to F is 9.

Trace back from F to A on the final diagram on page 202. The weight of each arc you go along must equal the difference in the final values of the vertices at each end of the arc.

b)

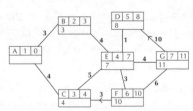

ABEG (11)

Exam Questions

1 a) DG (22) – add; EF (24) – add; BE (25) – add; DE (25) – add; EG (26) – don't add; AB (26) – add; BD (27) – don't add; BF (28) – don't add; FH (28) – add; CF (30) – add; AC (32) – don't add; GH (33) – don't add. Vertices of equal length can be considered in either order. *[3 marks available — 1 mark for edges in correct order, 2 marks for all added edges correct. Lose 1 mark for each error.]*

b) 22 + 24 + 25 + 25 + 26 + 28 + 30 = £180 *[1 mark]*

c)

[2 marks for correct edges. Lose 1 mark for each error.]

d) FC / CF.
Order edges added: EF, EB, ED, DG (or ED, DG, EB), BA, FH, FC.

[2 marks available. 1 mark for correct edge, 1 mark for evidence that Prim's algorithm has been applied.]

e) E.g. Prim's algorithm can be applied to data in matrix form; you don't have to check for cycles using Prim's; the tree grows in a connected way using Prim's.
[2 marks — 1 mark for each.]

2 a) Bipartite graph *[1 mark]*

b) 3 (A2, B1, B3) *[1 mark]*

c) 1 (e.g. AD) *[1 mark]*

d) 8 (2 × number of edges) *[1 mark]*

e) The sum of the orders is double the number of edges, so is always even *[1 mark]*. There are 5 vertices, and the sum of 5 odd numbers is always odd *[1 mark]*.

3 a)

	A	B	C	D	E
A	–	14	22	21	18
B	14	–	19	21	20
C	22	19	–	21	15
D	21	21	21	–	24
E	18	20	15	24	–

Order arcs added: AB, AE, EC, AD/BD/CD
(Arcs AD, BD and CD are interchangeable.)

[3 marks available — 2 marks for arcs in correct order (1 mark if one error). 1 mark for correct use of matrix.]

Each time you circle a number, write down which arc it represents by reading the row and column labels. Don't leave it until the end, or you'll have forgotten the order you added them in. Doh.

b) E.g.

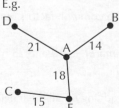

D may be connected to B or C instead of A *[1 mark]*.
weight = 68 *[1 mark]*.

c) 2 (using any of the alternatives AD, BD and CD) *[1 mark]*.

4 a) 8 (no. of vertices – 1) *[1 mark]*

b)

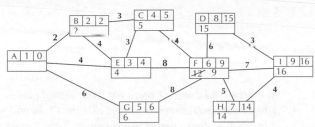

Fastest route = ABCFI, 16 minutes

[6 marks available — 1 mark for route, 1 mark for 16 minutes, 4 marks for all vertices correctly completed in diagram, lose 1 mark for each error.]

Find the fastest route by tracing back from the final destination. You know if a path is on the route because its weight is the difference between the final values at either end of it.

c) $6 + x + 4 < 16$, $x < 6$

Fastest route from A to G is 6. Fastest route from H to I is 4.
Total route AGHI must be less than 16 minutes.

[2 marks available — 1 mark for 6 + x + 4 as new route length, 1 mark for solving inequality for x.]

D1 Section 3
— The Route Inspection Problem
Warm-up Questions

1) a) Eulerian

b) semi-Eulerian

c) semi-Eulerian

d) neither

Eulerian. It's my new favourite word. In fact, I think I'll name my first-born child "Eulerian".

2) A, B, F, J
AB + FJ, AF + BJ, AJ + BF

3) a) It's Eulerian, so length = weight of network = 36.

b) It's semi-Eulerian, so length = 31 (weight of network) + 4 (distance AB) = 35

c) 4 odd vertices: A, B, D, F
AB + DF = 7 + 8 = 15
AD + BF = 5 + 3 = 8 (minimum)
AF + BD = 4 + 7 = 11
length = 42 (weight of network) + 8 = 50

4) a) 36, any vertex.

b) 31, start and end at A and B

c) BF is shortest distance between odd vertices, so start and end at A and D. Length = 42 + 3 = 45.

Exam Questions

1 a) Logo A = semi-Eulerian *[1 mark]*
Logo B = neither *[1 mark]*

b) (i) Logo A = 0 *[1 mark]*
Logo B = once *[1 mark]*

(ii) B or D *[1 mark]*

c) Logo A = 1 *[1 mark]*
Logo B = 2 *[1 mark]*

2 a) There are odd vertices *[1 mark]*

So it's not Eulerian. And there are four odd vertices, so it's not even semi-Eulerian.

b) Odd vertices are A, D, I, J *[1 mark]*
Pairings: AD + IJ = 180 + 100 = 280
AI + DJ = 440 + 310 = 750
AJ + ID = 490 + 340 = 830
[1 mark for pairings, 1 mark for lengths]
minimum pairing = AD + IJ *[1 mark]*
route length = 2740 + 280
= 3020 *[1 mark]* m *[1 mark]*

The question says that you must start and end at K — but as you've made the graph effectively Eulerian, it doesn't actually matter which vertex you start and finish at.

c) (i) IJ is minimum distance between odd vertices *[1 mark]*
Length of route = 2740 + 100 = 2840 m *[1 mark]*

(ii) A or D *[1 mark]*

3 a) Odd vertices are B, G, M, L *[1 mark]*
Pairings: BL + GM = 41 + 26 = 67
BG + LM = 30 + 29 = 59
BM + GL = 30 + 28 = 58
[1 mark for pairings, 1 mark for times]
minimum pairing = BM + GL *[1 mark]*
route time = 336 + 58 = 394 *[1 mark]* mins *[1 mark]*

b) (i) GM is minimum *[1 mark]* so end at L *[1 mark]*
(ii) 336 + 26 = 362 minutes *[1 mark]*

4 a) Odd vertices are A, D, E, F *[1 mark]*
Pairings: AD + EF = 12 + 15 = 27
AE + DF = 12 + 21 = 33
AF + DE = 18 + 6 = 24
[1 mark for pairings, 1 mark for distances]
minimum pairing = AF + DE *[1 mark]*
route length = 106 + 24 = 130 *[1 mark]* miles *[1 mark]*

b) Example route = ABFGEFBACBEDECDA *[1 mark]*
so 5 times *[1 mark]*.

Alternatively, you could say that C has four edges connected to it, so must be passed through twice. E has six edges connected to it (including the extra pass along DE) so must be passed through three times *[1 mark]*, so 2 + 3 = 5 times past an ice-cream shop *[1 mark]*.

c) 106 × 2 *[1 mark]* = 212 miles *[1 mark]*

You've effectively doubled the edges and made the graph Eulerian. You have to traverse it twice, so the distance is just double the network's weight.

Answers

D1 Section 4 — Critical Path Analysis

Warm-up Questions

1) a)

Activity	Immediate preceding activities
A	—
B	—
C	A
D	B
E	C, D

b)

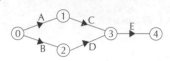

2) E.g.

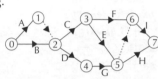

3) a) a = 0, b = 4, c = 9, d = 6, e = 5, f = 13, g = 14

b) (i) A series of activities running from the start to the end of the project (source node to sink node). The activities have zero total float (they are all critical) and if their start is delayed, the entire project time will be extended.

(ii) ACGH and ADJ

c) E's duration isn't equal to the difference between the time at the node before and the time at the node after.

d) 14 hours

e) B = 6 − 3 − 0 = 3 hours
E = 10 − 3 − 4 = 3 hours
I = 14 − 1 − 11 = 2 hours

f)

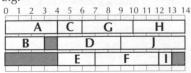

g) Sum of activity durations =
4 + 3 + 2 + 5 + 3 + 5 + 4 + 4 + 1 + 5 = 36
Critical time = 14
36/14 = 2.57, so lower bound = 3

h) E.g.

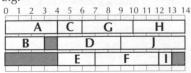

Exam Questions

1 a) E.g.

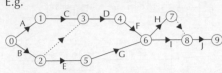

[5 marks available — 1 mark for 9 numbered nodes, 1 mark for each correctly placed dummy (2 dummies), 2 marks for other correct precedences. Lose 1 mark for each error.]

b) The dummy between nodes 2 and 3 is needed to show dependency (that D depends on B and C, but E depends on B only). **[1 mark]**

The dummy between nodes 7 and 8 is needed so that all activities are uniquely represented in terms of their events. **[1 mark]**

2 a)

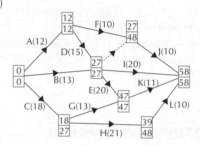

[4 marks available — 1 mark for every 4 correct numbers.]

b) Critical activities: ADEK **[1 mark]**
Length of critical path = 58 **[1 mark]** days **[1 mark]**

c) Total float on F = 48 − 10 − 12 = 26 days
Total float on G = 47 − 13 − 18 = 16 days
[3 marks available — 1 mark for each total float and 1 mark for showing correct working.]

d)

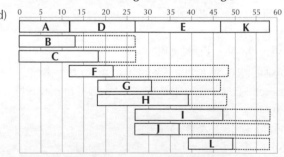

[4 marks available — 1 mark for every 3 correctly plotted activities.]

e) Lower bound = 173 ÷ 58 = 2.98, which rounds up to 3
[2 marks available — 1 mark for method, 1 mark for rounded answer.]

f) E.g.

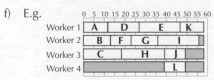

[3 marks available — 1 mark for each 4 activities correctly scheduled.]

Answers

g) It's seems that it isn't possible to complete the project in 58 days with the lower bound number of workers. / An extra worker is likely to be needed. *[1 mark]*

The lower bound often isn't enough workers to get the project done, so watch out for questions about this.

3 a)

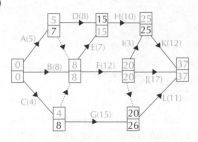

[3 marks available — 1 mark for every 2 correct numbers.]

b) BEHK *[1 mark]*, BFJ *[1 mark]*.

c) Lower bound = 112 ÷ 37 = 3.02 *[1 mark]* which rounds up to 4 *[1 mark]*

d) H (critical activity between day 15 and day 25) *[1 mark]* J (critical activity between day 20 and day 37) *[1 mark]*

e) From c) you know you need a minimum of 4 workers. E.g.

```
        0  5 10 15 20 25 30 35 40
Worker 1 | B | E | H |   K   |
Worker 2 | A |   F   |   J   |
Worker 3 | C | D |   I   |
Worker 4 |    G   |    L   |
```

[3 marks available — 1 mark for every 4 correctly scheduled activities.]

f) No. *[1 mark]*
D won't be finished until day 22. Critical activity H must start after 15 days, and depends on D being completed. *[1 mark]*

D1 Section 5 — Linear Programming

Warm-up Questions

1) a) Decision variables represent the quantities of the things being produced in a linear programming problem.

b) The objective function is the thing you're trying to optimise — an equation in terms of the decision variables that you want to maximise or minimise.

c) The optimal solution is a feasible solution that optimises the objective function (i.e. maximises or minimises it).

2) A dotted line represents a strict (< or >) inequality. You don't include this line in the feasible region.

3) An objective line is a line in the form $Z = ax + by$, which represents all solutions of the objective function that have the same value of Z.

4) E.g. A linear programming problem involving liquid doesn't need integer solutions — e.g. mixing vinegar and oil to make vinaigrette. You can have fractions of a litre. E.g. A linear programming problem about making musical instruments needs integer solutions — you can't make half a trumpet.

5) a) The decision variables are the number of large posters and the number of small posters, so let x = number of large posters and y = number of small posters.
The constraints are the different amounts of time available. Printing a large poster takes 10 minutes, so printing x large posters takes 10x minutes. Printing a small poster takes 5 minutes, so printing y small posters takes 5y minutes. There are 250 minutes of printing time available, so the inequality is $10x + 5y \leq 250 \Rightarrow 2x + y \leq 50$. Using the same method for laminating time produces the inequality $6x + 4y \leq 200 \Rightarrow 3x + 2y \leq 100$. The company wants to sell at least as many large posters as small posters, so $x \geq y$, and at least 10 small posters, so $y \geq 10$. Also, $x, y \geq 0$ (the non-negativity constraints).
The objective function is $P = 6x + 3.5y$ (profit) which needs to be maximised.

You could use x for the number of small posters and y for the number of large posters instead — so x and y in each inequality would just swap round.

b)

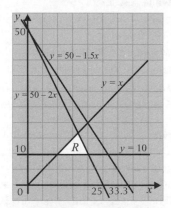

c) E.g. starting with the objective line for $P = 42$ (which goes through (0, 12) and (7, 0)) and moving it towards R gives:

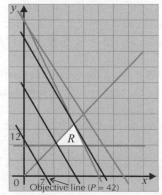

So the final point in the feasible region touched by the objective line is the point of intersection of the lines $y = x$ and $y = 50 - 2x$. Solving these simultaneous equations gives the point of intersection as $\left(\frac{50}{3}, \frac{50}{3}\right)$. Putting these values into the objective function gives a maximum profit of £158.33.

You can choose any value of P as a starting value — 42 makes it easy to draw the line. This answer uses the objective line method, but you could have used the vertex method instead.

d) From part c) above, the maximum profit is found at $\left(\frac{50}{3}, \frac{50}{3}\right)$. However, you can't have fractions of a poster, so an integer solution is required. The points with integer coordinates nearby are (16,16), (16,17), (17, 17) and (17, 16). (16, 17) doesn't satisfy the constraint $x \geq y$ and (17, 17) doesn't satisfy the constraint $y \leq 50 - 2x$. The value of the objective function at (16, 16) is £152 and at (17, 16) it's £158, so (17, 16) gives the maximum solution.

Exam Questions

1   The objective function is to minimise the cost *[1 mark]*, $C = 0.75x + 0.6y$ in £ (or $C = 75x + 60y$ in pence) *[1 mark]* (where x is the number of red roses and y is the number of white roses), subject to the constraints:

$x, y > 0$ *[1 mark]* (from the statement that she will sell both red and white roses — x and y can't be 0).

$x > y$ *[1 mark]* (from the statement that she will sell more red roses than white roses).

$x + y \geq 100$ *[1 mark]* (from the statement that she will sell a total of at least 100 flowers).

$x \leq 300$ *[1 mark]* and $y \leq 200$ *[1 mark]* (from the statement that the wholesaler has 300 red roses and 200 white roses).

Don't waste time trying to solve these inequalities — the question doesn't ask you to find a solution. Just write them down and run.

2   a) There are 6 sheets of foil in a gold pack, so in x gold packs there will be $6x$ sheets of foil. There are 2 sheets of foil in a silver pack, so in y silver packs there will be $2y$ sheets of foil. There is 1 sheet of foil in a bronze pack, so in z bronze packs there will be z sheets of foil. There are 30 sheets of foil available, so the inequality is $6x + 2y + z \leq 30$ *[2 marks — 1 mark for LHS, 1 mark for correct inequality sign and RHS]*. Using the same method for sugar paper produces the inequality $15x + 9y + 6z \leq 120$ *[1 mark]*, which simplifies to give $5x + 3y + 2z \leq 40$ *[1 mark]*. For tissue paper, the inequality is $15x + 4y + z \leq 60$ *[2 marks — 1 mark for LHS, 1 mark for correct inequality sign and RHS]*. Finally, the amount of foil used is $6x + 2y + z$, and the amount of sugar paper used is $15x + 9y + 6z$. The amount of sugar paper used needs to be at least three times the amount of foil, so the inequality for this constraint is $15x + 9y + 6z \geq 3(6x + 2y + z)$ *[1 mark]*
$15x + 9y + 6z \geq 18x + 6y + 3z$
$\qquad 3y + 3z \geq 3x$
$\qquad y + z \geq x$ *[1 mark]*

b) (i) If the number of silver packs sold is equal to the number of bronze packs, then $y = z$. Substituting this into the inequalities from part (a) gives:
$6x + 2y + y \leq 30 \Rightarrow 6x + 3y \leq 30 \Rightarrow 2x + y \leq 10$
$5x + 3y + 2y \leq 40 \Rightarrow 5x + 5y \leq 40 \Rightarrow x + y \leq 8$
$15x + 4y + y \leq 60 \Rightarrow 15x + 5y \leq 60 \Rightarrow 3x + y \leq 12$
$y + y \geq x \Rightarrow 2y \geq x$. *[3 marks available — 1 mark for making the correct substitution, 1 mark for correctly forming the inequalities and 1 mark for simplifying the inequalities.]*

(ii)

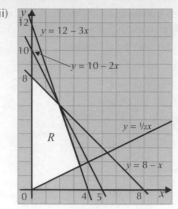

[5 marks available — 1 mark for each of the four inequality lines (with equations as shown on the graph) and 1 mark for correct feasible region]

(iii) Using the vertex method, the feasible region has vertices (0, 0) (the origin), (0, 8) (intersection of the y-axis and $y = 8 - x$), (2, 6) (intersection of $y = 8 - x$ and $y = 10 - 2x$) and $\left(\frac{24}{7}, \frac{12}{7}\right)$ (intersection of $y = \frac{1}{2}x$ and $y = 12 - 3x$) *[1 mark]*. The number of packs made on Monday is $x + y + z = x + 2y$, so the numbers made at each vertex are 0, 16, 14 and $\frac{48}{7} = 6\frac{6}{7}$, so the maximum number of packs made that day is 16 = 8 silver and 8 bronze *[1 mark]*.

You could have used the objective line method instead, using the line $Z = x + y + z = x + 2y$ to find the maximum.

(iv) The objective function is $P = 3.5x + 2y + z$ $= 3.5x + 2y + y = 3.5x + 3y$, which needs to be maximised. The value of P for each of the vertices found in part (iii) is £0, £24, £25 and £17.14 *[1 mark]*. The maximum value is £25 *[1 mark]*, which occurs at (2, 6), so the company needs to sell 2 gold packs, 6 silver packs and 6 bronze packs (as the number of bronze packs is equal to the number of silver packs) *[1 mark]*.

3   a) First, the non-negativity constraints are $x, y \geq 0$. The solid line that passes through (0, 6) and (3, 0) has equation $y = 6 - 2x$, and as the area below the line is shaded, the inequality is $2x + y \geq 6$. The dotted line that passes through (2, 0) has equation $y = x - 2$, and as the area below the line is shaded, the inequality is $x - y < 2$. The horizontal solid line that passes through (0, 4) has the equation $y = 4$, and as the area above the line is shaded, the inequality is $y \leq 4$. *[5 marks available — 1 mark for each line equation and 1 mark for all inequality signs correct]*

b) The coordinates of the vertices of R are (1, 4) (the intersection of the lines $y = 4$ and $y = 6 - 2x$) *[1 mark]* (6, 4) (the intersection of the lines $y = 4$ and $y = x - 2$) *[1 mark]* and $\left(\frac{8}{3}, \frac{2}{3}\right)$ *[1 mark]* (the intersection of the lines $y = 6 - 2x$ and $y = x - 2$) *[1 mark for solving the simultaneous equations]*.

c) The value of *C* at (1, 4) is 8, the value of *C* at (6, 4) is 28 *[1 mark if both are correct]* and the value of *C* at $\left(\frac{8}{3}, \frac{2}{3}\right)$ is $\frac{34}{3} = 11\frac{1}{3}$ *[1 mark]*. Hence the minimum value of *C* is 8, which occurs at the point (1, 4) *[1 mark]*.

This answer uses the vertex method, but you could also have used the objective line method to answer this question — pick whichever method you prefer.

D1 Section 6 — Matchings
Warm-up Questions

1) a)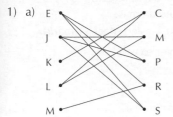

b)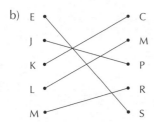

So E = S, J = P, K = C, L = M, M = R.
An alternative complete matching would start with E = P and J = S. Kitty, Lydia and Mary can't change.

These are both complete matchings — it shows that you can have more than one.

2) a)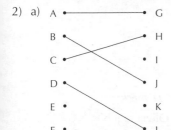

b) Find an alternating path starting from E:
E - - G — A - - I (where - - = not in, — = in).
(You could have done E - - H — C - - K instead — it reaches a breakthrough just as quickly.)
Changing the status of the arcs gives:
E — G - - A — I
Construct the improved matching:
A = I, B = J, C = H, D = L, E = G

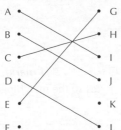

Now try and find an alternating path from F to K:

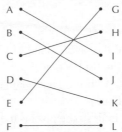

L — D - - K   Breakthrough
F
J — B - - I

The path F - - L — D - - K reaches a breakthrough first, so use this one (if you found a path from E to K in your first alternating path, the second path would be F - - J — B - - I). Changing the status of the arcs: F — L - - D — K
Construct the improved matching:
A = I, B = J, C = H, D = K, E = G, F = L

A •   • G
B •   • H
C •   • I
D •   • J
E •   • K
F •———————• L

There are no more unmatched nodes, so this is a complete matching.

An alternative complete matching would be A = G, B = I, C = K, D = L, E = H, F = J if you'd used the alternating paths in brackets.

Exam Question

1 a) Alternating path is M - - 4 — J - - 2 (where - - = not in, — = in) *[1 mark]*.
Change the status of the arcs: M — 4 - - J — 2 *[1 mark]*
So in the new matching, Mia will teach class 4 and James will teach class 2. Lee and Kelly are unchanged.
Construct the improved matching:
J = 2, K = 1, L = 3, M = 4 *[1 mark]*

J •   • 1
K •   • 2
L •———————• 3
M •———————• 4
N •   • 5

Make sure you list your matchings — you might not get marks for just drawing the graph.

b) Lee is the only person who can teach class 3 and the only person who can teach class 5, so both classes cannot be taught at the same time *[1 mark]*.

c) Finding a new alternating path from N:

4 — M - - 3 — L - - 5   Breakthrough
N
1 — K - - 2 — J - - 4

The path N - - 4 — M - - 3 — L - - 5 reaches a breakthrough quicker, so use this one *[1 mark]*.
Change the status of the arcs: N — 4 - - M — 3 - - L — 5
Construct the improved matching:
James = 2, Kelly = 1, Lee = 5, Mia = 3 and Nick = 4 *[1 mark]*

Answers

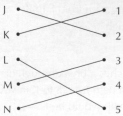

J   1
K   2
L   3
M   4
N   5

There are no more unmatched nodes,
so this is a complete matching *[1 mark]*.

D1 — Practice Exam 1

1   a)   There are 11 items in the list, so the pivot is the ½(11 + 1)
= 6th item = Nahla *[1 mark]*. The list becomes:
Frederick, Dexter, Kamui, Janet, Anthony, Luke, <u>Nahla</u>,
Vicky, Sophia, Paul, Samira *[1 mark]*.
The first small list has 6 names in it, so the pivot is the 4th
name = Janet. The second small list has 4 names in it, so
the pivot is the 3rd name = Paul. The list becomes:
Frederick, Dexter, Anthony, <u>Janet</u>, Kamui, Luke, <u>Nahla</u>,
<u>Paul</u>, Vicky, Sophia, Samira *[1 mark]*.
The first small list now has 3 names in it, so the pivot is the
2nd name = Dexter. The middle small list has 2 names,
so the pivot is the 2nd name = Luke. The third list has 3
names, so the pivot is the 2nd name = Sophia. The list
becomes:
Anthony, <u>Dexter</u>, Frederick, <u>Janet</u>, Kamui, <u>Luke</u>, <u>Nahla</u>,
<u>Paul</u>, Samira, <u>Sophia</u>, Vicky *[1 mark]*.
The remaining names that haven't been chosen as pivots
are all in lists by themselves, so the list is now in order
[1 mark].

*If you want, you can abbreviate the names (e.g. Fr, De, Ka etc.)
to speed things up a bit. It's dead easy to see if you've made
a mistake here — if your final list isn't in alphabetical order,
something's gone horribly wrong...*

  b)   The list to search is: 1. Anthony, 2. Dexter, 3. Frederick,
4. Janet, 5. Kamui, 6. Luke, 7. Nahla, 8. Paul, 9. Samira,
10. Sophia, 11. Vicky
From above, the 'middle' item is the 6th name = Luke.
Sebastian ≠ Luke and Sebastian is after Luke, so throw away
the first half of the list (including Luke) *[1 mark]*.
The remaining list is:
7. Nahla, 8. Paul, 9. Samira, 10. Sophia, 11. Vicky.
The 'middle' name is the ½(7 + 11) = 9th name = Samira.
Sebastian ≠ Samira and Sebastian is after Samira, so throw
away the first half of the list (including Samira) *[1 mark]*.
The remaining list is now: 10. Sophia, 11. Vicky.
The 'middle' name is the ½(10 + 11) = 10.5 = 11th name
= Vicky. Sebastian ≠ Vicky and Sebastian is before Vicky,
so throw away Vicky *[1 mark]*. The remaining name is
10. Sophia, and Sebastian ≠ Sophia, so Sebastian is not
in the list *[1 mark]*.

2   a)   Box 1: 1.8, 3.5   space left: ~~4.2~~ 0.7
       Box 2: 2.6, 1.2, 0.8   space left: ~~3.4~~ ~~2.2~~ 1.4

Box 3: 4.1   space left: 1.9
Box 4: 2.0, 2.4   space left: ~~4.0~~ 1.6
Box 5: 3.1   space left: 2.9   *[1 mark]*
5 boxes are used *[1 mark]* and there is:
0.7 + 1.4 + 1.9 + 1.6 + 2.9 = 8.5 kg wasted space *[1 mark]*.

  b)   In decreasing order, the weights of the items are:
4.1, 3.5, 3.1, 2.6, 2.4, 2.0, 1.8, 1.2, 0.8 *[1 mark]*.
Using the first fit method on this new list gives:
Box 1: 4.1, 1.8   space left: ~~1.9~~ 0.1
Box 2: 3.5, 2.4   space left: ~~2.5~~ 0.1
Box 3: 3.1, 2.6   space left: ~~2.9~~ 0.3
Box 4: 2.0, 1.2, 0.8   space left: ~~4.0~~ ~~2.8~~ 2.0
[1 mark] 4 boxes are used and there is:
0.1 + 0.1 + 0.3 + 2.0 = 2.5 kg wasted space *[1 mark]*.

  c)   The boxes in part (a) are lighter so will
be easier to carry *[1 mark]*.

3   a)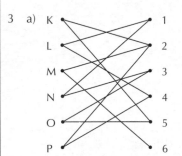

K   1
L   2
M   3
N   4
O   5
P   6

*[2 marks for all arcs correct, or 1 mark for 6 - 11 correct
arcs and no incorrect arcs]*

  b)   Initial matching:

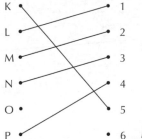

K   1
L   2
M   3
N   4
O   5
P   6   *[1 mark]*

Find an alternating path from Olivia to job 6:
O - - 5 — K - - 2 — M - - 6 *[1 mark]*
(the other route, which starts O - - 3 is longer, so shouldn't
be used). L, N and P remain unchanged.
The matching is now:

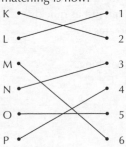

K   1
L   2
M   3
N   4
O   5
P   6

So K = 2, L = 1, M = 6, N = 3, O = 5, P = 4 *[1 mark]*.
There are no more unmatched nodes so the matching is
complete *[1 mark]*.

Answers

4 a) $0.4x + 0.5y \leq 8$ or $4x + 5y \leq 80$ *[1 mark]*.

b) Natalie can make no more than 8 blank cards *[1 mark]*. Also, the number of blank cards made must be no more than number of birthday cards made *[1 mark]*.

c)

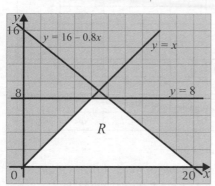

[4 marks available — 1 for each correct line (with equations as shown on graph), 1 for correct feasible region]

d) $P = 0.75x + 1.5y$ *[1 mark]*.

e) Using the vertex method, the vertices of the feasible region are (0,0), (8, 8) (the intersection of $y = x$ and $y = 8$), (10, 8) (the intersection of $y = 16 - 0.8x$ and $y = 8$) and (20, 0) *[2 marks for all 4 correct, or 1 mark for 2 correct vertices]*. Putting these values into the objective functions gives P values of £0, £18, £19.50 and £15 *[1 mark]*, so the maximum profit is £19.50 *[1 mark]*, which occurs at (10, 8), so Natalie must make 10 birthday cards and 8 blank cards *[1 mark]* to maximise her profit.

You could have used the objective line method instead, using the line $P = 0.75x + 1.5y$ to find the maximum.

5 a)

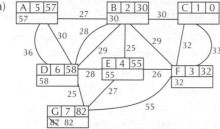

Route = CBEG, length = 82 miles

[7 marks available — 1 mark for route, 1 mark for length, 5 marks for all vertices correctly completed, alternatively 4 marks for 6 vertices correct, 3 marks for 5 vertices, 2 marks for 4 vertices, 1 mark for 2 or 3 vertices.]

b) (i) AB and BF *[1 mark]*

This is the shortest route between the other odd vertices, A and F.

(ii) 460 + 56 = 516 miles *[1 mark]*

c) Odd vertices are A, C, F, G *[1 mark]*
Pairings:　AC + FG = 57 + 53 = 110
　　　　　AG + CF = 55 + 32 = 87
　　　　　AF + CG = 56 + 82 = 138
[1 mark for pairings, 1 mark for lengths]
minimum pairing = AG + CF *[1 mark]*
route length = 460 + 87 = 547 miles *[1 mark]*

6 a) Order paths added:
GF, FB, BC, CA, GH/CH, HI, ID, IK, KL, KJ, IE

[5 marks for answer fully correct. Lose 1 mark for each mistake.]

b)

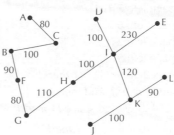

(CH is an alternative to GH.)
Length = 1200 metres
[3 marks available. 1 mark for 9 or 10 correct arcs, 2 marks for 11 correct. 1 mark for length of spanning tree.]

c) [AC – add; FG – add]; [BF – add; KL – add]; [BC – add; DI – add; HI – add; JK – add]; [AB – don't add; GH/CH – add; GH/CH – don't add]; IK – add; [HJ – don't add; HK – don't add; IL – don't add]; CD – don't add; EI – 230 – add; GJ – 250 – don't add; DE – 270 – don't add; EL – 280 – don't add; FJ – 300 – don't add. (The edges in square brackets can be considered in any order.)

The tenth and eleventh edges are IK and EI.

[3 marks available — 1 mark for each correct edge, 1 mark for evidence that Kruskal's algorithm has been applied.]

7 a)

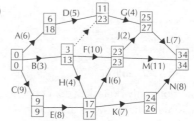

[5 marks available — 1 mark for each two vertices correctly completed.]

b) 23 – 5 – 6 = 12 days *[1 mark]*

c) 34 days *[1 mark]*

d) Lower bound = 90 ÷ 34 = 2.65 *[1 mark]*, which rounds to 3 *[1 mark]*

e)

	0 2 4 6 8 10 12 14 16 18 20 22 24 26 28 30 32 34
Worker 1	C　　E　　I　　M
Worker 2	B H　　F　　J L
Worker 3	A　D　G　K　N

Other answers are possible.
[4 marks available — 1 marks for each 3 activities scheduled suitably.]

To decide which activity to schedule next, look at the late event time for each activity and see which has to be finished soonest. (The late event time for an activity is the one in the bottom box at the end of the activity's arc.)

f) To show that G depends on both B and D, but F and H depend only on B *[1 mark]*.

Answers

D1 — Practice Exam 2

1 a) $9 + 10 + 6 + 14 + 7 + 11 + 3 = 60$. $60 \div 16 = 3.75$
   [1 mark], so the lower bound for the number
   of flash drives needed is 4 *[1 mark]*.

   b) Drive 1: 9, 6   space left: 7̶ 1
      Drive 2: 10, 3   space left: 6̶ 3
      Drive 3: 14   space left: 2
      Drive 4: 7   space left: 9
      Drive 5: 11   space left: 5   *[1 mark]*
      5 flash drives are used *[1 mark]*, and there is
      $1 + 3 + 2 + 9 + 5 = 20$ MB wasted space *[1 mark]*.

   c) By eye, $9 + 7 = 16$ and $10 + 6 = 16$ *[1 mark]*,
      so fill those drives first, then use the first fit algorithm:
      Drive 1: 9, 7   space left: 0
      Drive 2: 10, 6   space left: 0
      Drive 3: 14   space left: 2
      Drive 4: 11, 3   space left: 5̶ 2   *[1 mark]*
      This solution uses 4 drives, with $2 + 2 = 4$ MB wasted
      space. The lower bound is 4 so this is an optimal solution
      [1 mark].

2 a) Alternating path is E - - 3 — F - - 2 — D - - 1 *[1 mark]*.
      Switching status gives E — 3 - - F — 2 - - D — 1 *[1 mark]*.
      So in the new improved matching, Edgar will do task 3,
      Fatima will do task 2 and Diego will do task 1. Carlos and
      Gino remain unchanged. This is the improved matching:

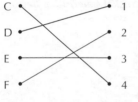

      C   1
      D   2
      E   3
      F   4
      G   •   5   *[1 mark]*

   b) Edgar and Gino can both only do task 3, so a complete
      matching is not possible (other answers are possible)
      [1 mark].

   c) Alternating path is G — 3 - - E — 5 *[1 mark]*.
      Switching status gives G - - 3 — E - - 5.
      So in the new matching, Gino does task 3 and Edgar does
      task 5. The improved matching is:
      C = 4, D = 1, E = 5, F = 2, G = 3 *[1 mark]*

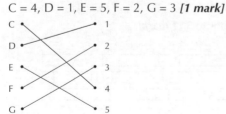

      C   1
      D   2
      E   3
      F   4
      G   5

      There are no more unmatched nodes,
      so this is a complete matching.

3 a)

A	B	B = 1 digit?	B = 9?	Output
977	23	No		
	5	Yes	No	
978	24	No		
	6	Yes	No	
979	25	No		
	7	Yes	No	
980	17	No		
	8	Yes	No	
981	18	No		
	9	Yes	Yes	981

   The result is 981.
   *[4 marks available in total. 4 marks if all
   correct, lose 1 mark for each mistake.]*

   b) 981 *[1 mark]*. This starting value would produce the same
      trace table as in part (a), but starting one row down.

   c) The next multiple of 9 is at most 8 numbers
      away from the starting value, so no more
      than 9 passes will be needed *[1 mark]*.

4 a)

   *[4 marks available — 1 mark each for 2 correct dummies.
   1 mark for 8 activities shown correctly, 2 marks for 10
   activities shown correctly.]*

   b) The dummy between events 1 and 2 allows activities A
      and B to be uniquely represented in terms of their events
      [1 mark]. The dummy between events 3 and 4 shows that
      F depends on C and D, but E depends only on C *[1 mark]*.

      *There are no marks for suggesting that it's a car factory and
      that crash test dummies are needed for safety testing.*

5 a) Odd vertices are B, D, F, G *[1 mark]*
      Pairings: BF + DG = 190 + 100 = 290
                 BD + FG = 180 + 110 = 290
                 BG + DF = 80 + 210 = 290
      [1 mark for pairings, 1 mark for lengths]
      minimum pairing = any of above *[1 mark]*
      route length = 2200 + 290
      = 2490 *[1 mark]* metres *[1 mark]*

   b) (i) D and F *[1 mark]*. B and G are the pair of odd vertices
           with the shortest path between them, so that's the path
           that should be repeated, which means the route must
           start and finish at the other odd vertices *[1 mark]*.
      (ii) $2200 + 80 = 2280$ m *[1 mark]*

Answers

c)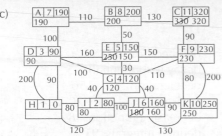

Distance = 320 metres
Route = HIGFC

[6 marks available — 1 mark for distance,
1 mark for route, 1 mark for correct order of labelling,
1 mark for correct final values, 1 mark for correct working
values at 9 or 10 vertices or 2 marks for correct working
values at all 11 vertices.]

6 a)

	Ⓐ	Ⓑ	Ⓒ	Ⓓ	Ⓔ	F
A	–	140	167	205	150	173
B	⑭⓪	–	145	148	159	170
C	167	⑭⑤	–	210	195	180
D	205	⑭⑧	210	–	155	178
E	⑮⓪	159	195	155	–	185
F	173	⑰⓪	180	178	185	–

Order of arcs = AB, BC, BD, AE, BF

[3 marks available — 1 mark for first 3 vertices in correct
order or 2 marks for all 5 vertices in correct order.
1 mark for correct crossing out and circling on matrix.]

Once you get into the swing of this — cross out, circle, cross out
circle — it's easy. Check back to D1 Section 2 if you've forgotten
what you're crossing out and circling.

b)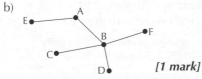

[1 mark]

753 metres *[1 mark]*

c) Cycles are only formed when joining vertices already in the
tree *[1 mark]*. Prim's algorithm only considers connections
to vertices not already in the tree, so a cycle can't form
[1 mark].

7 a) $w = 8$, $x = 8$, $y = 11$, $z = 12$
[2 marks available — 1 mark for each 2 values.]

For the top numbers, go from source to sink, taking the biggest
number each time. For the bottom number, go from sink to source,
taking the smallest number each time.

b) B, H, I, N *[1 mark]*

c) D = 13 – 6 – 4 = 3 hours *[1 mark]*
E = 10 – 2 – 4 = 4 hours *[1 mark]*
K = 13 – 3 – 8 = 2 hours *[1 mark]*

d)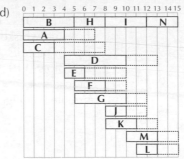

[4 marks available — 4 marks for all correctly
plotted, 3 marks for 12 or 13 activities correctly
plotted, 2 marks for 9, 10 or 11 correctly plotted,
1 mark for 6, 7 or 8 correctly plotted.]

e) Hour 4: A, B
Hour 10: D, G, I

[3 marks available — 1 mark for critical activities
B and I. 2 marks for activities A, D and G, or 1
mark for any two of these activities. Lose 1 mark
for each incorrect activity included in answer.]

8 a) (i) Number of tickets: $5x + 10y \leq 300 \Rightarrow x + 2y \leq 60$
[1 mark]. Number of bottles of wine:
$2x + 8y \leq 160 \Rightarrow x + 4y \leq 80$ *[1 mark]*.

(ii) $x \geq 5$, $y \geq 5$ *[1 mark]*.

There have to be at least 5 business class packages and at
least 5 premier packages sold

$x + y \geq 20$ *[1 mark]*.

The total number of packages sold has to be more than 20.

b) (i)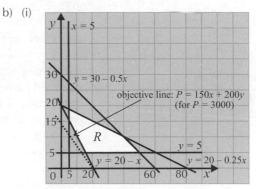

[6 marks available — 1 for each correct line
(with equations as shown on graph — 4 in total),
1 for correct feasible region and 1 for
objective line with correct gradient]

(ii) Using the objective line, the last point in the feasible
region it touches is the point (50, 5), the intersection of
the lines $y = 5$ and $y = 30 - 0.5x$. At this point,
$P = £8500$ *[1 mark]*, so this is the maximum profit
possible, achieved when 50 business class packages
and 5 premier packages are sold *[1 mark]*.

(iii) Using the objective line, the first point in the feasible
region it touches is the point (15, 5), the intersection of
the lines $y = 5$ and $y = 20 - x$. At this point,
$P = £3250$ *[1 mark]*, so this is the minimum profit
possible, achieved when 15 business class packages
and 5 premier packages are sold *[1 mark]*.

Index

Index

Index

Index